Modelling and Prediction Honoring Seymour Geisser

Springer
New York
Berlin
Heidelberg
Barcelona
Budapest
Hong Kong
London
Milan
Paris
Santa Clara
Singapore
Tokyo

Seymour Geisser
1996

Jack C. Lee
Wesley O. Johnson
Arnold Zellner
Editors

Modelling and Prediction Honoring Seymour Geisser

With 76 Figures

Springer

Jack C. Lee
National Chiao Tung University
Institute of Statistics
1001 Ta-Hseuh Road
Hsichu, Taiwan R.O.C.

Wesley O. Johnson
University of California, Davis
Department of Statistics
Davis, CA 95616
USA

Arnold Zellner
University of Chicago
Graduate School of Business
1101 East 58th Street
Chicago, IL 60637
USA

Library of Congress Cataloging-in-Publication Data
Modelling and prediction : honoring Seymour Geisser / edited by Jack C. Lee, Wesley O. Johnson, Arnold Zellner.
p. cm.
"Proceedings of the 'Conference on Forecasting, Prediction, and Modelling in Statistics and Econometrics' which was held at National Chiao Tung University, Hsinchu, Taiwan during the period December 12–14, 1994"—Pref.
Includes bibliographical references and index.
ISBN 978-1-4612-7529-9 (hardcover : alk. paper)
1. Bayesian statistical decision theory—Congresses. I. Geisser, Seymour. II. Lee, Jack C. III. Johnson, Wesley O. IV. Zellner, Arnold V. Conference on Forecasting, Prediction, and Modelling in Statistics and Econometrics (1994 : National Chiao Tung University)
QA279.5.M595 1996
519.5′4—dc20 96-19132

Printed on acid-free paper.

ISBN-13: 978-1-4612-7529-9 e-ISBN-13: 978-1-4612-2414-3
DOI: 10.10007/978-1-4612-2414-3

Softcover reprint of the hardcover 1st edition 1996

Production managed by Allan Abrams; manufacturing supervised by Jeffrey Taub.
Camera-ready copy prepared from the author's files.

9 8 7 6 5 4 3 2 1

Preface

This volume, which is dedicated to the work of Professor Seymour Geisser of the University of Minnesota, contains the refereed proceedings of the "Conference on Forecasting, Prediction and Modelling in Statistics and Econometrics" which was held at National Chiao Tung University, Hsinchu, Taiwan during the period December 12-14, 1994. The Conference was organized under the auspices of National Chiao Tung University and the University of Chicago with the generous support of ROC's National Science Council and Ministry of Education, the U.S. National Science Foundation, and the International Society for Bayesian Analysis.

The Conference was co-chaired by Jack C. Lee and Arnold Zellner with the help of the following two committees. The Program Committee members were C.K. Chu, M.N.L. Huang, G.C. Tiao, P.C. Wang and C.Z. Wei. The Local Arrangements Committee members were R.J. Chou, L.A. Chen, Y.J. Jou, H.H. Lu, N.F. Peng, J.J.H. Shiau, and F. Su. During the Conference, the generous support from Professor T.S. Chuu, Dean of Science College, President C.F. Den of NCTU and Dr. Y.S. Lee, Chairman of the Farmers Bank of China was crucial to its success. We would also like to express our thanks to Professor Alan Izenman, Director of NSF's Probability and Statistics Program, and Professor W.C. Huang, Director of NRC's Statistics Program, for their support in this endeavor.

The topics of forecasting and prediction, and their dependence on modelling are extremely important. In any statistical paradigm, it is essential that future outcomes be reasonably assessed by employing reliable models. The reliability of a statistical model is thus ultimately determined by the model's capacity for predicting/forecasting future observations. The failure of a model in this regard inherently leads to the consideration of other, hopefully more useful, models. In this way, the topics of forecasting/prediction and modelling form a natural union, and hence, a natural combination of topics for an international conference. Because of Professor Geisser's pathbreaking efforts in these areas, we dedicate this volume to him.

Professor Seymour Geisser received his B.A. in mathematics from the City College of New York in 1950. He then received his M.A. and Ph.D. in mathematical statistics from the University of North Carolina in 1952 and 1955, respectively. His dissertation advisor was Professor Harold Hotelling. After receiving his Ph.D., he worked as a mathematical statistician for the National Institute of Mental Health until 1961 when he became Chief of the Biometry section of the National Institute of Arthritis and Metabolic Diseases, a position he held until 1965. During the period 1960-1965, he also maintained the position of Professorial Lecturer at the George Washington University. In 1965, he became Professor and Chairman of the Department of Statistics at SUNY, Buffalo, where he stayed until 1970. Since 1971, Professor Geisser has been continuously appointed to the position of Director of the School of Statistics at the University of Minnesota. During his career, he has had visiting appointments at a number of universities including Iowa State, Stanford, Harvard and Carnegie-Mellon, and the Universities of Wisconsin, Tel-Aviv, Waterloo, Chicago, Warwick and the Orange Free State, and he was the Lady Davis Visiting Professor at the Hebrew University of Jerusalem.

In 1963, Professor Geisser was elected Fellow of ASA and Fellow of IMS in 1967. He was on the Board of Directors of the ASA from 1964-1965, member of the IMS Council from 1978-1981, Chair of SBSS from 1995-1996, Chair of the Council of Sciences for ISBA from 1992-1993, and a member of ISBA's International Advisory Board from 1994-1996. He was associate editor of JASA from 1968-1970 and from 1986-1988. He has been on numerous program review committees and advisory committees including the National Research Council Committee on National Statistics from 1968-1987. He has been a member and chair of a number of committees to select editors, fellows and other honorees of awards in Statistics.

Seymour Geisser has published nearly 100 papers in a wide variety of outlets. He has edited two volumes, written a monograph on prediction and published approximately 60 discussions, letters to the editor and book reviews in a broad spectrum of journals and volumes. His research efforts divide into a number of categories. His early work emphasized mathematical statistics and applications in psychology and biology. In the 1960's, he began making his extensive contributions to the theory of multivariate analysis, discriminant analysis and growth curves, and also developed strong interests in the foundations of Statistics and in prediction. He proceeded in the 1970's to develop his widely read theory of "predictive sample reuse", developed further interests in model selection, and continued with his contributions to classification, growth curves foundations and prediction, areas in which he continues to publish to this day. Major new interests in the 1980's were diagnostics and discordancy tests. In the 1990's he also began working in the area of interim analysis, and most recently, he has been publishing in the area of statistical issues related to forensics, and in particular, on controversial statistical issues related to DNA fingerprinting.

The 10 Ph.D. students supervised by Seymour Geisser are: (1) Peter Enis, Professor, Dept. of Statistics, SUNY, Buffalo; (2) Russell Kappenman, Statistician, Northwest and Alaska Fisheries Center, Seattle, WA; (3) Jack C. Lee, Professor and Director, Institute of Statistics, NCTU; (4) Wesley Johnson, Professor, Div. of Statistics, U.C. Davis; (5) James Hodges, Senior Research Associate, Dept. of Biostatistics, Univ. of Minnesota; (6) Robert McCulloch, Professor, Graduate School of Business, Univ. of Chicago; (7) Michael Lavine, Assoc. Professor, Inst. of Decision Sciences, Duke Univ.; (8) C.N. Chen, Member of Technical Staff, Bellcore; (9) L.S. Chen, Associate Professor, Institute of Statistics, National Cheng-Chi University (10) George D. Papandonatos, Asst. Professor, Dept. of Statistics, SUNY, Buffalo.

On behalf of all participants, we would again like to thank the sponsors for their support for this conference. The conference was extraordinarily successful, with superb talks, excellent opportunities for collaboration, and a broad balance in terms of expertise and interests of participants.

We would also like to express our gratitude to the many referees whose efforts improved the quality of the contributions to this volume, to the Spring Foundation of NCTU for their financial support, to Springer-Verlag for publishing this volume, and to Martin Gilchrist and John Kimmel for their support given to this endeavor. And our deepest gratitude goes to Seymour for his contributions to Statistics, for his mentoring and for his wit. Thanks Seymour.

After Dinner Remarks: On the Occasion of Seymour Geisser's 65th Birthday, Hsinchu, Taiwan, December 13, 1994

Marvin Zelen
Dana-Farber Cancer Institute and Harvard School of Public Health

I have been asked to make some after dinner remarks to commemorate the 65th birthday of Professor Seymour Geisser. My only qualification to stand before you is that I have known him for 44 years which is longer than anyone else in this room. Also perhaps many of you have not even reached that age.

Before beginning my talk, I think it appropriate to thank the organizers of this meeting for planning a very stimulating program which has served to motivate much discussion and I am sure will generate a great deal of future research. We especially should thank the co-chairmen of the meeting Professors Jack Lee and Arnold Zellner. Actually I have known both of them for more than 25 years–Professor Lee when he was a graduate student at the State University of New York at Buffalo and Professor Zellner when he was a junior faculty member at the University of Wisconsin. If nothing else, our acquaintanceship provides some evidence that statisticians have the genes for longevity.

The preeminent theme of this meeting is Bayesian Statistics. Although I have been interested in this area for many years, I have not been a very active practitioner of the subject. Let me relate an incident which happened to me about two years ago. At Harvard I had presented a seminar on Bayesian Methods in Case Control Studies. At about 3 A.M. the next morning I was sound asleep when the telephone rang by my bed. "Is this Dr. Zelen?" a rather inebriated male voice said. I was half asleep as I held the phone and sleepily said "Yes." In a voice which could be barely understood because he was so drunk the caller said that he was an epidemiologist and he and his friends had attended my seminar. They were at a party and were in an argument about Bayesian methods. They wanted some clarification between Bayes and frequency interpretations. I was exasperated and said, "That's something I can't explain over the telephone. Furthermore, I'm too sleepy and you're too drunk. Come around to my office in the morning and I'll be glad to discuss it with you." The voice at the other end said, "But Professor Zelen, we can't wait until tomorrow; we must know now." I was getting a bit angry and said, "Why can't you wait until tomorrow? Why do you have to know now?" Because the voice said in a patient manner, "Tomorrow we won't give a damn!"

I first met Seymour–I shall refer to him as Seymour as everyone else does, when he was a student at the University of North Carolina in Chapel Hill. He was fresh from the City College of New York, had spent a summer working in Washington and possessed bright red hair. He was an absolutely outstanding student there and was the last doctoral student advised by Professor Harold Hotelling.

As I remarked earlier, I have known Seymour for a very long time and we have long been friends. I knew his family quite well. Many years ago, I recall a conversation with Seymour's father. He was very proud of Seymour. He volunteered the information that when Seymour was about 12 years old he was regarded by everyone as being very clever for his age. I guess even now people are still making the same remark.

I remember an incident many years ago which I am sure he has long forgotten. We were attending a meeting and we met a very attractive young lady. She seemed to be interested in Seymour and asked what he did. With a twinkle in his eye he said that he was a Bayesian. She asked if that was some kind of religion. He replied "not quite" but it had to do with a scientific philosophy of data analysis and there were two points of view. At the meeting we were attending there were a series of debates of Bayesians versus the Frequentists. Not knowing very much the lady asked for some explanation. So Seymour with an amusing smile, that you can all picture, told her about the need for looking at posteriors. At which point she interrupted him and said she wasn't interested in posteriors. With her it was the frequency which counted!

Seymour's research for the past three decades has been primarily focused on *Bayesian Prediction*. The basic philosophy is that prediction is the most important part of inference. He regards the current practice of statistics as having too much emphasis on parameter estimation and testing. His recent book, published last year entitled, *Predictive Inference* is an attempt to change this over-emphasis. The book references 37 papers authored by Seymour on this subject. As a comparison, Fisher and Jeffreys are only referenced once in the bibliography. The book is a "tour de force" and essentially summarizes Seymour's contributions to this subject. It not only outlines the theory, but has many diverse examples of applications.

Some statisticians are convinced that the only way to think about statistical inference is through a Bayesian formulation. The issue is that it does not work for many problem settings. Seymour's book is an important book. Not only is there a self-contained presentation of predictive inference but it does attempt to deal with many different problem settings and applications. Our profession has been enriched, not only by his research, but by his many contributions to the discussion of invited papers. I have always enjoyed reading his discussions. They are erudite and sometimes quite "cutting." He has described current work on inference as consisting of assuming, averaging and approximating. After all this is accepted, claims are made for optimality.

One of his remarks I remember is particularly incisive. He wrote, "We are beset by principles involving gambling, entropy, divergence, repeated sampling, conditionality, sample re-use, predictive, likelihood, invariance, admissibility" (with many more in the hatching stage)..."Say what you like about statisticians (liars, damn liars, etc.) but we are certainly not unprincipled."

Seymour has been very active in forensic consultations. He told me that he was an expert witness with regard to potential flaws in the use of statistics in DNA forensics in many capital cases involving the death penalty. His vitae lists forensic consultations on 17 rape cases, 20 homicides, 13 paternity suits, 3 cases of discrimination, one incest case, and several commercial disputes.

My career and Seymour's have crossed many times since our graduate school days. He was a colleague of mine at the Statistical Engineering Laboratory of the National Bureau of Standards. He later transferred to the National Institute of Health where he became a colleague of Sam Greenhouse. Later in 1963 I too joined the NIH. At the time, the center of biostatistics in the United States and probably the world was at the NIH. Among the prominent statisticians there were Jerome Cornfield, Max Halperin, Nathan Mantel, Ed Gehan, Marvin Shneiderman as well as Sam Greenhouse and Seymour. It was particularly embarrassing to have outside seminar speakers as at the conclusion of the seminar, the speaker would always be challenged. Sometimes people couldn't wait for the seminar to end, but would interrupt the speaker with strident questions. The same scenario was repeated every few weeks. Some unsuspecting speaker would be the target and would take heavy hits from the audience. Seymour was one of the heavy hitters. I called it "murderers' row".

Seymour has produced one statistician the old fashioned way. Among his four children, one is a statistician.

We later became colleagues at the State University of New York at Buffalo where he chaired the Department of Statistics. I remember one evening he called me at home and said "Let's go down to the University to try to stop a riot," which was in the making. The students were demonstrating against the Vietnam War and had provoked the Buffalo police. Hundreds of police had suddenly invaded the campus. He wanted both of us to join some other faculty to stand between the angry students and the police who were looking for a fight. I told him he was out of his mind and to go to sleep early. Those of you who know him know that Seymour is absolutely fearless. He was determined to go and I trailed along. Actually only a few faculty had responded to an urgent call to try to prevent a confrontation. We tied handkerchiefs around our arms and glowered at the police. We did succeed in keeping the students and police apart. Well you know faculty. They were not prepared to stay all night. We left about mid-night–there were classes to be taught the next day. Once we left, the police and students fought a bloody one sided battle. However, I do have a memento of the evening. This picture (slide) was taken that evening.

Seymour later moved to the University of Minnesota where he assumed the position of Director of the School of Statistics. He still retains this post and under his leadership, the School has become one of the important centers of statistics in the world. At the time there were serious internal problems among the statisticians in Minnesota. They wanted a strong outside leader who would "crack heads". Seymour was their man. Senior faculty from Minnesota called me to ask what it would take to bring Seymour to Minnesota. My memory is a bit vague as to why they called me...probably because Seymour was out of the country at that time. In any event I suggested salary, authority and other perks. They then contacted Seymour and were successful in luring him away from Buffalo.

Time moved on and we did not see too much of one another. I ultimately moved to Harvard and became Chairman of Biostatistics in the School of Public Health. In 1981 I invited Seymour to be a visitor in our Department. It turned out I was incredibly busy during his visit and was out of town a great deal. At the time I had an administrator who was in charge of administering the Department of Biostatistics. She was very good at her job and we were quite friendly. I explained to her my plight with Seymour. We had planned his visit for some time, but I happened to be away and was not able to extend too much hospitality. She could do me a favor and make sure that Seymour was able to find his way in Boston. I asked her for tickets to concerts and plays, to which I was going to take Seymour, but could not go. I should add that my assistant was single, attractive and vivacious. She was obliging and agreed to do me a favor and look after Seymour. Of course I never consulted Seymour about the change of plans. To make a long story short, my assistant took her job so seriously that she married Seymour. She is here tonight...many of you know her...Anne, could you stand so people can identify you.

At Minnesota, Seymour has become a living legend. He has a reputation as an excellent mentor. Also as a no-nonsense professor with high standards. He has earned the reputation of being a witty lecturer. But his memory is not great! He is famous among the students for never knowing their names. Another student observation is that at seminars he takes a seat in the front, but never faces the speaker. He always sits sideways to observe which students are attending the seminar. If a graduate student has a poor attendance record their teaching assistantship is not likely to be renewed.

On Seymour's 60th birthday, his wife Anne composed a 12 verse ode. I would like to close by reading the last verse.

All can see, he has aged to perfection
Clearly a classic, upon inspection
It's now time to say
Happy Birthday
To Seymour a toast and ovation

Story About Cosimo di Medici

David Lane
University of Modena

My "Geisser story" isn't about Seymour—it's about Cosimo di Medici. Before telling Cosimo's story, I'll explain what it has to do with Seymour. It starts with a puzzle. When I arrived at Minnesota's School of Statistics, I quickly noticed that something was really different there than anything I had experienced at any of my previous academic stopping points. It was partly a matter of intellectual climate. There was a lot more openness at Minnesota, a willingness to look broadly at things statistical—discussions about almost anything kept bursting through the boundaries between mathematical statistics, probability theory, data analysis, and philosophy. Not to mention sociology—what the statistic community was currently up to was often interpreted in terms of the sociology of that community, not just in terms of ideas leading inexorably (or repudiatingly) to other ideas. All very invigorating stuff, in comparison to the usual American approach to doing science, which the Italians call the "hot needle" (very deep, and very, very narrow). Seymour played a part in these discussions, of course, but he wasn't usually the center of them. Even if (as I gradually discovered) he knew a lot more history and philosophy (not to mention statistics) than the rest of us, he typically spent more time listening than talking—and when he did join in, it was with utterances that were partly funny and partly delphic and meant a lot more later than we generally appreciated at the time.

So part of the puzzle was about why the discussions in the School had the character they had. But there was more to it than that. I noticed that almost all my colleagues LIKED the School, felt they were integral parts of its functioning, saw it as an "us" rather than a conglomeration of "thems".

Most everyone had something to complain about, of course; but the complaints always sounded as though they were coming from the inside, not from outside. This atmosphere—which I liked a lot, and which contrasted sharply from anything I'd ever experienced before—had to have something to do with how the School was run, but I couldn't figure out what the connection was. Seymour was the boss, but he didn't boss—he rarely told anyone what to do (though he could nag), he didn't articulate visions about what the School was or could be, he generally didn't take strong positions about the curricular or personnel issues that sometimes divided the faculty in our School meetings. Instead, there were those jokes again and the delphic utterances—and a lot of encouragement to everyone, even to people on opposing sides of some of the divisive issues to keep on their respective (contradictory) courses. Furthermore, when I'd discuss with my colleagues what was going on in the School, I generally found that no two people had the same interpretation of what Seymour was doing and why. So the puzzle was: if coherent leadership wasn't guiding the School along, why did the course the School was taking seem so positive to all of us (the occasional grumblings aside)?

Enter Cosimo. I suppose my ideas about leadership, coherence and rationality derived ultimately from Machiavelli's take on Cosimo and his grandson. Recently, my friend John Padgett of the University of Chicago's Political Science Department has developed a new interpretation of Cosimo and the rise of the Medici. John's interpretation starts with a puzzle. Cosimo was acknowledged by everybody— contemporaries and historians alike—as the head of the political faction that took power in 1434 and controlled Florentine politics for the next 60 years. Yet Cosimo never held any major political office and delivered few public addresses; those he did were most noteworthy for their nearly total lack of content. Every extant account of Cosimo's private audiences confirms that he rarely gave advice or orders, or commented on plans presented to him by associates, beyond a noncommittal but encouraging "Yes, my son". Cosimo seems to have existed mainly in the eyes of his beholders: the so-called New Men (whose families had only recently attained wealth and with it access to power) regarded him as the champion of the New Men; while the old aristocrats (to which class he belonged by birth) regarded him as the great bulwark of their threatened hegemony. This picture raises two principal puzzles: what was Cosimo doing that enabled him to control Florentine politics as no one had succeeded in doing before; and how could such contradictory attributions of who he was be sustained over the whole period of his ascendancy?

The key to John's explanation of these puzzles is his findings about the structure of the interaction network that became the Medicean party. Briefly, the Medici network had a star-like configuration, with the Medici at the center: the other families in the network connected to one another only through the Medici. The Medici married into other aristocratic families, and they did business with New Men. The New Men with whom they did business had no ties of any kind with the aristocratic families that the Medici married. All Cosimo's associates were free to interpret him as they pleased, without risk of contradiction by others. And any plans for concerted Medici action had to flow through Cosimo; his merest encouragement was necessary and nearly sufficient to initiate concerted action.

John's historical analysis makes clear that neither Cosimo nor anyone else designed the structure of the Medicean interaction network. It came about through the concatenation of a series of what can best be described as historical accidents. Moreover, it requires quite a bit of statistical analysis to make visible the network's structure, from the myriad of local interactions of which it is composed. John can "see" the network's structure, but I am fairly certain that Cosimo could not have done so.

On the other hand, Cosimo was, in John's suggestive terminology, a robust actor. He could feel the advantages his structural positioning in the network offered him, and he learned how to exploit the stream of opportunities that this positioning kept flowing in his direction. Cosimo reacted—or better, he let things from which he benefited happen.

There is no evidence that Cosimo engaged in "rational planning", based on an assessment of his goals and available options. Indeed, rationality requires that an individual construct for himself a coherent identity, based on a consistent set of goals and values. It is unlikely that a Cosimo so constructed could have maintained the contradictory set of attributions others held about him; action coming from rational planning would exhibit a coherence that would have given these others a window looking into the "real" Cosimo inside. There just is not a "real" person—that is, a coherent identity—inside the robust actor. Yet Cosimo's style, and the structural position in which he found himself, were mutually supporting, and led to more "intelligent" action than could possibly have resulted had he followed the injunctions of rational choice theory.

CONTENTS

Section I

General Methodological Issues

Some Statistical Issues in Forensic DNA Profiling

Seymour Geisser
School of Statistics
270 Vincent Hall
University of Minnesota
Minneapolis, MN 55455

Abstract

We present a discussion of some of the flaws and problems associated with the statistical methodology used by laboratories in DNA profiling and "matching".

1 Introduction

A rich source of highly polymorphic genetic markers based on recombinant DNA technology was described by Botstein et al. (1980) that would lead to a human genetic linkage map. These markers are called restriction fragment length polymorphisms (RFLP). A form of RFLP is generated by the presence of variable number tandem repeats (VNTR). A variation of the VNTR developed by Jeffreys et al. (1985a,b) used probes that recognized multiple loci was designated as DNA fingerprinting. Jeffreys then suggested its use for forensic identification. A description of its first "successful" use in a murder case in England known as the "Pitchfork" case became the subject of a popular book by Wambaugh (1989). Actually in this case it was used to exonerate one suspect and its potential use was sufficient to induce the murderer to confess without actually being DNA "fingerprinted". Starting in about 1988, in the United States the technology was mainly to use single locus probes for DNA profiling in forensic identification. The main laboratories were Lifecodes, Cellmark and slightly later the Federal Bureau of Investigation. Currently there are many state and county laboratories doing DNA profiling.

At each VNTR locus (a genetic location on a chromosome) an individual's genotype is expressed by two alleles (genetic expressions), one inherited from one's mother and the other from one's father. These alleles are basically a discrete number of tandem repeats that may vary greatly anywhere from 30 to well over 100, depending on the locus. Current technology probes a locus using electrophoresis to obtain a sizing of the fragment lengths (alleles). These alleles are sometimes referred to as "junk" DNA since they are presumed to be from noncoding regions of DNA and possibly not affected by selective forces. However, this is an unproven presumption rather than a fact.

A profile for an individual is then developed by probing several loci and reporting the two alleles at each locus. When two alleles are the same at a locus this is termed homozygotic and when different, heterozygotic.

The "revolutionary" aspects over previous blood groups profiled for forensic identification are the very large number of possible alleles at a locus and the fact that the profile can be obtained from various human tissues, including blood, semen and hair.

However, there are some problems associated with RFLP-VNTR profiling. The electrophoretic method does not yield a precise resolution of the VNTR values associated with the bands at the locus but is subject to measurement error unlike many of the older blood groups. Further, for the probes used, the number of different possible alleles and the interval in sizing between them is unknown. This requires statistical criteria for determining whether the two allelic band values can be considered similar or not. This is necessary to determine whether a crime scene profile "matches" either a suspect or victim. The second problem facing the forensicist is the so-called rarity of a profile consisting of several probes.

We shall present a critique of the statistical methods used by forensic laboratories.

2 The Initial Comparison (or "Match")

The major forensic laboratories—Cellmark, Lifecodes and the Federal Bureau of Investigation—all have somewhat different statistical approaches to the initial comparison because of the error in measuring a band value induced by the electrophoretic process. Lifecodes generally assumes that 2 band values match (really are similar) if the observed difference between them is not more than 1.8% of their average. According to Lifecodes, Baird et al. (1986), the standard deviation of the difference between two alleles measured on the same gel is .6% of the band size, hence the tolerance is 3 standard deviations. If the misclassification of the match/nonmatch procedure were due only to normally distributed measurement error and not to a host of other possibilities that can occur in a laboratory, then it is clear that theoretically the false exclusion rate (asserting that two bands do not match when in fact they do) is .0027 or 1/370. It is not possible to easily calculate the false inclusion rate (assuming that two bands match when in fact they don't). This can best be done by proficiency tests that are external and blind and of sufficient size to estimate the false exclusion rate. Such tests would include not only measurement errors but other kinds of errors as well. Apparently the only external test that Lifecodes has been subjected to was run by the California Association of Crime Laboratory Directors (CACLD) on a profile of 4 probes. In a total of 100 samples sent, 85 were analyzed without error. The others were inconclusive.

Cellmark changed from using standard deviations to a system of resolution limits on a band which it defines as ± 1 millimeter on the autoradiograph. If the two bands from a single locus probe are no more than a resolution limit apart they are declared to match. A resolution limit is also expressed as a varying percent of band size, varying from 1.15% to 5.15% of the band size. The larger the band size the larger the percent of the band size. It would appear from some studies of the variation of repeat measurements conducted by Cellmark that the one resolution tolerance on the difference between 2 bands translates approximately to between 3 and 4 standard deviations depending on the band size. Again the false exclusion rate due only to normally distributed measurement error is approximately between 1/370 and 1/15800 for a match between two bands.

As indicated before, the false inclusion rate can only be measured by external proficiency testing. This was also done by the CACLD and Cellmark's false inclusion rate over approximately 100 trials was 1/50.

The FBI asserts that 2 bands match if their difference is no larger than 5% of their average size, Budowle et al. (1991). A study of repeated measurements on pairs from the same individual on the same gel yields a standard deviation of the difference to be .744% of the band size. Hence the FBI's 5% tolerance translates into about 6.7 standard deviations. The false exclusion rate based solely on normally distributed measurement error on a match of two bands is less than 10^{-10}. No blind external proficiency tests have been reported for the FBI. The FBI's claim, Budowle et al. (1991), that this favors a defendant seems preposterous since even if 8 bands of 4 probes constituting a profile were independent, the theoretical false exclusion rate of the profile would still be less than 10^{-8}. However, in an examination of an unpublished study by the FBI of 225 agent-trainees DNA profiled on two occasions, it was ascertained that many did not match themselves by the FBI's own standards, United States vs. Yee, Thompson (1993). The FBI claimed that different conditions existed for the two occasions.

3 A Statistical Test for the Initial Match

The theoretical false exclusion rates due only to measurement error depend on the reasonable assumption that two bands are the same but differ only on normally distributed measurement error. For a probe consisting of 2 pairs of heterozygotic measurements (X_1, Y_1) and (X_2, Y_2) that require comparison, the assumption of bivariate normality of

$$Z = (X_1 - X_2), \quad W = (Y_1 - Y_2)$$

with highly correlated measurement error appears to be standard model for this situation, c.f. Berry et al. (1992) with estimate $\hat{\rho} = .9$. Given this situation, then under the null hypothesis Z and W represent measurement error and

$$(Z, W) \sim N(\mathbf{O}, \Sigma)$$

$$\Sigma = \begin{pmatrix} \sigma^2 & \rho\sigma\tau \\ \rho\sigma\tau & \tau^2 \end{pmatrix}$$

where

$$\rho \doteq .9, \quad \sigma \doteq .0074\left(\frac{x_1 + x_2}{2}\right), \quad \tau = .0074\left(\frac{y_1 + y_2}{2}\right).$$

Hence under the hypothesis of a match

$$Y = \frac{1}{(.0074)^2(1 - .9^2)}\left[\left(\frac{X_1 - X_2}{\overline{x}}\right)^2 - \frac{2 \times .9(X_1 - X_2)(Y_1 - Y_2)}{\overline{xy}} + \left(\frac{Y_1 - Y_2}{\overline{y}}\right)^2\right]$$

is approximately χ^2 with 2 degrees of freedom. At a given α then one would reject a match if $Y \geq y_\alpha$ where y_α is such that $\Pr[\chi_2^2 \geq y_\alpha] = \alpha$. This should result in a better criterion for a match than treating each Z and W separately, and resolve to a large degree the problem of bandshifting. Thus a simple approximate significance test is available for testing whether there is a match for a probe, that allows for the fact that it is known that measurement error is highly correlated and can result in bandshifting. The forensic laboratories disregard this correlated error by applying their criteria to the components of the pairs individually rather than jointly. At any rate, the plausibility of a match is not graded but decided yes or no.

4 The Relative Frequency of a Profile

Once a match is asserted, the next step for the forensicist is to determine how rare a profile is in a population or the prediction of the fraction of individuals that match the crime scene profile. This is framed as the probability that an unrelated individual chosen at random from an "appropriate" population will also match the crime scene profile. The logic for this calculation can be described as follows. Let A = accused, C = culprit, M = asserted match of crime scene profile and accused profile, or profile of victim and profile of evidence found on the accused. Now it is tacitly assumed that $\Pr[M|A = C] = 1$. Then adopting the common (though false) simplifying assumptions here, the likelihood ratio is

$$\frac{\Pr[M|A = C]}{\Pr[M|A \neq C]} = \frac{1}{\Pr[M|A \neq C]}. \tag{4.1}$$

The denominator in (4.1) is interpreted as the chance that a individual chosen at random from an appropriate population will also be declared a match. Then this is a quantity that the forensic laboratories will estimate and submit in their report to the court. Prosecutors, judges, jury etc. will generally interpret this as the odds that the accused is guilty. This is often termed the "prosecutors" fallacy or more generally transposing the conditional, c.f. Thompson and Schumann (1987). This is, of course, if correctly computed, a likelihood ratio and not the posterior odds of guilt. To turn this into posterior odds requires prior odds that the accused is the culprit, which then would result in the posterior odds ratio as

$$\frac{\Pr[M|A = C]\Pr[A = C]}{\Pr[M|A \neq C]\Pr[A \neq C]} = \frac{\Pr[A = C|M]}{\Pr[A \neq C|M]}. \tag{4.2}$$

Now suppose that if not for the genetic evidence there were N other suspects as likely as A to be the culprit, then it might be sensible to assume a priori

$$\frac{\Pr[A = C]}{\Pr[A \neq C]} = \frac{\frac{1}{N+1}}{\frac{N}{N+1}} = \frac{1}{N}, \tag{4.3}$$

and of course N may very well be the size of the entire population. The prosecutorial fallacy is to present the evidence as if $N = 1$ or that, a priori, the probability is .5 that the accused is the culprit. When the prosecutor implies this fallacy, as she often will, the defending attorney can counter with a more modest fallacy, and that is to ascertain the expected number of individuals K in a sufficiently large population, say of size T, such that

$$K = T \times \Pr[M|A \neq C]$$

and now assert that

$$\Pr[A = C|M] = \frac{1}{K+1}.$$

The potentially fallacious assumption here is that every other individual in the population who matches is as likely as the accused to be the culprit. Certainly where there is no other evidence other than the profile match this may be reasonable.

A weakness of this setup is the arbitrary match/nonmatch criterion and the tacit assumption

$$\Pr(M|A = C) = 1.$$

Actually,

$$\Pr(M|A = C) = 1 - \alpha$$

where α is the actual probability of a false exclusion whose properly calculated value varies with the varying laboratory criteria, joint distributional assumptions which in turn critically involve the independence of alleles both within a locus and between loci and errors other than measurement. These factors may also play an even more critical role in the estimation of

$$\Pr(M|A \neq C).$$

For a full discussion of other potential problems of this likelihood ratio, see Balding and Donnelly (1995).

5 Estimation of the Relative Frequency of a Profile

We assume first that there is a sample of profiles from a relevant population so that an estimate of this profile frequency can be made. We first describe the methods of estimation used by the three major laboratories.

For a single probe on N individuals, Lifecodes will order the $2N = n$ allelic values and for one of the profiled band values it will count all bands in the database that are within 1.8% of that value. That number will then be divided by n which results in (say) $\hat{p}_1$ for band 1. The same thing is done for band 2 resulting in $\hat{p}_2$. On the assumption of a heterozygote, i.e. two unequal bands, the estimate for the ith locus is $2\hat{p}_{1i}\hat{p}_{2i}$. If it is assumed that the initial band values are the same, or a homozygote, the estimate is $\hat{p}^2$, although a more conservative estimate $2\hat{p}$ is often used. This method assumes the independence of the 2 bands at the locus or in the genetic parlance the locus for the population is assumed to be in Hardy-Weinberg equilibrium (basically the Binomial theorem). The same procedure is then applied to the several other probes making up the profile and results in a final estimate for t probes

$$\hat{P} = 2^t \prod_{i=1}^{t} \hat{p}_{1i}\hat{p}_{2i} \tag{5.1}$$

for a heterozygote. The multiplication of the probabilities for the probes assumes the independence of loci, or in genetic parlance, linkage or gametic phase equilibrium depending on the chromosomal situation. We note first that the original match is on the same gel but the "matches" in the population are on different gels. The percent standard deviation between values on different gels is close to twice (1.82) that of the percent standard deviation on the same gel, as estimated by FBI data. Hence instead of using 1.8% they should be using 3.3%.

Cellmark uses basically the same idea but uses resolution limits, one for the initial match and two for the match in the sample database.

The FBI, which uses a 5% initial match window, is a bit schizophrenic about the match in the database. It has divided the sample database for a probe arbitrarily into 31 intervals or bins, then determines the bin in which the initial band falls, and reports the relative frequency of that bin, Budowle et al. (1991). When the band value falls close to the boundary between 2 bins, both relative frequencies are reported and often the larger one is used. However, experts familiar with this procedure (NRC report, 1992) advise that the number in the adjacent bins be summed and the relative frequency of the total be reported. Due to the controversy on the use of DNA profiling in forensics, the National Research Council formed a committee of experts to attempt to resolve the issues. They issued the above-mentioned NRC report. In many cases the FBI has used a floating bin of $\pm 2.5\%$ and $\pm 5\%$ about the band value to be matched, but neither of the two percentages are concordant with the fact that the intergel standard deviation is about 1.82 times the intragel standard deviation. This would require a $\pm 9\%$ window, which the FBI does not use. Assuming that the sampling procedures are adequate and mutual independence is appropriate (both issues will be discussed subsequently), it is of interest to assess the estimate $\hat{P}$ of $P = 2^t \prod_{i=1}^t p_{1i}p_{2i}$, the population value. That is tantamount to t independent trinomial distributions for the heterozygotic case. Since p_{ji} $j = 1, 2$, $i = 1, \ldots, t$ can all be considerably less than .5, it is clear that the sampling distribution of $\hat{P}$ will be skewed right. Now assuming that N individuals are all measured on t probes,

$$\text{cov}\ \hat{p}_{1i}\hat{p}_{2i} = E(\hat{p}_{1i}\hat{p}_{2i}) - \hat{p}_{1i}\hat{p}_{2i} = -\frac{p_{1i}p_{2i}}{n}$$

it is clear then that

$$E(\hat{p}_{1i}\hat{p}_{2i}) = p_{1i}p_{2i}\left(\frac{n-1}{n}\right),$$

which slightly underestimates each product within a probe. Since the probes are assumed independent (for the present),

$$E(\hat{P}) = \left(\frac{n-1}{n}\right)^t P$$

so the bias of underestimation becomes worse as the number of probes increase and the sample size decreases. Further and more importantly it is of interest to calculate

$$\Pr[\hat{P} \leq P].$$

This is analytically difficult so we resort to some Monte Carlo simulations to have an idea. This is presented in Table 1. Again we note that $\Pr[\hat{P} \leq P]$ can be quite large, close to 1 in fact in many cases but in addition to increasing with an increasing number of probes and decreasing sample size as the bias of $\hat{P}$ does, it also increases as the $k = 2t$ probabilities p_{ij} decrease. For the sake of ease the table is constructed using equal probabilities for each allele in each probe to obtain the values. These probabilities are then varied. This also, under the best of circumstances, puts to rest the claims of forensic laboratories that their estimates are conservative and favor the defendant. Moreover, sample databases as collected often have varying numbers of individuals on each probe. For example, Table 2 describes the number of individuals on the same probes in Cellmark's databases for three major U.S. groups, Blacks, Caucasians and Hispanics, this would require that

$$E(\hat{P}) = P \prod_{i=1}^{t} \left(\frac{n_i - 1}{n_i} \right),$$

stressing the dependence on the minimum n_i. Balding (1995), who raised the issue that the sampling distribution of $\hat{P}$ has substantial mass below P, has discussed methods for better estimation of P and has provided some shorter tables of $\Pr[\hat{P} \leq P]$ under the assumption of random samples where there is mutual independence within and between loci. We will discuss these assumptions subsequently.

6 Population Issues

Clearly, the appropriate reference population is the population of identifiable possible perpetrators, but this is almost always unknown. If the race or ethnicity of the culprit, who left the crime scene profile, has been established beyond any doubt, and the accused was a member of that group, then it would appear that it is the proper reference population. However, even this is a doubtful scenario. What is done in practice is to use the race (Black, Caucasian or Hispanic) of the accused to calculate the relative frequency and more generally to make the calculation for all three groups and report all of them. The NRC report (1992) strongly advised calculating a 95% upper confidence interval for each band or .1, whichever was larger, and then selecting the maximum over 3 or more major races as an interim ceiling principle that would be conservative when applying the product rule. (A more elaborate ceiling principle they suggested has never been put into practice.) Almost as soon as the interim ceiling principle was declared to be always conservative, it was demonstrated by Slimowitz and Cohen (1993) that when mutual independence was not assured, this principle need not be conservative, i.e. underestimate the true value. In that paper, dependence was theoretically violated in the form of a mixture of differing populations or substructuring, which is often the way that a genetic population is in disequilibrium.

7 Sampling Issues

In any event several populations are sampled. The major assumption that the product estimate (5.1) is based on requires a random sample of unrelated individuals that have been randomly mating for a sufficient number of generations to ensure independence. It turns out that the sampling is not random but at the laboratory's convenience and is actually referred to as "convenience" sampling, Roeder (1994). For example, Cellmark's entire Black database was obtained from a Detroit blood bank who in turn was recruiting rare donor volunteers for Black patients. The director of the blood bank indicated that the blood bank had no knowledge of how to obtain a simple random sample and no knowledge of possible familial relationships. The donors were asked for their racial category—Black, White, etc.—but not mixed. The Caucasian database was obtained much in the same manner from the Blood Bank of Delaware. These then purport to represent U.S. Blacks and Caucasians. Further, the sample sizes themselves and their distribution among the various loci are inadequate for proper testing of independence assumptions and precise estimation, see Table 2. The FBI samples tend to be about twice as large but fall far short of what would be required to properly test their fixed binning system. A proper analysis requires consideration of $496 = 31 \times 32/2$ cells in an upper triangular contingency table which is actually close to the number of individuals in their samples. This results in a rather sparse table. In reviewing the original FBI databases, it was discovered that there were about 25 apparent matches. Some of these were tracked down to duplicate submissions from the same individual. However, others not known to be duplicates were deleted based on a match criterion, Sullivan (1992). The use of the criterion to justify the criterion can bias results. Again, this is a result of careless "convenience" sampling. Attempts have been made to justify small relative frequencies for a profile by looking at all possible pairs, Risch and Devlin (1992), using a criterion (actually a 2.4% window rather than a 9% window).

8 Independence

The foundation for the so-called product rule rests on the assumption of mutual independence among all the alleles within and between loci. This issue was examined by Weir (1992a,b), who claimed that both the FBI's fixed bin and the floating bin approaches of Cellmark and Lifecodes indicated mutual independence and thence the propriety of the product rule. For the FBI, Weir basically admitted that a proper test of the 496 two-dimensional bins would require much larger sample sizes than had been collected, aside from the fact that the samples were not randomly collected. For Hardy-Weinberg equilibrium, he used intraclass correlation, but this is a measure of linear association and need not have power against other types of dependence. He also applied global chi-square and likelihood ratio tests, but because of the sparsity of cells, he applied bootstrap sampling methods to determine the p-values. He expected these to be powerful because they have "large degrees of freedom". That "large degrees of freedom" necessarily lead to a powerful test is a total non sequitur. In fact, those tests, given the sample sizes in relation to the number of cells, are not likely to be very powerful in detecting various forms of dependence. For the floating bin situation where bins are not determined in advance, he claimed he could not determine a global test of independence. He created 10,000 profiles and tested them using a chi-square statistic calculated from the actual databases, again claiming that the results were consistent with independence. These tests would also tend to have poor power to detect various possible dependencies.

Because of the paucity of observations in relation to the potential number of pairs of bins with regard to the FBI methods or the floating bins of Lifecodes and the resolution limit of Cellmark, Geisser and Johnson (1992) devised a quantile chi-square approach to the problem of testing. The method for testing the independence of (X, Y) when the form of the exchangeable bivariate distribution is unspecified is fairly simple. Assume that a random sample (X_i, Y_i), $i = 1, \ldots, N$ has been obtained. In addition, the parental alleles are not identifiable in these databases, i.e., it is not possible to ascertain which allele of the pair is X and which is Y. Hence, under independence all the values come from the same distribution. We order the $2N$ values into $Z_{(1)}, \ldots, Z_{(2N)}$ and divide them into q quantiles $Q_1, \ldots, Q_q$. We then form a $q \times q$ folded quantile table (Table 3) with sample entries n_{ij}, the number of pairs of (X_i, Y_i) that are in (Q_i, Q_j) $i \leq j$, since we can only observe $n^*_{ij} = n_{ij} + n_{ji}$, $i < j$. It was then shown that

$$Z = \frac{\left(N^{-1} \sum_1^q n_{ii} - \frac{1}{q}\right) \sqrt{N}}{\sqrt{\frac{1}{q}\left(1 - \frac{1}{q}\right)}} \to N(0,1) \tag{8.1}$$

and the statistic

$$X = \frac{q^2}{N} \sum_1^q \left(n_{ii} - \frac{N}{q^2}\right)^2 + \frac{q^2}{2N} \sum_{i>j} \left(n^*_{ij} - \frac{2N}{q^2}\right) \tag{8.2}$$

tends to χ^2 with $q(q-1)/2$ degrees of freedom. The basis for this test is that under independence $E(n_{ii}) \doteq \frac{N}{q}$ and $E(n^*_{ij}) \doteq \frac{2N}{q}$, $i = 1, \ldots, q$ and $i < j = 1, \ldots, q$.

The Z test is more useful for the substructuring alternative as a one-sided test. Since under the alternative one expects the diagonal entries to be larger than under the null hypothesis. This simple method was applied to the FBI sample databases by Geisser and Johnson (1993). It was applied to 6 different probes. For $q = 2$ it was determined that independence was rejected for D2S44, D17S79 and D1S7 at $q = 2$, and D14S13 at $q = 3$, for the Black database. For the Caucasian database, D17S79, D1S7 and D14S13 appeared to exhibit dependence, while D2S44, D17S79, D14S13 appeared to exhibit dependence for the Hispanic database.

Recently, the FBI has begun using a new probe D5S110. An analysis of this probe reveals that for Caucasians for $q = 2$, using Z the substructuring alternative $P = .05$, while for Blacks, $q = 3$ and using X, we reject at $P = .03$. The Hispanic data were divided into two groups by the FBI, Southeastern and Southwestern Hispanics. While in neither group were the tests for independence rejected at $P = .05$, in both groups there is a tendency for the diagonals to be less than expected under independence.

With regard to Cellmark, 5 probes were analyzed in a similar manner, and either for $q = 2$ or 3, only MS31 was not rejected both for Caucasians and Hispanics, and only MS43 was not rejected for Blacks. So for each major database, 4 out of 5 probes exhibited dependence.

For Lifecodes, only 2 probes—D2S44 and D17S79 on a Caucasian database—were available and both exhibited dependence, Geisser and Johnson (1993).

A rigorous derivation of the theory for these tests is presented by Geisser and Johnson (1995), who also include a quantile chi-square test for linkage equilibrium that takes advantage of the exchangeability of the alleles within a locus.

Now, it is clear that the quantile chi-square test sensitivity to dependence may depend critically on q, the number of quantiles. Depending on the configuration of dependence at a locus, certain values of q may be insensitive to detecting the dependence while others may be quite sensitive. Weir (1993), in faulting the test, apparently misunderstands this issue. Only if independence is not rejected for a series of different values of q can one have some confidence that dependence is not a critical issue. Another issue he miscontrues is his assumption that the quantiles should be the binning procedure itself The quantiles' major utility in testing independence is when a floating bin is used or when the sample sizes are inadequate for testing the FBI's fixed binning procedure.

It has also been proposed by Devlin and Risch (1993), that technical flaws in the electrophoretic process tend to make the quantile chi-square overly sensitive in detecting dependence. The first technical flaw is termed coalescence—a blurring on the autoradiograph such that presumably close but different bands are erroneously judged as the same, i.e. a homozygote. A second flaw is termed a null allele in which one of the band's size (or perhaps both) is too small to appear on the autoradiograph, and the band that appears is mistakenly judged as a homozygote. Further, the fact that the measurement error is highly correlated is also touted as a factor in making the test too sensitive.

Coalescence can only affect the test close to the intersection of the boundaries of the diagonal quantiles. Hence, only if q is large can there be an effect if the autoradiograph is unable to discriminate between adjacent alleles. For $q = 2$ or 3, any such effect is negligible. If it is known that one of the bands is an overly small null allele, there is absolutely no effect on the test. Similarly, the same holds if it is known to be overly large. If it is not known, the laboratory judges the two bands to be the same and is so entered into the database, which could lead to an excess along the main diagonal. All tests that have been proposed would be subject to some error in the presence of unknown null alleles. This would be confounded with the substructuring alternative. However, it is interesting to note the tendency to deficits along the main diagonal in the Southeastern and Southwestern Hispanic data for D5S110.

Since the intergel standard deviation of a band value is less than 1% of the band value, only a very small portion of the observable is subject to this measurement error correlation. This could have a minor effect for large q in some cases but only a negligible effect for modest q. At any rate, Devlin and Risch (1993) and Weir (1995) claim that the true alleles are independent and that any dependence disclosed by the test is due to the technical flaws of the electrophoretic procedure. Chakraborty et al. (1994) rehash the same arguments. The results on D5S110 reported tend to question these explanations.

But clearly even in the highly unlikely event the locus is in equilibrium, the observables, flawed or not, are used. If they exhibit sufficient dependence to be detected, then their use negates the product rule. It is unusual to argue that the virtues of a procedure are its technical flaws or that the best test is an insensitive one.

9 Summary

In DNA forensic profiling, the following problems and flaws are listed:

1. Technical flaws leading to coalescence and null alleles

2. False claim of favoring a defendant when the false exclusion rate appears to be orders of magnitude less than the false inclusion rate

3. Invoking or implying surreptitiously the prosecutor's fallacy

4. Lack of random sampling from well specified populations that exclude related individuals

5. Inadequate sample sizes
6. Omission of sampling error estimates of the profile relative frequencies
7. Estimates used of the relative frequency of a profile that are biased downward further belying the false claim of favoring the defendant even if all assumptions were valid
8. Reliance on the false assumption of mutual independence whether caused by substructuring or biased sampling
9. Refusal to engage in periodic, blind, external proficiency tests
10. Lack of implementation of many of the recommendations of the 1992 NRC report.

Clearly all of these problems/flaws should be resolved or corrected because DNA forensics are involved in very serious issues, mainly capital offenses and not infrequently DNA may be the only available evidence.

For some other concerns regarding the exclusion of close relatives that may also lead to a nonconservative relative frequency, see Donnelly (1992, 1994) and Balding and Donnelly (1995).

10 Acknowledgement

This work was supported in part by NIH Grant GM 25271 and the Lady Davis Trust.

References

Baird, M., Balazs, I., Giusti, A., Miyazaki, L., Nicholas, L., Wexler, K., Kanter, E., Glassberg, J., Allen, E., Rubenstein, P. and Sussman, L. (1986). Allele frequency distribution of two highly polymorphic DNA sequences in three ethnic groups and its application to the determination of paternity. *American Journal of Human Genetics* **39** 489–501.

Balazs, I., Baird, M., Clyne, M. and Meade, E. (1989). Human population genetic studies of five hypervariable DNA loci. *American Journal of Human Genetics* **44** 182–190.

Balding, D. J. (1995). Estimating products in forensic identification. *Journal of the American Statistical Association*, to appear.

Balding, D. J. and Donnelly, P. (1995). Inference in forensic identification. *Journal of the Royal Statistical Society, A* **158** 21–53.

Berry, D. A., Evett, I. W. and Pinchin, R. (1992). Statistical inference in crime investigations using DNA profiling: single locus probes. *Journal of the Royal Statistical Society, C* **41** 499–531.

Botstein, D., White, R. L., Skolnick, M. and Davis, R. W. (1980). Construction of a genetic linkage map in man using restriction fragment length polymorphisms. *American Journal of Human Genetics* **32** 314–331.

Budowle, B., Giusti, A. M., Wayne, J. S., Baechtel, F. S., Fourney, R. M., Adams, D. E., Presley, L. A., Deadman, H. A. and Monson, K. L. (1991b). Fixed bin analysis for statistical evaluation of continuous distributions of allelic data from VNTR loci for use in forensic comparisons. *American Journal of Human Genetics* **48** 841–855.

Chakraborty, R., Zhong, Y. and Budowle, B. (1994). Non-detectability of restriction fragments and independence of DNA fragments within and between loci in RFLP typing of DNA. *American Journal of Human Genetics* **55** 391–401.

Devlin, B. and Risch, N. (1993). Physical properties of VNTR data and their impact on a test of allelic independence. *American Journal of Human Genetics* **53** 324–328.

Donnelly, P. (1992). Discussion on "Statistical inference in crime investigations using deoxyribonucleic acid profiling" by D. A. Berry, I. W. Evett and R. Pinchin. *Applied Statistics* **41** 524–525.

Donnelly, P. (1995). The non-independence of matches at different loci in single-locus DNA profiles. *Heredity*, to appear.

Geisser, S. and Johnson, W. (1992). Testing Hardy-Weinberg equilibrium on allelic data from VNTR loci. *American Journal of Human Genetics* **51** 1084–1088.

Geisser, S. and Johnson, W. (1993). Testing independence of fragment lengths within VNTR loci. *American Journal of Human Genetics* **53** 1103–1106.

Geisser, S. and Johnson, W. (1995). Testing independence when the form of the bivariate distribution is unspecified. *Statistics in Medicine* (to appear).

Jeffreys, A. J., Wilson, V. and Thein, S. L. (1985a). Hypervariable "minisatellite" regions in human DNA. *Nature* **314** 67–73.

Jeffreys, A. J., Wilson, V. and Thein, S. L. (1985b). Individual-specific "fingerprints" of human DNA. *Nature* **316** 76–79.

National Research Council (1992). *DNA Technology in Forensic Science*. National Academy Press, Washington, D.C.

Risch, N. J. and Devlin, B. (1992). On the probability of matching DNA fingerprints. *Science* **225** 717–720.

Roeder, K. (1994). DNA fingerprinting: A review of the controversy. *Statistical Science* **9** 222–278.

Slimowitz, J. R. and Cohen, J. E. (1993). Violations of the ceiling principle: Exact conditions and statistical evidence. *American Journal of Human Genetics* **53** 314–323.

Sullivan, P. (1992). DNA fingerprint matches. *Science* **256** 1743–1744.

Thompson, W. C. (1993). Evaluating the admissibility of new genetic identification tests: Lessons from the "DNA" war. *Journal of Criminal Law and Criminology* **84** 1 22–104.

Thompson, W. C. and Schumann, E. L. (1987). Interpretation of statistical evidence in criminal trials. *Law and Human Behavior* **11** 167–187.

Weir, B. S. (1992a). Independence of VNTR alleles defined by fixed bins. *Genetics* **130** 873–887.

Weir, B. S. (1992b). Independence of VNTR alleles defined as floating bins. *American Journal of Human Genetics* **51** 992–997.

Weir, B. S. (1993). Tests for independence of VNTR alleles defined as quantile bins. *American Journal of Human Genetics* **53** 1107–1113.

Weir, B. S. (1994). Comment on DNA fingerprinting. *Statistical Science* **9** 266–267.

Table 1: Probability that the product estimate is less than the true product, based on 30,000 simulations

	k = 2	k = 4	k = 6	k = 8	k = 10
p = .01					
n					
5	0.9977667	1.0000000	1.0000000	1.0000000	1.0000000
10	0.9909333	1.0000000	1.0000000	1.0000000	1.0000000
20	0.9705667	0.9989667	1.0000000	1.0000000	1.0000000
50	0.8452000	0.9762333	0.9964000	0.9992000	0.9998000
100	0.6021000	0.8405000	0.9351000	0.9741333	0.9895000
200	0.5702333	0.7006000	0.7842667	0.8447667	0.8821333
400	0.5413000	0.6120000	0.6496667	0.6843333	0.7119000
p = .02					
n					
5	0.9921000	0.9999000	1.0000000	1.0000000	1.00000000
10	0.9691000	0.9993333	1.0000000	1.0000000	1.0000000
20	0.8936333	0.9884667	0.9987667	0.9999000	1.0000000
50	0.5983000	0.8382667	0.9375000	0.9741667	0.9891667
100	0.5693667	0.7021667	0.7887000	0.8467667	0.8849333
200	0.5400667	0.6231000	0.6582333	0.6914667	0.7172000
400	0.5355667	0.5808667	0.6105333	0.6316667	0.6526667
p = .05					
n					
5	0.9572000	0.9981000	0.9999333	1.0000000	1.0000000
10	0.8488667	0.9771333	0.9964000	0.9994000	0.9998667
20	0.7391667	0.8550333	0.9355667	0.9740333	0.9890667
50	0.6603333	0.7026000	0.7631000	0.8041667	0.8385000
100	0.5868333	0.6015333	0.6425333	0.6743333	0.6979000
200	0.5587333	0.5709667	0.5979000	0.6198667	0.6371333
400	0.5399000	0.5487667	0.5674333	0.5814333	0.5949667
p = .1					
n					
5	0.8511000	0.9777667	0.9964667	0.9995333	0.9999333
10	0.7394333	0.8522667	0.9332333	0.9733333	0.9891000
20	0.7011000	0.7396667	0.7856667	0.8406667	0.8760667
50	0.5885667	0.6008333	0.6486667	0.6829667	0.7041000
100	0.5578333	0.5680667	0.6026333	0.6221667	0.6405333
200	0.5456667	0.5578333	0.5734000	0.5877667	0.6044000
400	0.5355667	0.5375667	0.5511333	0.5602667	0.5712333
p = .2					
n					
5	0.7477333	0.8454667	0.9265333	0.9670333	0.9865333
10	0.7099333	0.7322000	0.7863333	0.8344333	0.8721000
20	0.6222333	0.6546333	0.6880667	0.7220000	0.7476333
50	0.5778333	0.5902333	0.6156667	0.6370333	0.6563000
100	0.5549000	0.5624000	0.5769333	0.5921333	0.6048333
200	0.5359000	0.5431667	0.5553000	0.5696333	0.5804000
400	0.5237333	0.5257000	0.5360667	0.5433333	0.5505000

Table 2: Number of individuals with values on Cellmark probes

1. Pairwise

Pairs	Blacks	Caucasians	Hispanics
MS1/G3	10	235	155
MS1/YNH24	91	154	104
MS1/MS43	128	177	178
MS1/MS31	155	210	154
MS31/YNH24	80	110	94
MS31/MS43	103	189	151
MS31/G3	10	171	133
MS43/YNH24	65	146	95
MS43/G3	10	253	142
YNH24/G3	16	154	93

2. Omit One Probe

Omitted Probe			
MS1	2	79	63
MS31	2	108	73
MS43	2	77	72
G3	31	91	76
YNH24	8	153	109

3. None Omitted

	2	75	59

Total on each Probe			
MS1	240	262	215
MS31	238	264	183
MS43	223	294	192
G3	200	325	168
YNH24	146	208	110

Table 3: Folded Quantile Contingency Table

	Q_1	Q_2		Q_q
Q_1	n_{11}	n^*_{12}	$\cdots$	n^*_{1q}
Q_2		n_{22}	$\cdots$	n^*_{2q}
			$\ddots$	$\vdots$
Q_q				n_{qq}

Statistical Practice as Argumentation: A Sketch of a Theory of Applied Statistics

James S. Hodges
Division of Biostatistics, University of Minnesota

Abstract

We have theories of statistics and we use statistical methods to solve subject-matter problems. One might expect that the theories would affect how the methods are used, but they do so only superficially, because all statistical theories are quite incomplete as descriptions of and prescriptions for statistical practice. This paper sketches an extension of Bayesian theory that might address this incompleteness. The extension is based on five ideas:

1. The product of a statistical analysis is an argument – not an HPD region, posterior distribution, decision, or other data summary, but the entire argument, including premises and logical steps.
2. Arguments come in several logically distinct types, with an argument's type being defined by the form of its conclusion. The paper catalogs the types of argument and identifies the main burden of each.
3. EDA and model-building activities establish a plausible, tractable baseline argument for a given problem and dataset.
4. Diagnostics and sensitivity analyses vary the premises of the baseline argument and display the resulting variation in the conclusion (as opposed to intermediate quantities).
5. An argument is strong to the extent that:
   - its premises are conclusions of strong arguments, or
   - the region of premises yielding the same conclusion is large.

A simple example demonstrates the mechanics of the extended theory. The example is then generalized to draw implications for statistical foundations, methods, and computing. This paper amounts to a research agenda, so it poses more problems than it solves.

Keywords: applied statistics, Bayesian statistics, diagnostics, exploratory data analysis, foundations of statistics, rhetoric, sensitivity analysis.

1 Introduction.

We have theories of statistics and we use statistical methods to solve subject-matter problems. One might expect that the theories affect how the methods are used and, superficially, they do. For example, Fisherians report P-values and Bayesians do not, at least, not to Bayesian audiences. But the theoretical veneer is thin. When we work in substantive fields, we all do the same things: learn something about the subject matter; on obtaining data, poke around in it and settle on a few models; compute data summaries; do diagnostics; and report conclusions, perhaps after repeated cycles through these steps. The theoretical styles lean toward different data summaries and models, but otherwise, foundational leanings have little affect on statistical activities.

This is not just historical inertia. Rather, it happens because all statistical theories are incomplete as descriptions of and prescriptions for statistical practice. This being a largely Bayesian event, I will focus on Bayesian theory. Section 2 makes a brief argument for the incompleteness of Bayesian theory. Section 3 argues that Bayesian theory can be extended and gives five ideas that frame the extension. A simple example in Section 4 illustrates the mechanics of the extended theory. Section 5 generalizes from the example to draw implications for foundations, methods, and computing. Section 6 discusses direct antecedents and inspirations for the extended theory. This theory is embryonic; the present paper amounts to a research agenda, and I beg the reader's indulgence.

2 Bayesian theory is incomplete.

All of us, Bayesians included, learn from data by means other than Bayes' Theorem, the obvious example being exploratory data analysis (EDA). EDA makes little or no use of the Bayesian formalism and prominent Bayesians have certified this as kosher (Smith [1], Hill [2], Berger [3]). Predictive calibration is also difficult to square with Bayesian ideology, but many Bayesians consider it essential. (Geisser and Zellner are the best-known advocates of this view, but see also West and Harrison [4], Chapter 10.) Bayesian theory brings no special facility to the problem of learning enough subject matter to avoid being a menace. And so on; the examples are easily multiplied.

What might we ask of a theory of applied statistics? Such a theory should name the *activities* of applied statistics and the *products* of those activities. It should provide a *rationale* for doing certain activities in specific situations and not doing other activities. Finally, it should explain *how the activities combine* to yield the products. Although Bayesian theory supplies some necessary language, the next few sections will show by construction that it provides none of these elements of a theory of applied statistics – nor, it should be obvious, does any other statistical theory.

Why is this a problem? For the fastidious, incompleteness is problem enough. For the more pragmatic, it is useful to be able to explain what we do when we take something from a substantive problem, build a little mathematical world, do operations in that world, and then claim to be better informed about the original problem. What is the basis of this latter claim? Subjective (or personalistic) Bayesian theory – to which most of this audience would pledge some allegiance – describes and prescribes how to use data to make changes in an individual's beliefs. With few exceptions, however, a statistician's problem is to analyze how data can and should change the beliefs of many or all individuals. If we sweep this under the rug of informality, we risk a variety of errors in practice.

3 But Bayesian theory can be extended.

The problem in extending Bayesian theory to a theory of applied statistics is to connect mathematical reasoning and verbal reasoning. The time-honored approach of axiomatics will not do, because axioms are within mathematics. What is needed, instead, is a rhetorical structure that rationalizes specific mathematical activities and explicitly connects them to subject-matter issues. This section proposes such a structure as an extended Bayesian theory of statistics. Five ideas frame the extended theory:

1. The product of a statistical analysis is an argument.
2. Arguments come in several logically distinct types. An argument's type is defined by the form of its conclusion.
3. EDA and model-building activities establish a plausible, tractable baseline argument.
4. Diagnostics and sensitivity analyses vary the premises of the baseline argument and display the resulting variation in the conclusion.
5. An argument is strong to the extent that:
   - its premises are conclusions of strong arguments, or
   - the region of premises yielding the same conclusion is large.

These ideas will now be described.

3.1 Idea #1: The product of a statistical analysis is an argument.

The product of a statistical analysis is *not* a highest-posterior-density region, posterior distribution, likelihood, predictive distribution, or decision. Instead,

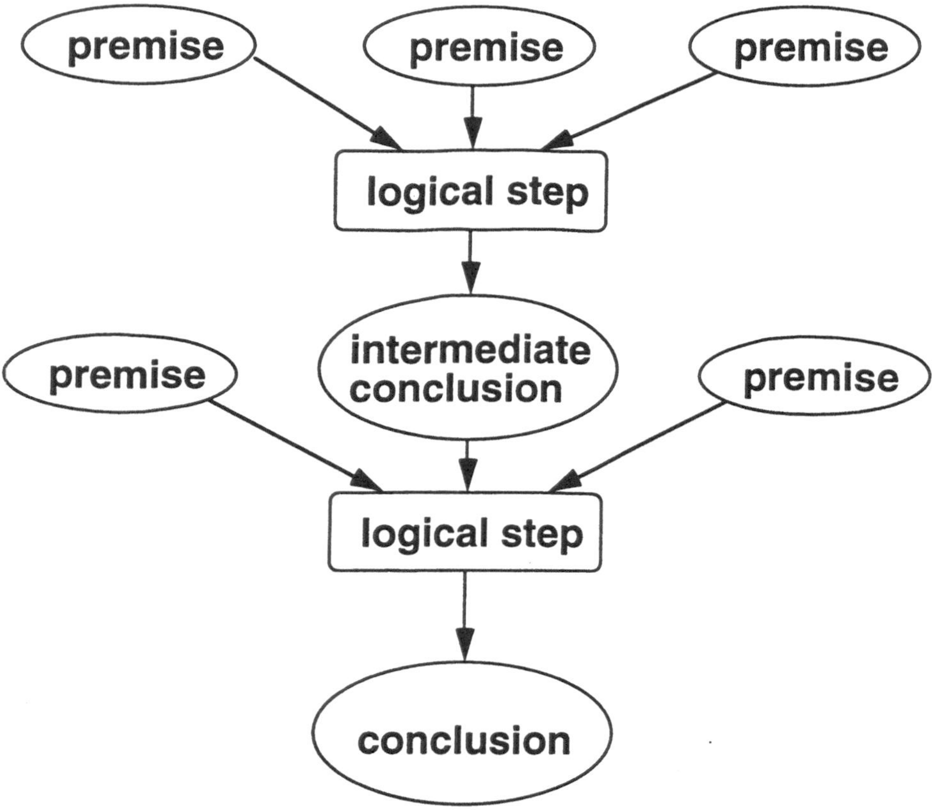

Figure 1: Schematic diagram of an argument

the product is an argument – the entire argument – parts of which may use data summaries like those just mentioned. An argument begins with premises, which are inputs to a logical step, which yields the conclusion. Sometimes premises are the conclusion of an argument that has its own premises and logical step. This is shown in Figure 1.

A premise is "a proposition upon which an argument is based or from which a conclusion is drawn" (*American Heritage College Dictionary*, 3rd. ed.). Statistical models are premises, as are assumptions like exchangeability or missing-at-random. The observed data are also a premise: different data could have occurred but did not, and an argument must use the data that did occur.

Binary logic provides many logical steps combining premises, such as *modus ponens*: A implies B, A is true, therefore B is true. So does the probability calculus, interpreted as an extension of binary logic. I also include arguments like Fisher's disjunction (Fisher [5], p. 42): if we observe a datum that is extremely unlikely under a postulated model, then *either* a rare event has occurred, *or* the posulated model is not true. In a future paper, I will argue that foundational approaches – of which I count three: randomization-theoretic, model-based frequentist, and personalistic Bayesian – differ in the explicitness and variety of premises they use. Randomization-theoretic arguments generally use the fewest premises and Bayesian arguments the most, while randomization-theoretic conclusions are least specific to the circumstances at hand and Bayesian conclusions the most specific (*cf* Efron [6]). No approach is correct or incorrect; rather, because they use different premises, they tend to produce different arguments. Further discussion of this topic is beyond the scope of this paper.

3.2 Idea #2: Arguments come in several logically distinct types. An argument's type is defined by the form of its conclusion.

This subsection discusses five logically distinct types of arguments, each having subtypes. There may be other types of arguments, but these five cover most uses of statistics. The arguments are defined below by the form of their conclusions. Each conclusion is stated non-probabilistically, but statements labeled B can be replaced by probability distributions in the personalistic fashion. The five types of arguments are:

- *Causal arguments*
  - inferential: "B *was* caused by A."
  - predictive: "A *will* cause B."
  - action: "B *was* caused by A, therefore action C is desirable," or "Taking action C in situation D *will* cause a desirable outcome."
- *Non-causal predictive arguments*
  - descriptive: "B *will* occur."
  - action: "Following rule C in situation D *will* yield a desirable outcome."
- *Description arguments*
  - direct: "We *summarize* the data as B."
  - inferential: "The unobserved constant C *takes the value* B."

  - action: "Constant C *takes the value* B, therefore, action D is desirable."

- *Existence arguments*
  - predictive: "B *has* occurred; therefore, it *can* occur."
  - action: "B *has* occurred, so it *can* occur; therefore, action C is desirable."

- *Hypothesis-generation arguments*
  - data-mining: "We ransacked the data and found B, but we have reason to fear that this finding is spurious."
  - innocuous identification: "We have found B, for which we have no explanation; explanations should be sought."
  - action: "On the basis of these data, we should invest in the study of A to make a more definitive argument."

All five types of arguments have "action" as a subtype. The non-action kinds of arguments will be considered first, followed by the action subtypes.

Before doing so, though, it is reasonable to ask why we should bother with a taxonomy like this. The basis of the taxonomy – and the strongest argument for using it – is that arguments can and should be distinguished according to their burdens of proof. The remainder of this subsection makes the case that the five types of arguments and their subtypes do, in fact, have distinct burdens of proof. Drawing on this material, Section 5 makes the case that the usual taxonomies of statistical problems – for example, inference vs. prediction vs. decision, or hypothesis testing vs. point or interval estimation vs. everything else, and so on – do not identify an argument's burden of proof and thus do not differentiate statistical problems usefully.

3.2.1 Non-action arguments.

Inferential causal arguments refer to the causation of an event that occurred in the past or was determined in the past, while predictive causal arguments refer to events, often counterfactual, that will or could occur in the future. The burden of an inferential causal argument is to rule out all but one causal agent. The additional burden of a predictive causal argument is to show that a postulated causal agent will produce a given effect in specified future situations.

For example, consider BW02, the clinical trial that justified US licensing of AZT to treat HIV infection (Fischl et al [7]). BW02 was a randomized, double-blind, placebo-controlled trial in patients who had had an AIDS-defining illness or who had AIDS-related complex (ARC). Competent use of randomization, double-blinding, and placebo controls are the crux of an argument that AZT caused the difference in the number of deaths between the two arms of the trial P an inferential causal argument.

It is harder to make the predictive causal argument that giving AZT to a general population with AIDS will extend their lifespans. This is because BW02:

- enrolled only people with ARC or people who had had exactly one episode of *Pneumocystis carinii* pneumonia (PCP) within six months of enrollment;
- enrolled only people with specific medication histories, in particular, with no prior use of antiretroviral drugs; and
- restricted concomitant medications, in particular, patients could not use prophylaxes for PCP, the most common AIDS-defining condition.

It might be possible to make a predictive causal argument in favor of AZT for patients meeting these conditions, but few such patients exist. A more ambitious predictive causal argument based on BW02 would require a subsidiary argument that AIDS-defining illnesses, medication history, and concomitant medications do not affect the efficacy of AZT. This is most unlikely; for example, lengthy use of AZT degrades its effectiveness (Kahn et al [8]).

For either type of causal argument, the burden of proof is met by a procedure – in this case, the data collection procedure – which is reflected either implicitly or not at all in formulating the statistical problem as a hypothesis test or interval estimation problem. This pattern recurs in the argument types to follow.

Non-causal predictive arguments refer to events that will occur in the future, but they make no assertion of causality. The burden of a non-causal predictive argument is showing that the predictive rule or procedure has the properties claimed for it. One example is economic predictions made with vector autoregressions, which capture relationships in economic time series without postulating causality (Litterman [9]). Another non-causal predictive argument is medical use of prognostic measurements. For example, dozens of studies have shown that in an unselected population of HIV-infected people, CD4+ lymphocyte count is a strong predictor of time to an AIDS-defining disease or to death. This is true even though *changes* in CD4+ are poor predictors of subsequent clinical events in interventional clinical trials (Fleming [20], De Gruttola et al [21], Choi et al [22]). Note that as with the causal arguments, the burden of proof of a non-causal predictive argument is provided by a procedure – in this case, an out-of-sample validation – that is not part of the mathematical formulation of a prediction problem.

Description arguments involve neither causality nor prediction. Direct description arguments also involve no uncertainty or sampling: all of the entities exist and have been measured. For example, certain vendors sell summaries of sales databases gathered by bar-code scanners in consumer-goods stores, mostly grocery stores [23]; in some regions, the sales data are exhaustive. Among other things, grocery stores use these data to answer the question "How am I doing?" compared to nearby grocery stores. The burden of proof in a direct description argument is showing that the summary actually describes the whole for the purpose at hand (Mallows [10], Draper et al [11]). An inferential description is almost a direct description – the entities involved exist and are ***potentially*** measurable – but some of the entities either were not or could not be measured. The obvious example is a census. Large censuses inevitably involve an inference, as the US census undercount controversy has made clear. Another example is estimating the number of birds killed in an oil spill (Carter, Page, and Ford [12]): some number of birds actually died, but it is impossible to count the number that died at sea and were not washed ashore, and if a census of beached carcasses is not exhaustive, it is impossible to count the number that washed ashore dead or died on shore. Any count of dead birds must be by inference. The burden of proof here is showing that the inferential statement has the properties claimed for it.

Estimating natural constants might seem to be inferential description, but it is not, necessarily. Some constants, such as the gravitational constant in Newton's theory of gravity, do not exist in any meaningful sense. Rather, because the theory is manifestly false, although useful, the gravitational constant is a mere tuning knob, like those on a radio, that is adjusted to make the theory fit the data as well as possible – that is, to facilitate predictions. Arguments estimating such constants must therefore be considered predictive. By contrast, other entities, such as the speed of light or the electron's mass, truly exist and thus admit of inferential description arguments. Estimating a given natural constant might initially be a predictive argument but later become an inferential description argument as the constant's *bona fides* are established.

Existence arguments are have a simple structure: some event *has* occurred; therefore, it *can* occur. The burden of existence arguments is showing that the event actually did happen. In medicine, case series are a popular but much-maligned form of existence argument. For example, Blakeman et al [13] described a series of patients who were on heart-transplant waiting lists because of severe congestive heart failure and coronary artery disease. Such patients have bleak prognoses but because of the risk of death during surgery, they are widely believed to be poor candidates for palliatives like coronary artery bypass grafts. Nonetheless, 17 of the 20 patients in Blakeman et al's series survived bypass grafting and 10 of the 17 were radically more able to perform ordinary activities. Thus, bypass grafts can be done on high-risk patients with good results and acceptable operative risk. The series does not identify ***which*** patients should be bypassed, but the result is significant nonetheless.

Another example indicates the possibilities of existence arguments. It is sometimes argued that the human immunodeficiency virus (HIV) cannot cause AIDS because lentiviruses such as HIV cannot work as the conventional view would have us believe. But the simian immunodeficiency virus (SIV) – which is similar in structure to HIV – has been shown experimentally to do in monkeys precisely what HIV is supposed to do in humans. Such viral behavior *has* happened, in monkeys; thus, it *can* happen. This argument does not imply that HIV behaves in humans in accordance with the conventional view; it does undermine the claim that no lentiviruses *can* work in accordance with the conventional view.

For an instance in which the burden of this argument's proof could not be met, consider Koech et al. [24] and Obel and Koech [25], who claimed that HIV-positive patients in their care became HIV-negative upon administration of low-dose oral α-interferon. At the time, no-one had claimed to induce so-called sero-deconversion, so these two reports were greeted with skepticism, and no investigator has been able to replicate the result. All but a few observers have concluded that sero-deconversion did not, in fact, occur. Note again that the burden of proof for existence arguments lies in procedures – external validation – that are not represented mathematically.

The foregoing notwithstanding, existence arguments are limited in scope. Suppose, for example, that an astronomer uses measurements to infer that an unobserved planet must exist. There is an existence argument here, but not a very interesting one: orbital irregularities (say) for this list of planets have occurred, therefore they can occur. The interesting argument is, instead, a predictive causal argument: the observed orbital irregularities were caused by an unobserved planet, which will cause this list of further observable irregularities.

Data-mining arguments are a loose end which is necessary because statisticians and others have not sorted out the relevant issues. The situation is this: you ransack a dataset, check subgroupings, delete apparent outliers, and so on, and find a nominally significant result. The traditional view is that such searches are extremely prone to spurious findings: they do not control Type I errors (Tukey [14]). Few Bayesians have commented explicitly on this issue. Leamer [15] can be interpreted as meaning that data-mining is essentially innocuous. With somewhat greater possibility of injustice, this view might be attributed to Lindley [16], [17]. On the other hand, Berry [18] and Hill [2] admit that data-mining affects the interpretation of a nominally significant result, although they do not give prescriptions.

The traditional view is a hypothesis-generation argument: this data-mining has suggested an interesting hypothesis; perhaps we can make a more compelling argument using another dataset. It captures the belief that, at the current state of theory at least, a causal argument cannot be based on a data-directed search because the chances of error are unknowable. The contrasting

Bayesian view is that data-directed searches can, in fact, be the basis of a predictive causal argument. An intermediate position is that data- directed searches can be the basis of an action causal argument, if not a predictive causal argument (Berry [18], discussed below).

Innocuous identification arguments present a result, usually from an observational study, refute uninteresting explanations like selection effects, and pose a challenge: "explain this result." For example, Neaton and Wentworth [19] used data on over 330,000 men screened for the MRFIT study between 1973 and 1975, along with the National Death Index and Social Security mortality data, to argue that in men between the ages of 35 and 57, low cholesterol and low blood pressure are risk factors for death from AIDS. Cross-sectional studies have shown that patients with HIV disease who are sicker, by various definitions, tend to have lower cholesterol, but it was unclear whether this was a consequence of the disease or a pre-existing condition. Neaton and Wentworth's result imples that it is at least partly a pre-existing condition and cannot be explained by misclassification of cause of death or by an association of sexual orientation with blood pressure and cholesterol. The result poses a challenge: show how cholesterol and blood pressure are causally related to death from AIDS or show that they are not. Neaton and Wentworth do not assert that raising cholesterol or blood pressure can reduce the risk of infection, so it is not a causal or predictive argument.

For both subtypes of hypothesis-generation argument, the burden is to find the result and to rule out as many uninteresting explanations as possible. Yet again, meeting this burden involves procedures not generally reflected in the explicit mathematical formulation of the statistical problem.

3.2.2 Action arguments.

Each type of argument has an action subtype, because the other subtypes need not imply action and vice versa. For example, one may be able to make a compelling action causal argument without being able to sustain either a predictive or inferential causal argument (*cf* Berry [18]). A living example of this is a colleague who has AIDS and uses an unproven anti-HIV drug, peptide-T. Peptide-T is an analog for the protein in the HIV virus's envelope that binds to the CD4 receptor in human cells; it blocks the binding of HIV to these receptors *in vitro*, and thus has been suggested for treating HIV. Observational studies suggest that peptide-T has a potent effect, but the sole controlled trial of peptide-T is measuring only neurological symptoms. Thus, it is difficult to sustain a predictive causal argument of more general clinical benefit. But peptide-T has no known side effects or interactions with other drugs, and my colleague gets his supply *gratis*. The cost and risk are small and the potential benefits immense, so an argument for action – for using peptide-T – is compelling.

Conversely, it is possible to make a cogent inferential or predictive causal argument without being able to sustain an action argument. For example, no-one questions that aerosolized pentamidine (AP) is an effective prophylaxis for PCP, i.e., it is easy to make a strong predictive causal argument that it reduces the chance of PCP. However, neither does anyone question that trimethoprim/sulfamethoxazole (TMS) is a much more effective prophylaxis for people who can tolerate it (Schneider et al [27], Hardy et al [26]). In this case, the action argument about taking AP ("don't take AP, take TMS if you can tolerate it") requires more information than the predictive causal argument about AP ("AP reduces the incidence of PCP").

Any action argument involves an implicit or explicit tabulation of costs and benefits associated with possible courses of action. In decision-theoretic views of statistics, the tabulation is explicit; in both of the arguments above, the tabulation is mostly implicit, although in each case it could be made explicit. (The value of such an exercise is a separate issue.) Whether the tabulation is implicit or explicit, it adds premises to the action arguments.

For the other argument types as for causal arguments, a non-action argument need not imply the corresponding action argument, and vice versa. Brief examples follow. Non-causal predictive: Litterman [9] and those following his lead make strong arguments for predictions, not arguments for actions; on the other hand, political and corporate actors must constantly rationalize action by predictive arguments that no well-informed person could find compelling. Descriptive: Counts of birds killed in oil spills have partly rationalized legal judgments against oil tanker companies; but less well-supported population counts have been used to draw inferences about sea-bird population dynamics. (Gross inferences, to be sure.) Existence: When Blakeman et al (1990) showed that patients with severe heart failure could be radically improved with bypass, avoiding a costly and hazardous heart transplant, it became necessary to act by identifying which patients could be so treated. But if someone discovered the remains of Noah's ark, this would confirm that it existed but the standards by which its existence was verified could not sensibly involve a calculus of costs and benefits. Hypothesis-generation: Had Neaton and Wentworth tried to use a specific calculus of gains and losses, it would have had only formal content. But a pharmaceutical company might reasonably use a subset analysis of a clinical trial, producing weaker evidence than Neaton/Wentworth's, to select among possible future trials in promising patient subgroups.

Now that the argument types have been differentiated according to their burdens of proof, a new diagram of arguments, Figure 2, can replace Figure 1. Commonly, researchers approach an investigation with the intention of eventually drawing a particular kind of conclusion, that is, of making a particular type of argument. The desired type of argument determines the mathematical

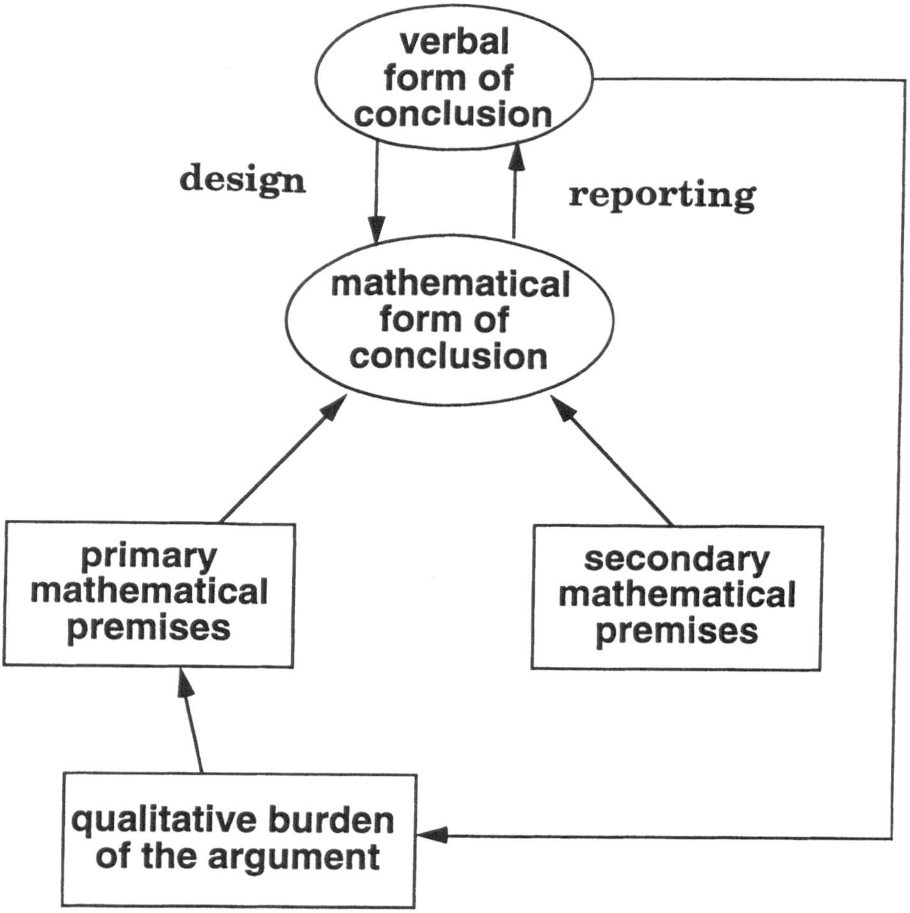

Figure 2: Revised schematic diagram of an argument, showing the relation of the argument type to the premises

3.3 Idea #3: Model-building activities establish a plausible, tractable baseline argument.

In the world portrayed in data analysis classes, certain activities precede model-fitting. First, the dataset is summarized until the analyst understands it as data, that is, to ensure that data items have not been garbled, to figure out coding schemes, and the like. Then, the dataset is summarized some more until the analyst has a few ideas of what it says about the subject-matter problem; for example, pictures are drawn to reveal trends, groupings, and other features that can be modeled. With luck, the result is a few models or data summaries and some tentative arguments that are tractable, have simple elements, and are not plainly inconsistent with the data.

But in some places where statistics is used – clinical trial organizations, for example – models and types of arguments are often specified before the data are collected. As soon as the data are entered into the computer system and run through basic edits, proportional-hazards models are fit to them like a bit and bridle.

Whether your world looks more like the data-analysis class or the clinical trial organization, what you have at this point is a baseline argument. It is an argument in the sense used above; it is a baseline in two senses. First, because it passed the EDA test (in the world of the data-analysis class) or because it has achieved the status of a convention (in the clinical trial world), it is the argument to which variations will be compared, in the hope that the variations have little effect and the baseline argument can be reported. Second, the term "baseline" commonly implies that attempts will be made to improve on it. In the extended theory of statistics, both senses are important. It is desirable to report the baseline argument because it is often simple and has conventional acceptability. On the other hand, it is important that the baseline argument not be accepted without substantial scrutiny.

3.4 Idea #4: Diagnostics and sensitivity analyses vary the premises of the baseline argument and display the resulting variation in the conclusion.

This idea has consequences for measuring the influence of perturbations and for choosing the class of perturbations to be examined.

Measuring the influence of perturbations. We do diagnostics and sensitivity analyses to see if a small change in the data, model, prior, or loss function induces a large change in . . . what? Cook's distance (Cook and Weisberg [28], p. 115) measures the change in the vector of regression coefficients against the inverse of its covariance matrix. Cook [29] assessed the effect of a perturbation by the change it induces in the parameters of the baseline model, measured by the associated change in the value of the baseline likelihood. Some Bayesian analogs use Kullback-Liebler distance to measure how perturbations change the posterior or predictive distributions (Johnson and Geisser [30], McCulloch [31], Carlin and Polson [32]).

These measures are helpful but we lack the words to describe how they help. What *does* one do if a point is influential? Cook and Weisberg ([28], p. 104) gave little advice, noting that "final judgments [about influential cases] must necessarily depend on context, making global recommendations impossible." They suggested some useful considerations – Is the influential point erroneous? Are more data available near the influential point? – but otherwise, they concentrated on methods for detecting influential cases.

The extended theory has a clear implication here: if a perturbation of the baseline argument has large influence according to some measure but does not change the outcome of the argument or suggest a need to change its structure, then . . . who cares? Influence measures should focus on conclusions of arguments, not intermediate steps.

The method of Kass, Tierney, and Kadane [33] allows this focus: it assesses the influence of perturbations on the posterior expectation of a general class of functions of the baseline model's parameters. The disadvantage of this method is that it measures influence locally, around the posterior mode. However, recent advances in Bayesian computing (e.g., Gelfand, Dey, and Chang [34]) should permit non-local influence measures for general functions.

Choosing the class of perturbations to be examined. Premises of a statistical argument often include exact mathematical specifications. Some specifications are explicit, such as probability density functions, while others are implicit, such as exchangeability assumptions (Draper et al [35]). It is rare to see a demonstration of strong support for the specifics of an argument's premises. More commonly, a case is made that the argument does not depend on the specifics, that is, the baseline argument is used for convenience and then shown to be *only* a convenience.

This is in the spirit of the extended theory, but one generally sees only modest variations on premises. Leave-one-out diagnostics are fairly common (Carlin [36]), as are checks of (say) logistic-regression-in-place-of-probit (Gibbons et al [37]) or if-we-used-this-different-measurement (Caulkins and Padman [38]). However, we see hardly any checks for failures of proportional hazards – and SAS even supplies a test! It is truly extraordinary to see the full range of an argument's premises bent to see where the argument breaks.

Why are we so timid? Why not bend all of an argument's premises? Tractability was an acceptable excuse ten years ago; perhaps it still is, but it will certainly not be in ten years. Why do we not use our computing power – raw CPU power and flexible modeling systems like generalized linear models – to put experimental design methods to work exploring variations of premises?

Of course, any applied statistician can supply plausible excuses. For example, we think we know which variations are important, and we examine those variations. But do we know which variations are important? For linear regressions with normal errors, we probably do. For other models, the research does not support such an assertion, especially not for the bigger, less comprehensible models that computing power has made possible. (For example, see Zaslavsky [39].) And when we do know which variations are important, do we examine them? Judging from the medical literature, almost never. It is harder to assess other fields, but based on editing experience and conferences, if variations are examined, they are not discussed.

Well, can we not handle such difficulties qualitatively? For example, editors of medical journals expect papers to report losses to follow-up, and if such problems are bad enough the editors will reject the paper. But there is a middle ground between rejecting papers and dropping standards entirely: if an argument can be shown to hold in spite of problems, then it has value. (Crawford et al [40] make an argument like this.)

At bottom, one senses a worry that if we bend premises too much, we will never sustain any results. But perhaps that is the right answer: maybe we need to find out how far we must bend premises to break a result, and then let people sort out their beliefs about the result according to their beliefs about how much the premises *should* be bent. This leads to the last of the five ideas.

3.5 Idea #5: An argument is strong to the extent that the premises are supported by strong arguments or the region of premises yielding the same conclusion is large.

A premise is strongly supported when it is the conclusion of a strong argument. (This appropriately raises the specter of infinite regress.) When a premise is strongly supported, there is no need to examine variations of it or, at least, not large variations. It is easy to think of strongly supported qualitative premises – for example, that a coefficient should be positive because it can't be negative – but difficult to find instances of strong support for mathematical specifics like normality or linearity. Ehrenberg and Bound [41] give some exceptions in the field of marketing; their exceptional quality suggests that we heed Ehrenberg and Bound's call to seek regularity across many sets of similar data.

It is more common to find arguments that are strong because the specifics of the premises do not matter, for example, the same result is obtained whether the errors are assumed to be normal or t. If the outcome of an argument is the same for all variations of its premises, the argument is strong. If the outcome of the argument varies as the premises vary, then you can argue about metrics on the space of premises, but ultimately the choice of a metric is subjective.

This is another way to reach the conclusion, sometimes attributed to D.F. Andrews, that "objectivity is a hoax": an argument's strength can only be defined with respect to a specific body of knowledge and belief. One cannot foreclose the possibility that ten years into the future, someone will invent a new way to bend premises that *does* change the outcome of the argument, at which point the old, sturdy argument falls or surrenders some of the ground it occupies, as Newtonian mechanics did a century ago.

4 A simple example illustrates the extended theory's mechanics.

This example is from Carlin and Louis [42] (henceforth CL), who were motivated by a clinical trial in which Bayesian monitoring was conducted in parallel to conventional monitoring (Carlin et al [43]). The trial compared pyrimethamine to placebo as prophylaxes against toxoplasmic encephalitis in HIV-infected people. For the time-to-event analysis below, the endpoint was death from any cause. The investigators wanted to make an inferential causal argument about the efficacy of pyrimethamine, which CL formulated as a Bayesian test of a point-null hypothesis against a composite alternative. (Other formulations might have made sense; this is discussed in Section 5.) The primary premises are that the treatment groups did not differ systematically in baseline characteristics, concomitant treatments, or in ascertainment of deaths. These are supported by the procedural features of randomization and blinding. CL focused on the secondary mathematical premises.

The baseline model – specified in the trial's protocol document – was a proportional-hazards regression. CL used two explanatory variables for each patient: a dummy indicating treatment group (1 for pyrimethamine, 0 for placebo) and the CD4+ count at entry into the trial. CL integrated the CD4+ coefficient out of the partial likelihood, leaving the marginal (partial) likelihood for the pyrimethamine coefficient. Call that coefficient θ and call the marginal likelihood $f(x|\theta)$, where x represents the data from the trial. The explicit mathematical form of the baseline argument was:

- Null hypothesis, $H_0 : \theta = 0$; the prior probability of H_0 is $\pi = 0.25$;
- Alternative hypothesis, $H_a : \theta \neq 0$, where the prior cumulative distribution function of θ under H_a is $G(\theta)$; and
- Test: if $P(H_0|x) > p$, choose H_0, otherwise choose H_a, where $p = 0.1$.

Applying the extended theory. To use the extended theory, one must first fix an aspect of the problem that is essential to its definition. For CL, that aspect was θ, the log relative hazard of patients receiving pyrimethamine compared to those receiving placebo. Any other aspect of the problem formulation is a premise that can be varied: exchangeability of the treatment groups, the proportional hazards assumption, the use of a partial instead of a full likelihood, $\pi, p, G(\theta)$, etc.

Next, the baseline model is expanded to represent the premise variations under consideration, with the baseline premises corresponding to a specific prior on the expanded space. The term "model expansion" may cause some confusion: in the extended theory, it includes not just familiar (and typically modest) model expansions, but any and all premise variations.

Having expanded the model, the next step is to examine priors on the expanded space to find ones that change the baseline argument's conclusion. If any are found, the last step is to understand why those variations change the conclusion.

CL considered only variations on G and constrained them by specifying quantiles. That is, they defined an indifference zone (an interval such that they are indifferent between θ in this region and $\theta = 0$; see Freedman and Spiegelhalter [44]) $(x_L, x_U) = (-0.288, 0)$ for θ and defined the constraints by $P_G(\theta \leq x_L) = a_L$ and $P_G(\theta > x_U) = a_U$, for specified $a_L, a_U \in [0, 1]$ satisfying $a_L + a_U < 1$. The maximum partial likelihood estimate of θ was 0.6, that is, pyrimethamine appeared to decrease survival time. Given the data, π, p, and the indifference zone, for any pair (a_L, a_U) the boundary of interest is linear: if $a_U > 0.145a_L + 0.273$, then a prior G exists that satisfies the constraints and results in the rejection of H_0. Otherwise, no prior permits rejection of H_0.

This is the boundary of variations at which the result changes. Its meaning is simple: if G puts enough probability above the indifference zone, it is possible to reject H_0; otherwise, it is not possible. This much is trivial; the non-trivial part is that CL quantified "enough" precisely, in the linear equation given above. (Sargent and Carlin [45] extended this result three ways, by using interval null hypotheses, including sample size considerations, and allowing π to vary.)

5 A generalization of the simple example suggests implications for foundations, methods, and computing.

The example in Section 4 can be generalized to derive five steps in applying the extended theory. In practice, it will often be necessary to iterate among the steps.

Step 1. Fix the aspect of the problem that is essential to its definition. There may be more than one way to do this. For example, suppose CL had varied the proportional hazards premise, replacing θ, the log relative hazard of pyrimethamine, by $\gamma + g(t)$, where t is time and $g(t)$ is a function of time. Are we interested in γ, or the whole function $\gamma + g(t)$, or some average of it? When the essential aspect of the problem has no single formulation, it may be necessary to vary the formulation in addition to varying baseline premises.

Step 2. Do EDA/model-building to form a baseline argument (Ideas #2 and #3). This is still art, not science, but the extended theory gives it a bottom line: get to an argument, or to the conclusion that no interesting argument can be made.

Step 3. Specify model expansions (Ideas #3 and #4). Each mathematical specific of the baseline premises is a *convenience* parameter: each specific is of no intrinsic interest but conveniently permits arguments to be built. Even the essential aspect of the problem is often just a convenience. This step requires a model expansion corresponding to each baseline premise that is not strongly supported.

Recall that "model expansion" includes *everything*. Whatever the premise variations, they can all be formulated as model expansions even if the various models cannot be nested. Sometimes this is awkward, but it can be done.

Step 4. Do a restricted search for priors on the convenience parameters that change the conclusion (Ideas #3 and #4). The search is restricted because it rarely makes sense to consider absolutely everything. One unhappy feature of robust Bayesian approaches generally and CL's approach in particular is that the G yielding extreme solutions are usually bizarre priors that nobody would advocate. Thus, it is desirable to impose smoothness constraints by means of, say, bounds on derivatives. Such constraints will partly address the concern mentioned in Section 3, that no result will ever stand up under the premise variations advocated here. The trick is to constrain priors in ways that are substantively innocuous but express genuine beliefs, like unimodality and continuity.

Step 5. On finding the boundary of such priors, figure out why the conclusion changes there (Idea #5). The baseline argument should be reported if it stands up under variations, but we also need to report where the baseline argument breaks in the expanded premise space, and why it breaks there. Does it break when we delete a few odd points? Is it so fragile that it breaks with some combination of modest deviations from several premises, or is a catastrophic failure of some premise needed? Such qualifications of the baseline argument are as important as the baseline argument itself.

5.1 Implications for foundations.

The most important implication is that the sensitivity analysis *is* the analysis, and that a sensitivity analysis must focus on the effects of perturbations on the *conclusion* of the baseline argument Sensitivity checks are not a mere nicety; they are central to the analysis.

Another implication is that the usual taxonomies of statistical problems – for example, inference vs. prediction vs. decision, or hypothesis testing vs. point or interval estimation vs. everything else – do not make useful distinctions among statistical problems. One might say, paraphrasing de Finetti, that inference does not exist: "inference" is not a separate type of argument, but several subtypes of different types of argument. Perhaps the futile quality of the foundational dispute over inference arises in part because there is no such thing as "inference", but, rather, qualitatively different types of inference. Similarly, there is no single kind of prediction or decision, but qualitatively different kinds of each. Finally, although it serves a mechanical purpose to formulate a problem as a hypothesis test or an interval estimate, it is clear that tests, estimates, and intervals play a role in most or all of the types of arguments.

The extended theory also makes it clear – if you are not already convinced – that it is not helpful to reduce all statistical arguments to exercises in decision theory. While non-action arguments can often be cast in decision-theoretic terms, it is sterile to do so. How could any meaningful loss function be constructed for the Neaton/Wentworth argument or for the causal arguments? The main burden of such arguments is not picking a particular estimate or posterior distribution, but sustaining the finding against alternative explanations. But fear not, purists: it is possible, after all, to use Savage's axioms for probability without using the axioms for utility.

A final implication, which cannot be treated here, is that past a certain point it is usually futile to try to express all variability and uncertainty as probability. For most argument types, it would be a waste of time to specify a probability distribution on the premise variations and integrate it out, because it would be just one more aspect of the problem to vary. (But perhaps not always; Hodges [46] discusses possible exceptions.)

5.2 Implications for methods.

Four of the five steps in applying the extended theory have immediate implications for methodological development.

Step 1. Each type of argument has a small group of characteristic problem formulations. For example, CL's inferential causal argument was formulated as a Bayesian hypothesis test. We need to catalog problem formulations for each argument type and develop CL-style setups friendly to perturbations of the generality discussed here.

Step 3. Step 2, the EDA/model-building step, produces a tractable baseline argument. Often this will involve a standard model, such as linear regression, so statisticians will routinely need to examine variations on the premises of these standard models. Thus, we will need standard "baskets" of model expansions for standard models. Of course, some users will need to go beyond the standard basket of model expansions, and one research challenge is to figure out how to allow them to do so without respecifying the standard model expansions.

Step 4. This step is a search through priors on the expanded model space for priors that change the conclusion of the baseline argument. To do this readily, we need friendly classes of priors on the standard baskets of model expansions, and constraints on them corresponding to conditions like continuity and unimodality.

Step 5. We need more ways to assess the effects of perturbations on conclusions of arguments. Kass, Tierney, and Kadane [33] suggests one approach; a computational idea will be suggested below. When the search is finished, we need ways to describe the boundary where the baseline argument's conclusion changes. CL drew a simple picture to summarize their search; more complicated model expansions will require more ingenious pictures.

5.3 Implications for computing.

It is not easy to search for the boundary where the conclusion changes. Explicit solutions are awkward for minor elaborations of CL's problem and for more general problems we cannot avoid computer-intensive searches.

One possible approach is an environment that searches premise variations stochastically, maps the boundary, and suggests diagnostics to elucidate why the conclusion changes there. Ordinarily, if we are using (say) a linear regression model, we have a dozen or so standard diagnostics that we apply, one by one, looking for problems. We may use all the standard diagnostics and find nothing, or a particular diagnostic may indicate a problem which we then pursue with other diagnostics. The computing system described above would invert this process by searching for places in premise space where the conclusion of the baseline argument changes and, by reference to the model expansion that seems to cause the change, displaying diagnostics that indicate what has gone wrong.

One might view the latter as an intelligent agent that does diagnostics for the human user. Other expert systems can help the human implement Step 2, the EDA/model-building step. Particularly in large data sets, it may not be cost-effective to have the human waiting while the computer does arithmetic; rather, it may be more efficient to have an intelligent agent do EDA/model-building and bring back a summary that the human can peruse at her convenience. The difficulty is designing an interface that allows the human to direct the agent so that it does not bring back mostly junk.

6 Antecedents.

The ideas in this paper have several immediate antecedents and, of course, innumerable less immediate predecessors. In a paper of this scope, only the former can be given their due; my apologies to the latter.

The nearest antecedent is a catalog of uses of simulation models in policy analysis (Hodges [47], Hodges and Dewar [48]) and Bankes' [49] notion of exploratory (as opposed to predictive) modeling which was followed by statistically-oriented unpublished work by Bankes and JL Adams. These streams merged and incorporated experimental design ideas in Dewar et al [50].

Carlin and Louis [42] is also a direct antecedent. Although their results are technically modest and related to other results in Bayesian robustness, they are a departure among statisticians in their focus on the conclusion of the analysis. (Others, such as decision analysts, have long focused on conclusions.)

Smith [51] suggested the idea of extravagant sensitivity analyses by noting that Bayes' theorem allows us to "report a rich range of the possible belief mappings induced by a data set"; he was referring to individual parameters, but the idea can readily be expanded. Cox and Snell [52] differentiated the primary aspect of a model, specifying the main question of interest, and the secondary aspects, which complete the model and indicate a suitable analysis. This is an ancestor of the notion of the essential aspect of a problem and of the distinction between primary and secondary premises. Cook and Weisberg (e.g., Cook and Weisberg [28], Weisberg [53]) advocated model expansion, which Weisberg credits to Cox [54]. Cook [29] and Ramsay and Novick [55] introduced methods for diverse perturbations, the latter in the form of PLU (prior-likelihood-utility) robust decision theory. Kass, Tierney, and Kadane [33] developed measures of case influence on general functions of parameters. Finally, the Bayesian robustness literature (Wasserman [56] is a recent survey) has focused on robustness to prior distributions, although some authors have considered sensitivity to other aspects of the baseline argument.

Other antecedents come from off the beaten track. Smith [1], Hill [2], and Berger [3] discussed Bayesian notions of data analysis, with the consensus that non-Bayesian methods are acceptable during model-building. Kadane and Schum [57] discussed Bayesian thinking within a scheme of diagramming legal arguments. Lindley and Singpurwalla [58] considered the amount and type of evidence needed to obtain agreement between adversaries with different prior information and loss functions. Rosenbaum (e.g., [59]) pushed randomization theory in an unusual direction with methods for sensitivity analyses in observational studies. These methods evaluate the strength of relationship between the mechanism treatment assignment and the outcome that is necessary to overturn an apparent causal relation between the treatment itself and the outcome. Finally, Leamer [15] incorporated model-searching into formal statistical inference and Leamer [60] catalogued distinct patterns of model-searching. I find it unhelpful to think of uses of statistics as specification searches, but the debt of the present paper to Leamer is clear.

Acknowledgements

The RAND blender mixed my ideas inextricably with those of John Adams, Steve Bankes, and Jim Dewar. Thanks also go to Brad Carlin, David Freedman, David Lane, Tom Louis, and Dan Sargent for helpful comments and discussion.

References

[1] Smith, A.F.M. 'Some Bayesian thoughts on modelling and model choice', *The Statistician*, **35**, 97–102 (1986).

[2] Hill, B.M. 'A theory of Bayesian data analysis', In *Bayesian and Likelihood Methods in Statistics and Econometrics: Essays in Honor of George A. Barnard*, eds. S Geisser, JS Hodges, SJ Press, A Zellner, Amsterdam: North-Holland, 49–74 (1990).

[3] Berger, J.O. 'Contributed discussion', In *Case Studies in Bayesian Statistics*, eds. C Gatsonis, JS Hodges, RE Kass, ND Singpurwalla. New York: Springer-Verlag, 302–303 (1993).

[4] West, M. and Harrison, J. *Bayesian Forecasting and Dynamic Models.* New York: Springer-Verlag (1989).

[5] Fisher, R.A. *Statistical Methods and Scientific Inference*, 3rd edition, New York: Hafner (1973).

[6] Efron, B. 'Why isn't everyone a Bayesian?', *American Statistician*, **40**, 1-11 (1986).

[7] Fischl, M.A., Richman, D.D., Grieco M.H., et al. 'The efficacy of azidothymidine (AZT) in the treatment of patients with AIDS and AIDS-related complex', *New England Journal of Medicine*, **317**, 185–191, (1987).

[8] Kahn, J.O., Lagakos, S.W., Richman, D.D., et al. 'A controlled trial comparing continued zidovudine [AZT] with didanosine in human immunodeficiency virus infection', *New England Journal of Medicine*, **337**, 581-587, (1992).

[9] Litterman, R.B. 'A statistical approach to economic forecasting', *Journal of Business and Economic Statistics*, **4**, 1–4, (1986).

[10] Mallows, C.L. 'Data description', In *Scientific Inference, Data Analysis, and Robustness*, eds. GEP Box, T Leonard, C-F Wu, Academic Press, 135-152 (1983).

[11] Draper, D.C., Hodges, J.S., Mallows, C.L., and Pregibon, D. 'Exchangeability and data analysis (with discussion)', *Journal of the Royal Statistical Society*, Series A, **156**, 9–37 (1993).

[12] Carter, H.R., Page, G.W., and Ford, R.G. 'The importance of rehabilitation center data in determining the impacts of the 1986 oil spill on marine birds in central California', *Wildlife Journal*, **10**, 9–14 (1987).

[13] Blakeman, B.M., Pifarre, R., Sullivan H., et al. 'High-risk heart surgery in the heart transplant candidate', *Journal of Heart Transplantation*, **5**, 468–472 (1990).

[14] Tukey, J.W. 'Some thoughts on clinical trials, especially problems of multiplicity', *Science*, **198**, 679–684 (1977).

[15] Leamer, E.E. 'False models and post-data model construction', *Journal of the American Statistical Association*, **69**, 122-131 (1974).

[16] Lindley, D.V. Discussion of Efron [6]

[17] Lindley, D.V. 'The 1988 Wald Memorial Lectures: The present position in Bayesian statistics (with discussion)', *Statistical Science*, **5**, 44-89 (1990).

[18] Berry, D.A. 'Subgroup analyses', *Biometrics*, **47**, 1227-1230 (1990).

[19] Neaton, J. and Wentworth, D. 'Relationship of serum cholesterol and blood pressure measured prior to HIV-infection with risk of death from AIDS', Unpublished manuscript, Division of Biostatistics, School of Public Health, University of Minnesota.

[20] Fleming, T.R. 'Surrogate markers in AIDS and cancer trials', *Statistics in Medicine*, in press (1995).

[21] De Gruttola, V., Wulfsohn, M., Fischl, M.A., and Tsiatis, A.A. 'Modeling the relationship between survival and CD4 lymphocytes in patients with AIDS and AIDS-related complex', *Journal of Acquired Immune Deficiency Syndromes*, **6**, 359–365 (1993).

[22] Choi, S., Lagakos, S.W., Schooley, R.T., and Volberding, P.A. 'CD4+ lymphocytes are an incomplete surrogate marker for clinical progression in persons with asymptomatic HIV infection taking zidovudine', *Annals of Internal Medicine*, **118**, 674–680 (1993).

[23] Schmitz, J. 'Massive Marketing Datasets', presented on 7 July 1995 at "Statistical Challenges and Possible Approaches in the Analysis of Massive Data Sets," a conference sponsored by the Committee on Applied and Theoretical Statistics; Washington, DC.

[24] Koech, D., et al. 'Low-dose oral alpha-interferon therapy for patients seropositive for human immunodeficiency virus type-1 (HIV-1)', *Molecular Biotherapy*, **2**, 91–95 (1990).

[25] Obel, A.O. and Koech, D. 'Outcome of intervention with or without low-dose oral alpha-interferon in 32 HIV-1 seropositive patients in a referral hospital', *East African Medical Journal*, **67(7)**, 71–76 (1990).

[26] Hardy, W.D., Feinberg, J., Finkelstein, D.M., et al. 'A controlled trial of trimethprim-sulfamethoxazole or aerosolized pentamidine for secondary prophylaxis of *Pneumocystis carinii* pneumonia in patients with the acquired immuneodeficiency syndrome', *New England Journal of Medicine*, **327**, 1842–1848 (1992).

[27] Schneider, M.M.E., Hoepelman, A.I.M., Schattenkerk, J.K.M.E, et al. 'A controlled trial of aerosolized pentamidine or trimethoprim-sulfamethoxazole as primary prophylaxis against *Pneumocystis carinii* pneumonia in patients with human immunodeficiency virus infection', *New England Journal of Medicine*, **327**, 1836–1841 (1992).

[28] Cook, R.D. and Weisberg, S. *Residuals and Influence in Regression*, New York: Chapman and Hall. (1982).

[29] Cook, R.D. 'Assessment of local influence (with discussion)', *Journal of the Royal Statistical Society*, Series B, **48**, 133–169 (1986).

[30] Johnson, W.O. and Geisser, S. 'A predictive view of the detection and characterization of influential observations in regression analysis', *Journal of the American Statistical Assn.*, **78**, 137–144 (1983).

[31] McCulloch, R.E. 'Local model influence', *Journal of the American Statistical Assn.*, **84**, 473–478 (1989).

[32] Carlin, B.P. and Polson, N.G. 'Monte Carlo Bayesian methods for discrete regression models and categorical time series', In *Bayesian Statistics 4*, eds. J.M. Bernardo, J.O. Berger, A.P. Dawid, A.F.M. Smith, Oxford University Press, 577–586, (1992).

[33] Kass, R.E., Tierney, L., and Kadane, J.B. 'Approximate methods for assessing influence and sensitivity in Bayesian analysis', *Biometrika*, **76**, 663–674, (1989).

[34] Gelfand, A.E., Dey, D.K., and Chang, H. 'Model determination using predictive distributions with implementation via sampling-based methods', In *Bayesian Statistics 4*, eds. J.M. Bernardo, J.O. Berger, A.P. Dawid, A.F.M. Smith, Oxford University Press, 147–167, (1992).

[35] Draper, D., Hodges, J.S., Mallows, C.L., and Pregibon, D. 'Exchangeability and data analysis (with discussion)', *Journal of the Royal Statistical Society*, Series A, **156**, 9–37 (1993).

[36] Carlin, B.P., Kass, R.E., Lerch, F.J., and Huguenard, B.R. 'Predicting working memory failure: A subjective Bayesian approach to model selection', *Journal of the American Statistical Assn.*, **87**, 319–327 (1992).

[37] Gibbons, R.D., Hedeker, D., Charles, S.C., and Frisch, P. 'A random-effects probit model for predicting medical malpractice claims', *Journal of the American Statistical Assn.*, **89**, 760–767 (1994).

[38] Caulkins, J.P. and Padman, R. 'Quantity discounts and quality premia for illicit drugs', *Journal of the American Statistical Assn.*, **88**, 748–757 (1993).

[39] Zaslavsky, A.M. 'Combining census, dual-system, and evaluation study data to estimate population shares', *Journal of the American Statistical Assn.*, **88**, 1092-1105 (1993).

[40] Crawford, S.L., Johnson, W.G. and Laird, N.M. 'Bayes analysis of model-based methods for nonignorable nonresponse in the Harvard Medical Practice Survey', In *Case Studies in Bayesian Statistics*, eds. C. Gatsonis, J.S. Hodges, R.E. Kass, N.D. Singpurwalla, New York: Springer-Verlag, 78–117 (1993).

[41] Ehrenberg, A.S.C. and Bound, J.A. 'Predictability and prediction (with discussion)', *Journal of the Royal Statistical Society*, Series A, **156**, 167–206 (1993).

[42] Carlin, B.P. and Louis, T.A. 'Identifying prior distributions that produce specific decisions, with application to monitoring clinical trials.', In *Bayesian Analysis of Statistics and Econometrics: Essays in Honor of Arnold Zellner*, eds. D.A. Berry, K.M. Chaloner, J.K. Geweke, New York: Wiley. (1996).

[43] Carlin, B.P., Chaloner, K.M., Louis, T.A., and Rhame, F.S. 'Elicitation, monitoring, and analysis for an AIDS clinical trial', In *Case Studies in Bayesian Statistics, Volume II*, eds. C. Gatsonis, J.S. Hodges, R.E. Kass, N.D. Singpurwalla, New York: Springer-Verlag (1995).

[44] Freedman, L.S. and Spiegelhalter, D.J. 'Application of Bayesian statistics to decision making during a clinical trial', *Statistics in Medicine*, **11**, 23–35 (1992).

[45] Sargent, D. and Carlin, B.P 'Robust Bayesian design and analysis of clinical trials via prior partitioning (with discussion)', Research Report 94-016, Division of Biostatistics, University of Minnesota, 1994. To appear in the IMS Lecture Note Series.

[46] Hodges, J.S. 'Uncertainty, policy analysis, and statistics (with discussion)', *Statistical Science*, **2**, 259–291 (1987).

[47] Hodges, J.S. 'Six (or so) things you can do with a bad model', *Operations Research*, **39**, 355–365 (1991).

[48] Hodges, J.S. and Dewar, J.A. 'Is it you or your model talking? A framework for model validation', RAND, R-4114-AF/A/OSD, Santa Monica, California (1992).

[49] Bankes, S.C. 'Exploratory modeling and the use of simulation for policy analysis', RAND, N-3093-A, Santa Monica, California (1992).

[50] Dewar, J.A., Bankes, S.C., Hodges, JS., et al. 'Credible uses of the Distributed Interactive Simulation (DIS) System' RAND, MR-607-A, Santa Monica, California (1995).

[51] Smith, A.F.M. 'Present position and potential developments: Some personal views of Bayesian statistics (with discussion)', *Journal of the Royal Statistical Society*, Series A, **147**, 245–259 (1984).

[52] Cox, D.R. and Snell, E.J. *Applied Statistics: Principles and Examples.* London: Chapman and Hall (1981).

[53] Weisberg, S. 'Some principles for regression diagnostics and influence analysis: Comments on "Developments in linear regression methodology: 1959-1982" by RR Hocking', *Technometrics*, **25**, 240–244 (1983).

[54] Cox, D.R. 'Nonlinear models, residuals, and transformations', *Math. Operationsforsch. Statist. Ser. Statistics*, **8**, 3–22 (1977).

[55] Ramsay, J.O. and Novick, M.R. 'PLU robust Bayesian decision theory: Point estimation', *Journal of the American Statistical Association*, **75**, 901–907 (1980).

[56] Wasserman, L. 'Recent methodological advances in robust Bayesian inference', In *Bayesian Statistics 4*, eds. J.M. Bernardo, J.O. Berger, A.P. Dawid, A.F.M. Smith, Oxford University Press, 483–502 (1992).

[57] Kadane, J.B. and Schum, D.A. 'Opinions in dispute: The Sacco-Vanzetti case', In *Bayesian Statistics 4*, eds. J.M. Bernardo, J.O. Berger, A.P. Dawid, A.F.M. Smith, Oxford University Press, 267–287 (1992).

[58] Lindley, D.V. and Singpurwalla, N.D. 'On the evidence needed to reach agreed action between adversaries, with application to acceptance sampling', *Journal of the American Statistical Association*, **86**, 933–937 (1991).

[59] Rosenbaum, P.R. 'Sensitivity analysis for certain permutation inferences in matched observational studies', *Biometrika*, **74**, 13–26 (1987).

[60] Leamer, E.E. *Specification Searches*, New York: Wiley (1978).

Individual Rationality and Social Efficiency in an Information Contagion Model

David Lane
Department of Political Economy
University of Modena
Email: lane@unimo.it

Abstract

In the Arthur-Lane information contagion model, agents choose sequentially between two competing products, basing their decisions upon information obtained from a sample of previous adopters. The market shares that each product obtains depend upon the true difference in performance between the products, but also on the number of previous adopters each agent samples and the way in which agents use the sample information to guide their product choice.

If an agent in the information contagion world were to consult a statistician, he would surely be advised to sample more previous adopters rather than less (assuming observations are costless) and to base his decision rule on the value of sufficient statistics for the products' unobservable performance characteristics. Bayesian statisticians might also recommend that the agent choose the product that maximizes his expected utility.

Surprisingly, these recommendations can lead to undesirable effects at the social level. First, giving individual agents access to more information can lead to smaller market share for the superior product. Second, a simple rule-of-thumb based on insufficient statistics always leads to an asymptotic market share of 100% for the better product. No rule based on sufficient statistics enjoys this property. In particular, Bayesian optimization can result in substantial market share for the inferior product.

1 Introduction

Consider the following simple decision problem. You have to choose between two new competing products, A and B. The value to you of these products depends on their performance characteristics, c_A and c_B respectively. If you knew these two numbers, you would select the product associated with the larger one of them. Unfortunately, the publicly available information about c_A and c_B is quite vague. To augment this information, you decide to query n previous adopters of the products. From each of these individuals, you will learn two things: which product she adopted, say X; and an estimate Y of c_X, which you believe to be normally distributed with mean c_X and variance 1.

Suppose you decide to consult a statistician to help you deal with this problem. Virtually all statisticians would concur in the following two pieces of advice:

R1. If observations are costless, the bigger is n, the better. After all, the more information you get about each product, the better you can estimate c_A and c_B, and so the surer you can be that the product that seems to have the higher performance characteristic really is better.

R2. Sufficiency principle: your decision should depend on the data through the values of the sufficient statistics for c_A and c_B (which, under the given model, are the sample averages of the estimates obtained for the two products). Ever since Fisher first formulated the concept of sufficiency, leading exponents of every major statistical ideology have subscribed to the sufficiency principle, in practice if not as explicit dogma.

If you consulted a Bayesian statistician,[1] you would likely receive in addition the following recommendation, which constitutes a fundamental cornerstone of Bayesian statistical faith that is buttressed by theoretical derivations due to Savage, de Finetti and many others:

R3. Maximize expected utility: assess prior distributions for c_A and c_B and a utility function on performance characteristics, calculate the expected utility for each product, and choose the product associated with the higher expected utility.

Recommendations R1-3 are aimed at an *individual* decision-maker. In this paper, I point out some surprising *social* implications of these recommendations. The relevant society consists of an infinite population of homogeneous agents, who will choose sequentially between A and B. All the agents in this population have access to the same public information about the products, which they each[2] supplement by randomly sampling n previous adopters. Thus, each agent faces the individual decision problem described above—and then, having made her choice, enters the pool of previous adopters, from which future adopters sample and learn. Since agents are homogeneous, each uses the same procedures to sample previous adopters and to determine *which* product to choose on the basis of the information they obtain from their samples, and all the agents share the same utility function.

In the world described above, the product with the higher performance characteristic is "truly" better than the other. The focal question for this paper is: what percentage of the agents end up adopting the better product? This clearly is not a question that individual agents can answer, at least with the information structure described above. After all, they cannot even know for sure which product is better, never mind its asymptotic market share! But what agents cannot know may nevertheless materially affect them: increasing the percentage of agents that adopt the better product can convey real, measurable benefits to a society (and hence to its members), including competitive advantages over other societies. For example, suppose that agents use the products to manufacture some commodity—say food or weapons—and that the performance characteristic measures how much commodity can be produced with a unit of factor input; then how much the society can produce—and consequently how well-fed or -armed it is—is an increasing function of the percentage of agents that adopt the better product.

[1] And most neoclassical economic theorists.

[2] After an initial "seed" of adopters, as described in the next section.

Given the true values c_A and c_B, the limiting market share each product attains depends on the procedures the agents use to gather their information and make their choice. I will call the asymptotic proportion of agents adopting the better product the *social efficiency* of a particular set of such procedures. Now I can pose the primary problem this paper addresses: is it necessarily advantageous at the *social* level for the *individual* agents comprising the society to follow recommendations R1-R3?

The paper proceeds as follows. Section 2 describes the product choice model on which the model is based. Sections 3 and 4 analyze the social efficiency associated with two types of information-processing and decision rules. In Section 3, agents are Bayesian optimizers. I assume particular parametric forms for their prior distributions and utility functions. The limiting market shares for the two products depends on these parameters and on the true values c_A and c_B in very complicated ways. I highlight two surprising aspects of this dependence. First, for a wide range of parameter values, social efficiency is *not* an increasing function of the number of previous adopters each agent samples. In some examples, a sample size of 2 guarantees that the superior product will capture the whole market asymptotically, while the inferior product will retain substantial market share if each agent samples dozens or even hundreds of previous adopters. Second, for some parameter values, the percentage of agents that adopt the inferior product does not go to 0, *no matter how great the difference between the true values* c_A *and* c_B.

In Section 4, I analyze what happens when agents use a simple rule-of-thumb: pick the product associated with the highest observed value in the sample. I prove the remarkable fact that, if all agents use this rule, the percentage of agents adopting the superior product converges to 1, *no matter how small the difference between* c_A *and* c_B. Under the model premised in Section 2, this rule depends on the data through the values of statistics that are not sufficient for c_A and c_B. Taken together, then, the results of Sections 3 and 4 imply that, whatever the merits of R1-R3 at the individual level, they may fail to produce desirable results at the social level. What is good for each may not be good for all.

Section 5 concludes the paper.

A note of caution: the analysis presented in this paper is restricted to the stylized case in which the number of adopters goes to infinity. In this case, the proportion of agents who adopt each product converges with probability one, for any posited information-processing and decision procedure. Moreover, the support of the asymptotic distribution can be calculated analytically. These facts allow succinct comparisons of the social effects of different individual information-processing and decision procedures. Needless to say, the gain in simplicity and conceptual clarity afforded by asymptotics may be offset by a loss in practical relevance. For example, the asymptotic results often turn out to be independent of posited initial conditions. This eliminates two extra dimensions in parameter space, but also obscures any "first-mover" effects, which can be critically important in real competitive situations.

2 THE INFORMATION CONTAGION MODEL

The product choice model, introduced in Arthur and Lane[2] and there called the *information contagion* model, depends on five numbers and a decision rule:

- c_A and c_B are real numbers, supposed unknown.
- n, r and σ are positive integers, with $n \leq r + s$.
- the decision rule D is a function from $\{\{A, B\}xR\}^n \rightarrow \{A, B\}$.

Begin with r and s agents who have adopted products A and B respectively. The next agent—called agent 1—selects n of these adopters at random, without replacement. For the ith agent in his sample, agent 1 observes two random variables: X_{1i} and Y_{1i}. X_{1i} takes values in $\{A, B\}$ and identifies the product that the ith sampled agent adopted. Y_{1i} is a normal random variable, with mean $c_{X_{1i}}$ and variance 1.[3] Given $\{X_{11}, \ldots, X_{1n}\}$, $Y_{11}, \ldots Y_{1n}$ are independent. Agent 1 then adopts a product of type $D((X_{11}, Y_{11}), \ldots, (X_{1n}, Y_{1n}))$.

For $j \geq 2$, agent j selects n of the $r + s + (j - 1)$ previous adopters. He observes (X_{ji}, Y_{ji}), which has the same stochastic structure as (X_{1i}, Y_{1i}), and then he determines which product type to adopt by calculating $D((X_{j1}, Y_{j1}), \ldots, (X_{jn}, Y_{jn}))$.

To complete the stochastic specification of the model, we suppose that $Y_1, Y_2, \ldots$ are independent random vectors, given $\{X_1, X_2, \ldots\}$.[4]

With this specification, the adoption process can be analyzed via the theory of generalized Polya urn schemes (Hill, Lane and Sudderth[5]). According to this theory, the limiting market share for product A converges with probability one, and its distribution can be calculated from the function

$$f(x) = \sum_{i=0}^{n} p(k) \binom{n}{k} x^k (1 - x)^{n-k} \tag{2.1}$$

where $p(k)$ is the probability that an agent who samples k previous purchasers of A—and hence $(n - k)$ previous purchasers of B—will choose product A. In particular, the support of the limiting distribution is the set $\{x : x \text{ in } [0, 1], f(x) = x \text{ and } f'(x) \leq 1\}$. Arguments for these claims are given in Arthur and Lane.[2]

Thus, the analysis in the next two sections reduces to calculating $p(k)$ for particular decision rules and parameter values, and then studying the zeros in $[0, 1]$ of the nth degree polynomial $f(x) - x$.

3 BAYESIAN OPTIMIZATION

3.1 The Bayesian Optimization Decision Rule

Bayesian agents start by (1) coding the publicly available information as a prior distribution on the performance characteristics c_A and c_B; and (2) determining their utility function. I will assume that these assignments depend on the parameters μ, σ and λ as follows:

---

[3]Thus, c_A and c_B represent scalar performance characteristics of products A and B respectively, and observations yield the values of these characteristics, perturbed by independent standard normal errors.

[4]Note that, according to this "observation error" assumption, two different aggents who sample the same previous adopter will receive independent estimates of the relevant performance characteristic (conditional on the true falue of the performance characteristic).

Prior distribution: c_A and c_B have independent, $N(\mu, \sigma^2)$ distributions. According to this assumption, the publicly available information does not distinguish between the performance characteristics associated with the two products.

Utility function: The agents' utility function μ for product performance belongs to the family of constant risk aversion functions, parameterized by the nonnegative constant l as follows:

$$u(c) = \begin{cases} -e^{-2\lambda c} & \text{if } \lambda > 0 \\ c & \text{if } \lambda = 0 \end{cases}.$$

When λ equals 0, agents are risk-neutral; the greater is the value of λ, the more risk averse are the agents.

Next, I assume that the agents use the information in X to classify observations by product type, but do not try to extract the information about product quality that is implied by the choices that the sampled previous adopters have made. The reason for this is simple: it is just not feasible to construct a model that extracts the missing information, since such a model would entail assumptions about how other agents make their decisions on the basis of their assumptions about how other agents make their decisions, and so on—and even if it were possible to construct such a model, it would be unrealistic to presume that it could be regarded as common knowledge and hence shared by all agents.[5] Thus, agents use only the information about product quality that is contained in the observation vector Y to update their distributions for c_A and c_B.

Finally, I suppose that agents know how the observations depend on the true, unknown performance characteristics c_A and c_B. That is, they know that they observe these characteristics perturbed by independent standard normal observation errors.

With these assumptions, the expected utility for a product, A for example, is

$$E(U_A) = -\exp[2\lambda(\lambda\sigma_A^2 - \mu_A)]$$

where μ_A is the mean of the current posterior distribution for c_A, and σ_A^2 is the variance of this distribution. Thus, the Bayesian optimization decision rule is: choose A if $\mu_a - \lambda\sigma_A^2 > \mu_B - \lambda\sigma_A^2$, and if the inequality is reversed, choose B.[6]

Using i to denote either A or B, let n_i represent the number of A adopters a particular agent samples and $\bar{Y}_i$ the average value of the observations obtained from the sampled agents. Then

$$\mu_i - \lambda\sigma_i^2 = \frac{1}{n_i + \sigma^{-2}}(n_i\bar{Y}_i + \sigma^{-2}\mu - \lambda).$$

It is now possible to calculate $p(k)$—the probability that an agent will adopt A if he samples k previous A adopters and $n - kB$ adopters, before he observes the associated Y-values—as follows:

$$\begin{aligned} p(k) &= P(E(U_A) > E(U_B)|n_A = k) \\ &= P(k\bar{Y}_A + \sigma^{-2}\mu - \lambda > \frac{k + \sigma^{-2}}{n - k + \sigma^{-2}}[(n-k)\bar{Y}_B + \sigma^{-2}\mu - \lambda]). \end{aligned}$$

[5] See Arthur[1] for a stimulating discussion of the futility of such a modeling enterprise.

[6] Suppose agents are only interested in selected the better product, so that if they choose i, their utility is 1 if i is the better product, and otherwise it is 0. In this case, the decision rule corresponds to the decision rule given in the text, with λ equal 0.

Hence,

$$p(k) = \Phi\left[\frac{(2k-n)(\sigma^{-2}(c_B-\mu)+\lambda] + k(n-k+\sigma^{-2})(c_A-c_B)}{\sqrt{k(\sigma^{-2}+n-k)^2+(n-k)(\sigma^{-2}+k)^2}}\right] \tag{3.1}$$

where Φ is the standard normal cdf.

The information contagion model under Bayesian optimization has eight parameters: c_A, c_B, n, r, σ, μ, and λ. Two of these parameters, r and s, do not figure explicitly in the conclusions to follow. Moreover, these conclusions depend on c_A, c_B, and μ only through the differences d $= c_B - c_A$ and $e = c_B - \mu$. That still leaves five parameters to worry about: d, e, n, σ and λ.

In Section 3.4, I consider a variant of the full Bayesian optimization model in which only three of these parameters play a role. This variant corresponds to improper "uniform" priors for c_A and c_B. As long as the sample contains at least one adopter of each product type, the formal Bayes algorithm yields proper posteriors for both c_A and c_B, and from these posteriors we can calculate

$$p(k) = \Phi\left[\frac{(2k-n)\lambda - k(n-k)d}{\sqrt{k(n-k)n}}\right]. \tag{3.2}$$

However, if the sample is all of one type, then the posterior for the other type is improper. I will suppose in this case that the agent adopts the only product type represented in the sample. That is, $p(0)$ equals 0 and $p(n)$ equals 1. I will call the resulting class of decision rules the *diffuse Bayes* rules. Notice that there are only three parameters for this class of rules: λ, d and n. They deal with three different aspects of the world: λ is about agent psychology, d is a technological fact, and n determines how much information underlies an agent's choice.

The asymptotic properties of the market-share allocation process under Bayesian optimization depend on the model parameters in surprisingly complicated ways. In Section 3.2, I will describe how the *number* of possible asymptotic market allocations depends on these parameters. Section 3.3 shows by means of an example that even arbitrarily large differences in the performance characteristics of the two competing products does not guarantee that the better product will dominate the market. Finally, Section 3.4 shows that more information available to individual agents need not increase social efficiency.

3.2 Path Dependence

How many possible asymptotic market shares are there for product A? I believe, but so far have not succeeded in proving, that there are only three possibilities: 1, 2 or 3. Which of these obtains depends on parameter values in a quite complicated way, as is illustrated in Figure 1 below. The figure portrays a typical phase-transition diagram, obtained by fixing the values d, e and σ (in this case, at 0.1, 2 and 1 respectively) and varying the values of n and λ. Five different zones of (n, λ)-space, marked I through V, correspond to different limiting market-share regimes. Zones I and V produce no path dependence: there is only one possible limiting share for the inferior product, which does not even depend on the starting distribution r and σ. In zone I, the superior product gains 100% market share asymptotically. In zone V, the asymptotic market share of the

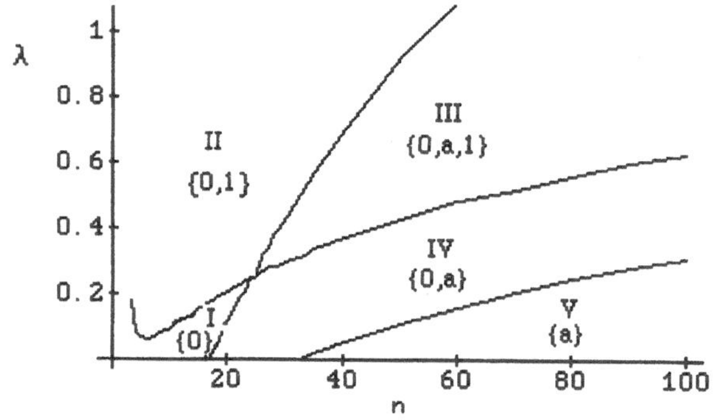

Figure 1: PATH DEPENDENCY REGIMES (for $d = 0.1, e = 2, \sigma = 1$).

inferior product, a, depends on the values of n and λ: for values of n less than 100, it is between 20% and 30%. In zones II and IV, there are two possible limiting values: in zone II, 0 or 1; and in zone IV, 0 or a, where for n less than 100, a is between 15% and 30%. In zone V, there are three possible limiting values, 0, 1 or a (where, for n less than 100 and λ less than 1.5, a is between 15% an 30%). Which limiting value is actually obtained in zones II-IV is path dependent, with a probability distribution that depends on the starting values r and σ.

For any fixed value of n, sufficiently large values of λ will lie in zone II: that is, the inferior product will attain either 0 or 100% of the market. This fact is true for any values of d, e and σ, and it has an easy intuitive explanation: when agents are sufficiently risk averse, they care much more about *how much* they know about each product than *what* they know. Consequently, when one product gets an edge in the market, agents are more likely to find out more about it—and hence to adopt it. But either product can get ahead, by chance sampling events early on. Hence, either may end up dominating the market. The probability that A dominates depends of course on the system parameters, including especially the initial distribution r and s and the difference between the true performance characteristics d.

Conversely, for any fixed value of λ, there will be only one possible asymptotic market share for the inferior product when n is sufficiently large. As n goes to infinity, this asymptotic market share goes to 0. This fact also holds for any values of the parameters e and σ and for nonzero values of d. The intuition behind it[7] is that if agents get access to enough information about both products, they are very likely to be able to distinguish the better product. When there is no real difference between the two products (that is, $d = 0$), sufficiently large values of n still result in zone V behavior, but in this case the only possible outcome is stable market sharing, with each product attaining 50% limiting market share.

[7]Which can be turned into a proof without much difficulty.

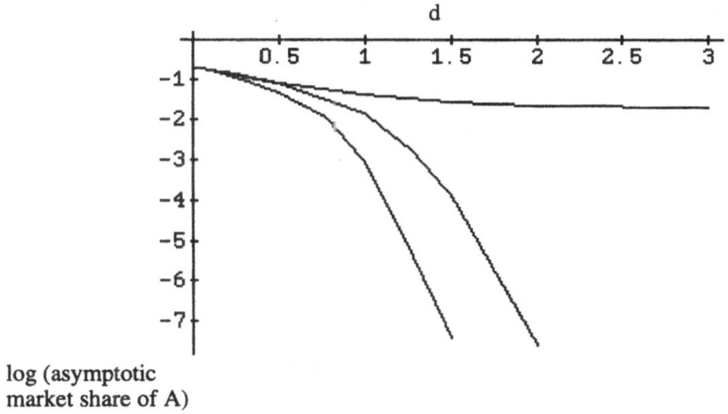

Figure 2: SOCIAL INEFFICIENCY OF BAYESIAN OPTIMIZATION RULES ($\lambda = 0, n = 5, \sigma^2 = 10$). For the upper curve, $\mu = c_B$; for the lower curve, $\mu = c_A$; and for the middle curve, $\mu = c_A + .5$.

3.3 Quality Domination Need Not Lead to Market Domination

What happens when one product is much better than the other? At first sight, one might suppose that if the difference between the products is sufficiently large, the superior product must end up dominating the market. Indeed, if we fix the values of the parameters n, d, λ, σ, c_A and the prior mean μ, and then let c_B go to infinity, the market share of the superior product A does go to 100%.

But suppose instead that the prior distribution is well calibrated for the superior product: that is, say, d is positive and ϵ is 0. In this case, the inferior product A need not be asymptotically shut out of the market, no matter how large the difference between the two products. The reason is simple. When the proportion of adopters of the superior product B becomes sufficiently large, an agent with high probability samples only B adopters. Then, his expected utility for A is just the prior mean $\mu = c_B$ (since $e = 0$), while his expected utility for B is a convex combination of c_B and the average of his observations, which will fall below c_B with probability 1/2. Thus, if he is not very risk averse, he is almost as likely to adopt A as he is to adopt B. This phenomenon is illustrated in Figure 2 below.

In the uppermost curve of Figure 2, the prior is well-calibrated for product B: that is, $\mu = c_B$. In the lower curve, the prior is well-calibrated for A, so $\mu = c_A$. The middle curve plots a case in which $\mu = c_A + 1/2$, so that the inferior product always performs worse than expected and the superior product does so for values of d less than 0.5, but for larger values of d it outperforms the adopters' expectations. For all three prior distributions, the inferior product A obtains substantial market share for small values of d. The dependence of limiting market share on the prior for larger values of d, though, is dramatic. If the prior is well-calibrated for A, the market share of A goes from 27% at $d = .5$ to 4% at $d = 1$ to 0.06% at $d = 1.5$. In contrast, no matter how large is d, the inferior product always gains an asymptotic market share of at least 18% if the prior is well-calibrated for the superior product B! This example underscores the fact that in the information contagion context the effects of the prior distribution do *not* disappear asymptotically, since the prior is "renewed" with each new adopting agent.

3.4 More Information Need Not Increase Social Efficiency

Suppose now that all the parameter values are fixed except n, the number of previous adopters each agent samples. The parameter n determines the quantity of information available to each agent. As the recommendation R1 implies, the value of additional information is always positive to an individual agent. Thus, as long as there are no costs to sampling, any individual agent ought to choose to increase n if he were allowed to do so. It is plausible that the same considerations should apply at the social level: the larger is n, the smaller ought to be the proportion of agents that end up adopting the inferior product. Indeed, as I claimed in the previous subsection, it is true that as n goes to infinity, the asymptotic proportion of agents adopting the inferior product does go to 0. However, as I shall show in this subsection, social efficiency does not in general increase monotonically with n. That is, over a wide range of parameter space, small values of n may result in a considerably smaller proportion of adoptions of the inferior product than much higher values of n.

In the diffuse Bayes case, maximal social efficiency is always achieved when n equals 2, in which case the asymptotic market share of the superior product is 100%. Figures 3 and 4 show how the social efficiency[8] varies with n and λ, for $d = 0.1$ and 0.5. For all values of λ, social efficiency decreases fairly rapidly with n, reaches a minimum value, and then slowly increases. The larger are d and λ, the greater is the value of n at which the social efficiency is minimized and the larger is the value of the minimal social efficiency attained. Note also that social efficiency is maximal, 100% asymptotic market share for the superior product, for an interval of integers, starting with 2; the larger are d and λ, the longer is this interval.

The dependence of social efficiency on n is more complicated under proper Bayes optimization. Some parameter values generate behavior similar to the diffuse Bayes rules. Another typical pattern is illustrated in Figure 5 below, where $d = 0.1$, $e = 0$ and $\sigma = 1$. Here, for low values of λ, social efficiency increases monotonically with n. At about $\lambda = 0.7$, the shape of the social efficiency function changes: the asymptotic market share of the inferior product at first decreases with n, then increases to a maximum (which after $\lambda = 1.2$ is the global maximum), and then slowly decreases (eventually to 0).

Even more complicated behavior is possible. Consider the following example, for which the parameter values are $\sigma^2 = 10$, $\lambda = 1.5$, $e = 0.5$ and $d = 0.1$. For $n = 2$, the asymptotic market share of the inferior product A is essentially 0. For $n = 3 - 6$, both 0 and 1 are possible limits. The probability that A dominates the market in these cases depends on the initial proportion of A adopters; if this proportion is sufficiently high, the probability that A dominates is also high for each n between 3 and 6, but for fixed initial concentration, the probability decreases as n increases. For $n = 7 - 11$, B again attains essentially 100% market share (regardless of the initial proportion of A adopters). In contrast, when $n = 12$ or 13, two limiting market shares are possible: either 0 or 18.9% for $n = 12$ and 22.8% for $n = 13$. Unless r is much less than s, the larger market share

[8]What is plotted is the asymptotic share of the *inferior* product, 100%—social efficiency.

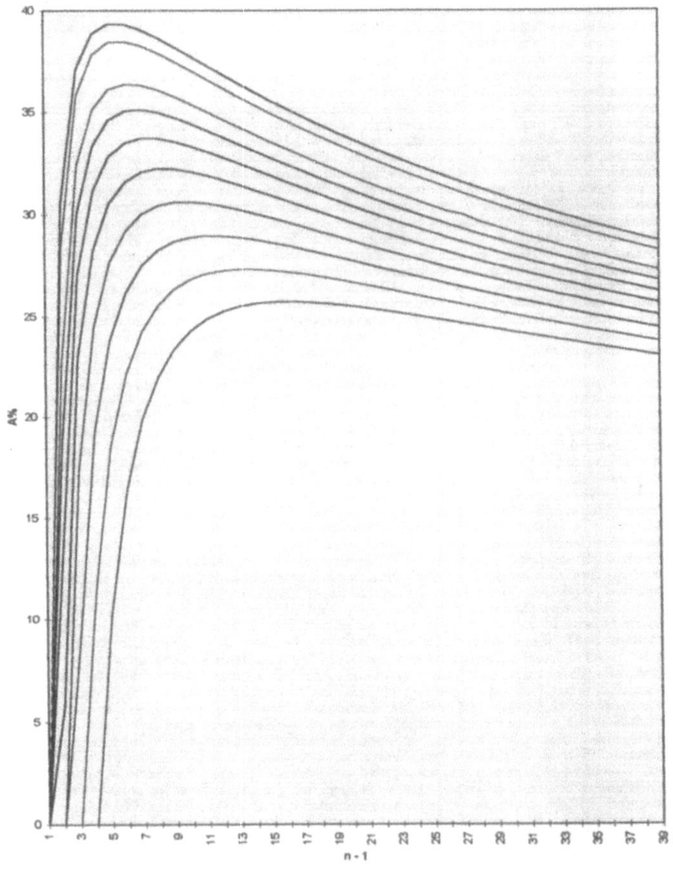

Figure 3: SOCIAL INEFFICIENCY FOR DIFFUSE BAYES RULES ($d = 0.2, \lambda$ from 0 to 1) For the upper curve, $\lambda = 0$; for successively lower curves, λ increments by 0.1 to 1.

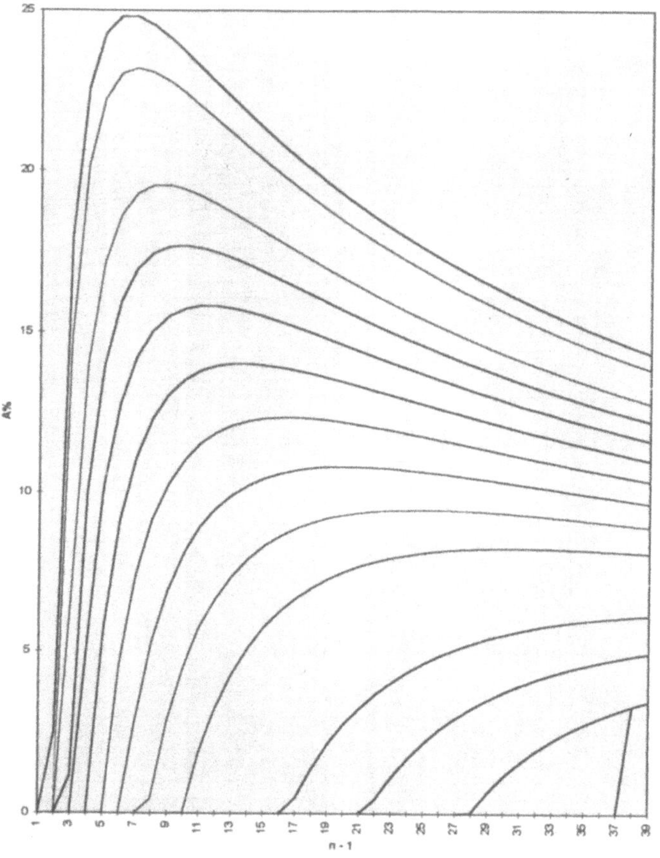

Figure 4: SOCIAL INEFFICIENCY FOR DIFFUSE BAYES RULES ($d = 0.5, \lambda$ from 0 to 1) For the upper curve, $\lambda = 0$; for successively lower curves, λ increments by 0.1 to 1.

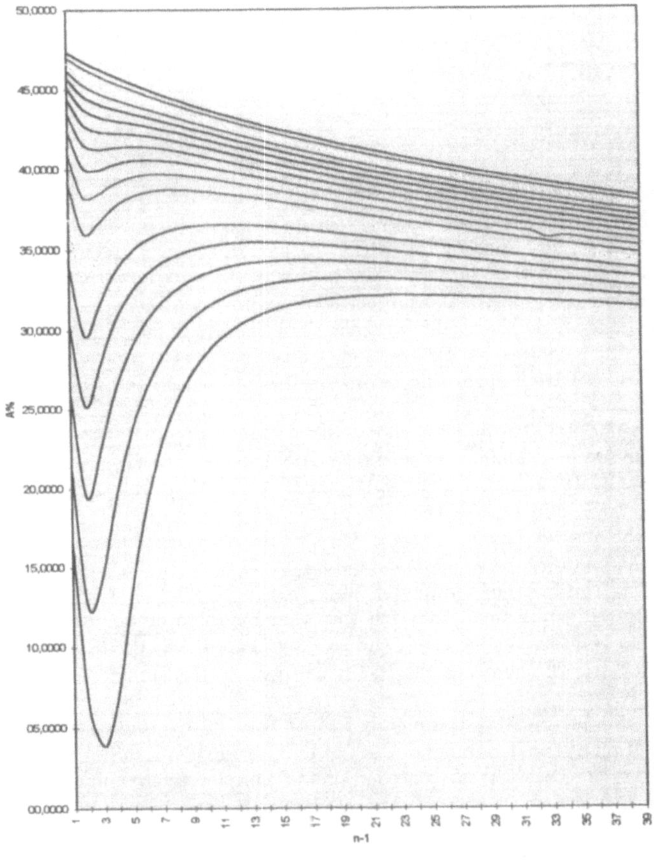

Figure 5: SOCIAL INEFFICIENCY FOR PROPER BAYES RULES ($d = 0.1, e = 0, \sigma = 1, \lambda$ from 0 to 1.6). For the upper curve, $\lambda = 0$; for successively lower curves, λ increments by 0.1 to 1.6.

for A is substantially more probable in both these cases. For $n \geq 14$, A has only one possible limiting market share. For $n = 14$, it is 25.4%, and it continues to increase up to a maximum value of 31.3% for $n = 30$! The market share of A decreases after 30, very slowly as usual: for $n = 100$, A gets 27.3% of the market, and for $n = 200$, its limiting market share is 23.1%.

4 The max rule

To investigate some of the predictions of the Arthur-Lane information contagion model, Warglien and Narduzzo[8] carried out an experiment with human subjects facing exactly the decision problem described by the model. As part of the experiment, they conducted a protocol analysis, in which subjects explained why they preferred one product to the other. Not surprisingly, none of the subjects claimed to reach their decision by Bayesian optimization. Instead, they typically invoked one or another of four simple rules-of-thumb to account for their choices. In this section, I discuss one of these rules, the max rule, which turns out to have a surprising optimality property from the standpoint of social efficiency .

This rule can be stated as follows: choose the product associated with the highest value observed in the sample. That is, denoting by max the index of $y(n) = \max(y_1, \ldots, y_n)$, the rule is given by $D((x_1, y_1), \ldots, (x_n, y_n)) = x_{\max}$. (Note that if all sampled agents have adopted the same product, then the max rule opts for that product, regardless of the magnitude of the observations $y_1, \ldots, y_n$.)[9] Clearly, the max rule is not based on sufficient statistics for the unknown parameters c_A and c_B!

The Warglien-Narduzzo experimental subjects who invoked the max rule justified its use with hearty self-confidence. They felt that they ought to be able to obtain product performance "as good as anyone else", and so they figured that the best guide to how a product would work for them was the *best* it had worked for the people in their sample who had used it.[10]

Of course, this justification completely fails to take into account sample size effects on the distribution of the maximum of a set of i.i.d. random variables. Since the current market leader tends to be overrepresented in agents' samples, its maximum observed value will tend to be higher than its competitors', at least as long as the true performance of the two products are about the same. Thus, it seems plausible that in these circumstances a market lead once attained ought to tend to increase. According to this intuition, the max rule should generate information contagion and thus path-dependent market domination.

This intuition turns out to be completely wrong. In fact, the max rule always leads to a maximally socially efficient outcome:

[9] $p(n) = 1$ and $p(0) = 0$. For $0 < k < n, p(k)$ is just the probability that the maximum of a sample of size k from a $N(c_A, 1)$ distribution will exceed the maximum of an independent sample of size $(n - k)$ from a $N(c_B, 1)$ distribution:

$$p(k) = \int \Phi[y - (c_B - c_A)]^{n-k} k \Phi[y]^{k-1} \phi[y] dy,$$

where φ is the standard normal density function.

[10] According to Warglien and Narduzzo,[8] this is an example of a "well-known bias in human decision-making", which they follow Langer[7] in calling the "illusion of control".

Proposition 1: Suppose $n \geq 2$. If the two products in reality have identical performance characteristics, the market share allocation process exhibits path-dependence: the limiting distribution for the share of product A is Beta(r, s). Suppose, however, that the two products are different. Then, the better product attains 100% limiting market share with probability 1.

Proof: The proposition follows from (4.1) and (4.2) below, using the results from Hill, Lane and Sudderth[5] and Arthur and Lane[2] cited in section 2 above:

(4.1) If $c_A = c_B$, then $p(k) = k/n$ and hence $f(x) = x$ for all x in [0,1]. In this case, $p(k)$ is just the probability that the maximum of the sample of the n i.i.d. random variables $Y_1, \ldots, Y_n$ is one of the k of them whose associated X-value is A. The value of f now follows from (3.1).

Since $f(x) = x$, the market allocation process is a Polya urn scheme, from which the Beta(r, s) limiting distribution follows.

(4.2) If $c_A > c_B$, then $p(k) \geq k/n$ for all k, and for $0 < k < n$ the inequality is strict; hence $f(x) > x$ for all x in (0,1). We now consider a sample of size k from a $N(c_A, 1)$ distribution and a sample of size $(n-k)$ from a $N(c_B, 1)$; $p(k)$ is the probability that the maximum of the combined sample comes from the first sample, which is clearly a non-decreasing (increasing for $0 < k < n$) function of $c_A - c_B$. The first assertion in (4.2) then follows from the first assertion of (4.1), which calculates $p(k)$ when $c_A - c_B = 0$. Since f is just a linear combination of the $p(k)$'s with positive coefficients in $(0, 1)$, the inequality $f(x) > x$ follows from the inequalities for each $p(k)$ and the second assertion in (4.1), which calculates f when $c_A - c_B = 0$.

The inequality $f(x) > x$, together with the evaluations $f(0) = 0$ and $f(1) = 1$, implies that 1 is the only value in the set $\{x : x \text{ in } [0, 1], f(x) = x \text{ and } f'(x) \leq 1\}$. Hence, the limiting market share of product A is 1 with probability 1.

5 SUMMARY

This paper has focussed on a simple instance of a deep and general problem in social science, the problem of micro-macro coordination in a multilevel system. Behavior at the micro-level in the system gives rise to aggregate-level patterns and structures, which then constrain and condition the behavior back at the micro-level. In the information contagion model, the interesting micro-level behavior concerns how and what individual agents choose, and the macro-level pattern that emerges is the stable structure of market shares for the two products that results from the aggregation of agent choices. The connection between these two levels lies just in the agents' access to information about the products.

The results described in Sections 3 and 4 show that mechanism at the micro-level has a determining and surprising effect at the macro-level. In particular, what seems good from a micro-level point of view can turn out to have bad effects at the macro-level, which in turn can materially affect micro-level agents. The paper highlighted two specific instances of this phenomenon. First, despite the fact that individual decision-makers "ought" to maximize their expected utility, a very large population of max rule followers will achieve a higher level of social efficiency than will a population of Bayesian optimizers. Second, although individual decision-makers should always prefer additional costless information, performance at the aggregate level can decrease as more information is available at the micro-level. I know of no other result that resembles the first of these findings, but an increasing body of examples suggests that the second may be a special case of some sort of general phenomenon (see, for example, Bassan and Scarsini[3]; Ellison and Fudenburg[4]; and Hutchins,[6] Chapter 5).

Acknowledgements

I would like the thank the Santa Fe Institute and the Italian Ministry of Research for support for this research. I benefitted from discussions with Massimo Warglien and Alessandro Narduzzo and my colleagues at the Santa Fe Institute (especially Brian Arthur and Bob Maxfield) and the University of Modena. Roberta Vescovini and Giordano Sghedoni contributed invaluable research assistance in the course of their baccalaureate thesis projects. Finally, I would like to thank Wes Johnson and an anonymous referee for a number of helpful suggestions on a previous draft of the paper.

I dedicate this paper to Seymour Geisser, whose good sense and persistent questioning have been an inspiration to me over the past two decades.

References

[1] Arthur, W. B. (1995). "Complexity in economic and financial markets." *Complexity* **1**, 20–25.

[2] Arthur, W.B. and D.A. Lane (1993). "Information contagion." *Economic Dynamics and Structural Change* **4**, 81–104.

[3] Bassan, B. and M. Scarsini (1995). "On the value of information in multi-agent decision theory." *Journal of Mathematical Economics* **24**, 557–576.

[4] Ellison, G. and D. Fudenberg (1995). Word of mouth communication and social learning. *Quarterly Journal of Economics* **110**, 93–126.

[5] Hill, B. M., D. A. Lane, and W. D. Sudderth (1980). "A strong law for some generalized urn processes." *Annals of Probability* **8**, 214–226.

[6] Hutchins, E. (1995). *Cognition in the Wild.* Cambridge: MIT Press.

[7] Langer, E. J. (1975). The illusion of control. *Journal of Personality and Social Psychology* **32**, 311-328.

[8] Warglien, M., and A. Narduzzo (1996). "Learning by the experience of others, an experiment on information contagion." To appear *Industrial and Corporate Change.*

Bayesian Method of Moments (BMOM) Analysis of Mean and Regression Models

Arnold Zellner*
University of Chicago

Abstract

A Bayesian method of moments/instrumental variable (BMOM/IV) approach is developed and applied in the analysis of the important mean and multiple regression models. Given a single set of data, it is shown how to obtain posterior and predictive moments without the use of likelihood functions, prior densities and Bayes' Theorem. The posterior and predictive moments, based on a few relatively weak assumptions, are then used to obtain maximum entropy densities for parameters, realized error terms and future values of variables. Posterior means for parameters and realized error terms are shown to be equal to certain well known estimates and rationalized in terms of quadratic loss functions. Conditional maxent posterior densities for means and regression coefficients given scale parameters are in the normal form while scale parameters' maxent densities are in the exponential form. Marginal densities for individual regression coefficients, realized error terms and future values are in the Laplace or double-exponential form with heavier tails than normal densities with the same means and variances. It is concluded that these results will be very useful, particularly when there is difficulty in formulating appropriate likelihood functions and prior densities needed in traditional maximum likelihood and Bayesian approaches.

1 INTRODUCTION

In the traditional likelihood and Bayesian approaches, it is usually assumed that enough information is available to formulate a likelihood function and, in the Bayesian approach, a prior density for the parameters of the selected likelihood function—see e.g. Jeffreys (1988), Box and Tiao (1973), Berger (1985), Geisser (1993), Press (1989), and Zellner (1971).

*Research financed in part by the National Science Foundation and by income from the H.G.B. Alexander Endowment Fund, Graduate School of Business, University of Chicago. Some of this work was reported to the U. of Chicago Econometrics and Statistics Colloquium in November, 1994, the Conference on Informational Aspects of Bayesian Statistics in Honor of H. Akaike, Japan, December 1993, the Department of Statistics Seminar, Iowa State U., April 1994, and the 1994 Conference in Honor of Seymour Geisser, Taiwan, December 1994. Thanks to participants and to a referee for helpful comments.

However, if not enough information is available to specify a form for the likelihood function, then clearly there will be problems in both the traditional likelihood and Bayesian approaches. In situations like this, some resort to non-likelihood based methods, say least squares regression, method of moments or boot-strap approaches that are "data-based." That is, least squares is usually justified by its producing the best fit to a given sample of data with no appeal to sampling properties. However, if appropriate sampling assumptions are made, unbiasedness and a variance-covariance matrix of the least squares estimator can be produced without specifying a complete likelihood function. Similarly, the well-known Gauss-Markov theorem's assumptions provide certain optimality properties for the least square estimator without introducing a likelihood function. These general results, obtained under certain sampling assumptions have been widely utilized. However, they do not yield conditional probability statements about possible values of parameters based on a *single set of given observations*. In the present paper, it will be shown how such conditional results can be obtained without introducing sampling assumptions, likelihood functions and prior densities. Moments of parameters and future values of variables will be derived and shown to have certain optimal properties based on just a single set of observed data.

Further, to obtain a posterior distribution for parameters given just their moments, a maximum entropy (maxent) approach, see e.g. Jaynes (1982a,b) and Cover and Thomas (1991) will be utilized since this provides a most conservative choice of density that incorporates information in moment side conditions. However, if enough information is available to formulate a tentative likelihood function and a prior density for its parameters, the results of a traditional Bayesian analysis can be compared and/or combined with those yielded by the BMOM approach. See Green and Strawderman (1994) for an application of the BMOM approach in the analysis of a natural resource model with an unknown likelihood function.

The plan of the paper is as follows. In Section 2, an analysis of a simple scalar mean process is presented since this is a central, important case and analysis of it reveals well the essential features of the BMOM approach. Then in Section 3, results for multiple regression and autoregressive models, are given. Section 4 includes a summary of results and indications of future research.

2 ANALYSIS OF A SCALAR MEAN PROBLEM

In this section, we assume that n given observations $\underline{y}' = (y_1, y_2, ..., y_n)$ have been obtained that relate to a scalar mean, θ, as follows:

$$y_i = \theta + u_i \qquad i = 1, 2, ..., n \tag{2.1}$$

where the u_i's are unobserved, ***realized*** error terms; see Chaloner and Brant (1988), Zellner (1975) and Zellner and Moulton (1985) for traditional Bayesian analyses of realized error terms. Note that we have made the important assumption that the errors, say measurement errors, are additive. Since θ and the u_i's are unobserved quantities with unknown values, we shall assume that we can view their possible values probabilistically. That is, given the data $\underline{y}$, we assume that the means of θ and the u_i's as well as other features of their distributions exist but have unknown values. Note that a ***realized*** error term, u_i, usually does not have a zero mean. Operations and assumptions introduced below will enable us to express these posterior moments and other quantities as functions of the given data.

2.1 First Order Posterior and Predictive Moments

Given the observations' values in (2.1) both θ and the realized error terms, the u_i's have fixed unknown values that we shall regard as subjectively random just as was done in Chaloner and Brant (1988), Zellner (1975), and Zellner and Moulton (1985). Here, in contrast to the work in these papers, we shall assume that not enough information is available to formulate a likelihood function and a prior density and thus it is not possible to use Bayes' theorem. From (2.1) we have

$$\bar{y} = \theta + \bar{u} \tag{2.2}$$

where $\bar{y} = \Sigma_{i=1}^{n} y_i/n$, a given sample mean and $\bar{u} = \Sigma_{i=1}^{n} u_i/n$, the mean of the realized error terms. The symbol E denotes a posterior expectation operator, so-called because it is utilized after the data have been observed. From (2.2), we have

$$\bar{y} = E\theta|D + E\bar{u}|D \tag{2.3}$$

where D denotes the given data, here $(y_1, y_2, ..., y_n)$.

We now introduce the following assumption.

Assumption I: $$E\bar{u}|D = \Sigma_{i=1}^{n} Eu_i|D/n = 0 \tag{2.4}$$

This assumption indicates that we believe that there is nothing systematic in the realized error terms and thus the expectation of their mean is equal to zero. If, for example, we believe that the i'th observation is an additive outlier, we would have $u_i = \eta + \varepsilon_i$ with the mean of the parameter η, $E\eta|D \neq 0$ and then $E\bar{u}|D \neq 0$.

Given Assumption I in (2.4), we have from (2.3),

$$E\theta|D = \bar{y} \tag{2.5}$$

that is the posterior mean of θ is the sample mean $\bar{y}$. Note that this result has been obtained without selecting a likelihood function and a prior density for its parameters as is done in many analyses.

Given the result in (2.5), we have on taking the expectation of both sides of (2.1), the posterior expectation of u_i, the i'th realized error term is:

$$Eu_i|D = y_i - E\theta|D = y_i - \bar{y} \tag{2.6}$$

which is the deviation of y_i from the mean $\bar{y}$. On summing both sides of (2.6) and dividing by n, we have $\Sigma_{i=1}^{n} Eu_i|D/n = \Sigma_{i=1}^{n} (y_i-\bar{y})/n = 0$, the "sample analogue" of Assumption I.

Further, note that if we seek an optimal point estimate, $\hat{\theta}$, relative to a quadratic loss function $L(\theta,\hat{\theta}) = (\theta-\hat{\theta})^2$, we have $EL(\theta,\hat{\theta}) = E(\theta-E\theta)^2 + (\hat{\theta}-E\theta)^2$ and as is well known, taking $\hat{\theta} = E\theta$ is optimal. Thus the estimate, $\bar{y}$ in (2.5) is optimal relative to squared error loss.

If y_{n+1} is a future, as yet unobserved value satisfying $y_{n+1} = \theta + u_{n+1}$, where u_{n+1} is a future, as yet unrealized error term, we can write $Ey_{n+1}|D = E\theta|D + Eu_{n+1}|D$ and assume that $Eu_{n+1}|D = 0$. Then

$$Ey_{n+1}|D = E\theta|D = \bar{y} \tag{2.7}$$

which is the mean of the future observation y_{n+1} given the data $\underline{y}$ and the information in Assumption I. $\bar{y}$ is an optimal point prediction for y_{n+1} relative to a squared error predictive loss function.

The first order moments in (2.5), (2.6) and (2.7) are identical to those obtained in a traditional normal likelihood function, diffuse prior analysis. However (2.5), (2.6) and (2.7) have been obtained without an iid normality assumption and without an improper diffuse prior.

2.2 Second Order Posterior and Predictive Moments

Having derived the posterior and predictive means in (2.5) and (2.7), we now turn to consider derivation of second order posterior and predictive moments.

From (2.1) in vector notation $\underline{y} = \underline{\iota}\theta + \underline{u}$ where $\underline{\iota}' = (1, 1, ..., 1)$ a $1 \times n$ vector of ones, we have $\hat{\underline{u}} = \underline{y} - \underline{\iota}\bar{y} = [I - \underline{\iota}(\underline{\iota}'\underline{\iota})^{-1}\underline{\iota}']\underline{y} = [I - \underline{\iota}(\underline{\iota}'\underline{\iota})^{-1}\underline{\iota}']\underline{u}$ and thus

$$\underline{u} - \hat{\underline{u}} = \underline{\iota}(\underline{\iota}'\underline{\iota})^{-1}\underline{\iota}'\underline{u} = \underline{\iota}(\underline{\iota}'\underline{\iota})^{-1}\underline{\iota}'(\underline{u}-\hat{\underline{u}}) \tag{2.8}$$

where $\underline{\iota}'\hat{\underline{u}} = 0$ has been employed in (2.8). Then,

$$V(\underline{u}|D) = E(\underline{u}-\hat{\underline{u}})(\underline{u}-\hat{\underline{u}})'|D = \underline{\iota}(\underline{\iota}'\underline{\iota})^{-1}\underline{\iota}'E(\underline{u}-\hat{\underline{u}})(\underline{u}-\hat{\underline{u}})'|D\underline{\iota}(\underline{\iota}'\underline{\iota})^{-1}\underline{\iota}' \tag{2.9}$$

is a functional equation that $V(\underline{u}|D)$ must satisfy, where D denotes given sample and prior information. Thus we introduce the following assumption.

Assumption II:
$$V(\underline{u}|\sigma^2,D) = E(\underline{u}-\hat{\underline{u}})(\underline{u}-\hat{\underline{u}})'|\sigma^2,D = \underline{\iota}(\underline{\iota}'\underline{\iota})^{-1}\underline{\iota}'\sigma^2 \tag{2.10}$$

with σ^2 a positive scalar parameter which satisfies (2.9). Note from $y_i = \theta + u_i$ and $y_i = \bar{y} + \hat{u}_i$ that $u_i - \hat{u}_i = \bar{y}-\theta$. Also, $u_j-\hat{u}_j = \bar{y}-\theta$ and thus $u_i-\hat{u}_i = u_j-\hat{u}_j$ for all i,j. Thus it is not surprising that in (2.10) all variances are the same and all correlations equal 1. Then from $\bar{y} - \theta = (\underline{\iota}'\underline{\iota})^{-1}\underline{\iota}'(\underline{y}-\underline{\iota}\theta) = (\underline{\iota}'\underline{\iota})^{-1}\underline{\iota}'(\underline{u}-\hat{\underline{u}})$, we have

$$\begin{aligned} V(\theta|\sigma^2,D) = E(\theta-\bar{y})^2|\sigma^2,D &= (\underline{\iota}'\underline{\iota})^{-1}\underline{\iota}'E(\underline{u}-\hat{\underline{u}})(\underline{u}-\hat{\underline{u}})'\underline{\iota}(\underline{\iota}'\underline{\iota})^{-1}|\sigma^2,D \\ &= (\underline{\iota}'\underline{\iota})^{-1}\sigma^2 = \sigma^2/n \end{aligned} \tag{2.11}$$

where Assumption II has been used for $E(\underline{u}-\hat{\underline{u}})(\underline{u}-\hat{\underline{u}})'|\sigma^2,D$ in the first line of (2.11). Thus σ^2/n is the posterior variance for θ given the parameter σ^2 and the data.

In addition, $E(\underline{u}-\hat{\underline{u}})'(\underline{u}-\hat{\underline{u}})|D,\sigma^2 = \mathrm{tr}\ \underline{\iota}(\underline{\iota}'\underline{\iota})^{-1}\underline{\iota}'\sigma^2 = \sigma^2$ from (2.10). Then $E(\underline{u}-\hat{\underline{u}})'(\underline{u}-\hat{\underline{u}})|D/n = E\underline{u}'\underline{u}/n|D - \hat{\underline{u}}'\hat{\underline{u}}/n = E\sigma^2|D/n$ where $E\underline{u}|D = \hat{\underline{u}}$ has been used. If we define $E\sigma^2|D = E\sum_{i=1}^{n}(u_i - E\bar{u})^2/n|D = E\underline{u}'\underline{u}/n|D$, since $E\bar{u}|D = 0$ by Assumption I, we have $E\sigma^2|D - \hat{\underline{u}}'\hat{\underline{u}}/n = E\sigma^2/n|D$ or

$$E\sigma^2|D = \hat{\underline{u}}'\hat{\underline{u}}/(n-1) \equiv s^2 \tag{2.12}$$

which is the posterior mean of σ^2 and thus $V(\theta|D) = s^2/n$.

For the future observation $y_{n+1} = \theta + u_{n+1}$, $Ey_{n+1} = \bar{y}$, given $E\theta = \bar{y}$ and $Eu_{n+1}|D,\sigma^2 = 0$. Also, $y_{n+1} - \bar{y} = \theta - \bar{y} + u_{n+1}$ and

$$E[(y_{n+1}-\bar{y})^2|\sigma^2,D] = E(\theta-\bar{y})^2|\sigma^2,D + Eu_{n+1}^2|\sigma^2,D = \sigma^2(1+1/n) \quad (2.13)$$

given that u_{n+1} and $\theta-\bar{y}$ are uncorrelated, (2.11) and $Eu_{n+1}^2|D,\sigma^2 = \sigma^2$.

A traditional diffuse prior, normal likelihood approach produces results identical to those in (2.11) and (2.13). However, the traditional approach yields $E\sigma^2|D = \nu s^2/(\nu-2)$ for $\nu > 2$ where $\nu = n-1$ rather than (2.12), $E(\sigma^2|D) = s^2$. Also $V(\theta|D) = s^2/n$ in the BMOM approach, whereas in the traditional Bayesian approach $V(\theta|D) = \nu s^2/(\nu-2)n$. Further, (2.1), (2.12) and (2.13) are obtained without an iid normality assumption and an improper prior density.

2.3 Derivation of Posterior and Predictive Densities

Above we have derived the first two posterior moments of θ given σ^2, namely $E\theta|D = \bar{y}$ and $Var(\theta|\sigma^2,D) = \sigma^2/n$. Also, $E\sigma^2|D = s^2 = \hat{u}'\hat{u}/(n-1)$, $V(\theta|D) = s^2/n$, $Ey_{n+1}|D = \bar{y}$ and $Var(y_{n+1}|\sigma^2,D) = (1+1/n)\sigma^2$. It is well known that maxent densities can be derived that incorporate the information in moment conditions; see e.g. Jaynes (1982a,b), Cover and Thomas (1991) and Zellner (1991, 1993). That is, we choose a density, say f(x), to maximize $H(f) = -\int f(x)\ln f(x)dx$ subject to $\int x^i f(x)dx = \mu_i$, $i = 0, 1, 2, \ldots$ with $\mu_0 = 1$, where we have utilized uniform measure in defining entropy, H(f). See Shore and Johnson (1980) for consistency, invariance and other desirable properties of this entropy-based procedure. Here we have the following results.

A. The proper maxent posterior density for θ given σ^2 is a normal density with mean $\bar{y}$ and variance σ^2/n, i.e. $g_N(\theta|\sigma^2,D) \sim N(\bar{y},\sigma^2/n)$.

B. The proper maxent predictive density for y_{n+1} given σ^2 and D with mean $\bar{y}$ and variance $\sigma^2(1+1/n)$ is a normal density $h_N(y_{n+1}|\sigma^2,D)$ with these moments, $N[\bar{y},\sigma^2(1+1/n)]$.

C. The proper maxent posterior density for σ^2 with $E\sigma^2 = s^2$ is the exponential density $g_e(\sigma^2|D) = (1/s^2)\exp\{-\sigma^2/s^2\}$, $0 < \sigma^2 < \infty$, with s^2 given in (2.12).

D. From A and C, the joint posterior density for θ and σ^2 is

$$f(\theta,\sigma^2|D) = g_N(\theta|\sigma^2,D)g_e(\sigma^2|D) \quad (2.14)$$

which is a maxent density given that θ/σ and σ are assumed independent.[1] On integrating over σ^2, $0 < \sigma^2 < \infty$, the marginal posterior density for θ is—see Appendix for the derivation,

$$g(\theta|D) = \sqrt{n/2s^2}\exp\left\{-\sqrt{2n}\,|\theta-\bar{y}|/s\right\} \qquad -\infty < \theta < \infty \quad (2.15)$$

This double-exponential or Laplace[2] marginal posterior density for θ is symmetric about $\bar{y}$, the mean, median and modal value, with variance equal to s^2/n and has thick tails relative to normal or Student-t densities. By a change of variable, $z = \sqrt{n}(\theta-\bar{y})/s$, a standardized form of (2.15) is: $g(z|D) = (1/\sqrt{2})\exp\{-\sqrt{2}\,|z|\}$, $-\infty < z < \infty$. See Appendix for further properties of this density.

E. From B and C, the joint density for y_{n+1} and σ^2 is

---

[1]A maxent density without this independence assumption has also been derived.

[2]See Stigler (1986, p.111) for a fascinating discussion of Laplace's derivation of this distribution using the "principle of insufficient reason."

$$h_N(y_{n+1}|\sigma^2,D)g_e(\sigma^2|D) \tag{2.16}$$

where $h_N \sim N[\bar{y}, \sigma^2(1+1/n)]$ and g_e is the exponential density shown in C. On integrating (2.16) with respect to σ^2, $0 < \sigma^2 < \infty$, the marginal predictive density of y_{n+1} is in the following double-exponential form,

$$h_e(y_{n+1}|D) = \left(1/s_e\sqrt{2}\right)\exp\left\{-\sqrt{2}\,|y_{n+1}-\bar{y}|/s_e\right\} \qquad -\infty < y_{n+1} < \infty \tag{2.17}$$

where $s_e^2 = (1+1/n)s^2$. The mean, median and modal value of (2.17) are all equal to $\bar{y}$ and its variance is s_e^2. As with the density in (2.15), the tails of the predictive density can be rather thick.

F. The proper maxent posterior density for $u_i = y_i-\theta$, the i'th realized error term with mean $\hat{u}_i$ and variance σ^2/n is given by

$$g_N(u_i|\sigma,D) = \sqrt{n/2\pi\sigma^2}\,\exp\left\{-(u_i-\hat{u}_i)^2n/2\sigma^2\right\} \qquad -\infty < u_i < \infty \tag{2.18}$$

G. The marginal posterior density for u_i, obtained by integration from the joint density, $g_N(u_i|\sigma,D)g_e(\sigma^2|D)$ is

$$g(u_i|D) = \sqrt{n/2s^2}\,\exp\left\{-\sqrt{2n}\,|u_i-\hat{u}_i|/s\right\} \qquad -\infty < u_i < \infty \tag{2.19}$$

a double-exponential density centered at $\hat{u}_i$, the i'th residual.

H. If it is known that $y_i > 0$ for all i, as with "time to failure" or "duration" data, $0 < \theta < \infty$ and the maxent proper density for θ subject just to $E\theta = \bar{y} > 0$ is the following exponential density

$$g_1(\theta|D) = (1/\bar{y})\exp\{-\theta/\bar{y}\} \qquad 0 < \theta < \infty \tag{2.20}$$

The posterior and predictive densities above can readily be implemented in practice. As indicated below, they can be compared and/or combined with posterior and predictive densities derived from assumed likelihood functions and prior densities using Bayes' theorem.

3 BMOM ANALYSIS OF MULTIPLE REGRESSION AND AUTO-REGRESSION MODELS

The BMOM analysis of multiple regression and autoregression models is quite similar to that utilized in analyzing the scalar mean model in the previous section. The given n × 1 observation vector $\underline{y}$ is assumed to satisfy the following n well known equations,

$$\underline{y} = X\underline{\beta} + \underline{u} \tag{3.1}$$

where X is a given n × k matrix of rank k, $\underline{\beta}$ is a k × 1 vector of regression coefficients with unknown values and $\underline{u}$ is an n × 1 vector of realized error terms. If (3.1) relates to an autoregression, it is assumed that the initial values of the output variable, say $(y_0, y_{-1}, \ldots, y_{-(q-1)})$, for a q'th order autoregression are known and are elements of the X matrix. Given that it is assumed that we do not have enough information to formulate a likelihood function with much confidence, the BMOM approach can be employed to derive posterior and predictive moments and densities.

3.1 Derivation of First Order Posterior and Predictive Moments

To derive first order moments, we multiply both sides of (3.1) by $(X'X)^{-1}X'$ and take the posterior expectation of both sides to obtain, as in connection with (2.2),

$$(X'X)^{-1}X'\underline{y} = E\underline{\beta}|D + (X'X)^{-1}X'E\underline{u}|D \tag{3.2}$$

Now paralleling Assumption I above in (2.4), the following Assumption I′ is introduced.

Assumption I′:

$$(X'X)^{-1}X'E\underline{u}|D = \underline{0} \tag{3.3}$$

If one of the columns of X is a column of ones, that is there is an intercept in (3.10), (3.3) includes Assumption I, namely $E\bar{u}|D = \sum_1^n Eu_i|D/n = 0$. Further, (3.3) implies that there is nothing "systematic" in $E\underline{u}|D$ that is correlated with the columns of X. For example, this would not be the case if it were believed that some important variable or variables had been omitted from the relation in (3.1) or if it were believed that the values of the independent variables in X were measured with error. In these two cases, the u_i's, would contain components that would be expected to be correlated with columns of X and (3.3) would not be expected to hold. However, if it is believed that (3.3) is valid, then clearly from (3.2) the posterior mean, $E\underline{\beta}$ is given by

$$E\underline{\beta}|D = (X'X)^{-1}X'\underline{y} = \hat{\underline{\beta}} \tag{3.4}$$

which is the least squares quantity, a rather simple result[3]. Further, given a quadratic loss function, $L(\underline{\beta},\hat{\underline{\beta}}) = (\underline{\beta}-\hat{\underline{\beta}})'Q(\underline{\beta}-\hat{\underline{\beta}})$, where Q is a given positive definite symmetric matrix, the value of $\hat{\underline{\beta}}$ that minimizes posterior expected loss is the posterior mean in general and given in (3.4) for this specific problem.

From (3.4), the mean of the realized error vector is given by

$$E\underline{u}|D = \underline{y} - XE\underline{\beta}|D = \underline{y} - X\hat{\underline{\beta}} = \hat{\underline{u}} \tag{3.5}$$

where $\hat{\underline{u}}$ is the least squares residual vector and we have $X'\hat{\underline{u}} = 0$, the sample analogue of Assumption I′ in (3.3).

As regards the predictive mean, we have for a future observation, y_f, assumed to satisfy $y_f = \underline{x}_f'\underline{\beta} + u_f$, with the $1 \times k$ vector $\underline{x}_f'$ given and u_f a random, as yet unrealized error term assumed drawn from a distribution with a zero mean (equal to $E\bar{u}|D = 0$, as assumed above in Assumption I′). Then the predictive mean is

$$Ey_f|D = \underline{x}_f'E\underline{\beta}|D = \underline{x}_f'\hat{\underline{\beta}} = \hat{y}_f \tag{3.6}$$

which is just the least squares point prediction. Given a squared error predictive loss function, the predictive mean in (3.6) minimizes the expectation of such a predictive loss function.

Having obtained the first order moments above, we now turn to derive second order

---

[3]If, for example, rather than (3.1), we have $\underline{y} = X\underline{\beta} + Z\underline{\gamma} + \underline{\varepsilon}$, that is $\underline{u} = Z\underline{\gamma} + \underline{\varepsilon}$, $\hat{\underline{\beta}} = (X'X)^{-1}X'\underline{y} = \underline{\beta} + (X'X)^{-1}X'Z\underline{\gamma} + (X'X)^{-1}X'\underline{\varepsilon}$ and $\hat{\underline{\beta}} = E\underline{\beta}|D + (X'X)^{-1}X'ZE\underline{\gamma}|D$ given $(X'X)^{-1}X'E\underline{\varepsilon}|D = \underline{0}$. Thus $E\underline{\beta}|D \neq \hat{\underline{\beta}}$ for $X'Z \neq 0$ and $E\underline{\gamma}|D \neq 0$.

moments.

3.2 Derivation of Second Order Posterior and Predictive Moments

From (3.1) and (3.4), we have $\underline{\beta} - E\underline{\beta}|D = (X'X)^{-1}X'(\underline{u}-\underline{\hat{u}})$, where $\underline{\hat{u}} = \underline{y}-X\underline{\hat{\beta}}$ with $X'\underline{\hat{u}} = 0$. Thus the posterior covariance matrix, denoted by $V(\underline{\beta}|D)$, is

$$V(\underline{\beta}|D) = E(\underline{\beta}-\underline{\hat{\beta}})(\underline{\beta}-\underline{\hat{\beta}})'|D = (X'X)^{-1}X'E(\underline{u}-\underline{\hat{u}})(\underline{u}-\underline{\hat{u}})'|DX(X'X)^{-1} \tag{3.7}$$

To evaluate (3.7) another assumption is needed, paralleling that in (2.10). Given that $\underline{\hat{u}} = (I - X(X'X)^{-1}X')\underline{y} = (I - X(X'X)^{-1}X')\underline{u}$, we have that

$$\underline{u} - \underline{\hat{u}} = X(X'X)^{-1}X'(\underline{u}-\underline{\hat{u}}) \tag{3.8}$$

and

$$E(\underline{u}-\underline{\hat{u}})(\underline{u}-\underline{\hat{u}})'|D = X(X'X)^{-1}X'E(\underline{u}-\underline{\hat{u}})(\underline{u}-\underline{\hat{u}})'|DX(X'X)^{-1}X' \tag{3.9}$$

is a functional equation that must be satisfied. Note that only k elements of $\underline{u}$ are free. Thus, we introduce the following assumption, the analogue of (2.10):

Assumption II': $$V(\underline{u}|\sigma^2,D) = E(\underline{u}-\underline{\hat{u}})(\underline{u}-\underline{\hat{u}})'|\sigma^2,D = X(X'X)^{-1}X'\sigma^2 \tag{3.10}$$

where σ^2 is a positive constant.[4] On inserting the expression in (3.10) in (3.9), it is seen that the functional equation is satisfied for any given value of σ^2. Substituting from (3.10) in (3.7), we have

$$V(\underline{\beta}|\sigma^2,D) = (X'X)^{-1}\sigma^2 \tag{3.11}$$

which is the posterior covariance matrix for $\underline{\beta}$ given σ^2 and D, the data.

To obtain a value for the posterior expectation of σ^2, from (3.10), $E(\underline{u}-\underline{\hat{u}})'(\underline{u}-\underline{\hat{u}})|\sigma^2,D = \mathrm{tr}\ X(X'X)^{-1}X'\sigma^2 = k\sigma^2$. Then $E\underline{u}'\underline{u}/n|D - \underline{\hat{u}}'\underline{\hat{u}}/n = kE\sigma^2|D/n$ and if we define $E\sigma^2|D = E\Sigma_{i=1}^n(u_i-E\bar{u})^2/n|D = E\underline{u}'\underline{u}/n|D$, since $E\bar{u}|D = 0$ from Assumption I', $E\sigma^2|D - \underline{\hat{u}}'\underline{\hat{u}}/n = kE\sigma^2|D/n$ or

$$E\sigma^2|D = \underline{\hat{u}}'\underline{\hat{u}}/(n-k) = s^2 \tag{3.12}$$

which is the posterior mean of σ^2.

For the future observation, $y_f = \underline{x}_f'\underline{\beta} + u_f$, $Ey_f|D = x_f'\underline{\hat{\beta}} \equiv \hat{y}_f$, as shown above. Then $y_f - \hat{y}_f = \underline{x}_f'(\underline{\beta}-\underline{\hat{\beta}}) + u_f$ and

$$E(y_f-\hat{y}_f)^2|\sigma^2,D = (1 + \underline{x}_f'(X'X)^{-1}\underline{x}_f)\sigma^2 \tag{3.13}$$

given that u_f and the elements of $\underline{\beta}-\underline{\hat{\beta}}$ have zero covariances and $Eu_f|D = 0$.

Given the above posterior and predictive moments, the following are maxent posterior and predictive densities that incorporate the information in these moments.

---

[4]If we write $\underline{y} = XHH^{-1}\underline{\beta} + \underline{u} = Z\underline{\theta} + \underline{u}$, where H is a square k×k non-singular matrix such that $H'X'XH = I_k$ and $\underline{\theta} = H^{-1}\underline{\beta}$, a k×1 vector. Then $Z'\underline{y} = \underline{\theta} + Z'\underline{u}$ and $E\underline{\theta}|D = Z'\underline{y}$ given $Z'E\underline{u}|D = 0$. Also, if we assume that the k error terms in $Z'\underline{u}$ have equal variances and are mutually uncorrelated, this implies (3.10) and (3.11).

A′. The proper maxent posterior density for $\underline{\beta}$ given σ^2 and the data with mean $\hat{\underline{\beta}} = (X'X)^{-1}X'\underline{y}$ and covariance matrix $(X'X)^{-1}\sigma^2$, is a multivariate normal density, $g_N(\underline{\beta}|\sigma^2,D) \sim MVN(\hat{\underline{\beta}}, (X'X)^{-1}\sigma^2)$.

B′. The proper maxent predictive density for y_f given σ^2, $\underline{x}_f$ and the data is a normal density, $h_N(y_f|\sigma^2,D) \sim N[(\hat{y}_f, (1 + \underline{x}_f'(X'X)^{-1}\underline{x}_f)\sigma^2)]$.

C′. The proper maxent density for σ^2 with $E\sigma^2 = s^2 = \hat{\underline{u}}'\hat{\underline{u}}/(n-k)$ is the exponential density $g_e(\sigma^2|D) = (1/s^2)\exp(-\sigma^2/s^2)$, $0 < \sigma^2 < \infty$.

D′. From A′ and C′, the joint posterior density for $\underline{\beta}$ and σ^2 is[5]

$$f(\underline{\beta},\sigma^2|D) = g_N(\underline{\beta}|\sigma^2,D)g_e(\sigma^2|D) \tag{3.14}$$

which is a maxent density given that $\underline{\beta}/\sigma$ and σ are independent. The marginal posterior density of a single element of $\underline{\beta}$, say β_i, can be obtained by integrating (3.14) with respect to the remaining elements of $\underline{\beta}$ and then with respect to σ^2. The result is the following double exponential density for β_i:

$$g(\beta_i|D) = \left(1/s_i\sqrt{2}\right)\exp\left\{-\sqrt{2}\,|\beta_i-\hat{\beta}_i|/s_i\right\} \qquad -\infty < \beta_i < \infty \tag{3.15}$$

where $\hat{\beta}_i$ is the i'th element of $\hat{\underline{\beta}}$ and s_i^2 is the (i,i)'th element of $(X'X)^{-1}s^2$. Also, from (3.14), the marginal distribution of $\eta = \underline{\ell}'\underline{\beta}$, where $\underline{\ell}$ is a given vector of rank one, is

$$g(\eta|D) = \left(1/s_\eta\sqrt{2}\right)\exp\left\{-\sqrt{2}\,|\eta-\hat{\eta}|/s_\eta\right\} \qquad -\infty < \eta < \infty \tag{3.16}$$

where $\hat{\eta} = \underline{\ell}'\hat{\underline{\beta}}$ and $s_\eta^2 = \underline{\ell}'(X'X)^{-1}\underline{\ell}\, s^2$

E′. From B′ and C′, the joint density for y_f and σ is

$$h_N(y_f|\sigma^2,D)g_e(\sigma^2|D) \tag{3.17}$$

where $h_N \sim N[\hat{y}_f, (1 + x_f'(X'X)^{-1}x_f)\sigma^2]$ and g_e is the exponential density shown in C′. On integrating (3.17) with respect to σ^2, the marginal predictive density for y_f is the following double-exponential density:

$$h_f(y_f|D) = \left(1/s_e\sqrt{2}\right)\exp\left\{-\sqrt{2}\,|y_f-\hat{y}_f|/s_e\right\} \qquad -\infty < y_f < \infty \tag{3.18}$$

The mean of this density is $\hat{y}_f$ and its variance is $s_e^2 = [1 + \underline{x}_f'(X'X)^{-1}\underline{x}_f]s^2$.

F′. The proper maxent posterior density for u_i with given mean $\hat{u}_i$ and given variance $v_i = \underline{x}_i'(X'X)^{-1}\underline{x}_i\sigma^2$ is

$$g_N(u_i|\sigma^2,D) = 1/\sqrt{2\pi v_i}\,\exp\{-(u_i-\hat{u}_i)^2/2v_i\} \tag{3.19}$$

G′. The marginal posterior density for u_i, obtained by integrating the joint density $g_N(u_i|\sigma,D)g_e(\sigma^2|D)$ with respect to σ is:

---

[5]To compute the joint density of the elements of $\underline{\beta}$, draw σ^2 from $g_e(\sigma^2|D)$ and insert the drawn value in $g_N(\underline{\beta}|\sigma^2,D)$ and draw $\underline{\beta}$ from this multivariate normal density. Thus draws from the joint density $f(\underline{\beta},\sigma^2|D)$ are obtained by repeated use of this procedure.

$$g_e(u_i|D) = \left[1/\sqrt{2v_i}\right]\exp\left\{-\sqrt{2}\,|u_i-\hat{u}_i|/\sqrt{v_i}\right\} \qquad -\infty < u_i < \infty \tag{3.20}$$

a double-exponential density.

H′. If it is known that $\theta = \underline{\ell}'\underline{\beta}$ is strictly positive, where $\underline{\ell}'$ is a given $1 \times k$ vector of rank one, the maxent density for θ subject just to $E\theta|D = \underline{\ell}'\underline{\hat{\beta}} > 0$, is the following exponential density:

$$g(\theta|D) = (1/\underline{\ell}'\underline{\hat{\beta}})\exp\{-\theta/\underline{\ell}'\underline{\hat{\beta}}\} \qquad 0 < \theta < \infty \tag{3.21}$$

3.3 Use of Additional Prior Information

If in addition to the sample information $\underline{y} = X\underline{\beta} + \underline{u}$, we represent additional prior information by use of a conceptual sample, $\underline{y}_c = X_c\underline{\beta} + \underline{u}_c$, where $\underline{y}_c$ is an $n_c\times 1$ vector of conceptual data points, X_c is a $n_c\times k$ given matrix, $\underline{\beta}$ is a $k\times 1$ vector of regression parameters and $\underline{u}_c$ is an $n_c\times 1$ vector of realized conceptual error terms. We can write

$$\begin{pmatrix} \underline{y}_c \\ \underline{y} \end{pmatrix} = \begin{pmatrix} X_c \\ X \end{pmatrix}\underline{\beta} + \begin{pmatrix} \underline{u}_c \\ \underline{u} \end{pmatrix} \tag{3.22a}$$

or

$$\underline{w} = W\underline{\beta} + \underline{\varepsilon} \tag{3.22b}$$

Then making Assumption I′ and II′ relative to the system in (3.22b), the posterior mean of $\underline{\beta}$ is

$$\begin{aligned} \underline{\hat{\beta}} \equiv E\underline{\beta}|D' &= (W'W)^{-1}W'w \\ &= (X_c'X_c + X'X)^{-1}(X_c'y_c + X'y) \\ &= (X_c'X_c + X'X)^{-1}(X_c'X_c\hat{\beta}_c + X'X\hat{\beta}) \end{aligned} \tag{3.23}$$

where, when X_c is of full column rank, $\underline{\hat{\beta}}_c = (X_c'X_c)^{-1}X_c'\underline{y}_c$ is a prior mean vector, $\underline{\hat{\beta}} = (X'X)^{-1}X'\underline{y}$, the least squares estimate and $X_c'X_c$ is assigned a value by the investigator. Then, with Assumption II′ relating to (3.22b), we have

$$E(\underline{\beta}-E\underline{\beta})(\underline{\beta}-E\underline{\beta})'|D,\sigma^2) = (W'W)^{-1}\sigma^2 \tag{3.24}$$

and

$$E\sigma^2|D = (\underline{w}-W\underline{\hat{\beta}})'(\underline{w}-W\underline{\hat{\beta}})/(n+n_c-k). \tag{3.25}$$

Also maxent distributions are available for this system that incorporates a conceptual sample.

Above are some results of applying the BMOM approach to analyze data assumed generated by a multiple regression or an autoregression model. In addition, posterior and predictive intervals can easily be computed from the above posterior and predictive densities by the procedures described in the Appendix. Generally these intervals are broader than corresponding intervals based on the conditional normal posterior and predictive densities, derived above, with $\sigma^2 = s^2$. Also they are broader than conventional intervals based on marginal Student-t densities based on posterior and predictive densities based on normal likelihood functions and diffuse prior densities for their parameters. See Appendix A for one such comparison.

4 Summary and Concluding Remarks

In the preceding sections, posterior and predictive moments for parameters and future, as yet unobserved observations have been derived using one given set of data and a few simple assumptions. There was no need to formulate a density function for as yet unobserved data as a basis for a likelihood function nor to introduce a prior density for its parameters. Also, no use was made of Bayes' Theorem in deriving posterior and predictive moments.

Then, proper posterior and predictive densities were derived by maximizing entropy subject to the derived moments. These densities for location or regression coefficients with doubly infinite ranges are in the double exponential or Laplacian form while the maxent posterior density for variance parameters are in the exponential form. Use of higher order moment constraints for variance parameters in deriving maxent posterior densities will be reported in future work. For location parameters with strictly positive values, the maxent posterior densities subject just to a first moment constraint are in the exponential form. Similar results were obtained for predictive densities for as yet unobserved observations.

It should be appreciated that the maxent predictive densities, in the double exponential or exponential form can serve as models for as yet unobserved data and employed in calculation of posterior odds in model comparison and selection problems using new data. In particular such models can be compared to those derived using a particular likelihood function, a prior density for its parameters and Bayes' Theorem. The two different posterior densities, based on analyses of a given sample of data can be employed as prior densities in the calculation of posterior odds using a second sample of data. The posterior odds can then be used to choose between the two models or to combine them using the approach described in Min and Zellner (1993) and Palm and Zellner (1992). In future work, such calculations will be reported. See Zellner (1995) for some BMOM results relating to multivariate regression and simultaneous equations models.

References

Berger, J.O. (1985), *Statistical Decision Theory*, 2nd ed., New York: Springer-Verlag.

Box, G.E.P. and G.C. Tiao (1973), *Bayesian Inference in Statistical Analysis*, Reading, MA: Addison-Wesley.

Chaloner, K. and R. Brant (1988), "A Bayesian Approach to Outlier Detection and Residual Analysis," *Biometrika*, 75, 651-659.

Cover, T.M. and J.A. Thomas (1991), *Elements of Information Theory*, New York: J. Wiley & Sons, Inc..

Geisser, S. (1993), *Predictive Inference: An Introduction*, New York: Chapman & Hall.

Gradshteyn, I.S. and J.M. Ryzhik (1980), *Table of Integrals, Series and Products*, (Corrected and enlarged edition prepared by A. Jeffrey), San Diego, CA: Academic Press.

Green, E.J. and W.E. Strawderman (1994), "A Bayesian Growth and Yield Model for Slash Pine Plantations," Department of Natural Resources and Department of Statistics, Rutgers U., New Brunswick, NJ.

Jaynes, E.T. (1982a), *Papers on Probability, Statistics and Statistical Physics*, Dordrecht, Netherlands: Reidel.

Jaynes, E.T. (1982b), "On the Rationale of Maximum-Entropy Methods," *Proc. of the IEEE*, 70, 939-952.

Jeffreys, H. (1988), *Theory of Probability* (3rd reprinted edition, 1st edition, 1939), Oxford: Oxford U. Press.
Min, C-k. and A. Zellner (1993), "Bayesian and Non-Bayesian Methods for Combining Models and Forecasts with Applications to Forecasting International Growth Rates," *Journal of Econometrics*, 56, 89-118.
Palm, F.C. and A. Zellner (1992), "To Combine or Not to Combine? Issues of Combining Forecasts," *Journal of Forecasting*, 11, 687-701.
Press, S.J. (1989), *Bayesian Statistics*, New York: J. Wiley & Sons, Inc.
Shore, J.E. and R.W. Johnson (1980), "Axiomatic Derivation of the Principle of Maximum Entropy and the Principle of Minimum Cross-Entropy," *IEEE Transactions*, Vol. IT-26, No. 1, 26-37.
Stigler, S.M. (1986), *The History of Statistics*, Cambridge: Harvard U. Press.
Zellner, A. (1971), "*An Introduction to Bayesian Inference in Econometrics*, New York: John Wiley & Sons, Inc., reprinted by Krieger Publishing Co., 1987.
Zellner, A. (1975), "Bayesian Analysis of Regression Error Terms," *Journal of the American Statistical Association*, 70, 138-144.
Zellner, A. (1991), "Bayesian Methods and Entropy in Economics and Econometrics," in W.T. Grandy and L.H. Schick (eds.), *Maximum Entropy and Bayesian Methods*, Kluwer, 17-32.
Zellner, A. (1993), "Prior Information, Model Formulation and Bayesian Analysis," paper presented at the Conference on Informational Aspects of Bayesian Statistics in Honor of H. Akaike, Fuji Conference Center, Japan, Dec. 1993. A short version of the paper is in the ASA's *1993 Proceedings of the Bayesian Statistics Science Section*, 202-207.
Zellner, A. (1995), "The Finite Sample Properties of Simultaneous Equations' Estimates and Estimators: Bayesian and Non-Bayesian Approaches," paper presented at the Conference in Honor of Professor Carl F. Christ, April 21-22, 1995 at Johns Hopkins U. and to appear in the conference volume.
Zellner, A. and B.R. Moulton (1985), "Bayesian Regression Diagnostics with Applications to International Consumption and Income Data," *Journal of Econometrics*, 29, 187-211.

Appendix
Derivation of Double Exponential Density in (2.15)

From (2.14), $f(\theta,\sigma^2|D) = g_N(\theta|\sigma^2,D)g_e(\sigma^2|D)$ with $g_N(\theta|\sigma^2,D) = \sqrt{n/2\pi\sigma^2} \times \exp\{-n(\theta-\bar{y})^2/2\sigma^2\}$ and $g_e(\sigma^2|D) = (1/s^2)\exp\{-\sigma^2/s^2\}$. Then, with $a \equiv 1/s^2$ and $b \equiv n(\theta-\bar{y})^2/2$

$$
\begin{aligned}
\int_0^\infty f(\theta,\sigma^2|D)d\sigma^2 &= \sqrt{n/2\pi}\, a \int_0^\infty \frac{1}{\sigma}\exp\left\{-[b/\sigma^2 + a\sigma^2]\right\}d\sigma^2 \\
&= \sqrt{2n/\pi}\, a \int_0^\infty \exp\left\{-[b/\sigma^2 + a\sigma^2]\right\}d\sigma \qquad \text{(A.1)} \\
&= \sqrt{2n/\pi}\, a \left[\frac{1}{2}\sqrt{\pi/a}\exp\left\{-2\sqrt{ab}\right\}\right]
\end{aligned}
$$

where the integral has been evaluated using a result given in Gradshteyn and Ryzhik (1980, p. 307, entry 3.325). On inserting the above values for a and b in (A.1), we have

$$
g(\theta|D) = \sqrt{n/2s^2}\,\exp\left\{-2\sqrt{n/2}\,|\theta-\bar{y}|/s\right\} \quad -\infty < \theta < \infty \qquad \text{(A.2)}
$$

Further, letting $w = \sqrt{n/2}(\theta-\bar{y})/s$, a "standardized" form of the density is

$$g(w|D) = \exp\{-2|w|\} \qquad -\infty < w < \infty \qquad (A.3)$$

Note that $\int_0^\infty \exp\{-2w\}dw = \left[-\tfrac{1}{2}\exp(-2w)\right]_0^\infty = 1/2$ and, given symmetry of g(w|D), $\int_{-\infty}^{\infty} g(w|D)dw = 1$. Also, the symmetry of g(w|D) implies that all odd order moments about zero are zero; that is, $\int_{-\infty}^{\infty} w^{2r+1}g(w|D)dw = 0$ for r = 0, 1, 2, ... The even order moments are given by

$$\begin{aligned}\mu_{2r} &= 2\int_0^\infty w^{2r}e^{-2w}dw \\ &= \int_0^\infty (x/2)^{2r}e^{-x}dx = \Gamma(2r+1)/2^{2r} \qquad r = 1, 2, \ldots \\ &= (2r)!/2^{2r}\end{aligned} \qquad (A.4)$$

where x = 2w and $\Gamma(\cdot)$ is the gamma function. For example, $\mu_2 = 1/2$, $\mu_4 = 3/2$, etc.

Note that $w = \sqrt{n}(\theta-\bar{y})/s\sqrt{2} = z/\sqrt{2}$, where $z = \sqrt{n}(\theta-\bar{y})/s$ is a standardized normal variable for the normal ***conditional*** posterior density for θ given $\sigma^2 = s^2$, $g_N(\theta|\sigma^2 = s^2,\bar{y}) \sim N(\bar{y}, s^2/n)$. For the conditional density, $E(z^2|\sigma^2 = s^2,D) = 1$, while for the unconditional density, $E(z^2|D) = 2E(w^2|D) = 1$ which, surprisingly, is equal to the conditional variance. However, $E(z^4|\sigma^2 = s^2,D) = 3$ in the conditional normal density for z while using the marginal density $Ez^4|D = 4Ew^4|D = 4 \cdot 3/2 = 6$, two times the value of that for the conditional density. The fourth moment of z divided by the squared second moment, denoted by $\mu_4/\mu_2^2 = 6$ and the "excess" over that for the conditional normal density for z is 6 - 3 = 3. Thus the marginal double-exponential density for z is quite leptokurtic.

Note that $z = \sqrt{n}(\theta-\bar{y})/s$ has a univariate Student-t posterior density with ν = n-1 degrees of freedom if a standard normal likelihood function for θ and σ and a diffuse prior $\pi(\theta,\sigma) \propto 1/\sigma$ were combined using Bayes' Theorem.[6] For this Student-t posterior density, we have: $Ez|D = 0$ for ν > 0, $Ez^2|D = \nu/(\nu-2)$ for ν > 2, and $Ez^4|D = 3\nu^2/(\nu-2)(\nu-4)$ for ν > 4. Thus the excess for the Student-t based posterior density for z is: $\mu_4/\mu_2^2 - 3 = 6/(\nu-4)$, for ν > 4, which for ν > 6 is considerably less than 3, the excess of the double-exponential density for z. Thus, for n - 1 = ν > 6 or n > 7, the double-exponential density is more leptokurtic. Also, as ν grows in value, the excess for the Student-t posterior density goes to zero whereas that for the double-exponential density has a constant value equal to 3.

Since $z = \sqrt{n}(\theta-\bar{y})/s = \sqrt{2}\,w$, the double-exponential posterior density for z is from (A.3).

$$p(z|D) = \left(1/\sqrt{2}\right)\exp\left\{-\sqrt{2}\,|z|\right\} \qquad -\infty < z < \infty \qquad (A.5)$$

Then for c > 0, $P_c \equiv \Pr(c < z < \infty|D) = \int_c^\infty f(z|D)dz = (1/2)\exp\left\{-\sqrt{2}\,c\right\}$ and thus $\ln 2P_c = -\sqrt{2}\,c$. For example for $P_c = .025$, $\ln .05 = -2.9957 = -1.4142c$ and c = 2.118. Thus .025 = Pr{2.118 < z < ∞} and from the symmetry of the density in (A.5), we have

$$\Pr\{-2.118 < z < 2.118|D\} = .95 \qquad (A.6)$$

[6]This well known result, probably first derived by Jeffreys, is presented in many works including Jeffreys (1988), Berger (1985), Box and Tiao (1973), Press (1989), and Zellner (1971).

This 95% interval for $z = \sqrt{n}\,(\theta-\bar{y})/s$, -2.118 to 2.118 implies that a 95% interval for θ is $\bar{y} \pm 2.118\ s/\sqrt{n}$. This interval is somewhat broader than a 95% interval based on the conditional normal posterior density for θ with $\sigma^2 = s^2$, namely $\bar{y} \pm 1.96\ s/\sqrt{n}$, the widths being $4.24\ s/\sqrt{n}$ in the former case and $3.92\ s/\sqrt{n}$ in the latter, about an 8% difference.

Section II

Prediction

On the Prediction of Growth Curves

Jack C. Lee*
National Chiao Tung University

Seymour Geisser*
University of Minnesota

Abstract

The subject of this paper is generalized linear growth curve models. In the past, these types of growth curves have been studied by authors including J. Wishart, G.E.P. Box and C.R. Rao. Further attention was directed towards this subject ever since the publication of the seminal paper on the subject by R.F. Potthoff and S.N. Roy in 1964 which modeled the growth curve as a generalized multivariate linear model. However, the emphasis up until 1970 was on the estimation and testing hypothesis about parameters. In this paper we mainly review work on prediction within the context of these growth models over the last 25 years.

1 Introduction

The subject of this paper is generalized linear growth curve models. In the past, these types of growth curves have been studied by many authors including Wishart (1938), Box (1950) and Rao (1958). Further attention was directed towards this subject since the publication of the seminal paper of Potthoff and Roy (1964) that modeled the growth curve as a generalized multivariate linear model. However, all of the attention up until Geisser (1970) was directed towards estimation and testing hypothesis about parameters. In this paper we mainly review work on prediction within the context of these growth curve models over the last 25 years.

In Section 2 we shall briefly review the estimation of parameters of growth curve models. In subsequent sections we turn our attention to various problems associated with prediction.

2 Covariance Structures and Parameter Estimation

Let Y_i be a $p_i \times 1$ vector of observations and

*Work supported in part by NSC grants 82-208-M009-054 and 84-2121-M009-008 and by NIGMS grant GM-25271, respectively. The authors wish to thank the referees and professor Arnold Zellner for some constructive comments.

$$Y_i = X_i\beta_a + \epsilon_i \tag{2.1}$$

$$\text{for } i = 1, \ldots, n_a^*;\ n_a^* = n_1 + \cdots + n_a;\ a = 1, \ldots, r$$

where X_i is a known $p_i \times m$ design matrix, β_a is an $m \times 1$ vector of unknown regression parameters, and ϵ_i is a $p_i \times 1$ vector of errors that are independently distributed as $N(0, \Omega_i)$. Let Σ_i be the covariance matrix of Y_i, then the form of Σ_i depends on the assumptions regarding Ω_i as well as β_a. For example, in the fixed effects model, β_a is assumed fixed but unknown, then $\Sigma_i = \Omega_i$. On the other hand, in the random effects model, if β_a is assumed to have $E(\beta_a) = \tau$, $\text{cov}(\beta_a) = \Gamma$, then $\Sigma_i = X_i\Gamma X_i' + \Omega_i$.

The linear growth function as specified by (2.1) allows for unbalanced situations, i.e., individual Y_i can have a specific design matrix X_i. However, we will restrict our attention to the balanced case, i.e., $X_i = X$, for the rest of the paper.

In the balanced case for the general linear growth curve, the model can be specified as

$$\underset{p\times N}{Y} = \underset{p\times m}{X}\ \underset{m\times r}{\tau}\ \underset{r\times N}{A} + \underset{p\times N}{\epsilon}, \tag{2.2}$$

where τ is the unknown matrix of growth function parameters, X and A are known design matrices of ranks $m < p$ and $r < N$, respectively. Furthermore, the columns of ϵ are each independent and p-variate normal with mean vector 0 and common covariance matrix Σ. Hence, Y is normally distributed with mean function $X\tau A$ and covariance matrix $\Sigma \otimes I$, where $\otimes$ denotes the Kronecker product. Usually p is the number of time points observed on each individual, m and r, which specify the degree of a polynomial in time and the number of distinct groups respectively, are assumed known. The design matrices X and A will therefore characterize the degree of the growth function and the distinct grouping out of N independent vectorial observations. For example, if each individual is observed at $t_1, \ldots, t_p$ and there are two different linear growth functions, then A consists of N_1 columns of (1,0) and N_2 columns of (0,1), $N_1 + N_2 = N$, and

$$X = \begin{bmatrix} 1 & t_1 \\ 1 & t_2 \\ \vdots & \vdots \\ 1 & t_p \end{bmatrix}. \tag{2.3}$$

The model as specified by (2.2) with Σ assumed to be an arbitrary unkown positive definite (p.d.) matrix and with $\Sigma = \sigma^2 C$, where $C = (\rho^{|a-b|})$, was proposed by Potthoff and Roy (1964). It was subsequently considered by many authors including Rao (1965, 1966, 1967, 1987), Khatri (1966, 1973), Geisser (1970, 1980, 1981) Lee and Geisser (1972, 1975), Fearn (1975), Zerbe (1979), Laird and Ware (1982), Lee and Tan (1984), Jenrich and Schluchter (1986), Lee (1988, 1991) and Keramidas and Lee (1990, 1995), among others. As indicated earlier, an arbitrary p.d. Σ arises when $\text{cov}(Y_i) = \text{cov}(\epsilon_i) = \Sigma$ in the model specified by (2.1). As for the serial covariance structure in which $\Sigma = \sigma^2 C$, $C = (\rho^{|a-b|})$, it arises from the situation in which the error terms $\epsilon_i = (\epsilon_{i1}, \ldots, \epsilon_{ip})'$

in the model specified by (2.1) follow the relationship $\epsilon_{ij} = \rho\epsilon_{i,j-1} + \eta_{ij}$ where η_{ij} are uncorrelated errors with common variance σ_η^2 and $\sigma^2 = \sigma_\eta^2/(1-\rho^2)$.

Although the focus of this paper is on the prediction of future observations assumed to have been generated from a growth curve model, we will still consider the estimation of parameters as they are needed in some instances. For an arbitrary p.d. Σ, the MLEs of the parameters are

$$\hat{\tau} = (X'S^{-1}X)^{-1}X'S^{-1}YA'(AA')^{-1}, \tag{2.4}$$

where

$$S = Y(I - A'(AA')^{-1}A)Y'$$

and

$$\hat{\Sigma} = (Y - X\hat{\tau}A)(Y - X\hat{\tau}A)'/N. \tag{2.5}$$

An alternative estimate for Σ is the posterior expectation of Σ, when the prior $p(\tau, \Sigma^{-1}) \propto |\Sigma|^{(p+1)/2}$ is assumed. Consequently, it can be shown that

$$\begin{aligned} E(\Sigma|Y) &= (N-p-1)^{-1}\{(Y - X\hat{\tau}A)(Y - X\hat{\tau}A)' \\ &+ (N-m-r-1)^{-1}X(X'S^{-1}X)X'[tr\, G^{-1}AA']\}, \end{aligned} \tag{2.6}$$

where

$$G^{-1} = (AA')^{-1} + T_2'(Z'SZ)^{-1}T_2$$

$$T_2 = Z'YA'(AA')^{-1}, \tag{2.7}$$

and $Z_{p\times p-m}$ is any matrix of rank $p-m$ such that $X'Z = 0$.

Comparison of (2.5) and (2.6) shows that the difference $E(\Sigma|Y) - \hat{\Sigma}$ is always non-negative definite.

The MLE of τ, as given in (2.4), was obtained by Khatri (1966) and was also derived by Rao (1967) as a covariance adjusted estimator. It can be shown that $\hat{\tau}$ is unbiased, i.e., $E(\hat{\tau}) = \tau$, although it is not a linear function of Y. Another unbiased estimator of τ, which is a linear function of Y, is

$$T_1 = BYA'(AA')^{-1}, \tag{2.8}$$

where

$$B = (X'X)^{-1}X'.$$

Since $E(T_1) = E(\hat{\tau}) = \tau$, the comparison of their covariance matrices would be instructive as to which would be a more desirable estimator for τ. Rao (1967) showed that T_1 is preferable if and only if the following condition holds

$$\Sigma = X\Gamma X' + Z\theta Z' + \sigma^2 I, \tag{2.9}$$

where Γ and θ are arbitrary unknown symmetric matrices. Such a situation arises when the following mixed model is considered,

$$\begin{aligned} Y_i &= X\alpha + X\beta + Z\gamma + \epsilon_i \\ \text{for } i &= 1, \ldots, n_a^*;\ a = 1, \ldots, r \end{aligned} \tag{2.10}$$

where α is a column of τ, and β, γ and ϵ_i are all uncorrelated random vectors such that $E(\beta) = 0$, $\text{cov}(\beta) = \Gamma$, $E(\gamma) = 0$, $\text{cov}(\gamma) = \theta$, $E(\epsilon_i) = 0$ and $\text{cov}(\epsilon_i) = \sigma^2 I$.

Geisser (1970) noted that

$$X(X'X)^{-1}X' + Z(Z'Z)^{-1}Z' \quad = \quad I$$

and hence claimed that, without loss of generality, the special structure for Σ as given by (2.9) can be defined as

$$\Sigma \quad = \quad X\Gamma X' + Z\theta Z', \tag{2.11}$$

which was later termed Rao's simple structure (RSS) by Lee and Geisser (1972). They also showed that the likelihood ratio criterion for testing

$$H_0 : \Sigma = X\Gamma X' + Z\theta Z' \;\; \text{vs} \;\; H_1 : \Sigma \text{ is p.d.}$$

is

$$\lambda \quad = \quad \frac{|(X'S^{-1}X)^{-1}|}{|BSB'|} \tag{2.12}$$

and is distributed under H_0 as $U_{m,p-m,N-p+m-r}$, a product of independent beta variates.

The two covariance structures considered so far are essentially the extremes for the growth curve model, one completely general and the other rather close to the independent situation. We will next consider a few other important structures.

The uniform covariance structure is defined as

$$\Sigma \quad = \quad \sigma^2[(1-\rho)I + \rho ee'], \tag{2.13}$$

where $e' = (1, \ldots, 1), -1/(p-1) < \rho < 1$. The uniform covariance structure arises from the situation when the error term $\epsilon_i = (\epsilon_{i1}, \ldots, \epsilon_{ip})'$ in the model specified by (2.1) has the property that $\text{cov}(\epsilon_{ij}, \epsilon_{ik}) = \sigma^2\rho$ for $j \neq k$, and $\text{var}(\epsilon_{ij}) = \sigma^2$. This indicates that the error components of ϵ_i are exchangeable. This covariance structure is very close to RSS. When $X = (e, X_2)$, that is, if there is a constant term in the growth function, then the MLEs of τ, σ^2, and ρ are, respectively, given by Lee (1988) as

$$\begin{aligned}
\hat{\tau} &= T_1 = BYA'(AA')^{-1}, \\
\hat{\sigma^2} &= tr\, S^*/pN \\
\hat{\rho} &= (e'S^*e - tr\, S^*)/(p-1)tr\, S^*,
\end{aligned} \tag{2.14}$$

where

$$\begin{aligned}
S^* &= S + ZDYA'(AA')^{-1}AY'D'Z', \\
D &= (Z'Z)^{-1}Z'.
\end{aligned} \tag{2.15}$$

Thus, as long as the constant term is in the growth function, which is typical in practice, the MLEs of the parameters are expressed in explicit form. Also, the MLE of τ is T_1, exactly as in the RSS case and consequently does not require covariance adjustment. However, according to our experience, this covariance structure may not be very practical, as least as far as the prediction of future observations is concerned. The likelihood ratio criterion for testing this particular covariance structure can be obtained easily but will be omitted here.

The covariance structure, which also caught the attention of Potthoff and Roy (1964), Lee and Geisser (1975) and Lee (1988, 1991), is the serial covariance structure which is defined as

$$\Sigma = \sigma^2 C, \tag{2.16}$$

where

$$C = (\rho^{|a-b|}).$$

The MLEs of the parameters are given by Lee (1988) as

$$\begin{aligned} \hat{\tau} &= (X'\hat{C}^{-1}X)^{-1}X'\hat{C}^{-1}YA'(AA')^{-1} \qquad (2.17) \\ \hat{\sigma^2} &= [tr\,(X'\hat{C}^{-1}X)^{-1}X'\hat{C}^{-1}S\hat{C}^{-1}X + tr\,(Z'\hat{C}Z)^{-1}Z'YY'Z]/pN \end{aligned}$$

where

$$\hat{C} = (\hat{\rho}^{|a-b|}),$$

and $\hat{\rho}$ is obtained by maximizing the profile likelihood function

$$L_{\max}(\rho) = (\hat{\sigma}^2(\rho))^{-pN/2}(1-\rho^2)^{-N(p-1)/2}, \tag{2.18}$$

and $\hat{\sigma}^2(\rho)$ is the $\hat{\sigma}^2$ given by (2.17) with $\hat{\rho}$ replaced by ρ. We thus see that the MLEs of τ and σ^2 depend on the MLE of ρ, which can be obtained by a one-dimensional search. No iterations are needed and the computations are relatively easy. A likelihood ratio test for testing

$$H_0 : \Sigma = \sigma^2 C \quad \text{vs} \quad H_1 : \Sigma \text{ is p.d.}$$

is $-2\ln\lambda$ which is distributed under H_0 as χ^2_ν, $\nu = p(p+1)/2-2$ when $N \to \infty$, where

$$\begin{aligned} \lambda &= N^{-pN/2}(\hat{\sigma}^2)^{-pN/2}(1-\hat{\rho}^2)^{-(p-1)N/2} \\ &\quad \times |(B'Z)|^{-N}|Z'YY'Z|^{N/2}|X'S^{-1}X|^{-N/2}. \end{aligned} \tag{2.19}$$

This result was obtained by Lee (1991).

The other covariance structures included in this paper are obtained from the different assumptions on $\Omega = \text{cov}(\epsilon_i)$ and β_a in the model specified by (2.1). For example, when $\Omega = \sigma^2 C$, and β_a is considered as a random vector with mean vector $E(\beta_a) = \tau$ and covariance matrix $\text{cov}(\beta_a) = \Gamma$, then

$$\Sigma = X\Gamma X' + \sigma^2 C \tag{2.20}$$

The estimation of the parameters for this model is quite complex as indicated by Keramidas and Lee (1995). Two somewhat more parsimonious models involve either the intercept random or the slopes random. The two resulting covariance structures are

$$\Sigma = \gamma ee' + \sigma^2 C \tag{2.21}$$

and

$$\Sigma = X_2 \Phi X_2 + \sigma^2 C, \tag{2.22}$$

where

$$X = (e, X_2).$$

The above three covariance structures were considered by Chi and Reinsel (1989).

In (2.20) if we set $\Omega = \sigma^2 I$, then we have the following covariance structure

$$\Sigma = X \Gamma X' + \sigma^2 I, \tag{2.23}$$

which was considered by Rao (1967) and later promoted by Fearn (1975) as a natural covariance matrix resulting from a two-stage hierarchical model in which each individual has a separate growth curve and the parameters associated with each curve are exchangeable, i.e., they are realizations of another normal distribution. It should be pointed out, however, that although the model started out with a separate curve for each individual, there is still a common growth curve at the end. The only difference is the final covariance structure of Σ which is different from $\mathrm{cov}(\epsilon_i)$, i.e., $\Sigma = X\Gamma X' + \sigma^2 I$ rather than $\Sigma = \mathrm{cov}(\epsilon_i)$, as assumed in the fixed effects model.

The MLEs of the parameters for the covariance structures (2.20) through (2.23) are not available in closed form. However, using BMDV(BMDP, 1988), one can obtain these estimates. Hence the MLEs of the parameters can be obtained in principle. Nevertheless, convergence can be a problem at times.

3 Prediction of Future Observations

In this paper we will consider three types of prediction for the generalized growth curve model as specified by (2.2) when the covariance matrix Σ has different structures. Let V be a set of $p \times K$ future observations drawn from the generalized growth curve model, i.e., the set of future observations is such that given the parameters τ and Σ,

$$E(V) = X\tau F, \tag{3.1}$$

where F is a known $r \times K$ matrix, usually formed by some columns of A, and the columns of V are independent and multivariate normal with a common covariance matrix Σ. Geisser (1970, 1980) and Lee (1982) considered prediction of V, given Y as the sample, from a Bayesian viewpoint.

The second type of prediction for the generalized growth curve model is to predict $V^{(2)}$ given $V^{(1)}$ and Y, if V is partitioned as

$$V = \begin{bmatrix} V^{(1)} \\ V^{(2)} \end{bmatrix}, \tag{3.2}$$

where $V^{(i)}$ is $p_i \times K (i = 1, 2)$ and $p_1 + p_2 = p$. If p is interpreted as the number of times being observed, then this problem is concerned with the growth curves for future individuals for subsets of size p_2, having observed subsets of size p_1. This type of prediction is called conditional prediction of $V^{(2)}$ given $V^{(1)}$ and Y. When $p_2 < p$ and $K = 1$, it is also called conditional prediction of the unobserved portion of a partially observed vector. This type of prediction has been considered by many authors including Lee and Geisser (1972, 1975), Fearn (1975), Rao (1975, 1987), Geisser (1981), Reinsel (1984), Lee (1988, 1991), and Keramidas and Lee (1995), among others.

The third type of prediction is somewhat different. It is concerned with predicting future values for the observed individuals. Let y, of dimension $q \times n$, be a set of $n (\leq N)$ future observations whose previous p-dimensional observations are a subset of Y. We are interested in predicting y given Y. This is time series type of prediction and thus is very important in practice as growth curve data are often time series in nature. However, Σ needs to be structured in order to predict y in a reasonable manner. For example, if Σ is assumed arbitrary p.d., then predictions can not be made simply because information is lacking in regard to the relationship between y and Y. This type of prediction is called extended prediction of y given Y, because the prediction is made beyond the observed time range of the sample Y. Extended prediction of y given Y has been considered by Rao (1977, 1987, 1984), Lee (1988, 1991) and Keramidas and Lee (1990, 1995).

Let x, of dimension $q \times m$, be a design matrix corresponding to y, $Y = (Y_1, \ldots, Y_N)$, $A = (A_1, \ldots, A_N)$, $y = (y_1, \ldots, y_n)$, and assume that for $i \leq n$,

$$E\begin{pmatrix} Y_i \\ y_i \end{pmatrix} = \begin{pmatrix} X \\ x \end{pmatrix} \tau A_i, \tag{3.3}$$

$$\Sigma^* = \operatorname{cov}\begin{pmatrix} Y_i \\ y_i \end{pmatrix} = \begin{pmatrix} \Sigma_{11}^* & \Sigma_{12}^* \\ \Sigma_{21}^* & \Sigma_{22}^* \end{pmatrix}. \tag{3.4}$$

We note in passing that $\Sigma_{11}^* = \Sigma$. From (3.4) it is clear that in order to make an inference about y, the form of Σ_{21}^* has to be known, i.e., the covariance structure for Σ has to be extendable to future values.

4 Σ is Arbitrary p.d.

From a Bayesian viewpoint Geisser (1970) considered the estimation problem for τ and predictive inference for V. Using a convenient diffuse prior, (Geisser and Cornfield, 1963; Geisser, 1965),

$$g(\tau, \Sigma^{-1}) \propto |\Sigma|^{(p+1)/2}, \tag{4.1}$$

Geisser (1970) showed that the predictive density of V given Y is

$$f(V|Y) \propto |Z'[YY' + (V - X\hat{\tau}F)(V - X\hat{\tau}F)']Z|^{-(N+K-m)/2}|G_1|^{m/2} \\ \times |I + G_1(V - X\hat{\tau}F)'S^{-1}X(X'S^{-1}X)^{-1}X'S^{-1}(V - X\hat{\tau}F)|^{-(N+K-r)/2}, \tag{4.2}$$

where

$$G_1^{-1} = [I - F'(HH')^{-1}F]^{-1} + (V_2 - Z'\hat{V})'(Z'SZ)^{-1}(V_2 - Z'\hat{V})$$

$$H = (A, F),\ V_1 = BV,\ V_2 = Z'\hat{V},\ \hat{V} = YA'(AA')^{-1}F.$$

It is clear that the predictive density of V given Y is a product of three general determinantal, or matrix T densities and is extremely complex. Drawings from the three general determinantal distributions can be accomplished as suggested by Zellner et al (1988). However, since the dimension of V is $p \times K$, the Monte Carlo method may be hard to use in practice for the predictive region of V. Instead, we will suggest the following. Geisser (1970) noted that $E(V|Y) = X\hat{\tau}F$. He also showed that $X\hat{\tau}F$ is not the mode of the predictive distribution of V given Y. However, numerical procedures, via two-dimensional search, for calculating the predictive mode when $K = 1$ were given by Lee and Geisser (1972). Geisser (1970) also showed that for

$$\begin{aligned} Q &= BYA'(AA')^{-1} + BSZ(Z'SZ)^{-1}Z'(V - YA'(AA')^{-1}F), \\ U_1 &= |I + (X'S^{-1}X)(BV - Q)G_1(BV - Q)'| \end{aligned} \tag{4.3}$$

is distributed as $U_{m,K,N-r}$ and is independent of

$$U_2 = |I + V'Z(Z'YY'Z)^{-1}Z'V| \tag{4.4}$$

which is distributed as $U_{p-m,K,N-m}$. Hence, $U_1 + U_2$ is distributed as the sum of two independent U variates. For an excellent approximation to this distribution and its special case, a linear combination of two independent F variates, see Lee and Hu (1995).

For conditional prediction of $V^{(2)}$ given $V^{(1)}$ and Y, we note that $f(V^{(2)}|V^{(1)}, Y) \propto f(V|Y)$ and hence the predictive distribution of $V^{(2)}$ given $V^{(1)}$ and Y is at least as complicated as the predictive distribution of V given Y. Therefore, some approximations are in order. Lee and Geisser (1972) showed that conditional on Σ^{-1} and Y, V is distributed as normal with mean

$$\mu_a = X(X'\Sigma^{-1}X)^{-1}X'\Sigma^{-1}YA'(AA')^{-1}F, \tag{4.5}$$

and covariance matrix

$$\begin{aligned} \Sigma_a &= X(X'\Sigma^{-1}X)^{-1}X' \otimes M^{-1} + [XB\Sigma Z(Z'\Sigma Z)^{-1}Z'\Sigma B'X' \\ &+ D'Z'\Sigma B'X' + XB\Sigma ZD + D'Z'\Sigma ZD] \otimes I_k, \end{aligned} \tag{4.6}$$

where

$$M = I - F'(AA' + FF')^{-1}F.$$

We thus see that the predictive distribution of V given Y is approximately $N(\hat{\mu_a}, \hat{\Sigma_a})$ where $\hat{\mu_a}$ and $\hat{\Sigma_a}$ are obtained by replacing Σ by its estimate, either the MLE or the posterior expectation as given by (2.5) or (2.6), respectively. Hence, a predictive region for V given Y as well as for $V^{(2)}$ given $V^{(1)}$ and Y can be obtained through standard normal theory, with appropriate rearrangement of V and the corresponding covariance matrix $\hat{\Sigma_a}$.

As for extended prediction of y given Y, the prediction can not be made because there is no information concerning the dependence structure between y and the observations in the sample. Hence, other covariance structures need to be assumed to facilitate extended prediction.

5 Σ has RSS

Using the convenient prior,

$$g(\Gamma^{-1},\theta^{-1},\tau) \quad \propto \quad |\Gamma|^{(m+1)/2}|\theta|^{(p-m+1)/2}, \tag{5.1}$$

Geisser (1970) showed that the predictive density of V given the sample Y is

$$f(V|Y) \propto |Z'[YY' + (V - XT_1F)(V - XT_1F)']Z|^{-(N+K)/2}$$
$$\times|I + (I - F'(HH')^{-1}F)(V - XT_1F)'B'(BSB')^{-1}B(V - XT_1F)|^{-(N+K-r)/2}. \tag{5.2}$$

It is clear that XT_1F is the mean and the mode of $f(V|Y)$. It can also be shown that a posteriori

$$U_1 \quad = \quad |I + (I - F'(HH')^{-1}F)(V - XT_1F)'B'(BSB')^{-1}B(V - XT_1F)|^{-1} \tag{5.3}$$

and

$$U_2 \quad = \quad |I + (V - XT_1F)'Z'(Z'YY'Z)^{-1}Z'(V - XT_1F)|^{-1} \tag{5.4}$$

are independently distributed as $U_{m,K,N-r}$ and $U_{p-m,K,N}$, respectively. Hence a predictive region for V can be obtained through $U_1 + U_2$ which is distributed as the sum of two independent U variates. For the special case in which $K = F = r = 1$, a predictive region for a future vectorial observation can be obtained from

$$Q \quad = \quad (V - XT_1)'[N(N+1)^{-1}X(X'SX)^{-1}X' + Z(Z'YY'Z)^{-1}Z'](V - XT_1) \tag{5.5}$$

which is distributed as a linear combination of two independent F variates. It is noted that Q can be written free of Z by use of the following identity

$$Z(Z'YY'Z)^{-1}Z' \quad = \quad (YY')^{-1} - (YY')^{-1}X[X'(YY')^{-1}X]^{-1}X'(YY')^{-1}. \tag{5.6}$$

Accurate approximations to the above distributions have been obtained by Lee and Hu (1995).

An alternative representation for the predictive density of V given Y, when $K = 1$, was obtained by Lee and Geisser (1972) as

$$f(V|Y) \quad = \quad \int_0^\infty f(V|Y,t)g(t|Y)dt \tag{5.7}$$

where

$$\begin{aligned} F(V|Y,t) &= St(.;XT_1F,(1+\gamma t)(2N+2-r-p)^{-1}J^{-1},2N+2-r), \\ J &= t\gamma MX(X'SX)^{-1}X' + Z(Z'YY'Z')^{-1}Z', \\ M &= 1 - F'(AA' + FF')^{-1}F, \\ \gamma &= (N+1-p+m)^{-1}(N+1-r-m), \\ G(t|Y) &= F(N+1-r-m,N+1-p+m), \end{aligned} \tag{5.8}$$

and $St(\mu, \Sigma, v+p)$ is a multivariate Student t distribution with density

$$f(T) = (\pi\nu)^{-p/2}\Gamma^{-1}(\nu/2)\Gamma(N/2)|\Sigma|^{-1/2}|1+(T-\mu)'(\nu\Sigma)^{-1}(T-\mu)|^{-N/2} \tag{5.9}$$

where T is $p \times 1$ and $\nu = N - p$. Thus, the predictive density of V given Y is expressed as an average of a multivariate Student t density over an F density.

As noted in Lyung and Box (1980), since F is nearly symmetric and well concentrated, a reasonable approximation for the predictive density of V given Y is

$$\begin{aligned} F(V|Y) &\doteq F(V|Y,\hat{t}) \\ &= St(XT_1F, (1+\gamma\hat{t})(2N+2-r-p)^{-1}J^{-1}(\hat{t}), 2N+2-r), \end{aligned} \tag{5.10}$$

where $\hat{t}$ is either the mean or the mode of the F dsitribution and they are $(N+1-p+m)/(N-1-p+m)$ or $(N-r-m-1)(N+1-p+m)/(N+1-r-m)(N+3-p+m)$, respectively, and $J(\hat{t})$ is the value of J evaluated at $t = \hat{t}$. From (5.10) we have

$$F(V^{(2)}|V^{(1)}, Y) \doteq St(\mu_{r2\cdot1}(\hat{t}), b(\hat{t})(2N+2-r-p_2)^{-1}J_{22}^{-1}(\hat{t}), 2N+2-r) \tag{5.11}$$

where

$$\begin{aligned} \mu_{r2\cdot1} &= X^{(2)}T_1F - J_{22}^{-1}J_{21}(V^{(1)} - X^{(1)}T_1F), \\ b &= 1+\gamma t + (V^{(1)} - X^{(1)}T_1F)'J_{11\cdot2}(V^{(1)} - X^{(1)}T_1F), \\ X &= \begin{pmatrix} X^{(1)} \\ X^{(2)} \end{pmatrix}, \quad X^{(i)} \text{ is } p_i \times m,\ p_1+p_2 = p, \\ J &= \begin{pmatrix} J_{11} & J_{12} \\ J_{21} & J_{22} \end{pmatrix}, J_{ij} \text{ is of dimension } p_i \times p_j, J_{11\cdot2} = J_{11} - J_{12}J_{22}^{-1}J_{21}. \end{aligned} \tag{5.12}$$

Thus, an approximate point estimate of $V^{(2)}$ given $V^{(1)}$ and Y is $\mu_{r_{2\cdot1}}(\hat{t})$ and an approximate predictive region for $V^{(2)}$ can be obtained through

$$(V^{(2)} - \mu_{r2\cdot1}(\hat{t}))'[b^{-1}(\hat{t})J_{22}(\hat{t})](V^{(2)} - \mu_{r2\cdot1}(\hat{t})) \tag{5.13}$$

which is distributed as $(2N+2-r-p_2)^{-1}p_2F(p_2, 2N+2-r-p_2)$. A better approximation for the predictive density of V given Y is

$$f(V|Y) \doteq \frac{1}{L}\sum_{i=1}^{L} f(V|Y, t^{(i)}) \tag{5.14}$$

where $f(V|Y,t^{(i)})$ is a multivariate Student t density as given in (5.10) and $t^{(i)}$ is the ith draw from $F(N+1-r-m, N+1-p+m)$. It is noted that (5.14) will clearly be better than the approximation given by (5.10) as $L = 1$ with

$t^{(i)} = \hat{t}$ will be a special case, see e.g., Liu (1995). However, an average of L multivariate Student t densities is no longer a multivariate Student t density and hence a predictive region based on the approximation (5.14) will be harder to obtain in practice.

For the conditional distribution of $V^{(2)}$ given $V^{(1)}$ and Y, we also have

$$f(V^{(2)}|V^{(1)},Y) = \int_0^\infty f(V^{(2)}|V^{(1)},r,t)g(t|V^{(1)},Y)dt, \tag{5.15}$$

where

$$\begin{aligned}
F(V^{(2)}|V^{(1)},Y,t) &= St(\mu_{s2\cdot 1}, b(2N+2-r-p_2)^{-1}J_{22}^{-1}, 2N+2-r),\\
\mu_{s2\cdot 1} &= X^{(2)}T_1F - J_{22}^{-1}J_{21}(V^{(1)} - X^{(1)}T_1F), \qquad (5.16)\\
b &= 1+\gamma t + (V^{(1)} - X^{(1)}T_1F)'J_{11\cdot 2}(V^{(1)} - X^{(1)}T_1F),\\
g(t|V^{(1)},Y) &= [f(V^{(1)}|Y)]^{-1}\pi^{p_2/2}t^{(N+1-r)/2-1}\\
&\times \Gamma[\frac{1}{2}(2N+1-r-p_2)]b^{-(2N+2-r-p_2)/2}C_0|J_{22}|^{-\frac{1}{2}},\\
&\qquad (5.17)\\
C_0 &= \pi^{-p/2}M^{m/2}|BSB'|^{-\frac{1}{2}}|DYY'D'|^{-\frac{1}{2}}\\
&\times \Gamma^{-1}[\frac{1}{2}(N+1-r-m)]\Gamma^{-1}[\frac{1}{2}(N+1-p+m)]\\
&\times \operatorname{mod}\left|\begin{pmatrix} B \\ D \end{pmatrix}\right| \qquad (5.18)\\
f(V^{(1)}|Y) &= \int_0^\infty f(V^{(1)}|Y,t)g(t|Y)dt, \qquad (5.19)\\
F(V^{(1)}|Y,t) &= St(X^{(1)}T_1F, (1+\gamma t)(2N+2-r-p)^{-1}J_{11\cdot 2}^{-1},\\
&\quad 2N+2-r-p_2), \qquad (5.20)\\
G(t|Y) &= F(N+1-r-m, N+1-p+m).
\end{aligned}$$

As in (5.10), a reasonable approximation for the conditional predictive distribution of $V^{(2)}$ given $V^{(1)}$ and Y is

$$\begin{aligned}
F(V^{(2)}|V^{(1)},Y) &\doteq F(V^{(2)}|V^{(1)},Y,t_0)\\
&= St(\mu_{s2\cdot 1}(t_0), b(t_0)(2N+2-r-p_2)^{-1}J_{22}^{-1}(t_0),\\
&\quad 2N+2-r) \qquad (5.21)
\end{aligned}$$

where t_0 maximizes $g(t|V^{(1)},Y)$. Similar to (5.14), a better approximation for the conditional predictive density of $V^{(2)}$ given $V^{(1)}$ and Y is

$$f(V^{(2)}|V^{(1)},Y) \doteq \frac{1}{L}\sum_{i=1}^{L} f(V^{(2)}|V^{(1)},Y,t^{(i)}) \tag{5.22}$$

where $t^{(i)}$ is the ith draw from $g(t|V^{(1)},Y)$. A comparison among (5.11), (5.15), (5.21) and (5.22), via a real data set, will be given in Section 12.

It is also noted that numerical prodecures are given for $K = 1$ by Lee and Geisser (1972) for obtaining the predictive mode of $V^{(2)}$ given $V^{(1)}$ and Y and an exact solution for the particularly interesting special case $p_2 = 1$. As in the arbitrary p.d. case, extended prediction for y given Y is not available for the RSS case. This is due to the fact that the dependence between y and its previous observations is not defined.

6 Σ has Uniform Covariance Structure

For the rest of this paper we will restrict our attention to the special case in which $K = 1$, i.e., there is one future vectorial observation to be predicted. When Σ has the uniform covariance structure, it has the form as given in (2.13). The predictive distribution of V given Y can be approximated as

$$F(V|Y) \doteq F(V|Y, \hat{\tau}, \hat{\rho}, \hat{\sigma}^2). \tag{6.1}$$

Thus, the predictive distribution of V given Y is approximately

$$F(V|Y) \doteq N(XT_1F, \hat{\sigma}^2[(1-\hat{\rho})I + \hat{\rho}ee'])$$

and the conditional distribution of $V^{(2)}$ given $V^{(1)}$ and Y is

$$F(V^{(2)}|V^{(1)}, Y) \doteq N(\mu_{u2\cdot 1}, \Sigma_{u22\cdot 1}), \tag{6.2}$$

where

$$\begin{aligned} \mu_{u2\cdot 1} &= X^{(2)}T_1F + \frac{\hat{\rho}}{1+(p_1-1)\hat{\rho}} e_{p_2}e'_{p_1}(V^{(1)} - X^{(1)}T_1F), \qquad (6.3) \\ \Sigma_{u22\cdot 1} &= \hat{\sigma}^2[(1-\hat{\rho})I_{p_2} + \frac{\hat{\rho}(1-\hat{\rho})}{1+(p_1-1)\hat{\rho}} e_{p_2}e'_{p_2}]. \end{aligned}$$

Hence, the approximate means, XT_1F and $\mu_{u2\cdot 1}$, can be used as the point predictors for V given Y, and $V^{(2)}$ given $V^{(1)}$ and Y, respectively, and the corresponding predictive regions can be obtained through standard normal theory.

Instead of the approximate mean as the point predictor for $V^{(2)}$ given $V^{(1)}$ and Y, an ad hoc predictor can be obtained by the weighted average of two independent predictors, one based on $V^{(1)}$ alone and the other based on Y, with the weights being the estimates of the relative precisions. This type of predictor was first proposed by Lee and Geisser (1975). The ad hoc predictor, when the uniform covariance structure is assumed, is

$$P_{ua} = (S_{m1}^{-1} + S_{m2}^{-1})^{-1}(S_{m1}^{-1}m_1 + S_{m1}^{-1}m_2), \tag{6.4}$$

where

$$\begin{aligned} m_1 &= X^{(2)}T_1F, \\ m_2 &= X^{(2)}\tau_vF + \Sigma_{u21}\Sigma_{u11}^{-1}(V^{(1)} - X^{(1)}\tau_vF), \\ \tau_v &= (X^{(1)'}\Sigma_{u11}^{-1}X^{(1)})^{-1}X^{(1)'}\Sigma_{u11}^{-1}V^{(1)}F'(FF')^{-} \qquad (6.5) \\ \Sigma_u &= \hat{\sigma}^2[(1-\hat{\rho})I + \hat{\rho}ee'] = \begin{pmatrix} \Sigma_{u11} & \Sigma_{u12} \\ \Sigma_{u21} & \Sigma_{u22} \end{pmatrix}, \end{aligned}$$

and

$$\begin{aligned}
S_{m1} &= \Sigma_{u22} + X^{(2)}B\Sigma_u B'X^{(2)'}F'(AA')^{-1}F,\\
S_{m2} &= \Sigma_{u22} + (b_1B_2 + \Sigma_{u21})'\Sigma_{u11}^{-1}(b_1B_2 + \Sigma_{u21})\\
&- (b_1B_2 + \Sigma_{u21})\Sigma_{u11}^{-1}\Sigma_{u12} + \Sigma_{u21}\Sigma_{u11}^{-1}(b_1B_2 + \Sigma_{u21})',\\
b_1 &= F'(FF')^{-}F,\\
B_2 &= (X^{(2)} - \Sigma_{u21}\Sigma_{u11}^{-1}X^{(1)})(X^{(1)'}\Sigma_{u11}^{-1}X^{(1)})^{-1}X^{(1)'},
\end{aligned}$$

where Σ_{uij} is $p_i \times p_j$, $p_1 + p_2 = p$, $(FF')^{-}$ is the generalized inverse of FF' and can be set as FF' in the special case $F = (1,0)'$ or $F = (0,1)'$. Note that $F'(AA')^{-1}F$ is a scalar and S_{mi} is the covariance matrix of the forecast error when m_i is the predictor for $V^{(2)}$. An alternative formula for S_{mi} was given by Lee and Geisser (1975). A justification for the ad hoc predictor is the fact that it is the optimal combination of two independent forecasts.

Next, consider extended prediction of y, given Y. This is a time series type of prediction. To make this type of prediction the covariance structure generally has to be extendable to future values of the individuals observed.

Let x, $q \times m$, be a design matrix corresponding to y, $Y = (Y_1, \ldots, Y_N)$, $A = (A_1, \ldots, A_N)$, $y = (y_1, \ldots, y_n)$, and assume that for $i \leq n$,

$$E\begin{pmatrix} Y_i \\ y_i \end{pmatrix} = \begin{pmatrix} X \\ x \end{pmatrix}\tau A_i, \tag{6.6}$$

and

$$\begin{aligned}
\Sigma^* &= \operatorname{cov}\begin{pmatrix} Y_i \\ y_i \end{pmatrix} = \begin{pmatrix} \Sigma_{11}^* & \Sigma_{12}^* \\ \Sigma_{21}^* & \Sigma_{22}^* \end{pmatrix}\\
&= \sigma^2\begin{pmatrix} V_{11} & V_{12} \\ V_{21} & V_{22} \end{pmatrix} = \sigma^2\begin{pmatrix} 1 & \rho & \cdots & \rho \\ \rho & 1 & \cdots & \rho \\ \vdots & \vdots & & \vdots \\ \rho & \rho & \cdots & 1 \end{pmatrix}.
\end{aligned} \tag{6.7}$$

Similar to (6.1), the predictive distribution of y_i given Y can be approximated as

$$F(y_i|Y) = F(y_i|Y, \hat{\tau}, \hat{\Gamma}, \hat{\sigma^2}). \tag{6.8}$$

Thus, the predictive distribution of y_i given Y is approximately

$$F(y_i|Y) \doteq N(\mu_{u2\cdot1}^*, \Sigma_{u22\cdot1}^*), \tag{6.9}$$

where

$$\begin{aligned}
\mu_{u2\cdot1}^* &= xT_1A_i + \frac{\hat{\rho}}{1+(p-1)\hat{\rho}}e_q e_p'(Y_i - XT_1A_i),\\
\Sigma_{u22\cdot1}^* &= \hat{\sigma}^2[(1-\hat{\rho})I_q + \frac{\hat{\rho}(1-\hat{\rho})}{1+(p-1)\hat{\rho}}e_q e_q'],
\end{aligned} \tag{6.10}$$

and T_1, $\hat{\rho}$, and σ^2 are given in (2.14).

7 Serial Covariance Structure for Σ

When Σ has the serial covariance structure, it takes the form $\Sigma = \sigma^2 C$ as given in (2.6). The predictive distribution of V given Y can be approximated as

$$F(V|Y) \doteq F(V|Y, \hat{\tau}, \hat{\rho}, \hat{\sigma}^2) \tag{7.1}$$

where $\hat{\tau}$, $\hat{\rho}$ and $\hat{\sigma}^2$ are given in (2.17). Thus, the predictive distribution of V given Y is approximately

$$F(V|Y) \doteq N(X\hat{\tau}F, \sigma^2 C(\hat{\rho})) \tag{7.2}$$

and the conditional distribution of $V^{(2)}$ given $V^{(1)}$ and Y is

$$F(V^{(2)}|V^{(1)}, Y) \doteq N(\mu_{s2\cdot1}, \Sigma_{s22\cdot1}) \tag{7.3}$$

where

$$\begin{aligned} \mu_{s2\cdot1} &= X^{(2)}\hat{\tau}F + (0, \hat{\eta})(V^{(1)} - X^{(1)}\hat{\tau}F), \\ \hat{\eta} &= (\hat{\rho}, \ldots, \hat{\rho}^{p_2})' \\ \Sigma_s &= \sigma^2 C(\hat{\rho}) = \begin{pmatrix} \Sigma_{s11} & \Sigma_{s12} \\ \Sigma_{s21} & \Sigma_{s22} \end{pmatrix}, \\ \Sigma_{s22\cdot1} &= \Sigma_{s22} - \Sigma_{s21}\Sigma_{s11}^{-1}\Sigma_{s12}. \end{aligned} \tag{7.4}$$

Hence, the approximate means, $X\hat{\tau}F$ and $\mu_{s2\cdot1}$, can be used as the point predictors for V given Y, and $V^{(2)}$ given $V^{(1)}$ and Y, respectively, and the corresponding predictive regions can be easily obtained by use of standard normal theory.

The ad hoc predictor, when the serial covariance structure holds, is

$$P_{sa} = (S_{m_1^*}^{-1} + S_{m_2^*}^{-1})^{-1}(S_{m_1^*}^{-1} m_1^* + S_{m_2^*}^{-1} m_2^*), \tag{7.5}$$

where m_1^*, m_2^* and $S_{m_2^*}$ are defined as in m_1, m_2 and S_{m_2}, respectively, except that T_1 is replaced by $\hat{\tau}$ and Σ_{uij} is replaced by Σ_{sij}, and

$$S_{m_1^*} = \Sigma_{s22} + X^{(2)}(X'\Sigma_s^{-1}X)^{-1}X^{(2)'}F'(AA')^{-1}F. \tag{7.6}$$

If the covariance structure is appropriate for the growth curve data at hand, the ad hoc predictor should be a very strong competitor as a point predictor for the unobserved portion, $V^{(2)}$, of the partially observed vector V. This is due to the fact that it utilizes all the available data in a very reasonable fashion. This observation has been borne out in studies such as Lee (1988) for several real data sets. One disadvantage of the ad hoc predictor is that the predictive region is not available in a natural manner. However, since the covariance matrix of the forecast error, when P_{sa} is the predictor for $V^{(2)}$, is approximately

$$\text{cov}(P_{sa}) \doteq (S_{m1}^{-1} + S_{m2}^{-1})^{-1}, \tag{7.7}$$

we have approximately,

$$F(P_{sa}|V^{(2)}, Y) \doteq N(V^{(2)}, (S_{m1}^{-1} + S_{m2}^{-1})^{-1}). \tag{7.8}$$

Hence, an approximate predictive region for $V^{(2)}$ given $V^{(1)}$ and Y can be obtained through use of standard normal theory.

We next consider extended prediction of y_i given Y. As in the previous section, assume for $i \leq n$,

$$E\begin{pmatrix} Y_i \\ y_i \end{pmatrix} = \begin{pmatrix} X \\ x \end{pmatrix} \tau A_i \tag{7.9}$$

and

$$\Sigma^* = \text{cov}\begin{pmatrix} Y_i \\ y_i \end{pmatrix} = \sigma^2 \begin{pmatrix} C_{11} & C_{12} \\ C_{21} & C_{22} \end{pmatrix} \tag{7.10}$$

where

$$\begin{pmatrix} C_{11} & C_{12} \\ C_{21} & C_{22} \end{pmatrix} = (\rho^{|a-b|}), a, b, = 1, \ldots, p+q, \tag{7.11}$$

C_{11} is $p \times p$, C_{12} is $p \times q$, C_{22} is $q \times q$, and $C_{21} = C_{12}'$. Similar to the predictive distribution of V given Y, the predictive distribution of y_i given Y can be approximated as

$$F(y_i|Y) \doteq F(y|Y, \hat{\tau}, \hat{\rho}, \hat{\sigma}^2). \tag{7.12}$$

Thus, the predictive distribution of y_i given Y is approximately

$$F(y_i|Y) \doteq N(\mu^*_{s2\cdot 1}, \Sigma^*_{s22\cdot 1}), \tag{7.13}$$

where

$$\mu^*_{s2\cdot 1} = X\hat{\tau}A_i + (0, \hat{\eta}^*)(Y_i - X\hat{\tau}A_i), \tag{7.14}$$

$$\hat{\eta}^* = (\hat{\rho}, \ldots, \hat{\rho}^q)',$$

$$\Sigma^*_{s22\cdot 1} = \Sigma^*_{s22} - \Sigma^*_{s21}\Sigma^{-1*}_{s11}\Sigma^*_{s12}, \tag{7.15}$$

$$\Sigma^*_s = \sigma^2 C(\hat{\rho}) = \begin{pmatrix} \hat{C}_{11} & \hat{C}_{12} \\ \hat{C}_{21} & \hat{C}_{22} \end{pmatrix}.$$

The approximate mean $\mu^*_{s2\cdot 1}$ can be used as the point predictor for y_i and the predictive region for y_i can be obtained through standard normal theory.

8 Other Covariance Structures

In this section we consider the situation in which the covariance structures are obtained from the consideration of random (and mixed) effects for the regression parameters. They include the covariance structures given in (2.20) through (2.23). In addition to the random effects for the regression parameters, the growth curve models associated with these covariance structures rely heavily on numerical solutions for the estimation of parameters. In other words, the MLEs are difficult to obtain numerically and convergence to the MLEs may not always be assured. Newer procedures such as Markov chain Monte Carlo maximum likelihood methods may alleviate the situation, Geyer and Thompson (1992).

Once the numerical estimates, $\hat{\tau}$ and $\hat{\Sigma}$, of the parameters, τ and Σ, are obtained, the predictive distributions of V given Y, and $V^{(2)}$ given $V^{(1)}$ and Y can be approximated as

$$F(V|Y) \doteq N(X\hat{\tau}F, \hat{\Sigma}), \tag{8.1}$$

and

$$F(V^{(2)}|V^{(1)}, Y) \doteq N(\hat{\mu}_{2\cdot 1}, \hat{\Sigma}_{22\cdot 1}), \tag{8.2}$$

where

$$\hat{\mu}_{2\cdot 1} = X^{(2)}\hat{\tau}F + \hat{\Sigma}_{21}\hat{\Sigma}_{11}^{-1}(V^{(1)} - X^{(1)}\hat{\tau}F), \tag{8.3}$$

$$\hat{\Sigma}_{22\cdot 1} = \hat{\Sigma}_{22} - \hat{\Sigma}_{21}\hat{\Sigma}_{11}^{-1}\hat{\Sigma}_{12}. \tag{8.4}$$

As for extended prediction of y given Y, the predictive distribution of y given Y can be approximated by

$$F(y|Y) \doteq F(y|Y, \hat{\tau}, \hat{\Sigma}). \tag{8.5}$$

Thus, the predictive distribution of y_i given Y is approximately

$$F(y_i|Y) \doteq N(\hat{\mu}^*_{2\cdot 1}, \hat{\Sigma}^*_{22\cdot 1}). \tag{8.6}$$

where

$$\hat{\mu}^*_{2\cdot 1} = X\hat{\tau}A_i + \hat{\Sigma}^*_{21}\hat{\Sigma}_{11}^{-1*}(V^{(1)} - X^{(1)}\hat{\tau}F), \tag{8.7}$$

$$\hat{\Sigma}^*_{22\cdot 1} = \hat{\Sigma}^*_{22} - \hat{\Sigma}_{21}\hat{\Sigma}_{11}^{-1*}\hat{\Sigma}^*_{12}, \tag{8.8}$$

and Σ^*_{ij} are defined in (3.4).

9 A Predictive Sample Reuse Approach for Conditional Prediction

This is a data analytic aparametric method which simulates the predictive process within the sample, given a complete lack of any distributional assumption. It is termed predictive sample reuse (PSR) by Geisser (1974, 1975) because each vector in the sample of size N will be utilized $(N-1)$ times in the prediction process.

In this section we are concerned with predicting one future vectorial observation, i.e., $K = 1$. Also, for ease of presentation let $V = Y_{N+1} = \begin{pmatrix} Y_{N+1}^{(1)} \\ Y_{N+1}^{(2)} \end{pmatrix}$ where $Y_{N+1}^{(i)}$ is $p_i \times 1$ and $p_1 + p_2 = p$. Suppose from the first N vectors $Y_1, \ldots, Y_N$, we generate a data-based predictor of $Y_{N+1}^{(2)}$, denoted by $\hat{Y}_{(N)}^{(2)}$. Further, suppose another predictor of $Y_{N+1}^{(2)}$, denoted by $\hat{Y}_{N+1}^{(2)}$, depends only on the observed $Y_{N+1}^{(1)}$. The two independent predictors are then combined to produce a new predictor

$$\dot{Y}_{N+1}^{(2)} = f(\hat{Y}_{(N)}^{(2)}, \hat{Y}_{N+1}^{(2)}; \theta) \tag{9.1}$$

for $\theta \in \Theta$, Θ being the admissible domain of θ and f an assumed predictive function. A particularly interesting case is

$$\dot{Y}_{N+1}^{(2)} = \theta\hat{Y}_{(N)}^{(2)} + (I-\theta)\hat{Y}_{N+1}^{(2)} \tag{9.2}$$

where θ is $p_2 \times p_2$ such that both θ and $I - \Theta$ are non-negative definite. Let

$$\dot{Y}_{\alpha}^{(2)} = \theta\hat{Y}_{(N-1,\alpha)}^{(2)} + (I-\theta)\hat{Y}_{\alpha}^{(2)} \tag{9.3}$$

where $\alpha = 1, \ldots, N$ and $\hat{Y}_{(N-1,\alpha)}^{(2)}$ is the predictor of $Y_{\alpha}^{(2)}$ based on $Y_1, \ldots, Y_{\alpha-1}$, $Y_{\alpha+1}, \ldots, Y_N$ and of the same functional form as $\hat{Y}_{(N)}^{(2)}$ and $\hat{Y}_{\alpha}^{(2)}$ is the predictor of $Y_{\alpha}^{(2)}$ and of the same functional form as $\hat{Y}_{N+1}^{(2)}$. Further, define the discrepancy measure

$$D(\theta) = \sum_{\alpha=1}^{N} d(\dot{Y}_{\alpha}^{(2)}, Y_{\alpha}^{(2)}) \tag{9.4}$$

which is then minimized w.r.t. $\theta \in \Theta$. If $\hat{\theta}$ is the unique solution then the final predictor is

$$\tilde{Y}_{N+1}^{(2)} = \hat{\theta}\hat{Y}_{(N)}^{(2)} + (I-\hat{\theta})\hat{Y}_{N+1}^{(2)}. \tag{9.5}$$

For an arbitrary quadratic discrepancy measure,

$$D(\theta) = \sum_{\alpha=1}^{N} (\dot{Y}_{\alpha}^{(2)} - Y_{\alpha}^{(2)})'\Lambda(\dot{Y}_{\alpha}^{(2)} - Y_{\alpha}^{(2)}), \tag{9.6}$$

of which a simplified form is to set $\Lambda = I$, Geisser (1975, 1980). With $D(\theta)$ given in (9.6) and $p_1 > m = 2$ and $p_2 = 1$, a solution for combining predictors based on simple least square predictors appears in Geisser (1975) where

$$\hat{Y}_{(N)}^{(2)} = X^{(2)}T_1 \text{ and } \hat{Y}_{N+1}^{(2)} = X^{(2)}(X^{(1)\prime}X^{(1)})^{-1}X^{(1)\prime}Y_{N+1}^{(1)}.$$

When Λ is known, a general solution may be obtained for other forms of $\hat{Y}_{(N)}^{(2)}$ and $\hat{Y}_{N+1}^{(2)}$, for $m < p_1$ and p_2 arbitrary,

$$\begin{aligned}\hat{\theta} &= \Big[\sum_{\alpha=1}^{N}(\hat{Y}_{(N-1,\alpha)}^{(2)} - \hat{Y}_{\alpha}^{(2)})\Lambda(\hat{Y}_{(N-1,\alpha)}^{(2)} - \hat{Y}_{\alpha}^{(2)})'\Big] \\ &\times \Big[\sum_{\alpha=1}^{N}(\hat{Y}_{(N-1,\alpha)}^{(2)} - \hat{Y}_{\alpha}^{(2)})\Lambda(\hat{Y}_{(N-1,\alpha)}^{(2)} - \hat{Y}_{\alpha}^{(2)})'\Big]^{-1},\end{aligned} \tag{9.7}$$

provided it exists and satisfies the constraints. When all of the components of Y_α are measured in the same units (as in the growth curve problem) it is often natural and convenient to use the simplified form $\Lambda = I$.

10 Transformations on Y

Motivated by real data considerations for forecasting technological subsitutions, Keramidas and Lee (1990) considered the following power transformations:

$$\underset{p\times N}{Y}^{(\lambda)} = \underset{p\times m}{X}\ \underset{m\times r}{\tau}\ \underset{r\times N}{A} + \underset{p\times N}{\epsilon}, \tag{10.1}$$

where

$$Y^{(\lambda)} = (Y_1^{(\lambda_1)}, \ldots, Y_N^{(\lambda_N)})$$

and

$$Y_i^{(\lambda_i)} = (Y_{1i}^{(\lambda_i)}, \ldots, Y_{pi}^{(\lambda_i)})',$$

with

$$\begin{aligned} Y_{ji}^{(\lambda_i)} &= \tfrac{(Y_{ji}+\gamma)^{\lambda_i}-1}{\lambda_i} \quad \text{when } \lambda_i \neq 0, \\ &= \ln(Y_{ji}+\gamma) \quad \text{when } \lambda_i = 0, \end{aligned} \tag{10.2}$$

and γ is assumed to be a known constant such that $Y_{ij}+\gamma > 0$ for all j and i, λ_i is an unknown parameter. Furthermore, the columns of ϵ are each independent and p-variate normal with mean vector 0 and common covariance $\Sigma = \sigma^2 C$, as defined in (2.16).

The MLEs of the parameters τ, σ^2, ρ and λ are

$$\hat{\tau} = (X'\hat{C}^{-1}X)^{-1}X'\hat{C}^{-1}Y^{(\hat{\lambda})}A'(AA')^{-1} \tag{10.3}$$

$$\hat{\sigma}^2 = \frac{1}{pN}[tr\,(X'\hat{C}^{-1}X)^{-1}X'\hat{C}^{-1}S\hat{C}^{-1}X + tr\,(Z'\hat{C}Z)^{-1}Z'(Y^{(\hat{\lambda})})(Y^{(\hat{\lambda})})'Z], \tag{10.4}$$

where

$$S = Y^{(\hat{\lambda})}(I - A'(AA')^{-1}A)Y^{(\hat{\lambda})'}, \tag{10.5}$$

$$\hat{C} = (\hat{\rho}^{|a-b|}), \tag{10.6}$$

and $\hat{\rho}$ and $\hat{\lambda}$ are obtained by maximizing the profile likelihood function

$$L_{\max}(\rho, \lambda) = (\hat{\sigma}^2(\rho,\lambda))^{-pN/2}(1-\rho^2)^{-N(p-1)/2}J, \tag{10.7}$$

where J, the Jacobian of the transformation from $Y^{(\lambda)}$ to Y, is defined as

$$J = \prod_{j=1}^{N}\prod_{i=1}^{p} Y_{ij}^{\lambda_i - 1}$$

and $\sigma^2(\rho, \lambda)$ is the $\hat{\sigma}^2$ given by (10.4) with $\hat{\rho}$ and $\hat{\lambda}$ replaced by ρ and λ, respectively.

We thus see that the MLEs of τ and σ^2 are expressed in explicit forms and require no iteration. The MLEs of ρ and λ can be obtained by a numerical search. Once $\hat{\rho}$ and $\hat{\lambda}$ are obtained, the joint MLEs of τ, σ^2, ρ and λ are established. Hence the most important step is to carry out the maximization of the profile likelihood function $L_{\max}(\rho, \lambda)$ as given in (10.7), which is an extension of (2.18). It is noted that in practice the maximization is easier than it looks, because the number of power transformations is less than N in practice. It was proposed by Keramidas and Lee (1990) that the number of power transformations be identical to the number of groups, that is, that there be only r different λ's, even though there are N independent vectors. This means that the same power transformation will be applied to each of the observations in the same group. Also, since p is usually quite small, the MLEs of λ are very hard to obtain if N different λ's are allowed. Thus, in (10.7) we are dealing only with a $(r+1)$-dimensional search .

For extended prediction of y given Y, we assume that for $i \leq n$,

$$E\begin{bmatrix} Y_i^{(\lambda)} \\ y_i^{(\lambda)} \end{bmatrix} = \begin{pmatrix} X \\ x \end{pmatrix} \tau A_i, \tag{10.8}$$

$$\operatorname{cov}\begin{bmatrix} Y_i^{(\lambda)} \\ y_i^{(\lambda)} \end{bmatrix} = \sigma^2 \begin{pmatrix} C_{11} & C_{12} \\ C_{21} & C_{22} \end{pmatrix} \tag{10.9}$$

where C_{ij} is specified in (7.11). From the conditional expectation of y_i given Y we obtained the following predictor for y_i,

$$\begin{aligned} \hat{y}_i &= \{e + \hat{\lambda}_i[x\hat{\tau}A_i + (0, \hat{\eta}^*)(Y_i^{(\hat{\lambda}_i)} - x\hat{\tau}A_i)]\}^{1/\hat{\lambda}_i} \quad \text{when } \lambda_i \neq 0 \\ &= \exp[x\hat{\tau}A_i + (0, \hat{\eta}^*)(Y_i^{(\hat{\lambda}_i)} - x\hat{\tau}A_i)] \quad \text{when } \lambda_i = 0, \end{aligned} \tag{10.10}$$

where $e' = (1, \ldots, 1)$, $\hat{\eta}^* = (\hat{\rho}, \ldots, \hat{\rho}^q)'$, and 0 is a $q \times (p-1)$ matrix with all 0 elements. In (10.10) we use the convention that $b^a = (b_1^a, \ldots, b_p^a)'$.

The model described in this section has been successfully applied to forecasting penetrations of telephone switching systems by Keramidas and Lee (1988, 1990).

11 Model Selection and Classification

The prediction problems considered so far can be useful in model selection for the growth curve data. Let $M_1, \ldots, M_g$ be the g possible growth curve models that could have generated the growth curve data at hand. Then the selected model, say M_{α^*}, is the one that yields the best predictive accuracy when the sample reuse procedure is used. This approach was utilized by Lee and Geisser (1975), Fearn (1975), Geisser (1981), Rao (1987), and Lee (1988, 1991), among others for the comparison of models based on the performance of the conditional prediction of $V^{(2)}$ given $V^{(1)}$ and Y. For more detailed discussion of model selection based on PSR, see Geisser and Eddy (1979). For model comparisons, it appears natural to use the same discrepancy measure. Then by the PSR method,

$$D_\alpha = \frac{1}{p_2 N}\sum_{j=1}^{N}(Y_j^{(2)} - \hat{Y}_{(j)}^{(2)}(\alpha))'(Y_j^{(2)} - \hat{Y}_{(j)}^{(2)}(\alpha)), \tag{11.1}$$

where

$$\hat{Y}_{(j)}^{(2)}(\alpha) = X^{(2)}\hat{\tau}_{(j)}(\alpha)A_j + \hat{\Sigma}_{21(j)}(\alpha)\hat{\Sigma}_{11(j)}^{-1}(\alpha)(Y_{(j)}^{(1)} - X^{(1)}\hat{\tau}_{(j)}(\alpha)A_j), \tag{11.2}$$

$$A = (A_1, \ldots, A_N),\ X = \begin{pmatrix} X^{(1)} \\ X^{(2)} \end{pmatrix},$$

$$Y_i = \begin{pmatrix} Y_i^{(1)} \\ Y_i^{(2)} \end{pmatrix},\ \Sigma = \begin{bmatrix} \Sigma_{11} & \Sigma_{12} \\ \Sigma_{21} & \Sigma_{22} \end{bmatrix},$$

$X^{(a)}$ is $p_a \times m$, $Y_i^{(a)}$ is $p_a \times 1$, Σ_{ab} is $p_a \times p_b$, and $\hat{\tau}_{(j)}(\alpha)$ and $\hat{\Sigma}_{ab}(j)(\alpha)$ are the estimates of τ and Σ_{ab} under the model M_α with sample $Y_{(j)} = (Y_1, \ldots, Y_{j-1}, Y_{j+1}, \ldots, Y_N)$. The model M_{α^*} corresponding to the minimum discrepancy is chosen as the most appropriate for the data. It is noted that in the prediction process, $Y_j^{(2)}$ is viewed as $V^{(2)}$ and $Y_j^{(1)}$ as $V^{(1)}$ and $Y_{(j)}$ as Y in the conditional prediction of $V^{(2)}$ given $V^{(1)}$ and Y.

This procedure can be applied to any number of competing models, nested or not. One possible drawback is that the choice of p_1 and p_2 may not be unique. A different choice of p_1 and p_2 could produce a different selection result. However, for most growth curve problems, the responses are time series in nature, and hence the most practical choice of p_2 is 1, i.e., $V^{(2)}$ represents the last component of the vector V (for $K = 1$). This choice of p_2 is particularly appealing if the ultimate goal in the modeling effort is the prediction of future values for each of the N vectors.

Instead of conditional prediction of $V^{(2)}$ given $V^{(1)}$ and Y, prediction can be made on the entire vector V. For this purpose, a natural discrepancy measure by the PSR procedure is

$$D_\alpha = \frac{1}{pN}\sum_{j=1}^{N}(Y_j - X\hat{\tau}_{(j)}(\alpha)A_j)'(Y_j - X\hat{\tau}_{(j)}(\alpha)A_j). \tag{11.3}$$

Again, the selected model M_{α^*} corresponds to the minimum discrepancy measure.

An advantage of this discrepancy measure is that the prediction is made on the whole vector. Hence there is no indeterminacy in the selection of p_2 as encountered in (11.1). However, the prediction of V given Y may not be as sensitive to the appropriateness of the covariance structure as conditional prediction of $V^{(2)}$ given $V^{(1)}$ and Y, or extended prediction of y given Y, which is the subject of the next discussion.

A discrepancy measure for extended prediction of y, when $q = 1$ and $n = N$, is

$$D_\alpha = (Y^{(2)} - \hat{Y^{(2)}})(Y^{(2)} - \hat{Y^{(2)}})'/N \tag{11.4}$$

where $\hat{Y^{(2)}} = X^{(2)}\hat{\tau}(\alpha)A + \hat{\Sigma}_{21}(\alpha)\hat{\Sigma}_{11}^{-1}(\alpha)(Y^{(1)} - X^{(1)}\hat{\tau}(\alpha)A)$. Σ_{ab} is defined as before with $p_1 = p - 1$, $p_2 = 1$, $\hat{\tau}(\alpha)$ and $\hat{\Sigma}_{ab}(\alpha)$ are estimates of τ and Σ_{ab} with the sample $Y^{(1)}$ under model M_α, $Y = (Y^{(1)'}, Y^{(2)'})'$, $Y^{(1)}$ is $(p-1) \times N$, $Y^{(2)}$ is $1 \times N$, $X = (X^{(1)'}, X^{(2)'})'$, $X^{(1)}$ is $(p-1) \times m$, and $X^{(2)}$ is $1 \times m$. The model M_{α^*} corresponding to the minimum discrepancy measure is chosen as the most appropriate for the data. It is noted that $Y^{(2)}$ is viewed as y and $Y^{(1)}$ as Y in extended prediction of y given Y. When $p > m+2$, the following pseudo-cross-validation procedure advocated by Keramidas and Lee (1990), which is prequential in nature (see Dawid, 1984), can be used. A discrepancy measure for this procedure is

$$D_\alpha = \frac{1}{(p-m-1)N} \sum_{j=m+2}^{p} (Y_j - \hat{Y}_j)(Y_j - \hat{Y}_j)'. \tag{11.5}$$

where $Y = (Y_1', \ldots, Y_p')'$, $\hat{Y}_j$ is the prediction of Y_j and is obtained in exactly the same manner as $Y^{(2)}$ in (11.4) with the sample $(Y_1', \ldots, Y_{j-1}')'$. Thus, the prediction is still extended in nature except it is done in a sequential manner. In the prediction of Y_j, $p_1 = j-1$, $p_2 = 1$, and $X^{(2)}$ is the first $j-1$ rows of X, if all the available data are used as the sample in the prediction process. In case only a sbuset of available data is used as the sample, the choice of the design matrix $X^{(2)}$, p_1 and p_2 should be self-evident. As in the previous case, the model M_{α^*} corresponding to the minimum discrepancy measure is chosen as the most appropriate for the data.

When the data involve samples from two or more populations, model selection can be based on the ability to classify the data at hand. The model selected is the one that produces the smallest probability of misclassification using the PSR procedure.

Here we consider the situation where g growth curves, as defined by (2.2), have been observed and a future observation matrix V, of dimension $p \times K$, is known to be drawn from one of g populations, $\pi_1, \ldots, \pi_g$, with prior probability $q_1, \ldots, q_g$, respectively. It is also assumed that

$$V \sim N(X\tau_i F_i, \Sigma_i \otimes I),$$

where F_i is a known design matrix formed by some columns of A_i, if V is from π_i. For the selection of a model using the PSR procedure, we will set $K = 1$.

Let $Y = (Y_1, \ldots, Y_g) = (y_1, y_2, \ldots, y_N)$, where $N = \sum_{j=1}^{g} N_j$ and θ is the collection of parameters. Then y_j is classified as π_α if

$$q_\alpha f_\alpha(y_j | Y_{(j)}, \hat{\theta}_{\alpha_{(j)}}, \pi_\alpha) > q_i f_i(y_j | Y_{(j)}, \hat{\theta}_{\alpha(j)}, \pi_i) \text{ for all } i \neq \alpha. \tag{11.6}$$

In case $g = 2$ and θ is known, (11.6) is equivalent to classifying y_j into π_1 if

$$\begin{aligned} &(y_j - X\tau_2 F_2)'\Sigma_2^{-1}(y_j - X\tau_2 F_2) + \log(|\Sigma_2|) \\ &> (y_j - X\tau_1 F_1)'\Sigma_1^{-1}(y_j - X\tau_1 F_1) + \log(|\Sigma_1|) + 2\log(q_2/q_1). \end{aligned} \tag{11.7}$$

If $\Sigma_1 = \Sigma_2 = \Sigma$, then (11.7) is further reduced to

$$[y_j - \frac{1}{2}(\tau_1 F_1 + \tau_2 F_2)]'X'\Sigma^{-1}X(\tau_1 F_1 - \tau_2 F_2) > \log(q_2/q_1) \tag{11.8}$$

which is an extension of equation (5) of Anderson (1984, p.205).

This procedure is quite practical, because the ability to classify the data correctly is very important. However, the procedure cannot be applied if there is no clear partitioning of the data into several distinct groups. Also, the sample size for each subgroup could be greatly reduced if the number of groups is large. For more detail about growth curve classification the reader is referred to Lee (1982).

12 Applications to Real Data

Several data sets have been used in the literature for the three types of prediction considered in this paper. The illustrative examples can be found in Lee and Geisser (1975), Fearn (1975), Rao (1977, 1984, 1987), Lee (1988,1991), Keramidas and Lee (1988, 1990, 1992). The data are all from biological applications with the exception of those in Keramidas and Lee (1988, 1990) in which the model was useful in technological substitutions.

Here we will restrict our attention to conditional prediction of $V^{(2)}$ given $V^{(1)}$ and Y for the dental data, as reported by Potthoff and Roy (1964). Dental measurements were made on 11 girls and 16 boys at ages 8, 10, 12 and 14 years. Each measurement is the distance (in mm) from the center of the pituitary to the pteryomaxillary fissure. As noted in Lee and Geisser (1975), individual 20 (the 9th boy) is suspected to be an aberrant observation and will be excluded in this illustration. Since the measurements are taken once every two years, the design matrix X is

$$X = \begin{pmatrix} 1 & 1 & 1 & 1 \\ -3 & -1 & 1 & 3 \end{pmatrix}' \tag{12.1}$$

The design matrix A is a 2×26 matrix composed of 11 (1,0) columns, followed by 15 (0,1) columns when both girls and boys are assumed to have a common covariance matrix. If the covariance matrices are distinct for girls and boys, the design matrix A is a 1×11 vector for girls and a 1×15 vector, both consisting of all 1s.

Let $Y = (Y_1, \ldots, Y_{11}, Y_{12}, \ldots, Y_{26})$ where Y_1 through Y_{11} represent the dental measurements of the girls and Y_{12} through Y_{26} are those of the boys. In conditional prediction of $V^{(2)}$ given $V^{(1)}$ and Y we will consider the special case in which $V = (V^{(1)'}, V^{(2)'})'$, $V^{(1)}$ is $(p-1) \times 1$ and $V^{(2)}$ is 1×1. A discrepancy measure given by (11.1) will be applied with $p_2 = 1$.

We will first apply the results (5.11), (5.15), (5.21) and (5.22) to the case in which $V^{(1)} = (26, 25, 29)'$ [corresponding to the first boy] and $V^{(1)} = (24.5, 25, 28)'$ [corresponding to the last girl]. The exact and approximate predictive densities of $V^{(2)}$ given $V^{(1)}$ and Y are shown in Figure 1. The difference between Figures 1(A) and 1(B) is in the treatment of the covariance matrix. In Figure 1(B) the dental data for both girls and boys are assumed to have an identical covariance matrix while in Figure 1(A) girls and boys are assumed to have different covariance matrices for their dental measurements. From Figures 1(A)-1(C) it is clear that the approximations given by (5.11) and (5.21) are quite adequate, at least as far as the predictive region is concerned. Meanwhile, even for the worst situation as shown in Figure 1(C), vast improvement can be accomplished via (5.22) as evidenced in Figure 1(D), which is a tremendous improvement over Figure 1(C) even for $L = 50$. As noted earlier, however, that an average of L multivariate Student t densities is no longer a multivariate

Student t density. Hence, an approximation such as (5.11) or (5.21) still has its place in our development unless the random variable of interest is one-dimensional in which (5.22) should be preferred.

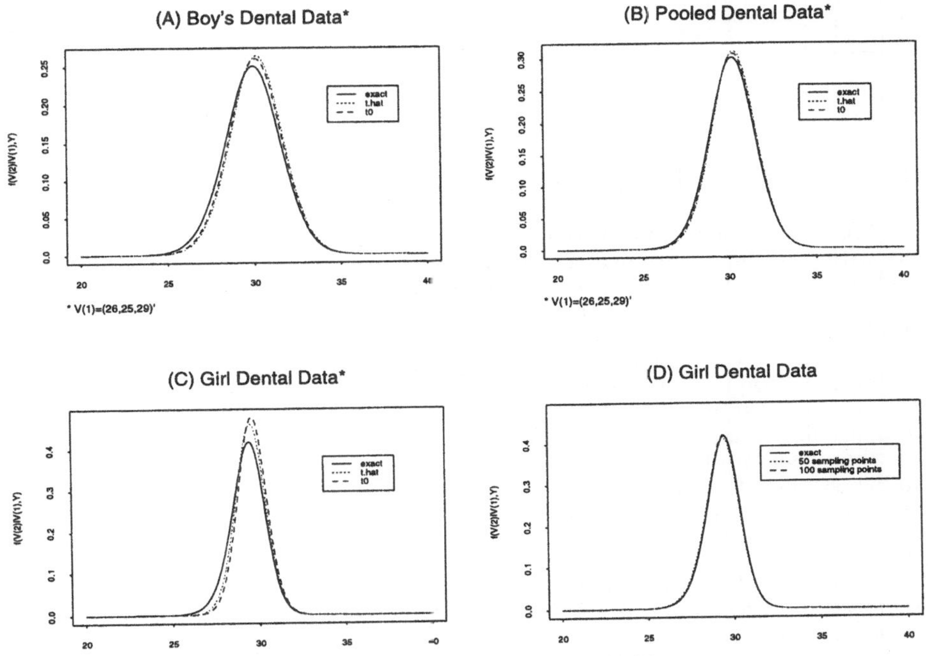

Figure 1. Predictive Densities of V(2) given V(1) and Y

We next compare the predictive accuracy for conditional prediction of $V^{(2)}$ given $V^{(1)}$ and Y by the approximate means under four different covariance structures : Arbitrary p.d., RSS, Uniform and Serial. The approximate mean for arbitrary p.d. Σ is obtained from (4.5) and (4.6) and the approximate means for RSS, Uniform and Serial structures are $\mu_{r2\cdot1}(\hat{t})$, $\mu_{u2\cdot1}$, and $\mu_{s2\cdot1}$, as given in (5.12), (6.3) and (7.4), respectively with $\hat{t}$ being the mode of the F distribution. The comparison is summarized in Table 1. The entries of this table are obtained from Lee and Geisser (1975) and Lee (1988). It appears that the serial structure is most appropriate for this data set.

Table 1. The MSD and MAD Between the Predicted and Actuals: Dental Data

	Arbitrary p.d.	RSS	Uniform	Serial
MSD	2.128	2.035	2.196	1.354
MAD	1.144	1.127	1.206	0.940

13 Concluding Remarks

We have reviewed predictive methods for many of the growth curve models with varying covariance structures. It is noted that none of these models will represent a true description of the underlying process for most rather complex data. However, for a particular data set what is usually required is that model be adequate for predictive purposes. Hence, estimates of predictive error can be made by randomly dividing the data into a construction sample and a validation sample and predicting the validation sample from the construction sample for a large sample. For a small sample, one can delete an observation and predict it from the rest and calculate the error and then cycle through all the observations yielding an average predictive error. Although this will underestimate the actual predictive error, it should prove useful as a gauge of the adequacy of the particular model for prediction in any real problem and serve to discriminate between alternative models. It is also to be noted that in many of these complex situations we have suggested use of "plug in estimates" and large sample normal approximations. Since the multivariate Student t distribution is somewhat more diffuse than the multivariate normal, it might be better to use the multivariate Student t distribution when the sample size is not overly large. Another inferential mode that may prove useful is the method of predictive likelihood, e.g., Butler (1986). However, it has not as yet been applied to growth curves.

References

Anderson, T. W. (1984). An Introduction to Multivariate Statistical Analysis, 2nd edition. New York: Wiley.

BMDP Statistical Software Manual, Vol. 2 (1988). University of California Press, Berkeley, California.

Box, G.E.P. (1950). Problems in the analysis of growth and wear curves. Biometrics, 6, 362-389.

Chi, E.M. and Reinsel, G.C. (1989). Models for longitudinal data with random effects and AR(1) errors. Journal of the American Statistical Association, 84, 452-459.

Dawid, A. P. (1984). Present position and potential development: Some personal views. Journal of the Royal Statistical Society, Series A 147, 278-292.

Fearn, T. (1975). A Bayesian approach to growth curves. Biometrika 62, 89-100.

Geary, D. N. (1989). Modeling the covariance structure of repeated measurements, Biometircs,45, 1183-1195.

Geisser, S. and Cornfield, J.(1963) Posterior distributions for multivariate normal parameters. Journal of the Royal Statistical Society, Series B 25, 368-376.

Geisser, S. (1965). Bayesian estimation in multivariate analysis. Ann. Math. Statist. 36, 150-159.

Geisser, S. (1970). Bayesian analysis of growth curves. Sankhya, Series A 32, 53-64.

Geisser, S. (1974). A predictive approach to the random effect model. Biometrika, 61, 101-107.

Geisser, S. (1975). The predictive sample reuse method with application. Journal of the American Statistical Association, 70, 320-328.

Geisser, S. and Eddy, W.F.(1979). A predictive approach to model selection . Journal of the American Statistical Association , 74, 153-160.

Geisser, S. (1980). Growth curve analysis. In Handbook of Statistics, Vol. 1,P.R.. Krishnaiah(ed.), 89-115. Amsterdam: North-Holland.

Geisser, S. (1981). Sample reuse procedures for prediction of the unobserved portion of a partially observed vector. Biometrika, 68, 243-250.

Geyer, C. J. and Thompson, E. A.(1992). Constrained Monte Carlo maximum likelihood for dependent data. Journal of the Royal Statistical Society, Series B 54,657-700.

Jenrich, R.I. and Schluchter, M.D. (1986). Unbalanced repeated-measures models with structured covariance matrices. Biometrics, 42, 805-820.

Keramidas, E.M. and Lee , J. C.(1988). Forecasting technological substitutions with concurrent short time series. Proceedings of the Business and Economic Statistics Section, American Statistical Association, 1-10.

Keramidas, E.M. and Lee, J. C.(1990). Forecasting technological substitutions with concurrent short time series. Journal of the American Statistical Association. 85, 625-632.

Keramidas, E.M. and Lee, J. C. (1995). Selection of a covariance structure for growth curves. Biometrical Journal,37,783-797

Khatri, C. G. (1966). A note on MANOVA model applied to problems in growth curve. Annals of the Institute of Statistical Mathematics, 18, 75-86.

Khatri, C. G. (1973). Testing some covariance structures under a growth curve model. Journal of Multivariate Analysis, 3, 102-116.

Laird, N. M. and Ware, J. H. (1982). Random effects models for longitudinal data. Biometrics,38, 963-974.

Lee, J. C. and Geisser, S. (1972). Growth curve prediction. Sankhya, Series A 34, 393-412.

Lee, J. C. and Geisser, S. (1975). Applications of growth curve prediction. Sankhya, Series A 37, 239-256.

Lee, J. C. and Hu, L. (1995). On the distribution of linear functions of independent F and U variates. Statistics and Probability Letters. (In Press)

Lee, J. C. (1982). Classification of growth curves. In Handbook of Statistics, Vol. II, P. r. Krishnaiah and L. W. Kanal(eds.), 121-137. Amsteram: North-Holland.

Lee, J. C. and Tan, W.Y. (1984). On the degree of polynomial in a general linear model. Communications in Statistics, Series A 13, 781-790.

Lee, J. C. (1988). Prediction and estimation of growth curves with special covariance structures. Journal of the American Statistical Association, 83, 432-440.

Lee, J. C. (1991). Test and model selection for the general growth curve model. Biometrics, 47,147-159.

Liu, S. I. (1995). Bayesian multiperiod forecasts for ARX models. Annals of the Institute of Statistical Mathematics(In press).

Lyung, G. M. and Box, G.E.P. (1980). Analysis of variance with autocorrelated observations, Scandinavian Journal of Statistics, 7, 172-180.

Potthoff, R.F. and Roy, S. N. (1964). A generalized multivariate analysis of variance model useful especially for growth curve problems. Biometrika, 51,313-326.

Rao, C. R. (1958). Some statistical methods for comparison of growth curves. Biometrics, 14,1-17.

Rao, C. R. (1965). The theory of least squares when the parameters are stochastic and its application to the analysis of growth curves. Biometrika, 52,447-458.

Rao, C. R.(1966). Covariance adjustment and related problems in multivariate analyses. In Multivariate Analyses, Vol. II, P. R. Krishnaish(ed.), 87-103. New York: Academic Press.

Rao, C. R.(1967). Least squares theory using an estimated dispersion matrix and its application to measurement of signals. Proceedings of the Fifth Berkeley Symposium on Mathematical Statistics and Probability, 1, 355-372.

Rao, C. R.(1977). Prediction of future observations with special reference to linear models. In Multivariate Analysis, Vol. IV, P. R. Krishnaiah(ed.), 193-208. Amsterdam: North-Holland.

Rao, C. R. (1984). Prediction of future observations in polynomial growth curve models. In Proceedings of the Indian Statistical Institute Golden Jubilee International Conference on Statistics: Applications and New Directions, 512-520. Calcutta: Indian Statistical Institute.

Rao, C. R.(1987). Prediction of future observations in growth curve models. Statistical Science, 2, 434-471.

Reinsel, G. (1984). Estimation and prediction in a multivariate random effects generalized linear model. Journal of the American Statistical Association, 79, 406-414.

Wishart, J. (1938). Growth rate determinations in nutrition studies with the bacon pig, and their analysis. Biometrika, 30,16-28.

Zellner, A., Bauwens, L and Van Dijk, H. K. (1988). Bayesian specification analysis and estimation of simultaneous equation models using Monte Carlo methods. J. of Econometrics, 38, 39-72.

Zerbe, G. O. (1979). Randomization analysis of the completely randomized design extended to growth and response curves. Journal of the American Statistical Association, 74, 215-221.

Predictive Influence in the Log Normal Survival Model

Wesley O. Johnson
Division of Statistics
University of California
Davis, CA 95616

Abstract

We discuss case deletion diagnostics for prediction of future observations in the log normal survival analysis model. The point of view taken is that prediction is the primary inferential goal in a survival analysis setting and that particular observations in the sample may be influential with regard to that goal. We thus consider the Kullback-Leibler divergence as a measure of the discrepancy between predictive densities based on full and case deleted samples. A large Kullback-Leibler number for a particular case is an indication that deletion of that case may result in substantially different predictive inferences.

1. Introduction

Our approach to prediction is Bayesian. Assuming censored log normal data and with a particular class of informative priors we derive the predictive density (PD) of a future vector of observations corresponding to specified covariate values. Denote the data as d, the covariate values as X_f, and the corresponding PD as $p(\cdot|X_f, d)$. Of particular interest is the PD for a single future value and its corresponding survival function. We use the same notation for the PD regardless of whether it is univariate or multivariate.

Johnson and Geisser (JG) (1982, 1983) introduced predictive influence measures for case deletion in the normal linear model. Let $d_{(i)}$ denote the data with case i deleted, and let $p(\cdot|X_f, d_{(i)})$ denote the corresponding predictive density. Then JG defined the predictive influence function (PIF)

$$KL_i = \int p(z|X_f, d)\ell n \frac{p(z|X_f, d)}{p(z|X_f, d_{(i)})} dz \qquad (1)$$

which is the Kullback-Leibler divergence between predictive densities based on full and case deleted data. There are of course numerous other possible discrepancy measures, but none appear to be obviously superior. We prefer to keep the exposition simple so we mainly discuss (1).

With censored log normal data the PD, $p(z|X_f, d)$, cannot be explicitly evaluated in general, and neither can the integral (1). However, current methods (Gelfand and Smith, 1990) allow for the straight forward numerical approximation of both the PD and KL_i. Importance sampling may be preferable if the dimension of the covariate vector is relatively small (Zellner and Rossi, 1984).

Due to the computational nature of approximating the exact KL_i, there is no insight into the nature of predictive influence. We thus consider prediction based on large samples in order to achieve greater tractability. Using a particular approximate PD and substituting it into (1), we obtain a simple approximation to KL_i in explicit form. In this instance, it is possible to make connections with Cook's distance (Cook; 1977, 1979). Using analogues to Pregibon's (1981) one step approximation to estimate parameters we are also able to obtain computationally efficient formulas that lend insight into the nature of predictive influence for survival models.

We model uncertainty about regression coefficients based on the work of Bedrick, Christensen and Johnson (1995). This specification results in a data augmentation prior cf Clogg et al. (1991), and allows for a "partial prior" specification in those instances when there are many covariates. It is also possible to assess the impact of various components of the prior specification as well as the impact of deleting data points.

In section 2, we discuss the log normal survival model, our class of prior distributions and we obtain posterior and predictive densities. In section 3 we discuss how to obtain approximate predictive influence and we give an illustration. In section 4 we discuss large sample inference and in section 5 the corresponding predictive influence problem as well as a further illustration. Section 6 gives final remarks and conclusions.

2. Background material

(2.1) The Model

Consider the log normal survival model

$$Y = \ell n T = X\beta + \sigma E \tag{2}$$

where $T = (T_1, ..., T_n)^T$ denotes a vector of n "survival" times and X is an $n \times p$ matrix of covariate values with first column all ones, $\beta = (\beta_0, \beta_1, ..., \beta_{p-1})$ is a vector of regression coefficients, σ is an unknown scale factor and $E = (\varepsilon_1, ..., \varepsilon_n)^T$ is a vector of independent standard normal errors. Assume $t = (t_1, ..., t_n)^T$ is the observed data.

Knowing that case i is right censored is equivalent to knowing that $T_i > t_i$ under the assumption that censoring and survival times are independent. Let n_u and n_c denote the number of uncensored and censored observations in the sample. Without loss of generality, partition Y, T and X so that the first n_u rows correspond to uncensored observations and the last n_c rows correspond to censored observations. Denote these components as $Y_u, Y_c, T_u, T_c, X_u, X_c$, etc. Then the data we see corresponds to

$$d = \{T_u = t_u, T_c > t_c\} \tag{3}$$

where $\{T_c > t_c\}$ denotes that the components of T_c are greater than the respective components of t_c.

Let $f_0(\cdot)$ and $S_0(\cdot)$ denote the pdf and survival function for ε_i respectively, and let x_i be the ith row of X. Then the pdf and survival function for T_i are respectively

$$f_{T_i}(t_i) = \frac{1}{\sigma t_i} f_0\left(\frac{\ell n t_i - x_i\beta}{\sigma}\right), \quad S_{T_i}(t_i) = S_o\left(\frac{\ell n t_i - x_i\beta}{\sigma}\right).$$

We thus obtain the likelihood function

$$L(\beta, \sigma|d) = \prod_{i=1}^{n_u} \frac{1}{\sigma t_i} f_0\left(\frac{\ell n t_i - x_i\beta}{\sigma}\right) \prod_{j=n_u+1}^{n} S_0\left(\frac{\ell n t_j - x_j\beta}{\sigma}\right), \tag{4}$$

where

$$f_0(\varepsilon) = \frac{1}{\sqrt{2\pi}} e^{-\varepsilon^2/2}, \quad S_0(\varepsilon) = \int_\varepsilon^\infty \frac{1}{\sqrt{2\pi}} e^{-u^2/2} du.$$

(2.2) The Prior

We require a model for our uncertainty about (β, σ). Since this is quite difficult to do directly, we consider an approach which is discussed in Bedrick, Christensen and Johnson (1995) for generalized linear models. We first note that the median of the log normal distribution for an observation with covariate vector x is just $e^{x\beta}$ since $S_0(0) = 1/2$. So we focus on specifying prior uncertainty about median survival times for individuals with various covariate combinations. Let

$$X_0 = \begin{pmatrix} x_{01} \\ x_{02} \\ \vdots \\ x_{0p'} \end{pmatrix}$$

be a full row rank $p' \times p$ matrix of covariate specifications with $p' \leq p$. Then

$$\tilde{m} = e^{X_0\beta} \tag{5}$$

denotes the vector of median survival times corresponding to p' individuals with these covariates. We select the x_0 's to be "widely spaced" but still within the set of vectors that are feasible, in order to make plausible the assumption that the uncertainty about $\tilde{m}_j$'s reflects independence of $\tilde{m}_j$'s. Bedrick, Christensen and Johnson (1995) elaborate upon this issue in great detail.

If $p' = p$ and a proper prior is specified for $\tilde{m}$, it is possible to induce a proper prior on β since the transformation (5) is $1-1$. With $p' = p$, we consider, conditional on σ^2,

$$\ell n(\tilde{m}) = X_0\beta \sim N_p\left(m_0, \sigma^2 W^{-1}\right)$$

with $W = diag(w_i)$ known, i.e., $\tilde{m}$ is log normal. The induced prior on β is

$$\pi(\beta|\sigma) \propto \sigma^{-p} e^{-\frac{1}{2}(X_0\beta - m_0)^T W (X_0\beta - m_0)/\sigma^2}. \tag{6}$$

Note that this is in the same form as a likelihood based on the assumption that $m_0 \sim N_p(X_0\beta, \sigma^2 W^{-1})$ is observed as data.

If p is large, it will often be quite difficult to specify a full prior for $\tilde{m}$. In this instance, we specify prior information for $p' < p$ median survival times. Our complete prior specification would be relatively noninformative about the remaining $p' - p$ medians. We thus assume that $(\tilde{m}_1, ..., \tilde{m}_{p'})^T$ is log normal as above and that $\tilde{m}_{p'+1}, ..., \tilde{m}_p$ have improper priors

$$\pi(\tilde{m}_i) \propto 1/\tilde{m}_i$$

for $i = p' + 1, ..., p$. This results in the induced partially informative but improper prior

$$\pi(\beta|\sigma) \propto \sigma^{-p'} e^{-\frac{1}{2}(X_0\beta - m_0)^T W (X_0\beta - m_0)/\sigma^2}, \tag{7}$$

where X_0 is $p' \times p$, m_0 is $p' \times 1$ and W is $p' \times p'$. This generalizes (6). Bedrick, Christensen and Johnson (1995) discuss this type of partial prior in some detail for generalized linear models.

2.3 Monte Carlo Approximations

The posterior for $\beta|\sigma, d$ is not a standard distribution unless $n_c = 0$, in which case it is multivariate normal. We thus consider the method of data augmentation introduced by Tanner and Wong (1987) and further discussed by Gelfand and Smith (1990). Recall that $Y_c = \ell n T_c$ is the set of log survival times that would have been observed except for censoring. Let the augmented data $y_A = \{y_u^T, y_{Ac}^T\}$ denote the actual log failure times for the uncensored data and the actual but unobserved log failure times for the censored data. Then

$$L(\beta, \sigma | y_A) \propto \prod_{i=1}^{n} \frac{1}{t_i \sigma} f_0 \left(\frac{y_A - x_i \beta}{\sigma} \right) \propto \sigma^{-n} e^{-\frac{1}{2}(y_A - X\beta)^T (y_A - X\beta)/\sigma^2} \tag{8}$$

and hence

$$\pi \left(\beta | y_A, \sigma^2 \right) \propto e^{-\frac{1}{2}(\tilde{y} - \tilde{X}\beta)^T \tilde{W} (\tilde{y} - \tilde{X}\beta)/\sigma^2} \propto e^{-\frac{1}{2}(\beta - \hat{\beta}_A)^T \tilde{X}^T \tilde{W} \tilde{X} (\beta - \hat{\beta}_A)/\sigma^2}$$

where

$$\tilde{y} = \begin{pmatrix} y_A \\ m_0 \end{pmatrix}, \quad \tilde{X} = \begin{pmatrix} X \\ X_0 \end{pmatrix}, \quad \tilde{W} = \begin{pmatrix} I_n & 0 \\ 0 & W \end{pmatrix}, \quad \hat{\beta}_A = \left(\tilde{X}^T \tilde{W} \tilde{X} \right)^{-1} \tilde{X}^T \tilde{W} \tilde{y}.$$

We employ the Gibbs sampling algorithm discussed by Gelfand and Smith (1990), which requires sampling from the conditional distributions (i) $y_{Ac} | y, \beta, \sigma$, (ii) $\beta | y_A, \sigma$ and (iii) $\sigma | y_A, \beta$. Clearly

$$Y_{Ac} | y, \beta, \sigma \sim TN \left(X_c \beta, \sigma^2 I_{n_c}, y_c \right)$$

where $TN(\cdot, \cdot, \cdot)$ denotes truncated normal with location $X_c \beta$, dispersion $\sigma^2 I_{n_c}$ and where the truncation is for $\{Y_{Ac} > y_c\}$. Furthermore,

$$\beta | y_A, \sigma^2 \sim N_p \left(\hat{\beta}_A, \left(\tilde{X}^T \tilde{W} \tilde{X} \right)^{-1} \sigma^2 \right). \tag{9}$$

We make the observation that the prior (7) is of data augmentation type (cf. Clogg et al. 1991). When combined with the likelihood, the resulting posterior, conditional on σ, is of the same form that would have been obtained had we seen the augmented data $(\tilde{y}, \tilde{X}, \tilde{W})$ and utilized the flat prior $\pi(\beta|\sigma) =$ constant, rather than the actual data (y, X, I_n) in conjunction with the prior (7).

In order to simplify the notation for the moment we let $\tau = 1/\sigma^2$ and assume

$$\pi(\tau) \propto \tau^{\frac{a}{2}-1} e^{-\frac{b\tau}{2}} \tag{10}$$

i.e., $\tau \sim \Gamma \left(\frac{a}{2}, \frac{b}{2} \right)$. Then utilizing (8)

$$\tau | y_A, \beta \sim \Gamma \left(\frac{n+a}{2}, \frac{(y_A - X\beta)^T (y_A - X\beta) + b}{2} \right). \tag{11}$$

The results (9) and (11) are particularly nice because they allow us to use the Gibbs sampler to obtain a Monte Carlo Sample, say $\left\{ \tau^{(j)}, \beta^{(j)}; \ j = 1, ..., M \right\}$ from the joint posterior $\pi(\tau, \beta | d)$, cf. Gelfand and Smith (1990). Given starting values $(\tau^{(0)}, \beta^{(0)})$, we sample

$$Y_{Ac}^{(k)} \sim TN \left(X_c \beta^{(k-1)}, I_{n_c} / \tau^{(k-1)}, y_c \right),$$

$$\beta^{(k)} \sim N_p \left(\hat{\beta}_A^{(k)}, \left(\tilde{X}^T \tilde{W} \tilde{X} \right)^{-1} / \tau^{(k-1)} \right),$$

$$\tau^{(k)} \sim \Gamma \left(\frac{n+a}{2}, \frac{\left(y_A^{(k)} - X\beta^{(k)} \right)^T \left(y_A^{(k)} - X\beta^{(k)} \right) + b}{2} \right)$$

for $k = 1, 2, \ldots$, and where $\hat{\beta}_A^{(k)}$ is calculated as $\hat{\beta}_A$ with $\tilde{y}^{(k)}$ substituted for $\tilde{y}$. After the initial "burn in" phase, the Markov Chain $\{\tau^{(j)}, \beta^{(j)}\}$ settles down and an ergodic theorem suggests that posterior expectations may be obtained by Monte Carlo approximation, namely

$$\int h(\beta, \tau)\, \pi(\beta, \tau|d)\, d\tau d\beta \simeq \sum_{j=1}^{M} h\left(\beta^{(j)}, \tau^{(j)}\right) / M$$

for M large (Tanner, 1994, p. 105).

Let

$$Y_f = \ell n T_f = X_f\beta + E_f/\sqrt{\tau} \tag{12}$$

be a vector of m independent future log normal survival times. The predictive density for T_f is thus approximated as

$$\begin{aligned} p(t_f|X_f, d) &= \int p(t_f|X_f, \beta, \tau)\, \pi(\beta, \tau|d)\, d\tau d\beta \\ &\simeq \sum_{j=M_0}^{M-M_0} p\left(t_f|X_f, \beta^{(j)}, \tau^{(j)}\right) / (M - M_0) \\ &= \left(\tfrac{1}{2\pi}\right)^{\frac{m}{2}} \sum_{j=M_0}^{M-M_0} \prod_{i=1}^{m} \left[\frac{\sqrt{\tau^{(j)}}}{t_{fi}} \exp\left\{-\frac{\tau^{(j)}}{2}\left(\ell n t_{fi} - x_{fi}\beta^{(j)}\right)^2\right\}\right] / (M - M_0) \\ &\equiv \tilde{p}(t_f|X_f, d), \end{aligned} \tag{13}$$

where M_0 is the "burn-in" sample size.

Finally note that we can numerically evaluate an integral of the form $\int g(t_f)p(t_f|X_f, d)dt_f$. We obtain the Monte Carlo sample

$$t_f^{(j)} \sim T_f|\beta^{(j)}, \sigma^{(j)}; \quad j = 1, \ldots, M,$$

i.e., generate log normal future vectors according to the model (12) for each given $(\beta^{(j)}, \sigma^{(j)})$ obtained from the Gibbs sampling procedure. Then the above integral can be approximated as $\sum_{j=1}^{M} g(t_f^{(j)})/M$. Alternatively, one could sample iid $\{t_f^{(j)},\ j = 1, \ldots, M^*\}$ from the approximate predictive distribution (13). We actually employ the latter approach.

Remark: The referee has made the correct observation that it is unnecessary to make the prior on β depend on σ. We could have simply assumed $\ell n(\tilde{m}) \sim N_p(m_0, W^{-1})$. The details of this section would be altered somewhat, but the same approach applies.

3. Predictive Influence via Monte Carlo Approximation

Our intention is to consider case influence for both the data and the prior. Accordingly, and to keep the notation simple, we now refer to the data, d, as the sample data and the "prior observations" for β; we refer to the case deleted data, $d_{(i)}$, as d minus case i, for $i = 1, \ldots, n+p'$. Then the exact predictive influence of deleting case i is defined as KL_i in expression (1). We approximate (1) by Monte Carlo to obtain a numerical approximation to exact predictive influence. But, first, we require an alternative representation of the posterior to facilitate the possibility of deleting prior observations.

Reparametrize with $\tau = 1/\sigma^2$, and let $\delta_j = 1$ for $j = 1, \ldots, n_u, n+1, \ldots, n+p'$ and $\delta_j = 0$ for $j = n_u + 1, \ldots, n$, and define $z_j = \tilde{w}_j^{1/2}(\ell n t_j - x_j\beta)\tau^{1/2}$ for $j = 1, \ldots n$ and $z_j = \tilde{w}_j^{1/2}(m_{0j} - x_{0j}\beta)\,\tau^{1/2}$ for $j = n+1, \ldots, n+p'$. Finally define

$$L_j(\beta, \tau) = \left\{\tilde{w}_j^{1/2}\tau^{1/2} f_0(z_j)\right\}^{\delta_j} \{S_0(z_j)\}^{1-\delta_j}, \tag{14}$$

the contribution to the posterior that is due to the jth sample point for $j = 1, \ldots, n$ and the contribution due to the jth "prior sample point" for $j = n+1, \ldots, n+p'$. Then

$$\pi(\beta, \tau|d) \propto \tau^{\frac{c}{2}-1} e^{-\frac{\tau b}{2}} \prod_{j=1}^{n+p'} L_j(\beta, \tau). \tag{15}$$

We see that case i can refer to either a data point or to one of the cases corresponding to the prior specification. We can thus ascertain the influence of particular components of the prior specification as well as that for components of the data on predictive inferences.

Now it is straight-forward to show that

$$p\left(t_f|X_f, d_{(i)}\right) = \frac{\int \frac{p(t_f|X_f,\beta,\tau)}{L_i(\beta,\tau)} \pi(\beta,\tau|d)\, d\beta d\tau}{\int \frac{1}{L_i(\beta,\tau)} \pi(\beta,\tau|d)\, d\beta d\tau}$$

where $L_i(\beta, \tau)$ was defined at (14). Then define the approximation

$$\tilde{p}\left(t_f|X_f, d_{(i)}\right) = \sum_{j=M_0}^{M_1} \frac{p\left(t_f|X_f, \beta^{(j)}, \tau^{(j)}\right)}{L_i\left(\beta^{(j)}, \tau^{(j)}\right)} \Big/ \sum_{j=M_0}^{M_1} \frac{1}{L_i\left(\beta^{(j)}, \tau^{(j)}\right)},$$

where we have used a subset of the previous Monte Carlo sample of size $M_1 - M_0$ which excludes the initial samples. Using the entire Monte Carlo sample is very time consuming.

When case i is implausible in the sense that $L_i\left(\beta^{(j)}, \tau^{(j)}\right)$ is often near zero, the denominator above will be unstable. However, as our goal is to find observations that are potentially influential, upon discovering such instability for a particular case, we would recommend study of the actual impact of deleting that case.

Finally, define

$$\tilde{KL}_i = \sum_{j=1}^{M^*} \ell n \left\{ \frac{\tilde{p}\left(t_f^{(j)}|X_f, d\right)}{\tilde{p}\left(t_f^{(j)}|X_f, d_{(i)}\right)} \right\} / M^*. \tag{16}$$

While larger M_1 will result in better approximations, all that is required is that influential cases be found, the exact value of this number is otherwise unimportant. We have found that $M_1 - M_0 = 200$ in conjunction with $M^* = 50$ is generally large enough to detect the most influential cases.

Remark: Alternatively, with extra expense, one could run different chains for each deleted observation in order to obtain a direct approximation to $p(t_f|X_f, d_{(i)})$. More generally, one one could consider the alternative Monte Carlo approach of importance sampling, cf. Zellner and Rossi (1984), which may be more efficient than Gibbs sampling for problems with a relatively small number of covariates and/or if there were many censored observations.

Illustration: Ovarian Cancer Data

We consider a censored data set, discussed by Collett (1994), which involves time until failure in days after surgical treatment of ovarian cancer. Survival time is related to treatment, age, extent of residual disease and performance status. Collett's analysis suggested that only age was important as a prognostic factor so we only consider the single covariate age. The data are listed in Table 1.

Table 1: Survival Times of Ovarian Cancer Patients

Case	Time	Cens	Age	Case	Time	Cens	Age
1	156	1	66	14	421	0	53
2	59	1	72	15	769	0	59
3	329	1	43	16	770	0	57
4	365	1	64	17	1227	0	59
5	268	1	74	18	1129	0	53
6	475	1	59	19	1206	0	44
7	464	1	56	20	1106	0	44
8	638	1	56	21	855	0	43
9	563	1	55	22	803	0	39
10	431	1	50	23	744	0	50
11	115	1	74	24	477	0	64
12	353	1	63	25	448	0	56
13	1040	0	38	26	377	0	58

We first considered a "flat" prior with $a = b = 0$ and with $\pi(\beta|\sigma) \propto$ const. We used a Monte Carlo sample size of $M = 1000$ to obtain inferences and the burn-in sample size $M_0 = 20$ was found to be reasonable. The diagnostics sample size $M_1 - M_0$ for our Figures was selected to be 500 and the MC sample size M^* for future observations was 250. Starting values for the Gibbs sampler were least squares estimates ignoring censoring. Estimates of (β_0, β_1, τ) and of several predictive probabilities were monitored to check on convergence of the Gibbs sampler.

It was decided to assess the impact of case deletion on 3 future cases with ages 46, 56 and 66 respectively. The mean age in the sample was 56 years with a standard deviation of about 10. The posterior estimate of the slope for age is $-.092$ and the standardized estimate is -3.45 indicating a negative association between survival and age. The posterior mean for τ is 1.30. Predictive survival curves for the 3 individuals of interest are given in Figure 1 and it is clear that age is a major prognostic factor for predictive survival. A second Monte Carlo run with $M = 1000$ gave virtually identical results.

Case deletion diagnostics are given in Figure 2. Uncensored case 3 is evidently the most influential followed by censored case 18. Case 3 corresponds to a 43 year old who only survived 329 days, an individual who died somewhat earlier than other individuals in the sample of similar ages. Case 18 corresponds to a censored time for an individual who lived longer than others of similar age. Figure 3 gives survival curve plots with case 3 deleted. The curves corresponding to the 56 and 66 year olds are basically unaffected by deleting case 3, but the effect on the curve for the 46 year old is dramatic. Deleting case 3 moves the median of the corresponding predictive survival curve from around 2000 days to around 3000 days. The slope estimate for age is $-.118$ and the standardized estimate is decreased to -3.85. The posterior mean for τ is increased to 1.72. Deletion of case 18 has no appreciable inpact on survival curves or on the slope or intercept estimates. The posterior mean for τ is increased to 1.6, which is likely the reason for it being indicated by the index plot in Figure 2.

Figure 4 indicates that new case 9 (old case 10) is now potentially influential, along with case 2 and new case 17 (old case 18). From Table 1, cases 10 and 18 correspond to individuals of similar age who died early and late respectively in the study. Case 3 evidently masked the potential impact of case 10 and diminished the potential impact of case 18. Case 2 is an "old" individual who died early in the study. All influential cases appear to correspond to deaths that were early or late relative to the remaining data. However, since there were only 12 deaths and only 26 observations, it is not surprising that potentially influential observations are showing up.

Figure 5 gives a predictive histogram of the log failure times of a sample of 1000 future values from the approximate PD (13) for a 56 year old. Exponentiating the x axis results

23.5cm

16 cm

Figure 1; Ovarian Data with Flat Prior

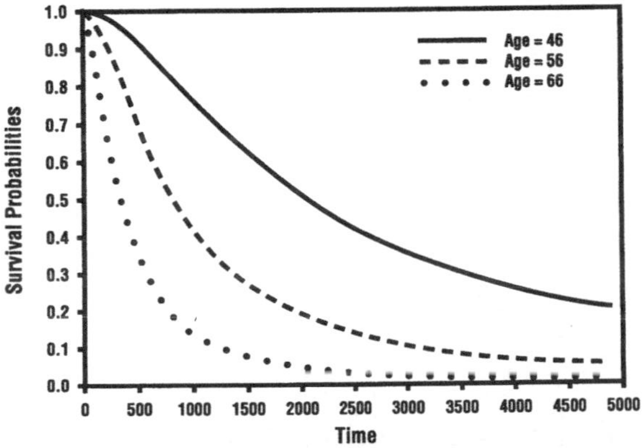

Figure 2; KL Divergence (16) versus Index
Ovarian Data; Flat Prior

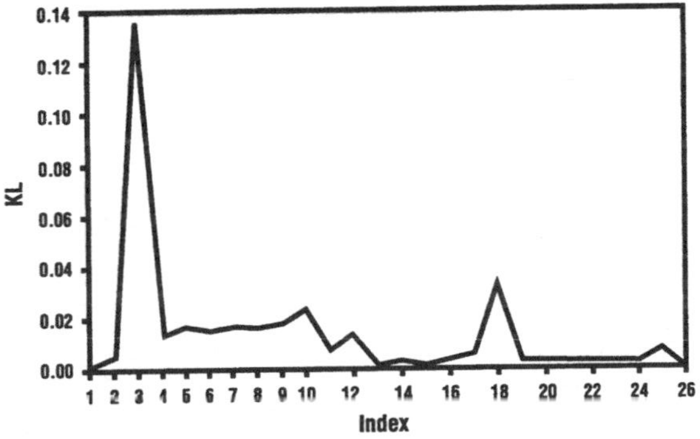

Figure 3; Ovarian Data with Flat Prior; Case 3 Deleted

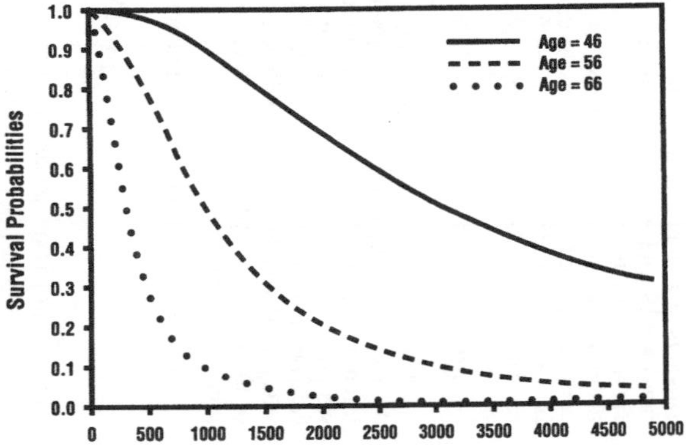

Figure 4; KL Divergence (16) versus Index
Ovarian Data minus Case 3; Flat Prior

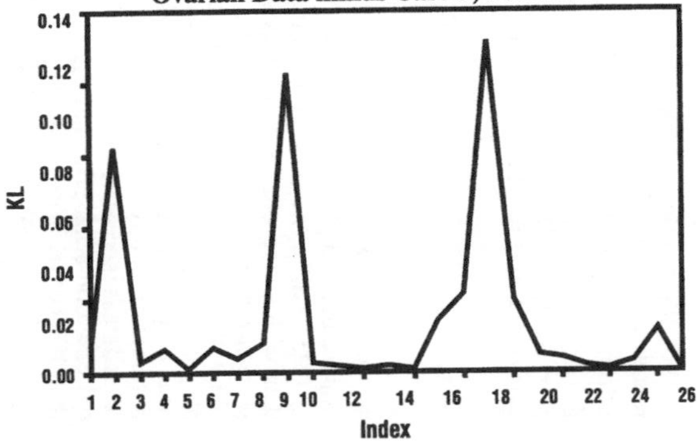

in a corresponding histogram for future times. Figure 6 gives a similar histogram of slope coefficients for age. The slope distribution is concentrated to the left of zero and is skewed left.

For illustration, we consider a partially informative prior on (β, τ). First note that the marginal prior for $\ell n(\tilde{m}_i)$ is $St[a, m_{0i}, b/aw_i]$ cf. DeGroot (1970, p. 170). We assume prior information about τ is reflected by a $\Gamma\left(\frac{5}{2}, \frac{2}{2}\right)$ pdf which corresponds to a prior mode of 1.5, a mean of 2.5, a median of 2.18, a 95th percentile of 5.5 and a 5th percentile of .57. Let $p' = 1$, $x_{01} = (1, 45)$ and assume that elicitation has resulted in $\Pr(\tilde{m}_1 \leq 1000) = .5$, and $\Pr(\tilde{m}_1 \leq 1400) = .95$. Then we must have $m_{01} = \ell n\, 1000$. Furthermore,

$$\sqrt{w_1}\sqrt{\frac{a}{b}}(\ell n\, \tilde{m}_1 - m_{01}) \sim t_a,$$

so we must have $.95 = \Pr(\tilde{m}_1 \leq 1400) = \Pr\left(t_5 \leq \sqrt{w_1}\sqrt{\frac{5}{2}}(\ell n\, 1400 - \ell n\, 1000)\right)$, which implies that $.53\sqrt{w_1} = 2.02$ or $w_1 = 14.5$. A 90% prior probability interval for $\tilde{m}_1$ is obtained as (715, 1400). We proceed to fit the model with this prior, which amounts to augmenting the data with a response of 1000 days at age 45 with a weight of 14.5. In the actual data there were two 44 year olds who survived 1106 and 1206 days before being censored and a 43 year old who survived 855 days before being censored. A 43 year old died after 329 days and a 50 year old died after 431 days. So the prior data is somewhat inconsistent with uncensored near neighbors and not overtly inconsistent with censored near neighbors. This single prior observation is obviously very informative when compared with the sample information. The prior data is inconsistent with the observed data in the sense that the survival curve for a 46 year old based on a flat prior indicates a 77% chance that a future 46 year old will survive longer than 1000 days. With such a large weight it is expected that the prior data will be influential. If $w_1 = 1$ were selected, the corresponding 90% prior confidence interval is (279, 3588). In this instance, the prior observation fails to be noticably influential.

Figure 7 gives an index plot with this prior. New case 1 corresponds to the prior observation, which is now indicated to be the most influential case, followed by new case 4 (old case 3) and new case 19 (old case 18). Figure 8 gives survival curves based on the partial prior. The median survival time for a 46 year old is now estimated to be around 1200 days. This is dramatically different than with the flat prior. All 3 curves in Figure 8 are shifted to the left from their counterparts in Figure 1, though the one for the 46 year old shifted the most. The slope estimate is $-.062$ and standardized, it is -4.62. A picture like Figure 6 for β_1 is centered at $-.06$ with no skewness, accounting for the smaller standardized slope. The posterior mean for τ is 1.98.

4. Moderate to Large Sample Inference

The above analysis is somewhat computationally intensive and provides no insight into the nature of why observations might be influential. We thus consider approximate predictive inferences based on a standard analytical approximation to the predictive density. In Section 5 we proceed to define approximate predictive influence as the KL divergence between full and case deleted approximate PD's. The KL divergence is tractable in this instance so we are able to study the derived formula. It will be seen to be a simple function of two statistics; one a measure of the difference in predictive location vectors and the other, a measure of the difference in predictive scales.

By first order Taylor approximation of the integrand of the PD, we obtain

$$p(t_f|X_f, d) \simeq p\left(t_f|X_f, \hat{\beta}, \hat{\tau}\right)$$

where $\left(\hat{\beta}, \hat{\tau}\right)$ is the posterior mode based on maximizing $L(\beta, \tau|d)\,\pi(\beta|\tau)\,\pi(\tau)$, with components defined at (4), (7) and (10). Large sample predictive inferences are made based on the

23.5cm

16 cm

Figure 5: PD of Log Failure Time for a 56 Year Old

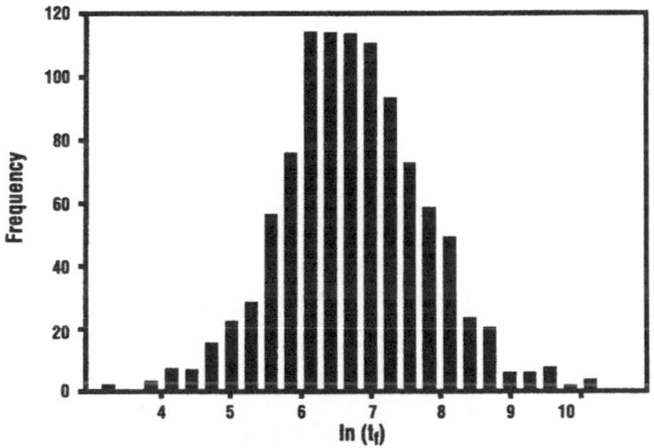

Figure 6: Histogram fro β_1

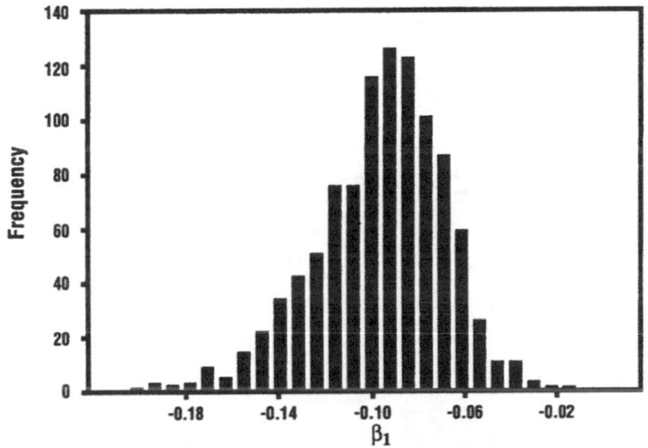

Figure 7: KL Divergence (16) vs. Index; Ovarian Data Partially Informative Prior

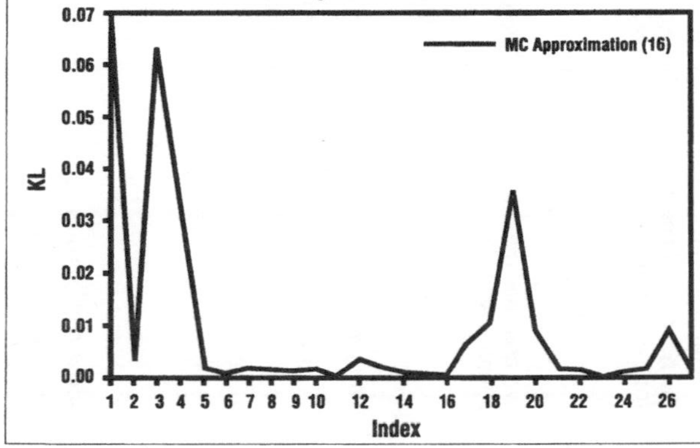

Figure 8; Ovarian Data with Partially Informative Prior

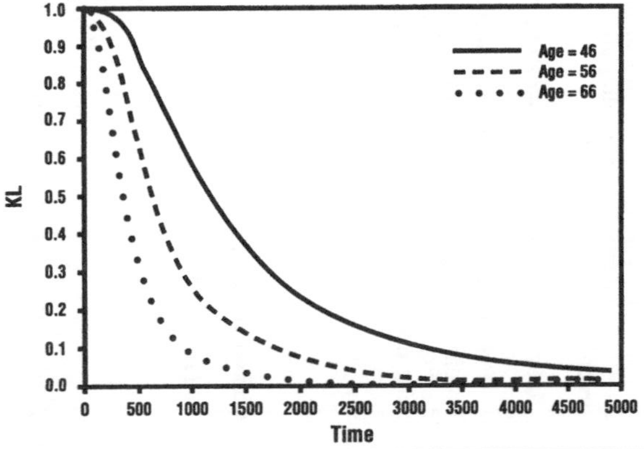

plug-in estimate $p\left(t_f|X_f,\hat{\beta},\hat{\tau}\right)$. With small n, this approximation may be appreciably less diffuse than $p\left(t_f|X_f,d\right)$ since it assumes β and τ are in fact the estimated values and consequently that there is no uncertainty about them. It furthermore fails to account for the correlation structure in a future vector.

5. Predictive Influence via Analytic Approximation

Even when large sample inferences are questionable, it may be cost effective to find observations that, upon deletion, have a large effect on $p(t_f|X_f,\hat{\beta},\hat{\tau})$. If a single case affects this density, it is likely that it would also affect the actual PD. The most influential cases can then be deleted and the ultimate effect on actual predictive inferences determined. We thus define

$$KL_i^* = \int p\left(t_f|X_f,\hat{\beta},\hat{\tau}\right) \ell n \left\{ \frac{p\left(t_f|X_f,\hat{\beta},\hat{\tau}\right)}{p\left(t_f|X_f,\hat{\beta}_{(i)},\hat{\tau}_{(i)}\right)} \right\} dt_f, \tag{17}$$

where $\left(\hat{\beta}_{(i)},\hat{\tau}_{(i)}\right)$ is the posterior mode with case i deleted.

Let $\ell n(V_i) \sim N\left(\mu_i,\sigma_i^2\right) \quad i=1,2$ and let $f_i(\cdot)$ be the pdf for V_i. It is well known that

$$\int f_1(v)\ell n \left\{\frac{f_1(v)}{f_2(v)}\right\} dv = \frac{\delta^2}{2} + \frac{1}{2}\left(\rho^2 - \ell n\rho^2 - 1\right),$$

where $\rho = \sigma_1/\sigma_2$ and $\delta = (\mu_1-\mu_2)/\sigma_2$. Thus the KL divergence between two log normal densities is a function of the squared normalized difference in location parameters and of the ratio of dispersion parameters.

Since we have assumed that the components of T_f are independent, conditional on $\left(\hat{\beta},\hat{\tau}\right)$, the KL divergence between joint predictive densities is simply the sum of the KL divergences for the marginal PD's. We thus obtain

$$KL_i^* = \frac{1}{2}\hat{\tau}_{(i)}(\hat{\beta}-\hat{\beta}_{(i)})^T X_f^T X_f(\hat{\beta}-\hat{\beta}_{(i)}) + \frac{m}{2}\left\{\hat{\tau}_{(i)}/\hat{\tau} - \ell n\left(\hat{\tau}_{(i)}/\hat{\tau}\right) - 1\right\}. \tag{18}$$

We consider the other directed divergence where the roles the full and case deleted PD's are interchanged. We obtain

$$KL_i^{**} = \frac{1}{2}\hat{\tau}(\hat{\beta}-\hat{\beta}_{(i)})^T X_f^T X_f(\hat{\beta}-\hat{\beta}_{(i)}) + \frac{m}{2}\left\{\hat{\tau}/\hat{\tau}_{(i)} - \ell n\left(\hat{\tau}/\hat{\tau}_{(i)}\right) - 1\right\}. \tag{19}$$

The symmetric KL divergence is obtained by taking the sum of KL_i^* and KL_i^{**}.

It is computationally inefficient to calculate $(\hat{\beta}_{(i)},\hat{\tau}_{(i)})$ exactly for each i, especially if n is large. Pregibon (1981) suggested approximating $(\hat{\beta}_{(i)},\hat{\tau}_{(i)})$ by taking one Newton-Raphson step towards $(\hat{\beta}_{(i)},\hat{\tau}_{(i)})$ starting from $(\hat{\beta},\hat{\tau})$. This has become standard practice. Let $\theta^T = (\beta^T,\tau)$ and $\hat{\theta}^T = (\hat{\beta}^T,\hat{\tau})$ etc. The one step approximation to $\hat{\theta}_{(i)}$ is

$$\hat{\theta}_{(i)}^{(1)} = \hat{\theta} - \ddot{\ell}_{(i)}^{-1}(\hat{\theta})\dot{\ell}_{(i)}(\hat{\theta}),$$

where $\ell_{(i)}(\cdot)$ is the log posterior based on $d_{(i)}$, $\dot{\ell}_{(i)}(\cdot)$ is the vector of first derivatives of $\ell_{(i)}(\cdot)$, and $\ddot{\ell}_{(i)}(\cdot)$ is the corresponding matrix of second derivatives. This approximation is not feasible for the problem we consider, as it will not be computationally efficient unless τ is assumed known.

However, we proceed to develop a simple variant which is feasible. Ultimately, we substitute our version of one step approximation to $(\hat{\beta}_{(i)}, \hat{\tau}_{(i)})$ into (18) and (19) to obtain very efficient approximations to KL_i^* and KL_i^{**}. Readers uninterested in the details of these one-steps may wish to skip to expression (27).

Defining $\ell_j(\beta, \tau) = \ell n L_j(\beta, \tau)$, where $L_j(\cdot, \cdot)$ was defined at (14), and $k(\tau) = -\ell n \pi(\tau)$, the log posterior can be expressed as

$$\ell(\beta, \tau) = -k(\tau) + \sum_{j=1}^{n+p'} \delta_j \left\{ \tfrac{1}{2} \ell n\, \tau + \ell n\, f_0(z_j) \right\} + (1 - \delta_j) \ell n S_0(z_j). \tag{20}$$

Furthermore, define $\dot{\ell}(\cdot), \dot{\ell}_j(\cdot)$ and $\dot{k}(\cdot)$ to be vectors of first derivatives with respect to θ, and define $\ddot{\ell}(\cdot)$, $\ddot{\ell}_j(\cdot)$ and $\ddot{k}(\cdot)$ to be second derivative matrices. We obtain the standard result

$$\hat{\theta}_{(i)}^{(1)} = \hat{\theta} + \left\{ \ddot{\ell}(\hat{\theta}) - \ddot{\ell}_i(\hat{\theta}) \right\}^{-1} \dot{\ell}_i(\hat{\theta}). \tag{21}$$

So we require $\ddot{\ell}_i(\cdot)$ and $\dot{\ell}_i(\cdot)$. We obtain

$$\dot{\ell}_i(\theta) = -\left\{ \delta_i z_i + (1 - \delta_i) h_0(z_i) \right\} \frac{\partial z_i}{\partial \theta} + \delta_i (\frac{1}{2\tau}) e_p, \quad \frac{\partial z_i}{\partial \theta} = \begin{pmatrix} -\tilde{w}_i^{\frac{1}{2}} \tilde{x}_i^T \tau^{\frac{1}{2}} \\ z_i / 2\tau \end{pmatrix}, \tag{22}$$

where $h_0(z_i) = f_0(z_i)/S_0(z_i)$ is the hazard function for the standard normal, $\tilde{w}_i$ is defined to be 1 for $i = 1, ..., n$, $\tilde{x}_{n+j} = x_{0j}$ for $j = 1, ..., p'$, and e_p is a vector with $p - 1$ zeros and a one in the p-th slot. Taking second derivatives

$$\begin{aligned} -\ddot{\ell}_i(\theta) &= \delta_i \left\{ \left(\tfrac{\partial z_i}{\partial \theta} \right) \left(\tfrac{\partial z_i}{\partial \theta^T} \right) + z_i \left(\tfrac{\partial^2 z_i}{\partial \theta \partial \theta^T} \right) + \left(\tfrac{1}{2\tau^2} \right) e_p e_p^T \right\} \\ &\quad + (1 - \delta_i) \left\{ \dot{h}_0(z_i) \left(\tfrac{\partial z_i}{\partial \theta} \right) \left(\tfrac{\partial z_i}{\partial \theta^T} \right) + h_0(z_i) \left(\tfrac{\partial^2 z_i}{\partial \theta \partial \theta^T} \right) \right\}, \end{aligned} \tag{23}$$

where

$$\begin{aligned} \left(\tfrac{\partial z_i}{\partial \theta} \right) \left(\tfrac{\partial z_i}{\partial \theta^T} \right) &= \begin{pmatrix} \tilde{w}_i \tilde{x}_i^T \tilde{x}_i \tau & -\tilde{w}_i^{\frac{1}{2}} z_i \tilde{x}_i^T / 2\tau^{1/2} \\ -\tilde{w}_i^{\frac{1}{2}} z_i \tilde{x}_i / 2\tau^{1/2} & z_i^2 / 4\tau^2 \end{pmatrix}, \\ \tfrac{\partial^2 z_i}{\partial \theta \partial \theta^T} &= \begin{pmatrix} 0 & -\tilde{w}_i^{\frac{1}{2}} \tilde{x}_i^T / 2\tau^{1/2} \\ -\tilde{w}_i^{\frac{1}{2}} \tilde{x}_i / 2\tau^{1/2} & -z_i / 4\tau^2 \end{pmatrix}, \\ \dot{h}_0(z_i) &= -z_i h_0(z_i) + h_0^2(z_i), \quad \ddot{k}(\tau) = \left(\tfrac{a}{2} - 1 \right) e_p e_p^T / \tau^2. \end{aligned} \tag{24}$$

We now sum the components of $-\ddot{\ell}_i(\theta)$. Starting with the upper block diagonal we obtain

$$\tau \left\{ \Sigma \tilde{w}_j \tilde{x}_j^T \tilde{x}_j \delta_j + \Sigma \dot{h}_0(z_j) \tilde{x}_j^T \tilde{x}_j (1 - \delta_j) \right\} = \tau \left\{ X_u^T X_u + X_c^T \dot{H}_0 X_c + X_0^T W X_0 \right\},$$

where $\dot{H}_0 = \text{diag}\left\{ \dot{h}_0(z_j) \right\}$. Then defining W^* to be

$$W^* = \begin{pmatrix} I_{n_u} & 0 & 0 \\ 0 & \dot{H}_0 & 0 \\ 0 & 0 & W \end{pmatrix}$$

we obtain the upper diagonal block to be $i_\beta(\beta,\tau) \equiv \tau\left(\tilde{X}W^*\tilde{X}\right)$. The lower diagonal is

$$i_\tau(\beta,\tau) \equiv n_u/2\tau^2 + \Sigma(1-\delta_i)\left\{z_i^2\dot{h}_0(z_i) - z_i h_0(z_i)\right\}/4\tau^2,$$

and the upper off diagonal vector is

$$i_{\beta\tau}(\beta,\tau) \equiv -\Sigma\delta_i\tilde{x}_i^T z_i\tilde{w}_i^{\frac{1}{2}}/\tau^{\frac{1}{2}} - \Sigma(1-\delta_i)\tilde{w}_i^{\frac{1}{2}}x_i^T\left\{z_i\dot{h}_0(z_i) + h_0(z_i)\right\}/2\tau^{1/2}.$$

Without censoring and with $p' = 0$, and $a = 2$, we obtain

$$-\ddot{\ell}(\theta) = \begin{pmatrix} \tau X^T X & -\sum_{i=1}^n z_i x_i^T/\tau^{1/2} \\ -\sum_{i=1}^n z_i x_i/\tau^{1/2} & n/2\tau^2 \end{pmatrix}$$

which has expectation

$$E\left\{-\ddot{\ell}(\theta)\right\} = \begin{pmatrix} \tau X^T X & 0 \\ 0 & n/2\tau^2 \end{pmatrix},$$

the usual Fisher expected information matrix for the above special case. The off diagonal terms will not have zero expectation in general. We have

$$-\ddot{\ell}(\theta) = \ddot{k}(\theta) + \begin{pmatrix} i_\beta & i_{\beta\tau} \\ i_{\beta\tau}^T & i_\tau \end{pmatrix} = \begin{pmatrix} i_\beta & i_{\beta\tau} \\ i_{\beta\tau}^T & i_\tau + \left(\frac{a}{2}-1\right)/\tau^2 \end{pmatrix} \equiv \begin{pmatrix} i_\beta & i_{\beta\tau} \\ i_{\beta\tau}^T & i_\tau^* \end{pmatrix} \equiv i_\theta.$$

It is not possible to obtain a simple one step approximation to $(\hat{\beta}_{(i)}, \hat{\tau}_{(i)})$ in the usual way unless the off block diagonal terms in the matrix $\ddot{\ell}_{(i)}(\hat{\theta})$ are zero. With $\tau = \hat{\tau}$ assumed known, however, it is easy to obtain a one-step approximation to the corresponding posterior mode $\hat{\beta}_{(i)}(\hat{\tau})$. We note that

$$-\ddot{\ell}_{(i)}(\beta) = \tau\tilde{X}_{(i)}^T W_{(i)}^*\tilde{X}_{(i)} = \tau\left\{\tilde{X}^T W^*\tilde{X} - \tilde{x}_i^T\tilde{x}_i w_i^*\right\} = -\left\{\ddot{\ell}(\beta) - \ddot{\ell}_i(\beta)\right\},$$

and thus

$$-\ddot{\ell}_{(i)}^{-1}(\beta) = \tau^{-1}\left\{(\tilde{X}W^*\tilde{X})^{-1} + \frac{w_i^*\left(\tilde{X}W^*\tilde{X}\right)^{-1}\left(\tilde{x}_i^T\tilde{x}_i\right)\left(\tilde{X}W^*\tilde{X}\right)^{-1}}{1 - w_i^*\tilde{x}_i\left(\tilde{X}W^*\tilde{X}\right)^{-1}\tilde{x}_i^T}\right\},$$

and from (22), we get

$$\dot{\ell}_i(\beta) = \tilde{w}_i^{1/2}\tau^{1/2}\left\{\delta_i z_i + (1-\delta_i)h_0(z_i)\right\}\tilde{x}_i^T.$$

Define $\hat{W}^*$ to be W^* with $(\hat{\beta},\hat{\tau})$ substituted for (β,τ), $\hat{A} = \tilde{X}\hat{W}^*\tilde{X}$ and $\hat{v}_i = \hat{w}_i\tilde{x}_i\hat{A}^{-1}\tilde{x}_i^T$. Note that $\tilde{w}_i = \hat{w}_i = w_i^*$ for $i = 1,\ldots,n_u,n+1,\ldots,n+p'$. Then we obtain

$$\hat{\beta}_{(i)}^{(1)}(\hat{\tau}) = \hat{\beta} - \frac{\hat{\tau}^{-1/2}\hat{w}_i^{1/2}}{1-\hat{v}_i}\left(\hat{A}^{-1}\tilde{x}_i^T\right)\left\{\delta_i\hat{z}_i + (1-\delta_i)h_0(\hat{z}_i)\right\} \equiv \tilde{\beta}_{(i)}^{(1)}, \tag{25}$$

after simple algebra since $\hat{\beta}(\hat{\tau}) = \hat{\beta}$. Similarly if $\beta = \hat{\beta}$ is fixed, we obtain

$$\hat{\tau}_{(i)}^{(1)}(\hat{\beta}) = \hat{\tau} + \hat{\tau}\left[(n_u - \delta_i)/2 + \sum_{(i)}(1 - \delta_j)\left\{\hat{z}_j^2 h_0(\hat{z}_j) - \hat{z}_j h_0(\hat{z}_j)\right\}/4 + \left(\tfrac{a}{2} - 1\right)\right]^{-1} \cdot$$
$$\left\{\delta_i(\hat{z}_i^2 - 1) + (1 - \delta_i)\,\hat{z}_i h_0(\hat{z}_i)\right\}/2 \equiv \tilde{\tau}_{(i)}^{(1)}, \tag{26}$$

since $\hat{\tau}(\hat{\beta}) = \hat{\tau}$. Accordingly, an uncensored case with $\hat{z}_i = 1$ will have $\tilde{\tau}_{(i)}^{(1)} = \hat{\tau}$, if $\hat{z}_i < 1$ the one-step is smaller and if $\hat{z}_i$ is larger the one-step is larger than $\hat{\tau}$. Since $\hat{z}_i h_0(\hat{z}_i) \to 0$ as $\hat{z}_i \to -\infty$ an observation that is censored near the beginning of the study will have no effect on the estimate of τ.

Our real interest is in

$$\begin{pmatrix} \hat{\beta}_{(i)}^{(1)} \\ \hat{\tau}_{(i)}^{(1)} \end{pmatrix} = \begin{pmatrix} \hat{\beta} \\ \hat{\tau} \end{pmatrix} + \left\{\ddot{\ell}\begin{pmatrix} \hat{\beta} \\ \hat{\tau} \end{pmatrix} - \ddot{\ell}_i\begin{pmatrix} \hat{\beta} \\ \hat{\tau} \end{pmatrix}\right\}^{-1} \dot{\ell}_i\begin{pmatrix} \hat{\beta} \\ \hat{\tau} \end{pmatrix}.$$

If we were to calculate $\hat{\tau}_{(i)}$ exactly, then $\hat{\beta}_{(i)}(\hat{\tau}_{(i)}) = \hat{\beta}_{(i)}$ and similarly $\hat{\tau}_{(i)}(\hat{\beta}_{(i)}) = \hat{\tau}_{(i)}$. If the off diagonal block components of $\left\{\ddot{\ell}\binom{\hat{\beta}}{\hat{\tau}} - \ddot{\ell}_i\binom{\hat{\beta}}{\hat{\tau}}\right\}^{-1}$ are near zero, then

$$\hat{\beta}_{(i)}^{(1)}(\hat{\tau}) \cong \tilde{\beta}_{(i)}^{(1)} \quad \text{and} \quad \hat{\tau}_{(i)}^{(1)}(\hat{\beta}) \cong \tilde{\tau}_{(i)}^{(1)}.$$

We note that if $\hat{\tau}_{(i)} \cong \hat{\tau}$, then $\hat{\beta}_{(i)} = \hat{\beta}_{(i)}(\hat{\tau}_{(i)}) \simeq \hat{\beta}_{(i)}(\hat{\tau})$, which in turn implies that $\tilde{\beta}_{(i)}^{(1)} \simeq \hat{\beta}_{(i)}^{(1)}(\hat{\tau})$, etc.

Our approach is to use the one step approximations $\tilde{\beta}_{(i)}^{(1)} = \hat{\beta}_{(i)}^{(1)}(\hat{\tau})$ and $\tilde{\tau}_{(i)}^{(1)} = \hat{\tau}_{(i)}^{(1)}(\hat{\beta})$. In general, these will not equal $\hat{\beta}_{(i)}^{(1)}$ and $\hat{\tau}_{(i)}^{(1)}$ respectively. However, all that we require is that influential cases be detected. Truly influential cases should be easily detected via these approximations.

We approximate (18) and (19) with our one-steps by substituting (25) and (26) for $\hat{\beta}_{(i)}$ and $\hat{\tau}_{(i)}$. Define $\hat{\gamma}_i = \tilde{\tau}_{(i)}^{(1)}/\hat{\tau}$. Then

$$KL_i^{*} \simeq \frac{\hat{\gamma}_i}{2}\left\{\delta_i\left(\frac{\hat{z}_i}{\sqrt{1-\hat{v}_i}}\right)^2\left(\frac{\hat{r}_i}{1-\hat{v}_i}\right) + (1 - \delta_i)\left(\frac{h_0(\hat{z}_i)}{\sqrt{1-\hat{v}_i}}\right)^2\left(\frac{\hat{r}_i}{1-\hat{v}_i}\right)\right\} + \frac{m}{2}\left\{\hat{\gamma}_i - \ell n\hat{\gamma}_i - 1\right\}, \tag{27}$$

$$KL_i^{**} \simeq \frac{1}{2}\left\{\delta_i\left(\frac{\hat{z}_i}{\sqrt{1-\hat{v}_i}}\right)^2\left(\frac{\hat{r}_i}{1-\hat{v}_i}\right) + (1 - \delta_i)\left(\frac{h_0(\hat{z}_i)}{\sqrt{1-\hat{v}_i}}\right)^2\left(\frac{\hat{r}_i}{1-\hat{v}_i}\right)\right\} + \frac{m}{2}\left\{\gamma_i^{-1} - \ell n\gamma_i^{-1} - 1\right\}, \tag{28}$$

where

$$\hat{r}_i = \tilde{w}_i\tilde{x}_i\hat{A}^{-1}X_f^T X_f\hat{A}^{-1}\tilde{x}_i^T.$$

These formulas are computationally efficient and lend themselves to interpretation.

The component of the approximation to KL_i^{**} corresponding to mean differences resembles Cook's distance (Cook, 1977). For an uncensored observation, we get the standardized residual $\hat{z}_i/\sqrt{1 - \hat{v}_i}$ squared times a term that involves the "leverage" $\hat{v}_i$ and another term $\hat{r}_i$, which involves $\tilde{x}_i$ and X_f. We proceed to obtain Cook's distance and to interpret it. Our measures can be similarly interpreted.

We obtain a standard simplification of the Cook distance analogue, namely

$$\begin{aligned} \tilde{D}_i^{(1)} &\equiv \hat{\tau}\left(\hat{\beta}-\tilde{\beta}_{(i)}^{(1)}\right)^T \tilde{X}\hat{W}^*\tilde{X}\left(\hat{\beta}-\tilde{\beta}_{(i)}^{(1)}\right) \\ &= \left\{\tfrac{\delta_i\hat{z}_i+(1-\delta_i)h_0(\hat{z}_i)}{\sqrt{1-\hat{\nu}_i}}\right\}^2\left(\tfrac{\hat{\nu}_i}{1-\hat{\nu}_i}\right)\tilde{w}_i/\hat{w}_i^* \qquad (29) \\ &= \delta_i\left(\tfrac{\hat{z}_i}{\sqrt{1-\hat{\nu}_i}}\right)^2\left(\tfrac{\hat{\nu}_i}{1-\hat{\nu}_i}\right)+(1-\delta_i)\left(\tfrac{h_0(\hat{z}_i)}{\sqrt{h_0(\hat{z}_i)}\sqrt{1-\hat{\nu}_i}}\right)^2\left(\tfrac{\hat{\nu}_i}{1-\hat{\nu}_i}\right). \end{aligned}$$

Cook's distance was designed to measure the effect of case deletion on the regression coefficient estimates. If $\hat{z}_i = 0$ and $\delta_i = 1, \tilde{D}_i^{(1)} = 0$. An observation with $\delta_i = 0$ and $\hat{z}_i = 0$ has $\tilde{D}_i^{(1)} = \hat{\nu}_i/(1-\hat{\nu}_i)^2$, so if the leverage $\hat{\nu}_i$ is large, a case censored at it's estimated median value can have a large impact according to Cook's influence measure. Clearly with $\hat{\delta}_i = 0$ and as $\hat{z}_i \to -\infty$, the effect of a censored observation diminishes to zero. This corresponds to censoring a time near zero and such a case should clearly not be influential. On the other hand, if we let $\delta_i = 0$ and $\hat{z}_i \to +\infty$,

$$\frac{h_0^2(\hat{z}_i)}{h_0(\hat{z}_i)} = \frac{h_0(\hat{z}_i)}{h_0(\hat{z}_i)-\hat{z}_i} = \frac{f_0(\hat{z}_i)}{f_0(\hat{z}_i)-\hat{z}_iS_0(\hat{z}_i)} \simeq \hat{z}_ih_0(\hat{z}_i) \simeq \hat{z}_i^2$$

by double application of L'Hopital's rule. Thus a large censored observation can be very influential according to this measure, and such a case will have approximately the same effect as deleting an uncensored case with the same $\hat{z}_i$.

It is of some interest to make a connection between Cook's distance CD and our predictive influence function. In order to do this, we need to let $X_f = X$ and consider a slightly different measure of predictive influence. First note from (19) and (28) that there is a clear similarity between the Cook statistic and the first component of our PIF when $X_f = X$. The difference is solely due to the placement of the matrix $\hat{W}^*$ in the center of expression (29). If we define the predictive influence function to be a weighted average of, say m, KL divergences for predicting future observations, one at a time, based on approximate marginal PD's of the type discussed in Section 4, if we let $X_f = X$, and if we let the weights be the $\hat{w}_i$'s, then the first component of (19) would be identical to (29). Johnson (1985) made a similar connection for logistic regression.

Ovarian Data Revisited:

Figures 9 and 10 give normalized values for the approximations (16) and (27) for our flat prior as well as the partially informative prior. In both instances, the underlying pattern and case influence is mimicked by the approximation (27). Case 3 is influential due to a small value for $\hat{z}_3 = -2.49$, and a moderate value $\hat{v}_3 = .21$. With the partially informative prior, $\hat{z} = -1.02$ and $\hat{v} = .66$ for new prior observation 1. This case has high leverage due to the combination of the relatively young age considered and the large weight. The standardized residual $\hat{z}$ is moderate. If a weight of 1 is attached to the prior observation rather than 14.5, the leverage $\hat{v} = .14$, $\hat{z} = -.80$ and this case is no longer influential. Note that old case 2, (new case 3) is now more influential than old case 3 (new case 4).

We finally consider a different choice of X_f. Johnson and Geisser (1983) used $X_f = X$ and we now compare our choice above with this one. Figure 11 gives normalized plots for both selections. Perhaps surprisingly, the prior observation is not indicated to be influential at all when prediction is to be at X, while old case 3 (new case 4) has the largest approximate KL number. The prior observation is not influential for predicting at X because the $(1,1)$ element of $X\hat{A}^{-1}X^TX\hat{A}^{-1}X^T$ is small while the corresponding element $X\hat{A}^{-1}X_f^TX_f\hat{A}^{-1}X$ is not. We remind the reader that the prior observation was in fact very influential for predicting survival for a future 46 year old.

23.5cm

16 cm

Figure 9: Normalized Approximate KL Divergences vs. Index; Ovarian Data; Flat Prior

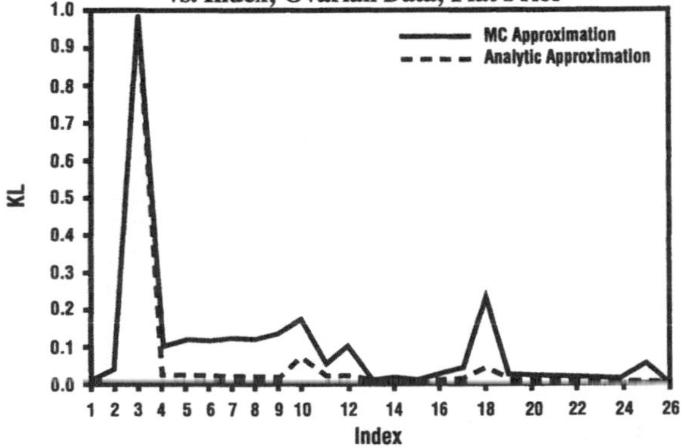

Figure 10: Normalized Approximate KL Divergences vs. Index; Ovarian Data; Partially Informative Prior

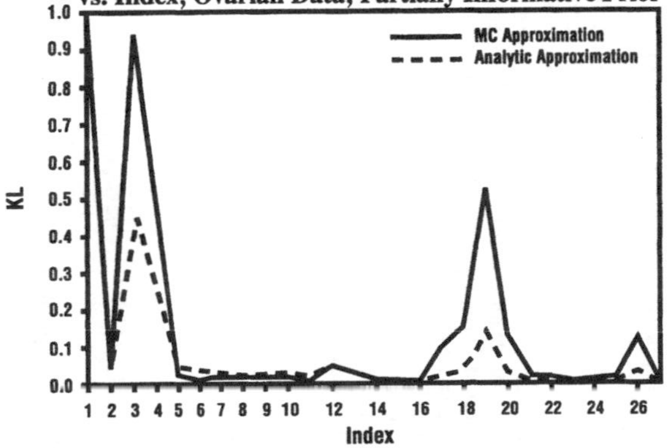

Figure 11: Approximate KL Divergences vs. Index; Ovarian Data; Partially Informative Prior

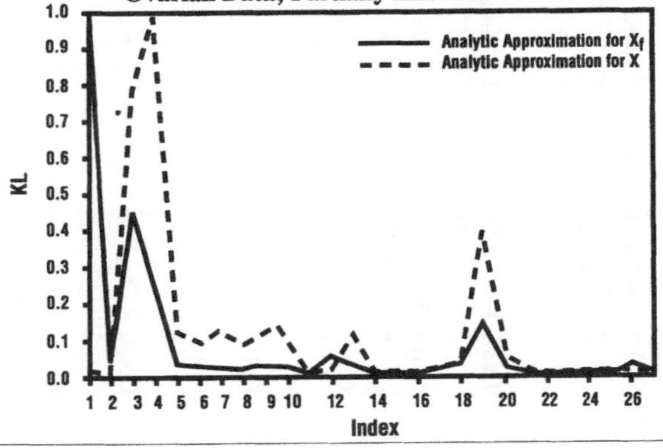

6. Conclusions

We have discussed a cost effective numerical approximation to an exact approach to predictive influence as well as a companion approach, suitable for large samples, which results in analytical formulas that give insight into why particular cases might be influential. The latter approach is virtually cost free once standard calculations have already been performed. We have noted the nature of the distinction between case influence for censored and uncensored data. Our approach also allows for the assessment of the impact of components of the prior on predictive inferences. Future work will focus on extending this work to the entire class of accelerated failure time models including the Weibull and log logistic. My Ph.D. student, Alex Exuzides, is currently working out all the frequentist angles of this approach when applied to accelerated failure time models and the Cox model. Related previous work was done by Thomas and Cook (1990) who considered local influence diagnostics for predicting in generalized linear models. Weissfeld and Schneider (1990) did standard diagnostics for the log normal model. Finally, Carlin and Polson (1991) utilized the Gibbs sampler to calculate utility based influence diagnostics.

Acknowledgement: I thank Seymour Geisser for shining his light on prediction and for making me look. I also thank him for his support through the years. I also thank the referee for helpful suggestions.

References

Bedrick, E.J., Christensen, R.R. and Johnson, W.O. (1995). A new perspective on priors for generalized linear models. preprint.

Carlin, B.P. and Polson, N.G. (1991). An expected utility approach to influence diagnostics. *J. Amer. Statist. Assoc.* **86**, 1013-21.

Clogg, C.C., Rubin, D.B., Schenker, N., Schultz, B. and Weidman, L. (1991). Multiple imputation of industry and occupation codes in census public-use samples using Bayesian logistic regression. *J. Amer. Statist. Assoc.* **86**, 68-78.

Collett, D. (1994). *Modelling Survival Data in Medical Research.* Chapman and Hall, London.

Cook, R.D. (1977). Detection of influential observations in linear regression. *Technometrics* **19**, 15-18.

Cook, R.D. (1979). Influential observations in linear regression. *J. Amer. Statist. Assoc.* **74**, 169-74.

Gelfand, A.E. and Smith, A.F.M. (1990). Sampling based approaches to calculating marginal densities. *J. Amer. Statist. Assoc.* **85**, 398-409.

Johnson, W.O. (1985). Influence measures for logistic regression: Another point of view. *Biometrika* **72**, 59-65.

Johnson, W.O. and Geisser, S. (1982). Assessing the predictive influence of observations. In *Statistics and Probability: Essays in Honor of C.R. Rao*, Eds. G. Kallianpur, PR. Krishnaiah and J.K. Gosh pp. 343-58. Amsterdam: North Holland.

Johnson, W.O. and Geisser, S. (1983). A predictive view of the detection and characterization of influential observations in regression. *J. Amer. Statist. Assoc.* **78**, 137-44.

Pregibon, D. (1981). Logistic regression diagnostics. *Ann. Statist.* **9**, 705-24.

Tanner, M.A. (1994) *Tools for Statistical Inference.* Springer-Verlag, New York.

Tanner, M.A. and Wong, W.H. (1987). The calculation of posterior distributions by data augmentation. *J. Amer. Statist. Assoc.* **82**, 528-40.

Thomas, W. and Cook, R.D. (1990). Assessing influence on predictions from generalized linear models. *Technometrics* **32**, 59-65.

Weissfeld, L.A. and Schneider, H. (1990). Influence diagnostics for the normal linear model with censored data. *Austrl. J. Statist.* **32**, 11-20.

Zellner, A. and Rossi, P. (1984). Bayesian analysis of dichotomous quantal response models. Journal of Econometrics, *J. Econometrics* **25**, 365-93.

Selecting the Form of Combining Regressions Based on Recur sive Prediction Criteria

Kuo-yuan Liang*
Department of Economics
National Tsing Hua University
Hsin-chu, Taiwan

Keunkwan Ryu
Department of Economics
University of California
Los Angeles, CA, U.S.A.

Abstract

This paper reformulates the basic Granger and Ramanathan's (1984) combining regression framework based on *post-sample* predictive accuracies. Using recursive regression techniques, this paper develops an algorithm to estimate combining weights. Under the new prediction criteria, we show that Granger and Ramanathan's (1984) preference ordering, Method C → Method A → Method B, breaks down. To overcome this lack of ordering, the paper suggests that, by using Akaike's (1973) information or Amemiya's (1980) prediction criterion, one can recursively select the best form of combining regressions. Empirical examples using macroeconomic forecasts of Taiwan are presented to illustrate the validity of the theoretical arguments.

1 Introduction

Probably the most widely used statistical procedures on composite forecasting were developed in an influential article by Granger and Ramanathan (1984, hereafter as G&R). Historically, their contribution is a generalization of Bates and Granger (1969) as well as Newbold and Granger (1974). More specifically, G&R's regression models proposed three different types of specifications that could be used in the linear combination of forecasts.[1] Among them, Method B is the most restrictive model with two constraints: the regression coefficients sum to one and no constant term is used. Method C is the least restrictive with no constraint on the regression coefficients and includes a constant term.

---

*Financial support from the ROC National Science Council under grants 83-0301-H007-024 & 84-2411-H007-019 is gratefully acknowledged.

[1]For related papers, see Liang (1992) as well as Liang and Shih (1994).

Method A is a compromise which possesses a constraint that no constant term is used but does not restrict regression coefficients to add up to one. The methods proposed by G&R have stimulated discussions on the choice of the "best" form of three combining regressions by using different model selection criteria (e.g. Gunter, 1992; Liang and Gau, 1994). Out of these criteria, G&R suggested that one may focus on the model with the smallest sum of squared errors (SSE). In particular, let Q_a, Q_b, Q_c denote SSE from Methods A, B, and C, respectively. It is a standard result that the following inequality holds:

$$Q_b \geq Q_a \geq Q_c \tag{1}$$

Based on this inequality relationship, G&R recommended Method C as the best way of combining rival forecasts.

The recommendation, using the smallest SSE criterion, is subject to a fundamental criticism. Notice that inequality (1) merely reflects the fact that as we have more free parameters in a regression model, we can always better fit the within sample observations. That is, in accordance with the *ex-post within-sample* criterion, inequality (1) always holds. However, from the *ex-ante post-sample* viewpoint, inequality (1) does not carry much information.[2] It is no doubt that *the very motivation in applying composite forecasts is to improve the predictive accuracy, not the within sample fitting performance.* Consequently, it is more meaningful to use *ex-ante post-sample* predictive accuracies as the selection criterion.[3] Following this line of arguments and using recursive methods under Akaike's (1973) information or Amemiya's (1980) prediction criterion, this paper develops a new framework for selecting an optimal combining formula among Methods A, B, and C.

Like Akaike's (1973) information and Amemiya's (1980) prediction criteria, the trade-off between parsimony and flexibility of alternative model-specifications can also be approached through the conventional mean squared error (MSE) criterion. Previously two notable studies using the MSE criterion in the study of composite forecasts were Clemen (1986) as well as Trenkler and Liski (1986). Specifically, Clemen (1986) argued that in deciding whether to choose Method B (the restricted model) or Method C (the unrestricted model), it can be shown that if the restrictions are correct, we can obtain more efficient forecasts. Otherwise, we have to deal with biased forecasts. If the constraints are "almost" correct, then by using such constraints we may reduce MSE when the efficiency gain is not fully offset by the bias. The direction of the inequality as in (1) based on the minimum MSE criterion may go either way, depending on whether the constraints are correct, "almost" correct, or incorrect. In other words, on the basis of the minimum MSE criterion Method C is not necessary a sure winner. Along this line of reasoning and applying the results from the pre-test literature (e.g. Judge and Bock, 1978), an extension of Clemen's (1986) work was done by Trenkler and Liski (1986). Their testable conditions offer forecasters a practical method for selecting the best form of the combining regression.

Although minimization of MSE is widely used as an optimal search criterion for the specification of the combining regression, however, it is by no means the

[2]In fact, it is not novel to observe a model that produces the best *within-sample* fit may not be the one that produces the best *out-of-sample* results. For example, in their discussion of the empirical results, G&R recognized that their within-sample ordering may not hold up for the *out-of-sample* situation.

[3]Similar idea of using past out-of-sample performance to weight forecast can be founded in Clemen and Winkler (1986), Gupta and Wilton (1987), as well as Makridakis (1990).

only possible candidate. An alternative choice of criterion that has advantageous attributes is advocated here; namely, maximization *ex-ante post-sample* predictive accuracies under prediction criteria. The outline of this paper is as follows. Section 2 describes the basic G&R framework and presents it in a new form based on *ex-ante* considerations. Section 3 develops an algorithm to estimate combining weights using recursive regression techniques. Under our new setup we show that G&R's ordering, Method C → Method A → Method B, breaks down. Section 4 then applies Akaike's information (1973) and Amemiya's (1980) prediction criteria to rank Methods A, B, and C. Using multiple data sets of Taiwan's macroeconomic forecasts, section 5 illustrates the performance of the proposed composite forecasting algorithm. Closing remarks follow in section 6.

2 Within-and Out-of-sample-based Combining

Following G&R, the three methods of estimating the combining weights and their corresponding SSE representations can be captured by a linear regression model. To begin, Method (A) is based on :

$$\mathbf{Z} = \mathbf{F}\beta + \mathbf{U} \tag{2}$$

$$Q = \min_{\beta}(\mathbf{Z} - \mathbf{F}\beta)'(\mathbf{Z} - \mathbf{F}\beta), \tag{3}$$

where $\mathbf{Z} = (z_1, ..., z_k)'$ is a $(T\times 1)$ vector of the variable being forecasted, $\mathbf{U} = (u_1, ..., u_T)'$ is a $(T\times 1)$ vector of the disturbance term, $\mathbf{F} = (\mathbf{f}_1, ..., \mathbf{f}_k)$ is a $(T\times k)$ matrix of forecast series with $\mathbf{f}_j (j = 1,...,k)$ being a $(T\times 1)$ vector of forecast series from the jth expert or forecasting model[4], and β is a $(k\times 1)$ vector of combining weights. Let $\ell = (1, ..., 1)'$ be a vector of ones. Method (B) modifies Method (A) with the linear constraint $\ell'\beta = 1$ in the minimization problem Q. Method (C) is the same as Method (A) except that β contains an additional constant term β_0.

Now suppose that at time t we have access to the following information set Ω_t :

$$\Omega_t = (z_1, ..., z_t, f_{1j}, ..., f_{tj}, f_{t+1,j}, j = 1, ..., k) \tag{4}$$

In other words, by time t we have observed all the realizations of the variable being forecasted and we have access to all one-period-ahead forecasts from k different sources.[5] Our purpose at time t is to predict Z_{t+1} based on all the information represented by Ω_t. The basic notion here is that the optimal form of the linear combination of forecasts[6] should be determined exclusively by using information available at time t: that is, the parameters of the linear combination of forecasts should be determined by Ω_t. In the context of G&R, one way to formulate this concept is as follows. Parallel to (2) and (3), Method (A') is based on :

$$\mathbf{Z}_t = \mathbf{F}_t\beta_t + \mathbf{U}_t \qquad t = p, ..., T-1 \tag{5}$$

[4]Based on encompassing principles, a constructive procedure concerning the selection of component forecasts (columns of F) in (2.1) is provided in Liang and Ryu (1994).

[5]If we are interested in $H(\geq 1)$ periods ahead forecast, then Ω_t will also include $f_{t+2,j}, ..., f_{t+H,j}, j = 1, ..., k$.

[6]For simplicity, we consider only the linear combinations of $(f_{t+1,1}, ..., f_{t+1,k})$.

$$\min_{\beta_t}(\mathbf{Z}_t - \mathbf{F}_t\beta_t)'(\mathbf{Z}_t - \mathbf{F}_t\beta_t) \tag{6}$$

where $\underset{t\times 1}{\mathbf{Z}_t} = (z_1, ..., z_t)'$, $\underset{t\times k}{\mathbf{F}_t} = \begin{bmatrix} f_{11} & f_{12} & \cdots & f_{1k} \\ f_{21} & f_{22} & \cdots & f_{2k} \\ \cdots & \cdots & \cdots & \cdots \\ f_{t1} & f_{t2} & \cdots & f_{tk} \end{bmatrix}$, $\underset{t\times 1}{\mathbf{U}_t} = (u_1, ..., u_t)'$

with f_{tj} being the forecast for z_t from the *jth* expert or forecasting model, and β_t is based on the first t observations. Here p is the number of observations used to compute the first set of combining weights and should be greater than or equal to k+1. In this way we are able to reformulate G&R's ***within-sample ex-post*** framework based on ***post-sample ex-ante*** considerations. Similarly, Method (B') modifies Method (A') with the linear constraint $\ell'\beta_t = 1$ in the minimization problem (6). Method (C') is the same as Method (A') except that β_t contains an additional constant term β_{0t}.

Let β_{at}; β_{bt}; β_{0t}, and β_{ct} (t = p,...,T-1) be the corresponding β's according to the three different methods based on the first t observations. Once we compute the values of these β's, we can compare the performance of the three new combining Methods (A'), (B'), and (C') using the following measures:

$$Q_a' = \sum_{t=p+1}^{T} (z_t - \mathbf{f}_{t.}\beta_{at-1})^2$$

$$Q_b' = \sum_{t=p+1}^{T} (z_t - \mathbf{f}_{t.}\beta_{bt-1})^2$$

$$Q_c' = \sum_{t=p+1}^{T} (z_t - \beta_{0t-1} - \mathbf{f}_{t.}\beta_{ct-1})^2,$$

where $\mathbf{f}_{t.} = (f_{t1}, ..., f_{tk})$ is an 1×k vector of one period ahead forecasts for z_t from k different sources. These quantities correspond to the one period ahead prediction error sum of squares when the combined forecast is obtained solely based on information available at each moment in time. Clearly, for prediction purposes, comparison of Q_a', Q_b', and Q_c' makes much more sense than the comparison of the three SSEs (Q_a, Q_b, and Q_c) grounded on (3). However, as we show in the next section, Q_a', Q_b', and Q_c' do not allow any specific ordering in advance.

3 Recursive Regression Framework

This section shows that under our new setup G&R's performance ordering, Method C → Method A → method B, breaks down. To this end, it is convenient to recognize that the linear least squares in (A'), (B'), and (C') of the previous section can be written in a general form as follows:

$$\mathbf{Y}_t = \mathbf{X}_t\beta_t + \mathbf{U}_t, \qquad t = p, ..., T-1 \tag{7}$$

where $\underset{t\times 1}{\mathbf{Y}_t} = (y_1, ..., y_t)'$, $\underset{t\times 1}{\mathbf{U}_t} = (u_1, ..., u_t)'$ and $\underset{t\times q}{\mathbf{X}_t} = (\mathbf{x}_1, ..., \mathbf{x}_t)'$, with $\mathbf{x}'_t$ being the $(1 \times q)$ vector of the *t-th* observations on each of q regressors.

By taking $(\mathbf{X}_t, \beta_t, \mathbf{Y}_t)$ as in (a'), (b'), and (c') below, we can represent all three different cases (A'), (B'), (C') using these general notations:

(a')

$$\underset{t\times k}{\mathbf{X}_t} = \mathbf{F}_t \tag{8}$$

$$\underset{k\times 1}{\beta_t} = \beta_{at} \tag{9}$$

$$\underset{t\times 1}{\mathbf{Y}_t} = \mathbf{z}_t \tag{10}$$

(b')

$$\underset{t\times(k-1)}{\mathbf{X}_t} = \begin{bmatrix} f_{11} - f_{1k}, & f_{12} - f_{1k}, & \cdots, & f_{1k-1} - f_{1k} \\ f_{21} - f_{2k}, & f_{22} - f_{2k}, & \cdots, & f_{2k-1} - f_{2k} \\ \vdots & \vdots & \vdots & \vdots \\ f_{t1} - f_{tk}, & f_{t2} - f_{tk}, & \cdots, & f_{tk-1} - f_{tk} \end{bmatrix} \tag{11}$$

$$\underset{(k-1)\times 1}{\beta_t} = (\beta_{b1_t}, \beta_{b2_t}, ..., \beta_{bk-1_t})' \tag{12}$$

$$\underset{t\times 1}{\mathbf{Y}_t} = (z_1 - f_{1k}, z_2 - f_{2k}, ..., z_t - f_{tk})' \tag{13}$$

(c')

$$\underset{t\times(k+1)}{\mathbf{X}_t} = (\ell : \mathbf{F}_t) \tag{14}$$

$$\underset{(k+1)\times 1}{\beta_t} = \left[\frac{\beta_{0t}}{\beta_{ct}}\right] \tag{15}$$

$$\underset{t\times 1}{\mathbf{Y}_t} = \mathbf{Z}_t \tag{16}$$

Of course, the error process, $\mathbf{U}_t$, will be different for each of the cases.

Using the method of ordinary least squares and starting from t=p, β_t can be estimated at each moment in time by:

$$\hat{\beta}_t = (\mathbf{X}_t'\mathbf{X}_t)^{-1}\mathbf{X}_t\mathbf{Y}_t, \qquad t = p, ..., T-1$$

It is well know that $\hat{\beta}_{t+1}$ is updated from $\hat{\beta}_t$ by:

$$\hat{\beta}_{t+1} = \hat{\beta}_t + \frac{[\mathbf{X}_t'\mathbf{X}_t]^{-1}\mathbf{x}_{t+1}[y_{t+1} - \mathbf{x}_{t+1}'\hat{\beta}_t]}{d_{t+1}}, \tag{17}$$

where

$$d_{t+1} = 1 + \mathbf{x}_{t+1}'(\mathbf{X}_t'\mathbf{X}_t)^{-1}\mathbf{x}_{t+1}.$$

Now let $\tilde{u}_{t+1}$ be the prediction error for y_{t+1} when we predict y_{t+1} using the information available at time t. Then T-p $\tilde{u}_{t+1}$ are recursively generated:

$$\tilde{u}_{t+1} = y_{t+1} - \mathbf{x}_{t+1}'\hat{\beta}_t \qquad t = p, ..., T-1.$$

Denoting $\underset{(T-p)\times 1}{\tilde{U}} = (\tilde{u}_{p+1}, \ldots, \tilde{u}_t)'$, and using a matrix notation, we can write $\tilde{U}$ as follows:

$$\tilde{U} = CY$$

where

$$\underset{(T-P)\times T}{C} = \begin{bmatrix} -x'_{p+1}[X'_pX_p]^{-1}X'_p, & 1, & 0, \ldots\ldots\ldots, & 0 \\ -x'_{p+2}[X'_{p+2}X_{p+1}]^{-1}X'_{p+1}, & 1, & 0, \ldots, & 0 \\ & & \ddots & \vdots \\ \ldots\ldots\ldots\ldots\ldots\ldots\ldots\ldots\ldots\ldots\ldots & & \ddots & 0 \\ -x'_T[X'_{T-1}X'_{T-1}]'X'_{T-1} & & , & 1 \end{bmatrix}$$

Here the matrix C satisfies
(i) $CX = 0$ and (ii) $CC' = \text{diag}(d_{p+1}, \ldots, d_T)$.
The sum of squared prediction errors is represented as :
$Q^* = \tilde{U}'\tilde{U} = (CY)'(CY) = Y'(C'C)Y = U'(C'C)U$, since $CX = 0$.
Here a measure of the magnitude of Q^* is $\text{tr}(C'C) = \text{tr}(CC') = \sum_{t=p+1}^{T} d_t$. On the basis of this measure, we cannot say in advance whether it increases or not as we increase the number of explanatory variables in X in a nesting fashion. In fact, if the $\{x_t\}$ series is stationary, we can expect that:

$$x'_t[X'_{t-1}X_{t-1}]^{-1}x_t \approx \frac{1}{t-1}$$

Then tr(C'C) is approximated as:

$$\text{tr}(C'C) \approx \sum_{t=p+1}^{T}(1+\frac{1}{t-1}) = \sum_{t=p+1}^{T}(\frac{t}{t-1}).$$

This is independent of the explanatory variables, X, and thus the number of explanatory variables in X.

As a result, under our new setup, we cannot specifically order three combining methods in advance. To solve this indeterminacy problem, in the next section we suggest that one applies Akaike's (1973) information or Amemiya's (1980) prediction criterion to select the best form of the combining regression.

4 Applications of Prediction Criteria to Composite Forecasts

Let $M'_t = I_t - X_t(X'_tX_t)^{-1}X'_t$, t=p,...,T-1. At each moment in time, the within sample sum of squared residuals can be written as:

$$Y'_tM_tY_t, \qquad t = p, \ldots, T-1.$$

To select a model among (A') (B') and (C') based on predictive accuracy, we can adopt either Akaike's (1973) information criterion (AIC) or Amemiya's (1980) prediction criterion (PC).

According to AIC, the model at each moment in time t which gives the minimum value of AIC_t will be chosen, where

$$AIC_t = \log(Y_t' M_t Y_t) + \frac{2q}{t} \tag{18}$$

with q being the number of regressors in the model. According to PC, we select the model at each moment in time t which gives the minimum value of PC_t, where

$$PC_t = \frac{t+q}{t-q}(Y_t' M_t Y_t). \tag{19}$$

In both AIC_t and PC_t, we can easily see that there is a trade-off between $Y_t' M_t Y_t$ and q; if we increase q then $Y_t' M_t Y_t$ decreases, but 2q/t and (t+q)/(t-q) increase at the same time.

If we take logarithm of PC_t, then

$$\log PC_t = \log(1 + \frac{2q}{t-q}) + \log(Y_t' M_t Y_t).$$

when 2q is much smaller than t-q,

$$\log(1 + \frac{2q}{t-q}) \approx \frac{2q}{t-q} \approx \frac{2q}{t},$$

then

$$\log PC_t \approx \log(Y_t' M_t Y_t) + \frac{2q}{t} = AIC_t.$$

Therefore, if the available number of observations is much larger than the number of regressors, then both criteria will result in the same selection.

Due to the recursive method of computing regression parameters, we can also compute the sequence of the sums of squared residuals $Y_t' M_t Y_t$ ($t = p,\ldots,T-1$) easily. As a result, this method of sequential model selection should not cause any computational difficulties.

5 EMPIRICAL RESULTS

In this section, we describe the results of an empirical study designed to illustrate some of the major features of the theoretical arguments presented in previous sections. To begin, we employ quarterly forecasts of Taiwan's GDP and its components to examine the preference ordering of Methods (A, B, C) as well as (A', B', C') under G&R's *ex-post within-sample* sum of squared errors and Liang/Ryu's *ex-ante out-of-sample* sum of squared prediction errors (see sections 2 and 3), respectively. We find that under the Liang/Ryu setup, the preference ordering of the three methods indeed breaks down. We thus proceed to study the performance of using Akaike's (1973) information or Amemiya's (1980) prediction criterion to guide one to select combining regression at each point in time, and compare the resulting prediction error sum of squares from each of Methods A', B', and C'.

5.1 The Data

Forecasts on real quarterly GDP and its components for Taiwan have been made on a regular basis by the Directorate-General of Budget Accounting and Statistics (DGBAS) and the Institute of Economics of the Academia Sinica (IEAS) since the early eighties. Both forecast series are obtained from the Keynesian-type macroeconometric models, with around 40 behavioral equations in each. While the DGBAS model contains a more detailed equation system of the government sector, the IEAS model emphasizes more on the foreign trade equations. Each forecasting time-series released by the two agencies, however, consists of two distinct sets. The first set (hereafter as *old data set*), covers observations 1980.3Q through 1987.4Q, and has been generated from the data compiled from the old 1952-based UN national accounting system. The second set (hereafter, as *new data set*), begins in 1988.3Q, and has been generated from the data compiled from the new 1968-based UN national accounting system. In the old data set, the DGBAS published its one-quarter ahead forecasts every quarter in its *Quarterly National Economic Trend, Taiwan Area, The Republic of China*, whereas the IEAS released its forecasts up to four-quarters ahead only once a year in its *Taiwan Economic Forecast and Policy*. In the new data set, again the DGBAS series is issued every quarter, whereas the IEAS series is released only once a year. Both series provide forecasts up to four-quarters ahead. Liang (1995) provides a brief description of the DGBAS and IEAS models and a critical evaluation of their forecast performances.

To enlarge the number of components in the combining regressions, a series of forecasts are obtained from a sequence of ARIMA models. These models are estimated using the data publicly available at the time for which the forecast is required. Thus, in the old data set, data from 1961.1Q to 1980.2Q are used to identify and estimate the first acceptable ARIMA model for each variable being forecasted. The model is then used to make a one-quarter ahead forecast (1980.3Q) and a set of forecasts up to four-quarters ahead (1980.3Q-1981.2Q) for each variable being considered. The procedures are repeated for the next one-quarter ahead forecast (1980.4Q, using the data set from 1961.1Q to 1980.3Q) and the next set of forecasts up to four-quarters ahead (1981.3Q-1982.2Q, using the data set covering 1961.1Q to 1981.2Q) and so on. Similarly, in the new data set, data from 1961.1Q to 1988.2Q are used to identify and estimate the first acceptable ARIMA model for each variable being forecasted. The model is then used to make a one-quarter ahead forecast (1988.3Q) and a set of forecasts up to four-quarters ahead (1988.3Q-1989.2Q) for each variable being considered. The procedures are repeated for the next one-quarter ahead forecast (1988.4Q, using the data set from 1961.1Q to 1988.3Q) and the next set of forecasts up to four-quarters ahead (1989.3Q-1990.2Q, using the data set covering 1961.1Q to 1989.2Q) and so on. Consequently, in both cases of old and new data sets, our ARIMA models generate both one-quarter ahead forecasts and sets of forecasts up to four-quarters ahead.

With DGBAS, IEAS and ARIMA forecasts available, we consider the following combinations. Specifically, for the old data set, we consider a combination of two series of one-quarter ahead forecasts (ARIMA & DGBAS) and another combination of two series of forecasts up to four-quarters ahead (ARIMA & IEAS); for the new data set, we run a combination of two series of one-quarter ahead forecasts (ARIMA & DGBAS), a combination of three series of forecasts up to four-quarters ahead (ARIMA, DGBAS & IEAS), and three combinations of two series of forecasts up to four-quarters ahead ((ARIMA & DGBAS),

(ARIMA & IEAS), (DGBAS & IEAS)).

5.2 Results

Our empirical results are summarized in Table 1 through Table 7. Each table contains six variables (real GDP (Y) and its five components, consumption (C), investment (I), government purchases (G), exports (X) and imports (M)), and presents three types of information obtained from the empirical study. In the upper part, SSE and the sum of squared prediction errors computed from various methods are recorded. Here, A_{GR}, B_{GR}, and C_{GR} refer to the *ex-post* sum of squared errors based on G&R's Method A, Method B, Method C, respectively. Similarly, A_{LR}, B_{LR}, and C_{LR} refer to the *ex-ante* sum of squared errors based on Liang/Ryu's Method A', Method B', and Method C', respectively. Finally, AIC and PC represent the *ex-ante* sum of squared prediction errors when Liang/Ryu's combining regressions are optimally selected according to Akaike's (1973) information criterion and Amemiya's (1980) prediction criterion, respectively. To see the advantage of using the information and prediction criteria, we can compare the sum of prediction errors obtained from these criteria with the corresponding error measures computed from other methods. To this end, we can compute the error reduction rate using, for instance, $[(A_{LR}\text{-}PC)/A_{LR}]100$.[7] In the lower part, selection frequency of each combining method when using Akaike's (1973) information criterion and Amemiya's (1980) prediction criterion is provided. Here, AGREE refers to the frequency that both the information and the prediction criteria indicated the same combining method should be used. AIC_A, AIC_B, and AIC_C refer to the frequencies that Method A', Method B', and Method C' have been selected under the information criterion, respectively. Similarly, PC_A, PC_B, and PC_C refer to the frequencies that Method A', Method B', and Method C' have been selected under the prediction criterion, respectively.

Tables 1 and 2 report the empirical results estimated from the old data set. In this situation, 30 *ex-ante* quarterly forecasts (1980.3Q through 1987.4Q) were initially available for each variable being considered. To implement the recursive regression technique of Liang/Ryu (see sections 2 and 3), we choose p=10. That is, the first combining weights have been generated based on the first ten observations (1980.3Q through 1982.4Q). Consequently, for each variable being considered, we can recursively produce 20 combined forecasts (1983.1Q through 1987.4Q). Additionally, because one purpose of this empirical study is to compare various performance measures in a comparable mode, we also use the data set covering the same sample period (1983.1Q through 1987.4Q) to compute G&R's *ex-post within-sample* sum of squared errors.

[7] Since these error reduction rates are easily attainable from the upper part of each table, they are not reported here.

Table 1. Combining ARIMA & DGBAS Forecasts (one-quarter ahead; 83.1Q - 87.4Q)

	SSE & sum of squared prediction errors (in millions of 1981 N.T. dollars)							
	A_{GR}	B_{GR}	C_{GR}	A_{LR}	B_{LR}	C_{LR}	AIC	PC
Y	1.193E9	1.566E9	1.191E9	1.839E9	1.517E9	2.055E9	8.262E8	8.262E8
C	1.533E8	1.596E8	1.413E8	1.527E8	1.473E8	1.754E8	1.243E8	1.243E8
I	5.756E9	5.879E9	5.596E9	4.965E9	4.759E9	5.630E9	3.744E9	3.744E9
G	6.074E7	6.316E7	6.010E7	5.415E7	5.374E7	6.258E7	4.783E7	4.783E7
X	5.399E9	6.254E9	5.369E9	6.457E9	6.768E9	7.878E9	4.703E8	4.703E8
M	6.808E9	6.917E9	6.284E9	5.789E9	4.515E9	7.605E9	2.625E9	2.625E9

	Frequency (T=20)						
	AGREE	AIC_A	AIC_B	AIC_C	PC_A	PC_B	PC_C
Y	20	4	8	8	4	8	8
C	20	4	4	12	4	4	12
I	20	2	10	8	2	10	8
G	20	5	5	10	5	5	10
X	20	7	5	8	7	5	8
M	20	5	2	13	5	2	13

Table 2. Combining ARIMA & IEAS Forecasts (up to four-quarters ahead; 83.1Q -87.4Q)

	SSE & sum of squared prediction errors (in millions of 1981 N.T. dollars)							
	A_{GR}	B_{GR}	C_{GR}	A_{LR}	B_{LR}	C_{LR}	AIC	PC
Y	4.171E9	6.711E9	4.057E9	1.810E11	6.428E9	2.592E11	3.279E9	3.279E9
C	4.301E8	5.244E8	3.870E8	2.131E8	1.547E8	4.741E8	9.880E7	9.880E7
I	8.277E9	9.274E9	8.207E9	7.583E9	7.963E9	8.381E9	5.594E9	5.594E9
G	1.305E8	1.306E8	1.295E8	1.177E8	1.109E8	1.292E8	9.564E7	9.564E7
X	1.947E10	2.037E10	1.813E10	2.085E10	2.027E10	2.475E10	1.459E10	1.459E10
M	2.139E10	2.177E10	1.864E10	2.147E10	1.733E10	2.303E10	1.047E10	1.047E10

	Frequency (T=20)						
	AGREE	AIC_A	AIC_B	AIC_C	PC_A	PC_B	PC_C
Y	20	6	7	7	6	7	7
C	20	5	3	12	5	3	12
I	20	4	7	9	4	7	9
G	20	7	1	12	7	1	12
X	20	7	2	11	7	2	11
M	20	8	1	11	8	1	11

Table 3 through 7 present the empirical results estimated from the new data set. In this case, 18 *ex-ante* quarterly forecasts (1988.3Q through 1992.4Q) were initially available for each variable being concerned. To implement the recursive regression technique of Liang/Ryu, we again run (3.1) using the first ten observations (1988.3Q through 1990.4Q) to generate the first combining weights. As a result, for each variable being considered, we can recursively create 8 combined forecasts (1991.1Q through 1992.4Q).

Table 3. Combining ARIMA & DGBAS Forecasts (one-quarter ahead; 91.1Q -92.4Q)

	SSE & sum of squared prediction errors (in millions of 1981 N.T. dollars)							
	A_{GR}	B_{GR}	C_{GR}	A_{LR}	B_{LR}	C_{LR}	AIC	PC
Y	1.186E9	1.189E9	1.182E9	1.578E8	1.370E8	3.566E8	1.313E8	1.313E8
C	1.416E9	1.422E9	1.406E9	4.534E8	3.082E8	5.881E8	2.897E8	2.897E8
I	3.956E9	4.013E9	3.585E9	2.157E9	1.884E9	2.054E9	1.558E9	1.558E9
G	1.941E8	1.941E8	1.332E8	2.044E9	2.001E9	1.948E9	1.981E9	1.981E9
X	1.783E9	2.190E9	1.781E9	1.125E9	1.210E9	1.862E9	1.004E9	1.004E9
M	1.933E9	1.937E9	1.919E9	1.816E9	1.291E9	2.473E9	1.210E9	1.210E9

	Frequency (T=8)						
	AGREE	AIC_A	AIC_B	AIC_C	PC_A	PC_B	PC_C
Y	8	1	1	6	1	1	6
C	8	1	2	5	1	2	5
I	8	2	1	5	2	1	5
G	8	3	1	4	3	1	4
X	8	1	2	5	1	2	5
M	8	2	3	3	2	3	3

Table 4. Combining ARIMA, DGBAS & IEAS Forecasts (up to four-quarters ahead; 91.1-92.4Q)

	SSE & sum of squared prediction errors (in millions of 1981 N.T. dollars)							
	A_{GR}	B_{GR}	C_{GR}	A_{LR}	B_{LR}	C_{LR}	AIC	PC
Y	2.087E9	2.573E9	7.341E8	1.335E9	1.020E9	8.336E8	1.210E8	1.210E8
C	1.989E9	1.995E9	1.646E9	1.380E9	1.128E9	1.126E9	8.687E8	8.687E8
I	3.279E9	3.284E9	3.073E9	2.033E9	1.670E9	1.986E9	1.539E9	1.539E9
G	8.857E7	1.532E8	6.591E7	4.843E7	1.093E8	2.837E7	1.632E7	1.632E7
X	5.559E9	5.925E9	4.773E9	6.917E9	6.355E9	7.604E9	5.449E9	5.449E9
M	3.980E9	4.099E9	3.900E9	3.693E9	2.542E9	4.715E9	2.312E9	2.312E9

	Frequency (T=8)						
	AGREE	AIC_A	AIC_B	AIC_C	PC_A	PC_B	PC_C
Y	8	2	1	5	2	1	5
C	8	2	1	5	2	1	5
I	8	1	1	6	1	1	6
G	8	4	2	2	4	2	2
X	8	2	1	5	2	1	5
M	8	0	2	6	0	2	6

Table 5. Combining ARIMA & DGBAS Forecasts (up to four-quarters ahead; 91.1Q- 92.4Q)

	SSE & sum of squared prediction errors (in millions of 1981 N.T. dollars)							
	A_{GR}	B_{GR}	C_{GR}	A_{LR}	B_{LR}	C_{LR}	AIC	PC
Y	2.732E9	2.781E9	2.590E9	9.268E8	8.256E8	2.261E9	5.880E8	5.880E8
C	2.074E9	2.078E9	1.875E9	1.185E9	8.549E8	1.272E9	6.231E8	6.231E8
I	4.171E9	4.284E9	3.635E9	2.744E9	2.058E9	2.445E9	1.346E9	1.346E9
G	2.485E8	2.486E8	7.049E7	1.579E8	7.136E7	2.956E7	8.778E6	8.778E6
X	7.862E9	8.243E9	7.728E9	1.012E10	9.272E9	1.254E10	9.330E9	9.330E9
M	4.397E9	4.437E9	4.359E9	3.345E9	2.754E9	4.883E9	2.745E9	2.745E9

	Frequency (T=8)						
	AGREE	AIC_A	AIC_B	AIC_C	PC_A	PC_B	PC_C
Y	8	3	2	3	3	2	3
C	8	3	0	5	3	0	5
I	8	2	2	4	2	2	4
G	8	5	0	3	5	0	3
X	8	2	2	4	2	2	4
M	8	1	0	7	1	0	7

Table 6. Combining ARIMA & IEAS Forecasts (up to four-quarters ahead; 91.1Q -92.4Q)

SSE & sum of squared prediction errors (in millions of 1981 N.T. dollars)

	A_{GR}	B_{GR}	C_{GR}	A_{LR}	B_{LR}	C_{LR}	AIC	PC
Y	2.266E9	2.602E9	8.281E8	8.317E8	7.534E8	4.716E8	1.347E8	1.347E8
C	2.025E9	2.027E9	1.675E9	1.200E9	8.653E8	1.273E9	6.370E8	6.370E8
I	3.337E9	3.340E9	3.049E9	2.028E9	1.641E9	1.867E9	1.464E9	1.464E9
G	5.514E8	6.136E8	2.711E8	3.171E8	3.217E8	5.986E7	4.176E7	4.176E7
X	9.046E9	1.131E10	8.848E9	5.429E9	7.015E9	8.396E9	4.743E9	4.743E9
M	5.516E9	5.659E9	5.415E9	4.138E9	2.264E9	5.692E9	2.262E9	2.262E9

Frequency (T=8)

	AGREE	AIC_A	AIC_B	AIC_C	PC_A	PC_B	PC_C
Y	8	2	1	5	2	1	5
C	8	3	0	5	3	0	5
I	8	2	1	5	2	1	5
G	8	4	2	2	4	2	2
X	8	2	2	4	2	2	4
M	8	0	1	7	0	1	7

Table 7. Combining DGBAS & IEAS Forecasts (up to four-quarters ahead; 91.1Q -92.4Q)

SSE & sum of squared prediction errors (in millions of 1981 N.T. dollars)

	A_{GR}	B_{GR}	C_{GR}	A_{LR}	B_{LR}	C_{LR}	AIC	PC
Y	7.445E9	7.451E9	2.460E9	4.446E8	2.060E8	2.828E9	1.271E8	1.271E8
C	3.875E9	3.878E9	2.691E9	5.951E8	4.625E8	1.800E9	4.186E8	4.186E8
I	3.364E9	3.366E9	3.223E9	1.650E9	1.422E9	1.656E9	1.385E9	1.385E9
G	2.718E8	2.811E8	9.719E7	7.320E7	5.976E7	5.693E7	7.894E6	7.894E6
X	9.026E9	9.987E9	7.566E9	8.351E9	8.154E9	7.136E9	6.571E9	6.571E9
M	4.801E9	5.259E9	4.774E9	4.446E9	3.450E9	5.666E9	2.872E9	2.872E9

Frequency (T=8)

	AGREE	AIC_A	AIC_B	AIC_C	PC_A	PC_B	PC_C
Y	8	0	3	5	0	3	5
C	8	1	1	6	1	1	6
I	8	1	1	6	1	1	6
G	8	2	3	3	2	3	3
X	8	2	2	4	2	2	4
M	8	0	3	5	0	3	5

These tables offer the following results:

1. Using the *ex-post within-sample* sum of squared residuals, the preference ordering based on G&R's method still follows the pattern of Method C → Method A → Method B, an apparent result. Nevertheless, there is no systematic pattern of the preference ordering when using the sensible *ex-ante post-sample* sum of squared prediction errors. Furthermore, except for two rare cases in predicting government purchases (G) in Table 3 and exports (X) in Table 5, the variable combining method using the AIC or PC criterion reduces the *ex-ante* sum of squared prediction errors. In other words, the *variable* combining method using the AIC or PC criterion is preferred to any *fixed* recursive regression technique, A', B', or C'. These findings can be seen more vividly by examining the error reduction rates mentioned previously. For example, in Table 1, by comparing the PC criterion with the Method C', the sum of squared prediction errors for Y is reduced by 59.80%. Accordingly, these empirical results are consistent with our theoretical predictions.

2. In the upper part of each table, we find that the AIC and PC criteria do equally well. That is, in all cases under study, both criteria produced the same sum of squared prediction errors in all cases. This result is also reflected in the lower part of the table which shows that AGREE = T-p, implying that

both AIC and PC criteria picked the same combining method throughout the forecast periods.

Overall, Tables 1-7 lead us to conclude that using recursive methods under information or prediction criterion helps one to select the best form of combining regression and thus to reduce the prediction error. Accordingly, using *ex-post* samples to run G&R's regressions and then utilizing their SSE estimates to rank Methods A, B, or C is inappropriate. Judging by our theoretical and empirical results, our new proposal clearly provides a promising alternative in specifying and selecting the best regression form in the study of combining forecasts.

6 Closing Remarks

The central notion presented in this paper is that, the *very motivation in applying composite forecasts is to improve the predictive accuracy*, not *the within sample fitting performance.* In this connection, this paper reformulates the basic G&R's *within-sample* combining regression models in a new form based on the considerations of *out-of-sample* predictive accuracies. Using recursive regression techniques, an algorithm to estimate combining weights is developed. Under the new setup, we show that G&R's specific preference ordering, Method C → Method A → Method B, breaks down. To overcome this lack of ordering, this paper suggests one to use Akaike's (1973) information or Amemiya's (1980) prediction criterion to select the best form of combining regressions recursively. Empirical results, obtained by using multiple data sets of macroeconomic forecasts for Taiwan, confirm the validity of the theoretical arguments. Consequently, judging by our theoretical and empirical results, our new proposal clearly provides a promising alternative in specifying and selecting the best regression form in the study of combining forecasts.

References

[1] Akaike, H. Information Theory and an Extension of the Maximum Likelihood Principle, pp. 267-281 in B.N. Petrov and F. Csaki, (eds)., *Second International Symposium on Information Theory*, Budapest: Akademiai Kiado, 1973.

[2] Amemiya, T. Selecting of Regressors. *International Economic Review*, 21:331-345, 1980.

[3] Bates, J.M. and C.W.J. Granger. The combination of Forecasts. *Operations Research Quarterly*, 20:451-468, 1969.

[4] Clemen, R.T. Linear Constraints and the Efficiency of Combined Forecasts. *Journal of Forecasting*, 5:31-38, 1986.

[5] Clemen, R.T. and R.L. Winkler. Combining Economic Forecasts. *Journal of Business & Economic Statistics*, 4:39-46, 1986.

[6] Granger, C.W.J. and R. Ramanathan. Improved Method of Combining Forecasts. *Journal of Forecasting*, 3:197-204, 1984.

[7] Gunter, S.I. Nonnegativity Restricted Least Squares Combinations. *International Journal of Forecasting*, 8:45-59, 1992.

[8] Gupta, S. and P.C. Wilton. Combination of Forecasts: An Extension. *Management Science*, 33:356-372, 1987.

[9] Judge, G.G. and M.E. Bock. *The Statistical Implications of Pre-test and Stein-rule Estimators in Econometrics*, Amsterdam: North-Holland. 1978.

[10] Liang, K.-Y. On the Sign of the Optimal Combining Weights under the Error-variance Minimizing Criterion. *Journal of Forecasting*, 11:719-723, 1992.

[11] Liang, K.-Y. and J.S. Gau. Specification of the Combining Regressions under Stochastic Regressors. *Journal of the Chinese Statistical Association*, 32:367-388 (in Chinese with English Abstract), 1994.

[12] Liang, K.-Y. and K. Ryu. On the Forecast-encompassing Test and its Application to the Combination of Forecasts. *Manuscript*, Department of Economics, National Tsing Hua University and University of California-Los Angeles. 1994.

[13] Liang, K.-Y. and Y.C. Shih. Bayesian Composite Forecasts: An Extension and A Clarification. *Journal of Quantitative Economics*, 10:105-122, 1994.

[14] Liang, K.-Y. A Critical Evaluation of Macroeconomic Forecasting in Taiwan. *Taiwan Economic Review*, 23:43-82 (in Chinese with English Abstract), 1994.

[15] Makridakis, S. Sliding Simulations: A New Approach to Time Series Forecasting. *Management Science*, 36:505-512, 1990.

[16] Newbold, P. and C.W.J. Granger. Experience with Forecasting Univariate Time Series and the Combination of Forecasts (with discussions). *Journal of the Royal Statistical Society*, Series A 137:131-149, 1974.

[17] *Quarterly National Economic Trends, Taiwan Area, The Republic of China*, Taipei: Directorate-General of Budget, Accounting and Statistics, (various issues, in Chinese).

[18] *Taiwan Economic Forecast and Policy*, Taipei: Institute of Economics, Academia Sinica, (various issues, in Chinese).

[19] Trenkler, G. and E.P. Liski. Linear Constraints and the Efficiency of Combined Forecasts. *Journal of Forecasting*, 5:197-202, 1986.

Multiperiod-Ahead Predictive Densities and Model Comparison in Dynamic Models

Chung-ki Min
George Mason University, Fairfax, VA 22030

Abstract

This study proposes a new model comparison method which uses multiperiod-ahead predictive densities. Use of predictive densities allows us to incorporate the uncertainties associated with point forecasts in comparing the forecasting performances of various models. In the special case where one-period-ahead predictive densities are used, the method is equivalent to the Bayesian posterior odds. This new model-comparison method is contrasted to other measures of forecasting performance, such as the mean squared error, which don't consider the uncertainties associated with point forecasts. To evaluate the multiperiod-ahead predictive densities in dynamic models, this study uses simulation methods.

1 Introduction

It is widely agreed that model comparison methods should be determined by the intended use of models. In many applications, prediction is a primary purpose for the chosen models, and it is therefore desirable to use prediction criteria for model comparison. Use of prediction criteria has been suggested by many researchers including Friedman and Schwartz (1991), Geisser (1975), Gelfand et al. (1992), and Zellner (1994).

Many prediction criteria in the literature calculate a single summary number, such as the mean squared forecast error, using point forecasts obtained from each model. However, use of only point forecasts may not be appropriate for model comparison since they may not represent well what models tell about future outcomes. Point forecasts are decision variables which are determined to minimize the expected value of a certain loss function, and could be different depending on loss functions even with the same model (Zellner (1986)). Therefore, it would be more appropriate if we base our model comparison on the models' information about future outcomes, rather than on the optimal point forecasts determined by loss functions. Predictive densities are one form of the summarized models' information about future outcomes, and their use is thus proposed for model comparison in this study.

Predictive densities can help us determine which model produces more useful information for our decision making, considering the uncertainties of the models' predictions as well. Suppose that there are two models whose predictive densities for a future outcome have the same mean, but different variances. For a squared-error loss function, their optimal point forecasts are the predictive means which are the same in this example. Then, the two models will be evaluated the same if we use criteria which are based on point forecasts only. However, the two models provide users with different information about future outcomes, and should therefore be evaluated as different. For example, if the actual value happens to be near the predictive mean, the model with a smaller variance should be favored because it accurately predicted the actual value with a tight predictive density. On the other hand, if the actual value is far away from the predictive mean, the model with a larger variance should be favored because its flatter predictive density would correctly warn users against possible large deviations from the predictive mean. This model comparison method would be found useful when correct prediction of the uncertainty about future outcome is required, particularly when speculative prices are modeled.

The implementation of the proposed method requires evaluation of predictive densities. However, multiperiod-ahead predictive densities in dynamic models may not be derived analytically (Chow (1973), and Thompson and Miller (1986)). Alternatively, this study employs simulation methods for the calculations required by the model comparison method.

In sum, the objective of this study is to propose a model comparison method which can help us determine which model, among a given set of models, produces the most useful information about future outcomes. It is suggested that predictive densities be used for model comparison to consider the uncertainties, or variances, of the models' predictions. To implement the method, simulation methods would be useful for evaluating predictive densities.

This article is organized in the following way. A model-comparison method is proposed in Section 2, and simulation methods for performing the required calculations are explained in Section 3. In Section 4, the method is applied to simulated and real data. Finally, brief concluding remarks are made in Section 5.

2 Model Comparison using Predictive Densities

Use of predictive densities for model comparison has been suggested in the literature. In Geisser (1975) and Gelfand et al. (1992), cross-validation versions of predictive densities are used for model comparison.[1] In Geweke (1994) and Min and Zellner (1993), time-series models are compared using the Bayesian posterior odds which can be expressed as the product of the ratios of one-period-ahead predictive densities. This study focuses on time-series models and thus extends the methods used in Geweke (1994) and Min and Zellner (1993) to multiperiod-ahead forecasting problems.

For any two time-series models, M_1 and M_2, the Bayesian posterior odds, K_{12}, may be expressed as the product of the ratios of one-period-ahead predictive densities (Min and Zellner

[1]Geisser (1975) chooses one among a given set of models which maximizes the product of cross-validation predictive densities. That is, the best model maximizes $\Pi_{\tau=1}^{N} p(y_\tau | Y_{(\tau)}, I_i, M_i)$ where $Y_{(\tau)}$ represents a data set with the τth observation omitted, i.e., $(y_1, \cdots, y_{\tau-1}, y_{\tau+1}, \cdots, y_N)$, and I_i represents the prior information under a model M_i. However, the cross-validation predictive densities are not well-defined in dynamic models because of lagged values of y_τ present in subsequent periods. Therefore, for comparison of dynamic models this study uses k-period-ahead predictive densities instead of the cross-validation predictive densities, replacing $Y_{(\tau)}$ with $(y_1, \cdots, y_{\tau-k})$. In this sense, this model-comparison method may be viewed as an application of the cross-validation idea to comparison of dynamic models.

(1993)):

$$
\begin{aligned}
K_{12} &= \frac{\Pr(M_1|y_t, y_{t-1}, \cdots, y_1, I_1)}{\Pr(M_2|y_t, y_{t-1}, \cdots, y_1, I_2)} \\
&= \pi_{12} \times \frac{p(y_t, y_{t-1}, \cdots, y_1|I_1, M_1)}{p(y_t, y_{t-1}, \cdots, y_1|I_2, M_2)} \\
&= \pi_{12} \times \prod_{\tau=1}^{t} \frac{p(y_\tau|y_{\tau-1}, \cdots, y_1, I_1, M_1)}{p(y_\tau|y_{\tau-1}, \cdots, y_1, I_2, M_2)}
\end{aligned}
\tag{1}
$$

where π_{12} is the prior odds, and I_1 and I_2 represent the prior information for the two models, respectively. The last expression indicates that the posterior odds uses one-period-ahead predictive densities for model comparison. In other words, it compares models using the ordinates of one-period-ahead predictive densities evaluated at the realized values of y_τ's, thereby having the advantage of considering the uncertainties of models' predictions as well.

The above method for model comparison may be extended to a case of comparing multi-period-ahead forecasting models, suggesting that k-period-ahead predictive densities be used for model comparison. Suppose that we are interested in comparing two k-period-ahead forecasting models, M_1 and M_2. We first calculate the k-period-ahead predictive ratio, $r_{t,k}$, which is the ratio of the two predictive densities for y_t conditional on the information available in period $t - k$:

$$
r_{t,k} = \frac{p(y_t|D_{t-k}, I_1, M_1)}{p(y_t|D_{t-k}, I_2, M_2)}
\tag{2}
$$

where $D_{t-k} = (y_1, \cdots, y_{t-k})$ is the data information. After obtaining the k-period-ahead predictive ratios for all the periods of our interest, e.g., from $t_0 + k$ to T, we calculate the k-period-ahead cumulative predictive ratio as follows:

$$
\begin{aligned}
R_{T,k} &= \prod_{t=t_0+k}^{T} r_{t,k} \\
&= R_{T-1,k} \times r_{T,k}
\end{aligned}
\tag{3}
$$

where the cumulative ratio is calculated using the ratios from period $t_0 + k$ to period T. In this calculation, the observations up to period t_0 are used only for deriving posterior densities for model parameters, and may be called a 'training sample' (Berger and Pericchi (1992), Geweke (1994), and Min and Zellner (1993)). The use of a training sample ensures that the posterior densities for model parameters are proper even with improper diffuse priors, and that predictive densities are therefore proper. Concerning the choice of training-sample size, we may use the smallest number of data points such that the posterior densities are proper. If we are more interested in later periods, e.g., from $t_0 + k$ to T, we would consider using the first t_0 data points as a training sample in comparing k-period-ahead forecasting models. By increasing the size of the training sample, we will be able to reduce the effects of prior information on the posterior densities.

The cumulative predictive ratio measures the relative likelihood of realized values under the two models considered.[2] In a special case of using one-period-ahead densities, i.e., k=1,

[2]When there are more than two models to be compared, we can calculate the cumulative predictive ratio for all possible pairs and choose the best model.

the cumulative predictive ratio becomes the Bayes factor of the posterior odds. It is easy to update the cumulative predictive ratio with new observations. Suppose we have just observed the value of y_T in period T. With this new observation, we calculate $r_{T,k}$, the ratio of k-period-ahead predictive densities evaluated at the realized value of y_T. Note that the k-period-ahead predictive densities for y_T are conditioned on the information available in period $T-k$. Then, we obtain an updated k-period-ahead cumulative predictive ratio, $R_{T,k}$, by multiplying the previously-calculated cumulative predictive ratio, $R_{T-1,k}$, by the newly-calculated predictive ratio $r_{T,k}$. For future updating, all we need to save in period T is the most recent cumulative predictive ratio, i.e., $R_{T,k}$.

This model-comparison method suggests that appropriate predictive densities be used depending on the forecasting horizon of our interest. For example, if we are interested in forecasting k-period-ahead outcomes, we need to use k-period-ahead predictive densities in calculating the cumulative predictive ratio. It is therefore possible that a different model is selected for each forecasting horizon. Since no true model is assumed to exist in the Bayesian framework, this doesn't cause any contradiction. In fact, it has been reported in the literature on forecasting models that time-series models performed better for short horizons, while structural models performed better for long horizons, implying that use of a single model for all forecasting horizons might not be an optimal strategy. However, if we want to select one model for several forecasting horizons together, it is suggested to use joint predictive densities for the forecasting horizons, e.g., $p(y_{t+1}, y_{t+2}, \ldots, y_{t+k}|D_t, I_i, M_i)$, to evaluate the cumulative predictive ratio.

The method proposed above is designed to select one from a given set of models which produces the most useful information about future outcomes, but not to determine optimal point forecasts. Optimal point forecasts are decision variables which minimize the expected values of loss functions under a chosen model. Therefore, once a model is selected by the proposed method, optimal forecasts may be determined for any loss functions.

Caution should be exercised because there can be cases in which a few bad predictions influence model selection.[3] By examining the k-period-ahead predictive ratios, $r_{t,k}$'s, for the entire period of our interest, we would be able to identify bad predictions, if any. Good models are expected to produce large values of $r_{t,k}$ for the entire period, and to maximize the cumulative predictive ratio. Very small values of $r_{t,k}$ for certain periods indicate that a model needs to be improved for those periods. Thus, the information about $r_{t,k}$'s can be used for a model diagnostic.

3 Evaluation of Predictive Densities Using Simulation Methods

The calculations of the k-period-ahead cumulative predictive ratio require evaluations of k-period-ahead predictive densities, $p(y_{t+k}|D_t, I_i, M_i)$. Consider the following dynamic model, M_1, for a three-period-ahead predictive density,

$$\begin{aligned} y_{t+3} &= \beta_0 + \beta_1 y_{t+2} + \epsilon_{t+3} \\ &= \beta_0(1 + \beta_1 + \beta_1^2) + \beta_1^3 y_t + \beta_1^2 \epsilon_{t+1} + \beta_1 \epsilon_{t+2} + \epsilon_{t+3} \end{aligned} \tag{4}$$

[3] I am grateful to an editor and a referee for pointing out this issue.

Figure 1
Predictive Densities in a Dynamic Model

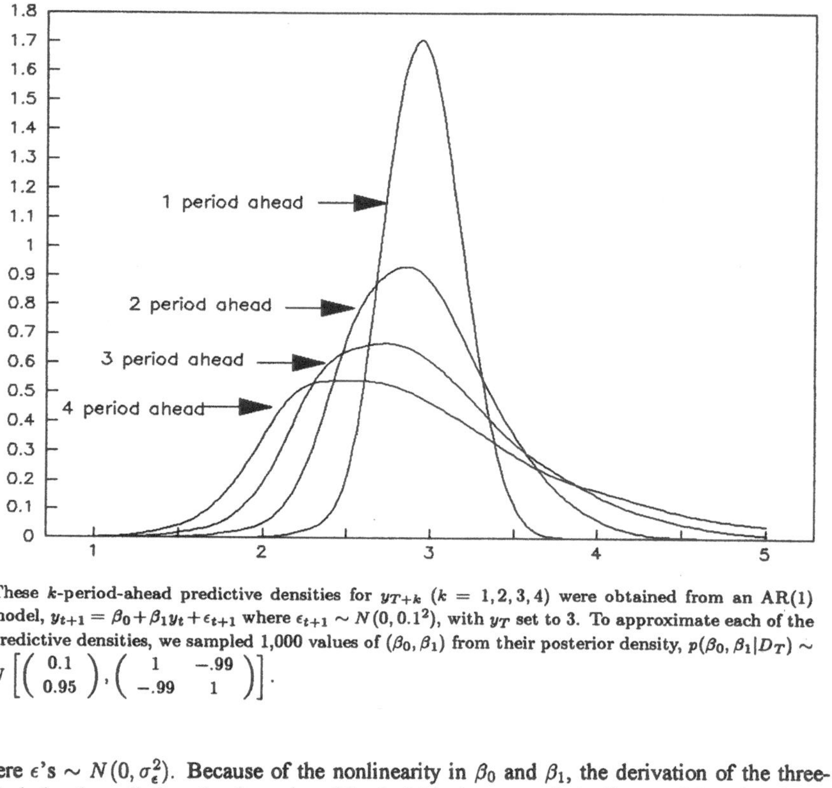

These k-period-ahead predictive densities for y_{T+k} ($k = 1, 2, 3, 4$) were obtained from an AR(1) model, $y_{t+1} = \beta_0 + \beta_1 y_t + \epsilon_{t+1}$ where $\epsilon_{t+1} \sim N(0, 0.1^2)$, with y_T set to 3. To approximate each of the predictive densities, we sampled 1,000 values of (β_0, β_1) from their posterior density, $p(\beta_0, \beta_1 | D_T) \sim N\left[\begin{pmatrix} 0.1 \\ 0.95 \end{pmatrix}, \begin{pmatrix} 1 & -.99 \\ -.99 & 1 \end{pmatrix}\right]$.

where ϵ's $\sim N(0, \sigma_\epsilon^2)$. Because of the nonlinearity in β_0 and β_1, the derivation of the three-period-ahead predictive density, $p(y_{t+3}|D_t, I_1, M_1)$, is not analytically possible. As shown in Figure 1, multiperiod-ahead predictive densities in dynamic models are skewed and not well-known distributions (Chow (1973), and Thompson and Miller (1986)). Alternatively, this study employs a simulation method for the calculations required by the model comparison method.

The simulation method used in this study exploits the ease of drawing a sample from k-period-ahead conditional predictive densities, $p(y_t|\underline{\beta}, D_{t-k}, I_i, M_i)$. To evaluate (unconditional) predictive densities $p(y_t|D_{t-k}, I_i, M_i)$, we need to integrate out model parameters $\underline{\beta}$, i.e.,

$$p(y_t|D_{t-k}, I_i, M_i) = \int p(y_t|\underline{\beta}, D_{t-k}, I_i, M_i) \cdot p(\underline{\beta}|D_{t-k}, I_i, M_i)\, d\underline{\beta}. \tag{5}$$

Since this integration cannot be done analytically, we alternatively draw a sample from the posterior density for $\underline{\beta}$, $p(\underline{\beta}|D_{t-k}, I_i, M_i)$, and then evaluate the conditional predictive density with that sample,

$$\hat{p}(y_t^a|D_{t-k}, I_i, M_i) = \sum_{j=1}^{N} p(y_t^a|\underline{\beta}^{(j)}, D_{t-k}, I_i, M_i)/N \tag{6}$$

where y_t^a is the observed value of y_t, and $\underline{\beta}^{(j)}$ are drawn values from the posterior density for $\underline{\beta}$, $p(\underline{\beta}|D_{t-k}, I_i, M_i)$. In many cases including linear and nonlinear dynamic models, the conditional predictive density $p(y_t^a|\underline{\beta}, D_{t-k}, I_i, M_i)$ is easy to evaluate. Although posterior densities for $\underline{\beta}$ cannot be derived analytically in some cases, simulation methods such as the Gibbs sampler allow us to draw a sample from posterior densities, and therefore expand our ability to apply the model-comparison method.

Concerning the convergence checking of the simulation methods, this study plots calculated ordinates of predictive densities over sampling iterations. If these calculated ordinates show little variation with additional iterations, we may conclude that the simulation method has converged to produce an accurate estimate of the ordinates of predictive densities.

4 Applications

4.1 Simulated Data

This application to simulated data is to show that the cumulative predictive ratio can detect data-mining problems, thereby keeping us from choosing wrong models. Belsley (1988) notes that a close fit to existing data provides no guarantee for good forecasts in novel situations. As illustrated below, arbitrarily fitted models can produce very tight predictive densities, implying that the models are unreasonably confident about future outcomes.

Data were generated using an autoregressive model of order one, AR(1), for $t = 1, \cdots, 100$,

$$y_t = \beta_0 + \beta_1 y_{t-1} + \epsilon_t \tag{7}$$

with $y_0 = 0$, $\beta_0 = 0.1$, $\beta_1 = 0.95$, and $\epsilon_t \sim N(0, 0.1^2)$. Suppose we need to select one among a given set of models which can provide the most useful information about 4-period-ahead values of y_t for the period from $t = 84$ to 100. The first 80 observations (t=1 to 80) were used as a training sample; in other words, they were used only for deriving the posterior densities for (β_0, β_1)[4] And, the model comparison was based on the cumulative predictive ratio calculated using the observations from $t = 84$ to 100. Since 80 observations were used as a training sample, prior densities had no significant influence on the posterior and predictive densities, and therefore no significant influence on the model comparison as well.[5]

Two models were compared: one is an AR(1), the model used for data generation, and the other is a 'data-mining' (DM) model which is a time trend model,

$$y_t = \alpha_0 + \alpha_1 \cdot t + u_t \qquad u_t \sim N(0, \sigma_u^2) \tag{8}$$

[4]Use of the first 80 observations as a training sample might be arbitrary. If our interest is not limited to the forecasting performance for the period of t=84 to 100, we may alternatively use several training samples of different sizes and select a model which performs well for all the training samples considered.

[5]This application used a diffuse prior $p(\beta_0, \beta_1, \sigma) \propto 1/\sigma$.

Figure 2
Four-period-ahead Predictive Densities for Simulated Data

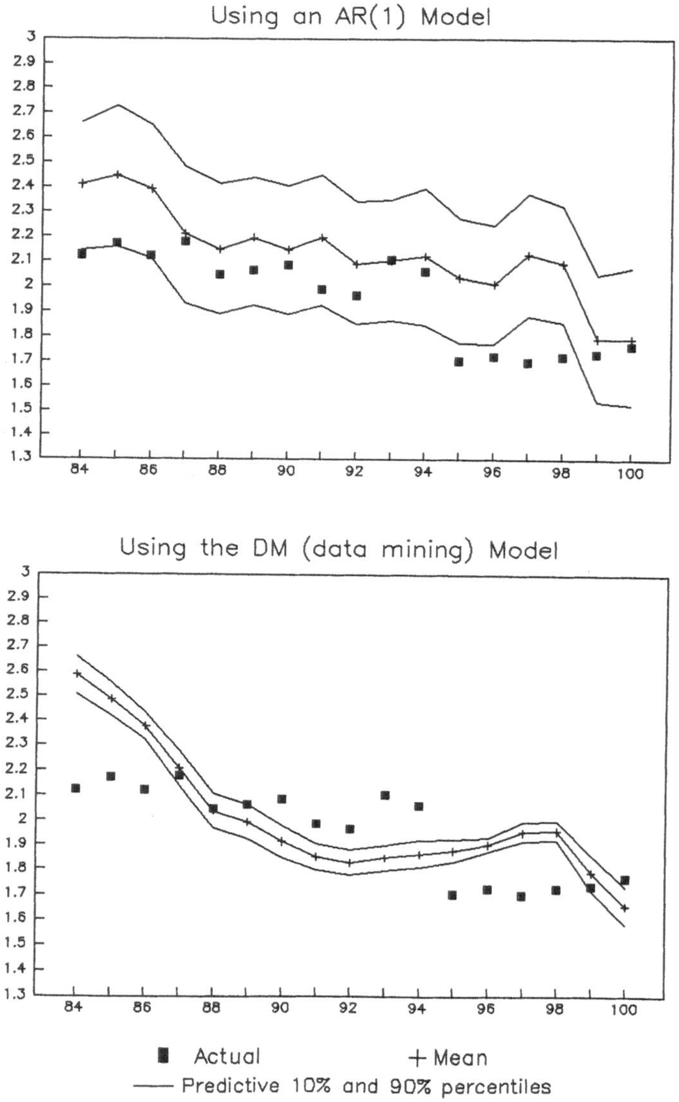

Table 1
Predictive and Cumulative Predictive Ratios of Four-period-ahead Predictive Densities for Simulated Data

Period	Ordinates of Predictive Densities[a]		Predictive Ratio($r_{t,4}$)[c]
(t)	AR(1) Model	DM Model[b]	DM/AR(1)
84	0.79	8.37×10^{-14}	1.07×10^{-13}
85	0.84	1.51×10^{-9}	1.79×10^{-9}
86	0.81	3.10×10^{-9}	3.83×10^{-9}
87	1.95	7.10	3.64
88	1.74	9.28	5.33
89	1.62	2.58	1.59
90	1.90	2.23×10^{-3}	1.18×10^{-3}
91	1.16	3.38×10^{-3}	2.93×10^{-3}
92	1.68	2.61×10^{-3}	1.56×10^{-3}
93	2.01	8.06×10^{-11}	4.02×10^{-11}
94	1.94	4.27×10^{-6}	2.20×10^{-6}
95	0.51	1.11×10^{-6}	2.18×10^{-6}
96	0.69	5.47×10^{-15}	7.90×10^{-15}
97	0.19	5.33×10^{-16}	2.79×10^{-15}
98	0.34	7.43×10^{-14}	2.17×10^{-13}
99	1.86	3.74	2.02
100	1.91	0.58	0.30

Cumulative Predictive Ratio ($R_{100,4}$): 6.76×10^{-101}

[a] These are the ordinates of four-period-ahead predictive densities, $p(y_t|D_{t-4}, I_i, M_i)$, evaluated at the actual values.

[b] It represents the 'data-mining' model, explained in the text.

[c] The predictive ratio is the ratio of ordinates of predictive densities, i.e., $p(y_t|D_{t-4}, I_{DM}, DM)/p(y_t|D_{t-4}, I_{AR(1)}, AR(1))$. Small ratios favor the AR(1) against the DM model.

with a moving window of width 10 periods.[6] Figure 2 shows the actual values and the four-period-ahead predictive densities obtained from the two models. As expected, the DM model produced unreasonably tight predictive densities due to data mining, thereby failing to provide us with useful information about future outcomes. In contrast, the AR(1) model produced reasonably wider predictive densities, thereby having us prepare for the uncertainty about future outcomes. These differences are well reflected in the the cumulative predictive ratio. As reported in Table 1, the AR(1) model was favored against the DM model, the cumulative predictive ratio being 6.76×10^{-101}. It is contrasted to the root mean squared error (RMSE) criterion which favored the DM model (RMSE= 0.211) against the AR(1) model (RMSE=0.224).[7]

[6] It means that the predictive densities are derived using the 10 most recent observations only.

[7] Since the predictive and cumulative predictive ratios consider both the means and variances of predictive densities by evaluating the ordinates, a model is not necessarily favored even though it produces predictive means closer to realized values than other models. For example, the DM model was not favored in periods 95 through 98 even though its predictive means were closer to the realized values than the AR(1) model's. In periods 89 and 99 the predictive means of the AR(1) model were as close to the realized values as those of the DM model.

Table 1 also reports the ordinates of the four-period-ahead predictive densities for each period. The ordinate for each period may be used for a model diagnostic since unusually small values indicate that the observations are improbable under the model. In the simulated data, the DM model had very low ordinates for many periods, indicating that the DM model was not a good description of the data. Further, the period-by-period predictive ratios show that the model selection conclusion was not influenced by a few bad predictions.

4.2 Diffusion Models

In the marketing literature, many theoretical and empirical models have been developed to forecast the number of adoptions of new durable products. Among them, the diffusion theory proposed by Bass (1969) has been widely employed. Focusing on interpersonal communication effects, the diffusion theory says that the probability that an individual will adopt one product in period t given that no adoption has yet been made is proportional to the number of previous adoptions in the market,

$$\frac{X_t}{M - S_{t-1}} = p + \frac{q}{M} S_{t-1} \tag{9}$$

where X_t is the number of adoptions made in period t; S_{t-1} is the accumulated number of adoptions made until $t-1$; and M is the total potential adopters in the market. The following statistical model was derived from (9) and used for estimation and forecasting,

$$X_t = \alpha_0 + \alpha_1 S_{t-1} + \alpha_2 S_{t-1}^2 + u_t \tag{10}$$

where $\alpha_0 = Mp$, $\alpha_1 = q - p$, $\alpha_2 = -q/M$, and u_t is the disturbance term of a normal distribution $N(0, \sigma_u^2)$.

An alternative model was proposed in Min (1995) to consider the impact on the adoption timing of its price variable, P_t, which is omitted in the Bass model, in addition to the interpersonal communication effects. The model also allows the coefficients to follow a vector random walk process, and is called the time-varying coefficient (TVC) model,

$$\frac{S_t}{M_t} = \beta_{0t} + \beta_{1t}\frac{S_{t-1}}{M_{t-1}} + \beta_{2t}\log P_t + e_t, \qquad e_t \sim N(0, \sigma^2) \tag{11}$$

$$\begin{bmatrix} \beta_{0t} \\ \beta_{1t} \\ \beta_{2t} \end{bmatrix} = \begin{bmatrix} \beta_{0,t-1} \\ \beta_{1,t-1} \\ \beta_{2,t-1} \end{bmatrix} + \begin{bmatrix} \eta_{0t} \\ \eta_{1t} \\ \eta_{2t} \end{bmatrix}, \qquad \begin{bmatrix} \eta_{0t} \\ \eta_{1t} \\ \eta_{2t} \end{bmatrix} \sim N(\underline{0}, \phi\sigma^2 I)$$

where M_t represents the number of potential adopters which may change over time.[8] When the hyperparameter ϕ is equal to zero, this model becomes a fixed-coefficient version of the model. With the time-varying coefficients, the model discounts the weights of old observations, thereby being able to adapt quickly to new information. In addition, since forecasts need to be made when only a few observations are available, models have to be simple and therefore include only a small number of characteristics that are believed to be central to adoption timing. In our model, variables other than price and cumulative adopters are omitted. Use

However, the AR(1) model had large predictive variances, and was not favored by the predictive ratio criterion.

[8] In this application, M_t was proxied by the total number of households in period t.

Figure 3
One-year-ahead Predictive Densities for Clothes Dryers

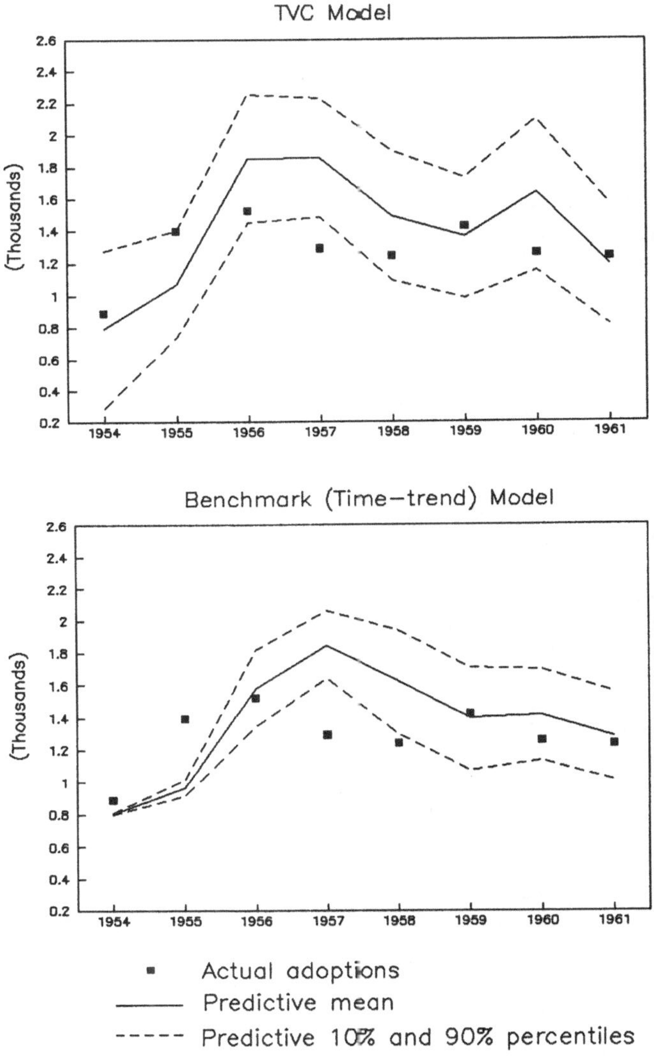

Figure 3 (continued)

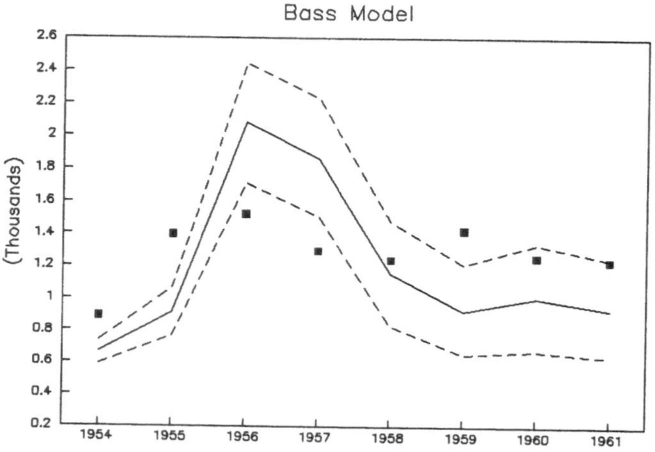

Table 2
Predictive and Cumulative Predictive Ratios of One-year-ahead Predictive Densities for Clothes Dryers

Year	Ordinates of Predictive Densities ($\times 10^{-4}$)[a]			Predictive Ratio ($r_{t,1}$)	
(t)	TVC[b]	Benchmark[c]	Bass[d]	TVC/Benchmark	TVC/Bass
1954	12.3440	0.0002	0.3640	57953.05	33.91
1955	4.6730	0.0004	0.0980	11682.50	47.68
1956	6.3350	23.0110	1.4740	0.28	4.30
1957	2.0010	0.1680	1.3780	11.91	1.45
1958	9.6420	3.9220	16.4700	2.46	0.59
1959	14.4870	17.6320	1.0040	0.82	14.43
1960	3.7830	13.8270	8.9830	0.27	0.42
1961	13.2300	19.7900	6.4160	0.67	2.06
	Cumulative Predictive Ratio ($R_{1961,1}$):			8.20×10^{8}	7.40×10^{4}

[a] These are the ordinates of one-year-ahead predictive densities, $p(y_t|D_{t-1}, I_i, M_i)$, evaluated at the actual values.

[b] It represents the TVC model, as in (11).

[c] It represents the time-trend model, as in (12).

[d] It represents the Bass model, as in (10).

of time-varying coefficients would alleviate the effects of omitted variables and specification errors.

Along with the above two models, the following time-trend model is also used as a benchmark model,

$$X_t = \gamma_0 + \gamma_1 t + \gamma_2 t^2 + \xi_t, \qquad \xi_t \sim N(0, \sigma_\xi^2) \tag{12}$$

where X_t is the number of adoptions made in period t.

Suppose that we need to select one among the three models which can provide the most useful information about one-year-ahead adoptions of clothes dryers.[9] Given the data for the period of 1949-1961, we began to produce one-year-ahead predictive densities after five years of observations were made. In other words, we used the first five observations as a training sample.[10] In the following years, models were re-estimated as additional observations were added.

When we used a predictive mean as a point forecast, the mean squared errors were 850 for the TVC, 1,159 for the Bass, and 817 for the benchmark model. Similar forecasting performance was observed when the mean absolute percentage forecast error was used; 19.4% for the TVC, 28.4% for the Bass, and 16.9% for the benchmark model. Based on these criteria, the benchmark model was the best, followed by the TVC model and then by the Bass model. However, the cumulative predictive ratio, $R_{1961,1}$, resulted in a different ranking of the three models. The ratio between the TVC and the benchmark model was 8.20×10^8, and the ratio between the TVC and the Bass model was 7.40×10^4, thus favoring the TVC model as the best and the Bass as the second.

Figure 3 shows why the cumulative predictive ratio favored the TVC as the best. As compared to the TVC model, the benchmark model produced very tight predictive densities in early periods. Even though its predictive means were close to the actual values, the benchmark model was heavily penalized and the ordinates of the tight predictive densities were very low. This is because the benchmark model was overly confident in its predictive means, while the TVC model provided users with useful information about possible large deviations from them. Similarly, the Bass model produced very tight predictive densities and failed to include the actual adoptions in the 80% predictive intervals in many periods. In this sense, the cumulative predictive ratio ($R_{T,1}$) and the yearly-calculated predictive ratios ($r_{t,1}$'s) can help us select models which produce useful information for correct decision making.

The selection of the TVC model is not sensitive to the choice of training sample. In Table 2, it appears that the benchmark model is not selected because of two poorly predicted observations for 1954 and 1955. However, when we add the two observations to the training sample and calculate the cumulative predictive ratio for the period of 1956-1961, the TVC model is still selected with the cumulative predictive ratio of 1.22.

Some might suggest that the benchmark model be used for forecasting adoptions beyond 1961 because it predicts better for 1959-1961 than the TVC model.[11] However, the benchmark model poorly predicts for earlier periods, particularly for 1954, 1955 and 1957. Unless we understand why its prediction is poor and expect no more poor predictions, the benchmark model should not be used. In contrast, the TVC model provides us with useful information

[9]The model comparison method can be applied for multi-year-ahead adoptions in the same way.

[10]Diffuse priors were used: $p(\alpha_0, \alpha_1, \alpha_2, \sigma_u) \propto 1/\sigma_u$ for the Bass model (10), $p(\beta_{0,t'}, \beta_{1,t'}, \beta_{2,t'}, \sigma) \propto 1/\sigma$, where $t' = 1948$ (the initial period), for the TVC model (11), and $p(\gamma_0, \gamma_1, \gamma_2, \sigma_\xi) \propto 1/\sigma_\xi$ for the time-trend model (12). See Min and Zellner (1993) for updating procedures for predictive density of the TVC model.

[11]It is equivalent to the using of observations up to 1958 as a training sample.

about the uncertainty associated with its prediction. It is therefore suggested that in order for us to better understand models, we need to use several training samples and to examine predictive ratios for each period.

5 Conclusion

This study has proposed a model selection method which uses predictive densities. It seems that the method can help us determine which model, among a given set of models, produces the most useful information about future values. While it was applied only to dynamic models in this study, the method can be applied to any other models in the same way. Further, with the help of simulation methods such as the Gibbs sampler, the proposed method finds an expanded area of application.

References

Bass, F. M. (1969), 'A New Product Growth Model for Consumer Durables,' *Management Science*, 15, 215-227.

Belsley, D. (1988), "Modelling and Forecasting Reliability," *International Journal of Forecasting*, 4, 427-447.

Berger, J.O., and L.R. Pericchi (1992), "The Intrinsic Bayes Factor," Purdue University Department of Statistical Report.

Chow, G.C. (1973), "Multiperiod Predictions from Stochastic Difference Equations by Bayesian Methods," *Econometrica*, 41, 109-118.

Friedman, M., and A. Schwartz (1991), "Alternative Approaches to Analyzing Economic Data," *American Economic Review*, 81, 39-49.

Geisser, S. (1975), "The Predictive Sample Reuse Method with Application," *Journal of the American Statistical Association*, 70, 320-328.

Gelfand, A.E., D.K. Dey, and J. Chang (1992), "Model Determination using Predictive Distributions with Implementation via Sampling-Based Methods," *Bayesian Statistics* 4, J.M. Bernardo, J.O. Berger, A.P. Dawid, and A.F.M. Smith, eds., Oxford: Clarendon Press, 147-167.

Geweke, J. (1994), "Bayesian Comparison of Econometric Models," Manuscript.

Min, C. (1995), "Forecasting the Adoptions of New Consumer Durable Products," Manuscript.

Min, C., and A. Zellner (1993), "Bayesian and Non-Bayesian Methods for Combining Models and Forecasts with Applications to Forecasting International Growth Rates," *Journal of Econometrics*, 56, 89-118.

Thompson, P.A., and R.B. Miller (1986), "Sampling the Future: A Bayesian Approach to Forecasting From Univariate Time Series Models," *Journal of Business & Economic Statistics*, 4, 427-436.

Zellner, A. (1986), "Biased Predictors, Rationality and the Evaluation of Forecasts," *Economics Letters*, 21, 45-48.

Zellner, A. (1994), "Time Series Analysis, Forecasting and Econometric Modeling: The Structural Econometric Modeling, Time Series Analysis (SEMTSA) Approach," *Journal of Forecasting*, 13, 215-233.

Bayesian Prediction Under Asymmetric Linear Loss:
Forecasting State Tax Revenues in Iowa

Charles H. Whiteman*
108 PBAB Suite 230; The University of Iowa; Iowa City, IA 52242

For the past five years, forecasts of regional economic conditions and state tax revenues under the Iowa Economic Forecasting Project have been produced using a Bayesian approach to multivariate time series analysis. This paper provides an assessment of that effort and addresses issues which have arisen in the specification of the time series models, collection of data, specification of prior distributions, and the production and interpretation of forecasts under symmetric and asymmetric loss.

1. INTRODUCTION

Virtually every public policy debate includes a common question: what resources will be available when the expenditures must be made? At the state level, answering this question requires a sensible tax revenue forecast. For the past five years, as Director of The Institute for Economic Research at The University of Iowa, I have produced tax revenue forecasts which have been used in statewide discussions of the course of available revenues. The approach I have taken involves the specification and use of Bayesian vector autoregressive (BVAR) models for regional economic variables of interest.

The primary focus is on general tax receipts, a variable for which a coherent time series exists only since the early 1980's. The strong prior views necessary to proceed in such a situation come from exclusion of many time series of potential interest, and in the use of very short lag lengths. Aside from this, and for computational convenience, a noninformative prior over VAR parameters was specified. At each forecast date, quarterly predictive densities for the ensuing two fiscal years are calculated by Monte Carlo methods, and forecasts are obtained from the moments and quantiles associated with these densities.

Revenue growth predictions produced by this method have been considered by the state's official revenue estimating body, the Revenue Estimating Conference, in the formulation of its estimates. The Bayesian predictions have in fact been markedly more accurate than the official forecasts.

A number of issues regarding the forecasts have arisen. First, huge data revisions plague state-by-state income data two or more times each year, and induce volatility in the quarterly forecasts. Second, recently developed numerical methods make the computational simplicity of uninformative priors relatively less important, and informative priors (e.g., a "Minnesota" prior) coupled with Gibbs sampling methods could facilitate the production of

* Support from the National Science Foundation under SBR-9422873 is gratefully acknowledged.

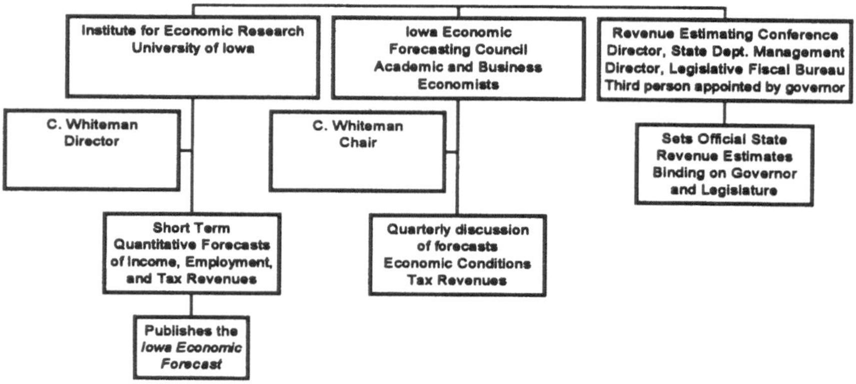

Figure 1. The Forecasting Structure in Iowa.

predictive densities which incorporate alternative views about the evolution of the time series in question. Finally, the forecasts have been used as input into what is actually a political discussion. Over the few years during which the forecasts have been produced via the Bayesian approach, policy makers' interpretation of the forecasts has progressed from "maneuvering room" (because predictive densities are more diffuse than the point forecasts politicians often seem to prefer) to a "hedging" device ("our forecast needs to be more conservative than the model's").

2. FORECASTING STRUCTURE AND POLITICAL CONSIDERATIONS

There are two officially constituted bodies responsible for economic forecasts in the state of Iowa. The Iowa Economic Forecasting Council is responsible for forecasts of economic conditions. The Revenue Estimating Conference is responsible for setting the official state revenue estimates. Figure 1 displays an organization chart of forecasting in Iowa.

The Forecasting Council consists of academic and business economists; since 1990, I have served as chair. The Council meets quarterly to consider short-term quantitative forecasts of economic conditions and state tax revenues. The forecasts considered at these meetings are produced by The Institute for Economic Research using methods alluded to above and described in detail below. The forecasts, together with a summary of the Council's discussion, are published by the Institute in the quarterly *Iowa Economic Forecast*.

The Revenue Estimating Conference consists of three individuals appointed by the Governor: the Director of the State Department of Management (which is analogous to the

U.S. Office of Management and Budget), the Director of the Legislative Fiscal Bureau (which is analogous to the U.S. Congressional Budget Office), and a third person typically not associated with state government. The Conference meets quarterly; the fourth quarter meeting is required by statute to occur no later than December 15. This requirement stems from the binding nature of the Conference estimates.

Historically, both the timing of the Revenue Estimating Conference decisions and political considerations in the state have influenced revenue forecasts. Until fiscal year 1993 (which ended June 30, 1993), the timing of revenue estimates and spending decisions was as follows: in December, the Revenue Estimating Conference would set the official forecast for revenue growth in the current fiscal year and the ensuing one. In January, the Governor would propose a budget for the ensuing fiscal year, which could involve "supplemental" spending for designated programs during the current fiscal year. The Governor could recommend spending 100% of the revenue projected to accrue over the eighteen month period until the end of the next fiscal year.

In March, the Revenue Estimating Conference would meet to reconsider the December estimate. If the revenue estimate was revised upward, additional spending became feasible. If the revenue estimate was revised downward and legislation had already passed which called for spending essentially all expected revenue (a common practice during and following the midwest "mini-recession" in the mid 1980's), spending cuts or layoffs of state workers became necessary.

Because the Democratic Party controlled the Legislature and the Republican Party controlled the Executive Branch of state government, the perverse incentives induced by this timing introduced political considerations into the revenue estimation process.[1] In particular, a high estimate in December would benefit the Governor, who could then propose and be credited with new and exciting spending programs. Members of the opposition party in the Legislature, wishing to see the Governor hamstrung, would want a low estimate. By March, the incentives would cause a role reversal. Since the Governor's budget proposal had already been made, the Legislature benefited from an upward revision of the revenue estimate because then *it* could initiate and be credited with exciting new spending programs not previously contemplated by the Governor. The Governor, on the other hand, mindful of his Constitutional responsibility to prevent budget deficits, would wish to see the *estimate* (though not the ultimate revenues) be low in March so that no new spending would occur.

The perverse incentives of the system seemed to script the experience of fiscal year 1991. The standing revenue estimate in the fall of 1990 was for growth of 5.5% during fiscal year 1991. At the December meeting, despite revenues running below projections, an impending national recession, and an incipient war, the Revenue Estimating Conference *raised* the revenue estimate to 6.1%.[2] Actual revenue growth for the fiscal year turned out to be 4.7%. The forecast made using the procedures to be described below was for 4.3% growth.

---

[1] There is an alternative interpretation of this situation which is based on time-dependent asymmetric loss functions–in the language of the loss functions introduced in Section IV below, deficits would appear less costly in December than in March to the party in power, and the reverse would be true for the opposition.

[2] Subsequently, the NBER dating committee placed the peak during the third quarter of 1990, so the recession was nearly six months old at the time of the meeting. Many economists believed that the recession had in fact begun, but Michael Boskin, Chairman of the Council of Economic Advisors, stated on December 14 (the day before the Revenue Estimating Conference) that the recession had not yet begun. He changed his mind ten days later, but added that the recession was by

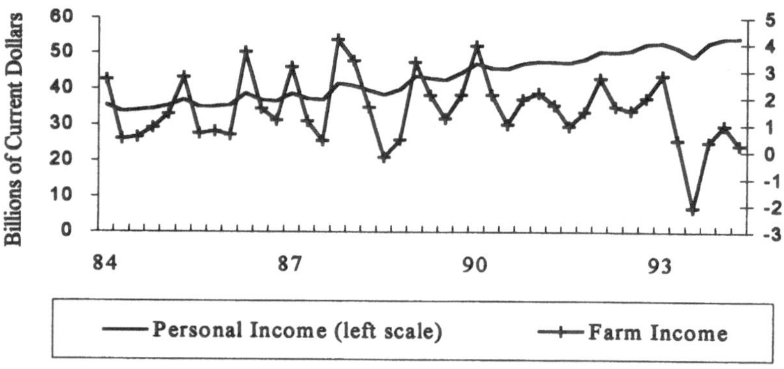

Figure 2. Economic Conditions in Iowa.

After the experience in fiscal year 1991, the players in this game recognized that the state would benefit by the removal of some of the perverse incentives. Thus beginning in fiscal year 1993, two changes were made to the process. First, the Governor and the Legislature were statutorily prevented from proposing to spend more than 99% of available revenue. Second, the December Revenue Estimating Conference estimate became binding for the entire year, so that upward revisions to the official forecast made in March would not be available for spending during the spring legislative session.

These changes eliminated some, but not all, of the political considerations from official revenue forecasts. For example, the binding nature of the official estimates and the 99% provision build in a simple way to press for smaller government--excessively conservative revenue forecasts. Indeed, the state budget deficit caused in part by the overly optimistic forecast of December 1990 was "paid off" by the Republican Administration a year ahead of schedule in 1994 because of a (purposefully?) conservative revenue forecast in December 1993--the official growth forecast: 5.0%; my forecast and the actual value: 6.9%.

While the nature of the forecasting process permits and perhaps encourages the politicization of the official forecast, the effects of this are difficult to pinpoint accurately. The reason is that there is more than enough volatility in Iowa economic conditions and revenue growth to swamp forecasters' incentives.

---

then over. To most observers, there seemed little room for optimism--recall that the Iraqi invasion of Kuwait was four months old by this time, and international tensions were high.

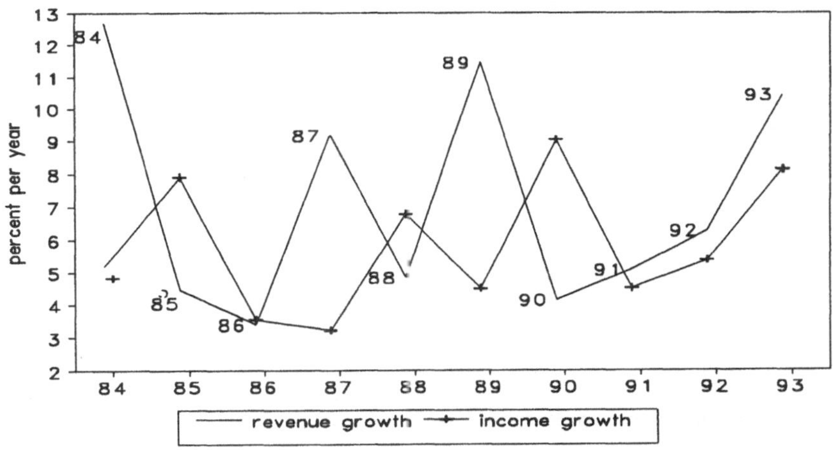

Figure 3. The record of revenue and income growth in Iowa (calendar years.)

3. ECONOMIC CONDITIONS IN IOWA

As recently as fifteen years ago, the Iowa economy was dominated by agriculture--either directly by farming, or indirectly through the production of farm equipment and raw materials--seed stock, feed additives for livestock, etc. But the farm crisis of the 1980's drove many marginal producers out of business and essentially forced the economy to diversify. Figure 2 demonstrates that since the mid 1980's farm income has accounted for less than 10% of personal income in the state. It also shows how volatile farm income can be. During the flood quarters of 1993--the third and fourth--farm income (the level!) was *negative*.

Unfortunately for forecasters, the volatility carries over to revenue growth. Figure 3 shows how wide the swings in income and revenue growth can be, and that comovement between them is not as simple as one might imagine. Figures 4 and 5 make this even more apparent: when fiscal year income and fiscal year revenue growth are considered, it turns out that the least squares regression of revenue growth on income growth yields a *negative* slope coefficient. The expected positive slope coefficient (though certainly not unity) is obtained when calendar year income and fiscal year revenues are compared. Still, this exercise suggests that the *timing* issues are important, and that a forecasting scheme would do well to include some dynamic features. Indeed, dynamics are the defining feature of the procedures I use--Bayesian Vector Autoregressions.

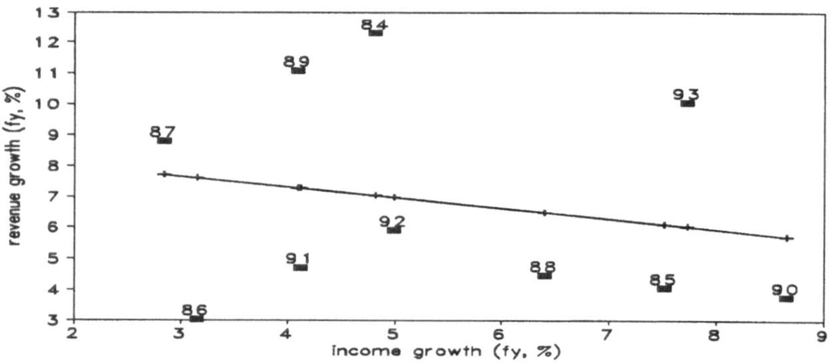

Figure 4. Revenue and Income Growth in Iowa (fiscal years).

4. INSTITUTE FOR ECONOMIC RESEARCH FORECASTING PROCEDURES

The *Iowa Economic Forecast* is prepared using the Iowa Forecasting Model which has been developed and is maintained by The Institute for Economic Research at The University of Iowa. The model is estimated each quarter using the latest available National and State historical data series. The data are obtained from commercial and government electronic data services such as Citicorp's Citibase and the Census State Data Center of the U.S. Department of Commerce. Forecasts produced by the Institute use publicly available data and are not judgementally adjusted.

The basic economic construct underlying the model is the export-base model. In the export-base model, economic activity in the region is closely linked to demand from the other regions (the export "base") which buy goods and services produced in the region of interest. Iowa's export base is taken to be the rest of the national economy, and thus how Iowa fares is tied to growth elsewhere in the U.S. (Foreign trade has become more important for Iowa in recent years, but getting reliable and timely data from other countries is problematic.) Thus Iowa's "share" in the national economy is of substantial importance, and developments nationally which affect demand for Iowa products are felt quickly.

The basic statistical construct underlying the forecasts is the Bayesian Vector Autoregression (BVAR). The BVAR procedure, pioneered at the Federal Reserve Bank of Minneapolis (Litterman, 1986) and currently in widespread use, has been called a "standard of comparison for other forecasts" (McNees, 1986). Judged against several large national forecasting models, forecasts produced using the BVAR maintained at Minneapolis (on which the Iowa model is based) "were generally the most accurate or among the most accurate for

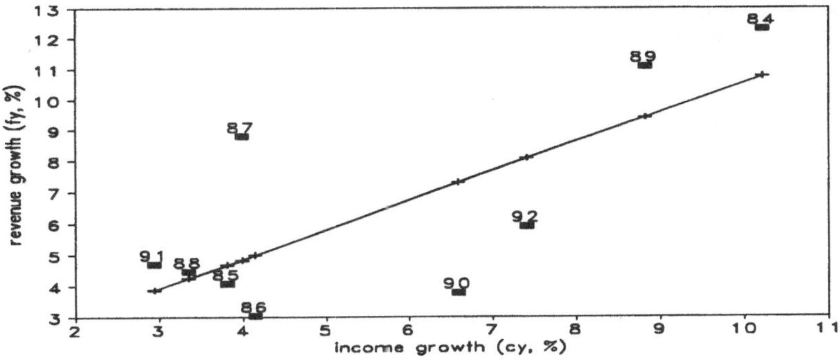

Figure 5. Revenue and Income Growth in Iowa (calendar year income, fiscal year revenues.)

real gross national product (GNP), the unemployment rate, and real nonresidential fixed investment" (McNees, 1986). Other methods could of course be substituted for the BVAR. An informal search quickly established that BVARs were superior to univariate methods in forecasting accuracy in Iowa. Two other alternatives, vector autoregressive-moving average (VARMA) models, and structural econometric models, were also considered. The VARMA approach is more general than a pure autoregression, but the generalization to moving average components (especially in the vector case) introduces computational difficulties deemed excessive given the requirement that forecasts of dozens of variables be produced quarterly in the face of continual data revisions (necessitating re-estimation of the model). Finally, there is not a widely embraced structure for regional economies (the export-base notion being just that--a notion rather than a fully articulated theoretical construct.) These considerations, and the absense of any evidence suggesting that the alternatives were more accurate or robust, led to the use of McNees's "standard", the BVAR.

A technical description of an L-lag VAR proceeds as follows. Let z_t denote an n×1 vector of variables of interest at time t. The equation for the ith component of z_t is

$$z_{it} = H_{i1}z_{t-1} + H_{i2}z_{t-2} + \dots H_{iL}z_{t-L} + \Gamma_i + \Delta_i S_t + \varepsilon_{it}, \quad t=1,\dots,T; \quad i=1,\dots,n$$

where H_{ij} denotes the coefficient matrix on lag j, Γ_i is the intercept, S_t denotes the 3×1 vector of seasonal dummies, and Δ_i denotes the coefficients on the seasonal dummies. Let Z_i denote the T×1 vector of observations on z_{it}, let X denote the T×(nL+4) matrix consisting of observations on $z_{t-1}, \dots z_{t-L}$, constant, and three seasonal dummies, let e_i denote the vector of residuals, and let B_i denote the matrix $[H_{i1}' \dots H_{iL}' \; \Gamma_i \; \Delta_i']'$. Then the equation can be written as

$$Z_i = XB_i + e_i, \qquad i = 1,\dots,n.$$

The residual e_i is assumed to be normally distributed with mean zero and variance σ_i. Letting $Z = [Z_1 \ldots Z_n]$, $B = [B_1 \ldots B_n]$, and $E = [E_1 \ldots E_n]$, the entire system can be written

$$Z = XB + E.$$

Applying the vec operator yields

$$\text{vec}(Z) \equiv z = (I_n \otimes X)\text{vec}(B) + \text{vec}(E)$$

or

$$z = x\beta + e \qquad\qquad e \sim N(0, \Sigma \otimes I_n)$$

where $x = I_n \otimes X$, $\beta = \text{vec}(B)$, $e = \text{vec}(E)$, and the (i,j) element of Σ is Ee_ie_j. The likelihood function can then be written

$$\ell(z,x|\ \beta,\Sigma) \propto |\Sigma|^{-1/2} \bullet \exp\{-0.5(z-x\beta)'\ \Sigma^{-1} \otimes I_n\ (z-x\beta)\}.$$

The BVAR permits the combination of powerful, reproducible statistical methods with economic theory such as the export-base model. It therefore permits the use of historical correlations in the data among Iowa and national employment and income series together with prior theories of how these variables interact. The theoretical restrictions are placed on the model through the use of Bayesian priors. For example, agriculture is treated as an export sector, and trade as a non-export sector, by specifying different patterns of pass-through from the associated national sectors--agricultural employment is affected more quickly by national developments than is trade employment.

The model comprises four quarterly VARs. Three VARs focus on economic conditions, and one is used to generate the revenue forecast. The economic conditions VARs are estimated using an equation-by-equation “mixed estimation” procedure under "Minnesota"-type random walk priors discussed, e.g., in Doan, Litterman, and Sims (1984): other things equal, each variable in the system is assumed to grow at its most recent historical rate. The export-base notion is implemented in two ways. First, whenever an Iowa variable appears, its national analogue also enters. Second, for each Iowa and national variable in the system the relative weight each of the other variables in the system has in its determination is specified. In this respect the model is similar in spirit to the Ninth Federal Reserve District regional model maintained at the Federal Reserve Bank of Minneapolis (Todd, 1984).

The Personal Income VAR involves total Iowa Personal Income, Wage and Salary Disbursements, Property Income, Transfers, and Farm Income, together with the parallel national economic variables. The Real Income VAR comprises the same 10 time series, deflated by the GDP Deflator. The employment VAR involves total Iowa nonagricultural employment, population, and employment in durable and nondurable goods manufacturing, services, wholesale trade, and retail trade, together with the parallel national economic time series. All VARs use four lags, and include constants and seasonal dummy variables. Priors for the latter are uninformative.

Tightness Parameter: 0.2
Harmonic Lag Decay With Parameter: 1.0
Means and Standard Deviations As Percentage of Own Lag
Listed Under The Dependent Variable

	YP_IA	**WSD_IA**	**YPROP_IA**	**V_IA**	**YPF_IA**
YP_IA	1.00	0.50	0.50	0.50	0.50
WSD_IA	0.50	1.00	0.10	0.10	0.10
YPROP_IA	0.10	0.10	1.00	0.10	0.10
V_IA	010	0.10	0.10	1.00	0.10
YPF_IA	0.10	0.10	0.10	0.10	1.00
YP	1.00	0.05	0.05	0.05	0.05
WSD	0.05	1.00	0.05	0.05	0.05
YPROP	0.05	0.05	1.00	0.05	0.05
V	0.05	0.05	0.05	1.00	0.05
VPF	0.05	0.05	0.05	0.05	1.00
MEAN	1.00	1.00	1.00	1.00	1.00
	YP	**WSD**	**YPROP**	**V**	**YPF**
YP_IA	0.01	0.01	0.01	0.01	0.01
WSD_IA	0.01	0.01	0.01	0.01	0.01
YPROP_IA	0.01	0.01	0.01	0.01	0.01
V_IA	0.01	0.01	0.01	0.01	0.01
YPF_IA	0.01	0.01	0.01	0.01	0.01
YP	1.00	0.50	0.50	0.50	0.50
WSD	0.50	1.00	0.10	0.10	0.10
YPROP	0.10	0.10	1.00	0.10	0.10
V	0.10	0.10	0.10	1.00	0.10
YPF	0.10	0.10	0.10	0.10	1.00
MEAN	1.00	1.00	1.00	1.00	1.00

Table 1. The prior used in the economic conditions forecast.

Forecasts must be made in March, June, September, and December. Employment data are released monthly with about a two-month lag, so employment data through the end of the previous quarter are available for each forecast. Income data are released quarterly, in January, April, July, and October, with a four month lag for state data, one month for national. Thus income data are available up to two quarters prior to the forecast date. To make maximum use of the available nonsynchronous data, the state income data are shifted forward one quarter for the economic conditions forecast, and two for the revenue forecast. (The last month of the revenue quarter is filled out by assuming growth for that single month will equal the year-to-date growth rate.)

The estimation procedure for the economic conditions VARs involves combining the normal likelihood function with a normal prior to obtain a mixed estimate of the posterior mean. Specifically, priors on lag coefficients in the three VARs are implemented using the RATS (Doan, 1993) "General" procedure; Table 1 provides a description of the prior in the income VAR. In each equation of the VAR, the prior mean on the own first lag coefficient is unity and all other prior means on lag coefficients are zero (in the prior, that is, all elements of B_i are zero except for the element corresponding to the first lag of z_i). The associated prior standard deviation is "Overall tightness" times the empirical innovation standard deviation from a univariate autoregression for that variable. Prior standard deviations for higher lag coefficients decay harmonically, so by lag J, the prior standard deviation is a fraction 1/J of that for the first lag. Standard deviations for coefficients on other variables are as listed in the table; for example, the prior standard deviation on the first lag of Iowa Wage and Salary Disbursements in the total Iowa Personal Income equation is one-half that of the first lag of Iowa Personal Income itself. Thus letting $\overline{B}_1$ denote the prior mean for B_1 and Ω_1 its prior covariance matrix, we have for the mixed estimate of B_1[3]

$$\hat{B}_1 = [\frac{X'X}{\sigma_1^2} + \Omega_1^{-1}]^{-1}(\frac{X'Z_1}{\sigma_1^2} + \Omega_1^{-1}\overline{B}_1).$$

Values for the prior means and standard deviations (the coefficients are independent in the prior) were determined in a forecasting experiment conducted in February 1990 using the most current data then available. The objective was forecast accuracy at a horizon of one to eight quarters; the procedures used form the basis for the quarterly computation of forecast error statistics and will be described presently.

Prior to producing the forecast, the VARs are estimated using data from the beginning of the sample through 38 quarters prior to the end of the sample. This VAR is used to forecast (using the chain rule of forecasting) up to 12 quarters ahead. Then one additional observation is added, and the one- through twelve-step forecasts are regenerated. Forecast statistics are collected, and the process is continued through the end of the sample. Table 2 provides an example of the results of this pseudo-real-time forecasting experiment. (The 1990 experiments to fix parameters of the prior distribution involved the production of tables like this for various different specifications of prior mean values, standard deviations, decay factors, etc.) The root mean squared errors (RMSEs) from this sort of experiment are reported along with the forecast (an example is provided in Table 3) as a measure of the uncertainty which should be attached to the forecasts at various horizons. Note, for example, that the forecast of 1994 Iowa Personal Income in Table 3, which was made in December 1994 using data through the second quarter of 1994, is assigned an RMSE of 1.59--the value for the two-quarter horizon forecast from the experiment reported in Table 2.

The revenue VAR consists of only two equations--total Iowa Personal Income, and General Receipts (revenues.) Logarithms of the series are taken, and two lags of each (quarterly) variable are used, as are constants and seasonal dummies. The primary reason for restricting attention to such a small system involves data availability. A time series for general receipts based on current accounting conventions exists only back to 1982. Attempts to reconstruct general receipts data (and a principal component, sales tax receipts) from historical components but using current accounting conventions failed to produce series which

---

[3] See Theil (1971), pp. 347-9.

Forecast Statistics For Series YP_IA

Step	Mean Error	Mean Abs. Error	RMS Error	Theil U	N. Obs
1	-0.22296121	1.0269801	1.2827061	0.72469577	38
2	-0.46802981	1.2273632	1.5853295	0.64743572	37
3	-0.72495514	1.1951002	1.6661997	0.67263407	36
4	-0.90984980	1.3773689	1.9485540	0.76609891	35
5	-1.04653230	1.5760876	2.0417895	0.65376933	34
6	-1.18804860	1.6326288	2.2267338	0.59591063	33
7	-1.43813410	1.7725910	2.3853949	0.58398948	32
8	-1.38169410	1.7133071	2.0302542	0.44756340	31
9	-1.41348760	1.7011489	2.0232481	0.38684837	30
10	-1.47744840	1.5979857	1.9742418	0.34016068	29

Forecast Statistics For Series YPF_IA

Step	Mean Error	Mean Abs. Error	RMS Error	Theil U	N. Obs
1	-0.23837244	0.64202075	0.83149702	0.52787965	38
2	-0.16113428	1.00550220	1.30953020	0.64382580	37
3	-0.4705E-01	0.97160019	1.31570090	0.76715093	36
4	-0.6791E-01	0.88637441	1.22632560	0.86182559	35
5	0.12962722	0.95395063	1.30128120	0.75514491	34
6	0.22154457	1.02976730	1.35721780	0.70767188	33
7	0.20470196	0.95136283	1.26724680	0.78258155	32
8	0.23935455	0.87907718	1.23326130	0.85556678	31
9	0.27260848	0.92430208	1.30399620	0.70973488	30
10	0.24675068	0.91093029	1.28747190	0.68719573	29

Table 2. Forecast statistics for the period 1984:III-1994:II, using the 1994:II data set.

could be spliced to the modern series without violating (my) prior views regarding smoothness of the splice.[4]

The revenue forecast is produced using a fully Bayesian procedure under the usual uninformative prior. (See Zellner, 1971) That is, the mixed estimation shortcut is not taken since the posterior distribution is tractable. In fact, under the uninformative prior, the posterior distribution of VAR parameters factors into the product of a conditional normal distribution for the coefficients given the innovation covariance matrix, and an inverted Wishart distribution for the covariance matrix. That is, with prior distribution

[4]Furthermore, as we shall see presently, it is difficult to argue that the forecasting procedure is "broken"--until it is, herculean efforts to gather historical data of questionable value are probably not warranted.

Annual Indicators Of The Iowa Economy

Real Iowa Personal Income (Billions of 1987 Dollars)

	1991	1992	1993	1994	1995	1996
YP_IA	40.574	42.143	41.745	43.549	45.041	46.111
RMSE	NA	NA	NA	1.537	1.508	1.370
% CHA	0.928	3.865	0.943	4.321	3.426	2.376
YP_IA	40.558	42.367	41.771	44.244	45.602	46.708
RMSE	NA	NA	NA	1.419	1.246	1.400
% CHA	0.967	4.460	1.407	5.920	3.070	2.426

Iowa Personal Income (Billions of 1987 Dollars)

	1991	1992	1993	1994	1995	1996
YP_IA	47.714	50.953	51.564	55.014	58.769	62.212
RMSE	NA	NA	NA	1.585	2.227	1.974
% CHA	2.888	6.788	1.199	6.690	6.827	5.858
YP_IA	47.696	51.225	51.597	55.985	59.775	63.452
RMSE	NA	NA	NA	1.682	1.983	1.960
% CHA	2.848	7.399	0.728	8.504	6.770	6.150

Iowa Non Farm Employment (Thousands)

	1991	1992	1993	1994	1995	1996
EEA	1238.117	1252.617	1277.092	1308.507	1344.405	1382.647
RMSE	NA	NA	NA	10.498	32.198	67.215
% CHA	0.960	1.171	1.954	2.430	2.743	2.845
EEA	1238.117	1252.617	1277.092	1317.352	1364.963	1408.098
RMSE	NA	NA	NA	10.959	38.168	75.894
% CHA	0.960	1.171	1.954	3.152	3.614	3.160

Table 3. An example forecast produced December 1994.

$$\Pi(B,\Sigma) \propto |\Sigma|^{-(n+1)/2},$$

the posterior distribution is

$$P(B,\Sigma) \propto P(B|\Sigma)P(\Sigma)$$

with

$$P(B|\Sigma) \propto |\Sigma|^{-(nL+4)/2} \exp\{-\frac{1}{2}(\beta-\hat{\beta})'\Sigma^{-1} \otimes X'X((\beta-\hat{\beta})\}$$

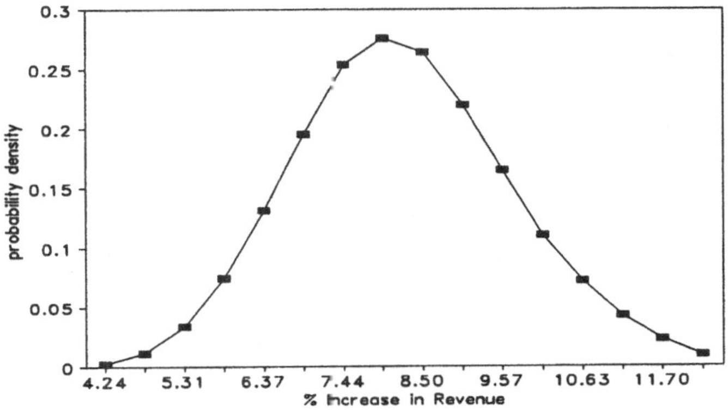

Figure 6. The predictive density for fiscal year 1995 revenue growth; forecast made December 1994.

and

$$P(\Sigma) \propto |\Sigma|^{-(T-(nL+4)+n+1)/2} \exp(-\frac{1}{2}\mathrm{tr}\Sigma^{-1}S)$$

where $\hat{\beta}$ and S denote the least squares estimates of the coefficient vector and the covariance matrix.

Because it is simple to sample from the posterior distribution, the predictive densities and CDFs are then easily computed by Monte Carlo integration. The procedure is as follows: draw a covariance matrix from the inverted Wishart distribution and use it to draw a sequence of innovations from the appropriate Normal distribution. Also, conditional on the covariance matrix, draw from the multivariate Normal distribution for the coefficients. Then using the just realized innovations, the just realized coefficients, and the historical data, generate a simulation of (log) general receipts through the next fiscal year, exponentiate, accumulate to annual figures, and record the result. Repeat this process a large number of times (1000 seems to be adequate and has been used in all computations reported in this paper and *The Iowa Economic Forecast*), and draw histograms of the densities and cumulative distributions. An example from the December 1994 forecast of fiscal year 1995 revenue growth is provided in Figures 6 and 7.[5]

Each quarter, a document is produced for the Revenue Estimating Conference which reports the predictive mean and standard deviation. As is well known, the predictive mean is an optimal forecast under quadratic loss. But because of the balanced budget provision of the

[5] The predictive density in the figure was smoothed using a local-quadratic smoother described in DeJong and Whiteman (1991). The smoothed density was accumulated to produce the CDF.

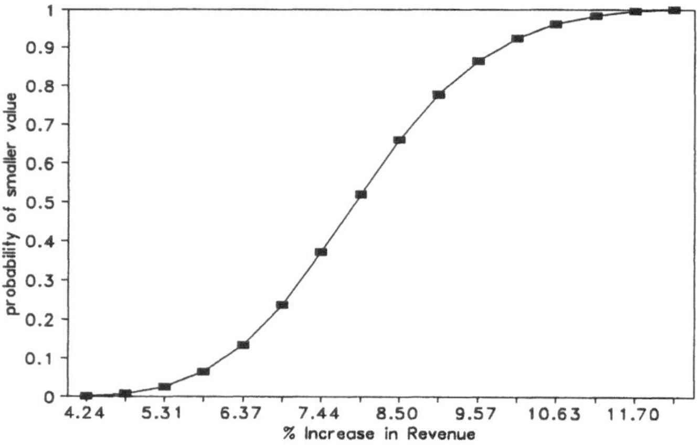

Figure 7. The predictive CDF for fiscal year revenue growth; forecast made December 1994.

state Constitution, in state revenue forecasting the symmetric loss function seems inappropriate. For example, a \$40 million surplus (about 1%) means that either taxpayers should not have been taxed or \$40 million in worthy expenditures were unrealized; a \$40 million shortfall means across the board cuts (generally long after the year's expenditures have been planned), layoffs of workers (as happened in 1991), or some other drastic measure.

Thus from the beginning I have also produced a set of asymmetric loss forecasts. The simplest asymmetric loss function is piecewise linear: let "F" denote the forecast and "A" the actual value. Then the loss function is

$$L(F,A) = \begin{cases} s|F-A| & \text{if } F \leq A \\ d|F-A| & \text{if } F \geq A \end{cases}$$

where s and d are parameters (s for "surplus", d for "deficit"). To capture the notion that shortfalls are more costly than surpluses, we typically imagine that d > s. Figure 8 displays the d=3, s=1 case. More generally, it is easy to show that the optimal forecast in this case is the s/(d+s) quantile of the predictive density. To see this, let p(.) denote the predictive density and P(.) the associated CDF. Then expected loss is

$$EL(F,A) = d\int_{-\infty}^{F}(F-A)p(A)dA + s\int_{F}^{\infty}(A-F)p(A)dA$$

Minimizing expected loss requires

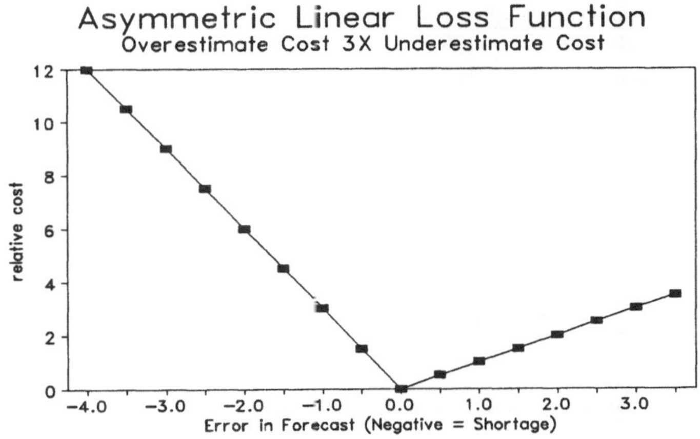

Figure 8. An example asymmetric loss function.

$$d(EL)/dF = 0 = dP(F) + dFP(F) - dFP(F) - sFP(F) - s[1-P(F)] + sFP(F)$$

$$= dP(F) - s + sP(F)$$

so that the optimal forecast satisfies $P(F) = s/(d+s)$.

Thus Table 4 provides the forecasts for s=1 and a host of values for d. Following custom and tastes of members of the Forecasting Council, the d=3 forecasts are often the focus of Council discussion. They are also the forecasts which I "recommend" to the Revenue Estimating Conference.

5. THE TRACK RECORD

While the accuracy of forecasts of state revenues, which are not subject to revision, is straightforward, assessing the accuracy of economic conditions forecasts is another matter.[6] Personal income and employment data are revised frequently, and the revisions are often quite large, especially for income. For example, over the roughly twenty quarter period 1990-1994, the mean square quarterly *revision* to Iowa Personal income from the first to the second data release was $864 million (more than 1.5%) The mean square revision from the first release

[6] There is one caveat to the statement about revenues. At the end of fiscal 1993, at the discretion of the Department of Management, funds were transferred from the general fund to the Liquor fund. The transfer occurred during the last week or so of the fiscal year, and was unprecedented. The footnote in Table 5 thus indicates that the "actual" figure for fiscal year 1993 does not deduct these transfers from general receipts.

Cost Ratio = Cost of Shortage : Cost of Surplus

If the Cost Ratio is this	Forecast this (FY 1995)	Forecast this (FY 1996)
1	7.9	4.8
2	7.3	3.9
3	7.0	3.4
4	6.8	3.0
5	6.6	2.8
6	6.5	2.6
7	6.4	2.5
8	6.3	2.3
9	6.2	2.2
10	6.1	2.1

Table 4. Asymmetric loss forecasts made December 1994.

to the release most recent as of December 1994 was nearly twice that amount: $1.6 billion (about 3%).

Faced with data revisions commonly in the 1.5-3% range, it is perhaps mildly comforting that treating the first release as data, the mean square error of Institute's one-quarter ahead forecast of personal income was $1 billion over the period 1990:III-1994:II. The most relevant test of the procedures, though, is in the revenue forecasts, because they matter most for public policy.

Table 5 presents the revenue forecast track record; the column marked "CW" denotes the forecasts made as described above (the 3:1 loss ratio forecasts), and "REC" denotes the official forecast. The fiscal 1991 episode is perhaps the most dramatic. It certainly garnered the most public attention. The Revenue Estimating Conference estimate set at the meeting was above the 90% quantile of my predictive density. After checking to ensure the correct terminology, David Yepsen, the chief political correspondent for the statewide *Des Moines Register*, widely reported that there was "a 90% chance that the REC estimate is too high." Subsequent events did much to enhance the credibility of Institute forecasts. By the fall of 1992, the Director of the Department of Management remarked that the Department intended its forecasts to be more conservative than mine.[7]

[7]In the first few months of 1991, the state entered in to collective bargaining with its three major unions. The state contended that it did not have the money to pay for raises in the 4-6% range. Impasses were reached in each negotiation, and in binding arbitration, state called me to testify (on the basis of the 4.3% revenue growth forecast) that the revenue situation was dire. Under cross-examination, the Director of the Department of Management (one of the principals in the 6.1% REC forecast) would then testify that although the state had no money, the 6.1% official forecast was likely to be realized. Each of the arbitrators decided in favor of the unions--each cited the contradictory stance of the state as decisive.

Date	CW	REC	Actual
3/90	4.3%	5.0%	3.8%
12/90	4.3%	6.1%	4.7%
3/91	4.5%	5.6%	4.7%
12/91	6.7%	6.8%	5.9%
3/92	5.7%	6.0%	5.9%
12/92	11.5%	10.3%	10.3%*
3/93	10.6%	10.3%	10.3%*
12/93	6.9%	5.0%	6.9%
3/94	7.0%	5.9%	6.9%

* Includes liquor transfers

Table 5. Track record of Iowa revenue forecasts.

6. CONCLUDING REMARKS

The Bayesian procedures I have used in forecasting economic conditions and tax revenues in Iowa have established a credibility which I think is well deserved. A former State Comptroller and Revenue Estimating Conference member once remarked that "you're doing well if your growth forecast is within three percentage points." The forecasts I have produced have been markedly more accurate, and noticeably more accurate than the official forecasts (though political considerations account for some aspects of the official forecasts.) More general prior distributions might permit the production of even more useful forecasts, and this is the subject of ongoing research.

Further, the Bayesian procedures have injected science into the forecasting process in two ways. First, because the official state forecasters see predictive densities instead of point forecasts, all of the unquantifiable expertise they bring to the forecasting effort can be brought to bear on the exercise. Second, the predictive densities and distributions have made probabilistic assessments of revenue forecasts possible, and have contributed to the recognition by all those involved that not only are shortfalls and surpluses not equally costly, but that it is possible to do something about it.

References

DeJong, D.N., and C.H.Whiteman, (1991). "Reconsidering 'Trends and Random Walks in Macroeconomic Time Series'," *Journal of Monetary Economics.*

Doan, T., (1993). *RATS 4.10.* Evanston: Estima.

Doan, T.; Litterman, R.; and C. Sims, (1984). "Forecasting and Conditional Projection Using Realistic Prior Distributions," *Econometric Reviews* 3:1-100.

Litterman, R.B. (1986). "Forecasting With Bayesian Vector Autoregression--Five Years of Experience," *Journal of Business and Economic Statistics* 4:25-38.

McNees, S., (1986). "Forecasting Accuracy of Alternative Techniques: A Comparison of U.S. Macroeconomic Forecasts," *Journal of Business and Economic Statistics*.

Theil, H. (1971). *Principles of Econometrics*. New York: Wiley.

Todd, R., (1984). "Improving Economic Forecasting With Bayesian Vector Autoregression," Federal Reserve Bank of Minneapolis *Quarterly Review* 8 (Fall):18-29.

Zellner, A., (1971). *An Introduction to Bayesian Inference in Econometrics* . New York: Wiley.

Section III

Design of Experiments and Classification

On small sample Bayesian inference and sequential design for quantal response curves

Tom Leonard[†] and John S.J. Hsu[‡]

† Department of Statistics, University of Wisconsin-Madison, and Department of Mathematics and Statistics, University of Edinburgh

‡ Department of Statistics and Applied Probability, University of California, Santa Barbara

Abstract

It is important in many econometric and biological applications to evaluate the effective dose (ED) points in the tails of quantal response curves and the curves according to a sensible criterion. Following Geisser (1971), it seems most sensible to evaluate the posterior mean value function of the response curve, since this also gives predictive probabilities of positive responses across the design region. While plausible, it can be insufficient for small samples to base the calculations upon standard multivariate normal likelihood approximations. Exact determinations, for example, via importance sampling (see Zellner and Rossi, 1984) are needed. Extending Leonard (1982a) it is now also possible to compute the exact posterior distribution of the ED points. These are proposed as "design measures", since they can be used to sequentially generate further design points. Related procedures (see Leonard, 1982b) yield excellent frequency properties. For example, a total of 40 observations (10 fixed in advance and 30 chosen sequentially) can assess a response curve to 6% accuracy for all design points between the ED60 and ED90 points, with an average of over 90% of the sequentially generated design points falling between the ED60 and ED90 points.

KEYWORDS: Biossay, quantal response curve, posterior mean, effective dose, design measure.

1. BACKGROUND AND INTRODUCTION

Consider a quantal response curve $\theta(x)$, for $x \in D = (a, b)$. Consider independent binomial responses $y_1, \ldots, y_m$ with respective probabilities $\theta_1 = \theta(x_1), \ldots, \theta_m = \theta(x_m)$, and sample sizes $n_1, \ldots, n_m$, where each specified design point $x_1, \ldots, x_m$ falls in D. Convenient link functions involve either the logits α_i, satisfying

$$\theta_i = \frac{e^{\alpha_i}}{1 + e^{\alpha_i}} \qquad (i = 1, \ldots, m), \tag{1.1}$$

or alternatively the probits, also denoted by the α_i, but satisfying

$$\theta_i = \Phi(\alpha_i) \qquad (i = 1, \ldots, m), \tag{1.2}$$

where Φ denotes the standard normal distribution function. In either case, let

$$\alpha_i = \mathbf{x}_i^T \boldsymbol{\beta}, \qquad (i = 1, \ldots, m),$$

with $\mathbf{x}_i = (1, x_i)^T$ and $\boldsymbol{\beta} = (\beta_0, \beta_1)^T$. Whenever, the maximum likelihood vector $\hat{\boldsymbol{\beta}}$ of $\boldsymbol{\beta}$ exists, and the likelihood information matrix is non-singular, the posterior distribution of $\boldsymbol{\beta}$ can, in the absence of prior information, be regarded as approximately multivariate normal with mean vector $\hat{\boldsymbol{\beta}}$ and covariance matrix equal to the likelihood dispersion matrix $\mathbf{Q}$, which satisfies

$$\mathbf{Q}^{-1} = \sum_{i=1}^{m} n_i \hat{\theta}_i (1 - \hat{\theta}_i) \mathbf{x}_i \mathbf{x}_i^T,$$

where $\hat{\theta}_i = \Phi(\mathbf{x}_i^T \hat{\boldsymbol{\beta}})$, for the probit model, and $\hat{\theta}_i = \exp(\mathbf{x}_i^T \hat{\boldsymbol{\beta}}) / \{1 + \exp(\mathbf{x}_i^T \hat{\boldsymbol{\beta}}\}$, for the logit model. Remember that $\hat{\boldsymbol{\beta}}$ and $\mathbf{Q}$ change according to the choice of link function, e.g., (1.1) or (1.2). Furthermore, in the presence of prior information the hierarchical Bayes techniques due to Hsu and Leonard (1995) can alternatively be employed.

For the probit model, many statisticians estimate the quantal response curve by

$$\hat{\theta}(x) = \Phi(\mathbf{x}^T \hat{\boldsymbol{\beta}}), \tag{1.3}$$

with $\mathbf{x} = (1, x)^T$ and where $\hat{\boldsymbol{\beta}}$ denotes the maximum likelihood vector of $\boldsymbol{\beta}$. However, Bayesians frequently prefer the posterior mean value function of $\theta(x)$, since this also provides predictive probabilities of positive responses for future observations. Under the above normal approximation, a couple of algebraic tricks can be used to demonstrate that the posterior mean value function of $\theta(x)$ is

$$\theta^*(x) = \Phi(\mathbf{x}^T \hat{\boldsymbol{\beta}} / \sqrt{1 + v}) \qquad (x \in D), \tag{1.4}$$

where

$$v = \mathbf{x}^T \mathbf{Q} \mathbf{x}.$$

The curve (1.4) is flatter then (1.3) and hence, for example, can provide ED90 or ED99 points in the more extreme tails of D. There are similar considerations under the logit model (1.1). In this case, $\theta^*(x)$ should instead be expressed as the expectation of $e^u/(1 + e^u)$, when, u is normally distributed with mean $\mathbf{x}^T \hat{\boldsymbol{\beta}}$ and variance v.

The data in Table 1 were introduced by Bliss (1967, p. 540) and indicate the fertility or otherwise of 57 rats, each of which has been subjected to one of six dose levels of a drug. The dose levels are described as fractions of 25 milligrams.

Table 1: Rat Fertility Data

Dosage level	Fertile	Sample size
0.15	0	5
0.20	2	10
0.25	4	10
0.30	8	10
0.40	11	11
0.60	11	11

Curve (a) of Figure 1 is the maximum likelihood curve (1.3) for the probit model. Curve (b) is the approximation (1.4) to the posterior mean value function of the response curve. This is however, surprisingly different from the exact posterior mean value function (c) of the response curve, under a uniform prior for $\boldsymbol{\beta}$. The later was calculated via the Zellner-Rossi (1984) importance sampling techniques. We have therefore demonstrated that the standard approximations, based upon approximate normality of the likelihood function, may not be adequate for small samples. For interpretation purposes, we also report curve (d), which replaces $\hat{\boldsymbol{\beta}}$ in (1.3) by the exact posterior mean vector of $\boldsymbol{\beta}$. Note that, as would be expected from two applications of Jensen's inequality, curve (d) is substantially steeper than curve (c). Also, for the current example, the maximum likelihood curve (a), is nearly identical to our recommended curve (c). However, an advantage of our Bayesian approach, when compared with the maximum likelihood procedures employed on standard statistical packages, is that the posterior standard deviation of the quantal response curve can be calculated exactly. See curve (e) of Figure 1. Note that, the observed proportions are represented by the crosses.

Nearly identical curves were obtained for the logit model (1.1). Table 2 provides numerical comparisons of the estimated ED95 and ED99 points, obtained by taking

Table 2: Numerical Comparisons of the Estimated ED95 and ED99 points

Model	Point	(a)	(b)	(c)	(d)
LOGISTIC	ED95	0.348	0.365	0.349	0.337
LOGISTIC	ED99	0.399	0.453	0.412	0.382
PROBIT	ED95	0.344	0.359	0.347	0.336
PROBIT	ED99	0.380	0.428	0.398	0.369

Key: (a) Maximum likelihood;
(b) Approximate posterior mean function;
(c) Exact posterior mean function;
(d) Replaces $\hat{\boldsymbol{\beta}}$ in (a) by exact posterior mean vector of $\boldsymbol{\beta}$.

projections from the exact curves. Note, for example, that approximation (1.4) suggests a value of 0.428 for the ED99 point, compared with the exact answer of 0.398. The differences can become greater for higher ED points.

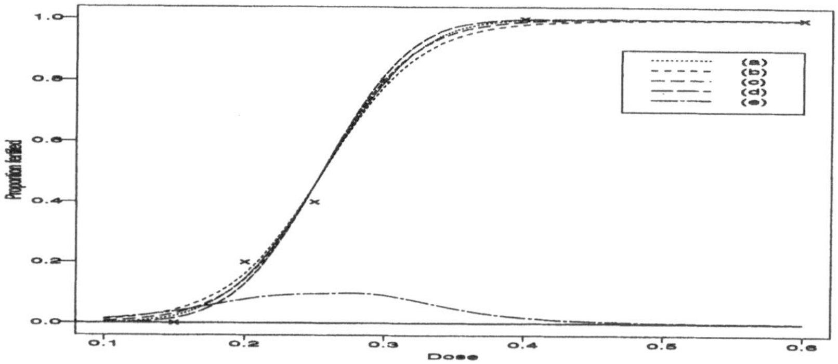

Figure 1: Fitted Response Curves for Probit Model. (a) maximum likelihood; (b) approximate posterior mean function; (c) exact posterior mean function; (d) Replaces $\hat{\boldsymbol{\beta}}$ in equation (1.3) by exact posterior mean vector of $\boldsymbol{\beta}$; (e) exact posterior standard deviation function.

2. SEQUENTIAL DESIGN

Consider the logit model (1.1). Then the ED99 point is

$$x_{99} = \frac{q_{99} - \beta_0}{\beta_1}, \tag{2.1}$$

where $q_{99} = \log(0.99) - \log(0.01)$, with similar definitions for the other ED points. Leonard (1982a) obtained a rough approximation to the posterior distribution of x_{99} and proposed this as a sequential design measure. This was developed in the context of suggesting the velocity needed for a missile to possess 99% probability of piercing its target. Note, the exact density of x_{99} can be derived when β_0 and β_1 are taken to possess a bivariate normal distribution.

The histogram in Figure 2 describes the exact posterior density of x_{99} for the Bliss data, calculated by importance sampling. This can be used to suggest the choice of next design point, i.e., the dosage level needed to give a rat 99% probability of fertility. Either the posterior median or mode might be employed. Clearly, the procedure can be completed sequentially. Alternatively, if a batch of several further design points is needed for the purpose of evaluating x_{99}, these could be equally spaced along approximate percentiles of our posterior distribution. Another possibility would be to randomly generate design points from this "design measure". Curve (b) describes the Laplacian approximation (e.g., Leonard et al, 1989) to the posterior density of the effective dose. The approximation

$$\pi^*(x_{99}|\mathbf{y}) \propto [r(x_{99})]^{-\frac{1}{2}} \bar{\pi}[\tilde{\beta}_1(x_{99})|\mathbf{y}] \tag{2.2}$$

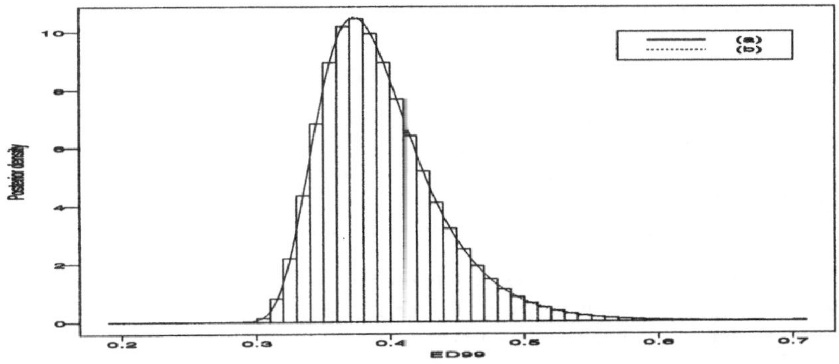

Figure 2: Posterior Density of x_{99} for the Bliss Data. (a) histogram; (b) Laplacian approximation.

is remarkably accurate, even in its tails. Here $\tilde{\beta}_1(x_{99})$ conditionally maximizes

$$\bar{\pi}(\beta_1|\mathbf{y}) = \frac{|\beta_1| \exp[(q_{99} - x_{99}\beta_1)\sum_{i=1}^{m} y_i + \beta_1 \sum_{i=1}^{m} x_i y_i]}{\prod_{i=1}^{m}\{1 + \exp[q_{99} + \beta_1(x_i - x_{99})]\}^{n_i}}, \tag{2.3}$$

with respect to β_1 for each fixed x_{99}, and $r(x_{99})$ evaluates

$$\sum_{i=1}^{m} \frac{n_i(x_i - x_{99})^2 \exp[q_{99} + \beta_1(x_i - x_{99})]}{\beta_1^2\{1 + \exp[q_{99} + \beta_1(x_i - x_{99})]\}^2} \tag{2.4}$$

at the conditional maximum $\tilde{\beta}_1(x_{99})$.

If it is required to estimate the response curve across a range, say between the ED60 and ED90 points, then we recommend a choice of design measure which averages all 31 posterior distributions of the effective dose, from ED60, ED61, up to ED90. Leonard (1982b) used frequency simulations to evaluate this procedure, when performed sequentially, and when the next design point was always the mode of his approximation to our design measure. He referred to the criteria,

(a) Estimation accuracy = average [$\max_{0.6 \le \theta(x) \le 0.9} |\theta^*(x) - \theta(x)|$], where $\theta^*(x)$ is defined in (1.4), and

(b) Design percentage efficiency = long run percentage of x values satisfying $0.6 \le \theta(x) \le 0.9$.

Here "average" and "long run percentage" refer to replications of the whole sequential experiment, and the x-values are confined to those chosen via the design measure. In a situation where the true response curve was chosen to be

$$\theta(x) = \frac{e^{-4.39+8.79x}}{1+e^{-4.39+8.79x}},$$

ten badly chosen initial design points $0, 0.05, 0.1, 0.15, \ldots, 0.45$ were selected corresponding to θ values ranging between 0.012 and 0.392. Note that, the initial values can in practice be chosen much more sensibly. However, even with these poor initial values, computer simulations demonstrated that only 30 further observations are needed to obtain the above defined (a) estimation accuracy = 6.1% and (b) design percentage efficiency = 91.3%.

Alternative Bayesian design approaches in the literature refer to formal criteria. These include Freeman (1970) and Kuo (1983). We regard the suggestion of a "sequential design measure" as proposing a largely inferential approach to this problem, and with a justification which does not depend upon a specific design criterion. Other Bayesian procedures for drawing inferences regarding effective doses are presented by Disch (1981) and Ramsey (1972) and some non-Bayesian procedures are described by Cox (1970, Chapter 4).

ACKNOWLEDGEMENTS

Thanks are due to Seymour Geisser for suggesting that predictive distributions can be used for estimating sampling distributions, to Irwin Guttman for further suggestions, and to Michael Hamada for his substantial help.

References

[1] Bliss, C.I. (1967). *Statistics in Biology: Statistical Methods for Research in the Natural Sciences.* New York: McGraw-Hill.

[2] Cox, D.R. (1980). *Analysis of Binary Data.* London: Chapman and Hall.

[3] Disch, D. (1981). Bayesian nonparametric inference for effective doses in a quantal-response experiment. *Biometrics*, **37**, 713-722.

[4] Freeman, P.R. (1970). Optimal Bayesian sequential estimation of the median effective dose. *Biometrika*, **57**, 79-89.

[5] Geisser, S. (1970). The inferential use of predictive distributions. In *Foundations of Statistical Inference*, V.P. Godambe and D.A. Sprott (Eds.). Toronto: Holt, Rinehart, and Winston.

[6] Hsu, J.S.J. and Leonard, T. (1995) Hierarchical Bayesian semi-parametric procedures for logistic regression. Unpublished manuscript.

[7] Kuo, L. (1983). Bayesian bioassay design. *Ann. Statist.*, **11**, 886-895.

[8] Leonard, T. (1982a). An inferential approach to the bioassay design problem. MRC Technical Report # 2416, Mathematics Research Center, University of Wisconsin-Madison.

[9] Leonard, T. (1982b). A small sample evaluation of a Bayesian design method for quantal response models. MRC Technical Report # 2478, Mathematics Research Center, University of Wisconsin-Madison.

[10] Leonard, T., Hsu, J.S.J. and Tsui, K. (1989). Bayesian marginal inference. *J. Am. Statist. Assoc.*, **84**, 1051-1057.

[11] Ramsey, F.L. (1972). A Bayesian approach to bioassay. *Biometrics*, **28**, 841-858.

[12] Zellner, A. and Rossi, P. (1984). Bayesian analysis of dichotomous quantal response models. *J. Econ.*, **25**, 365-393.

Some Simulation Results for the Performance of DNA Classification of Noisy Image Data

by

S.James Press[1]
University of California
Riverside, CA

Abstract

The Directional Neighborhoods Approach (DNA) to classifying pixels and reconstructing images from remotely sensed noisy data is a newly proposed, computer intensive, Bayesian-type procedure. It uses the observational data to select an optimal, generally asymmetric, but relatively homogeneous, neighborhood for contextually classifying pixels. We provide Monte Carlo simulations for a 2-population image and compare DNA results with those from a reference Bayesian contextual classification. We show that DNA improves substantially upon the reference classification procedure.

Key Words: Spatial, Bayes, Classification, Contextual, Neighborhood.

1 INTRODUCTION

This paper presents some simulation results that demonstrate the effectiveness of a new statistical procedure for classifying pixels from images reconstructed from remotely sensed noisy data. The new procedure, called DNA, for Directional Neighborhoods Approach, was described in Press, 1996, and is summarized briefly in Section 2. Section 3 discusses a simulation study of DNA compared to a reference Bayesian classification procedure. The paper ends with Section 4 where we present some conclusions.

We conceptualize the problem, as follows. There is a satellite equipped with a multispectral sensor that can receive signals reflected from the earth, in each of several frequencies. There is an area of interest on the ground called the "scene". Within the scene there may be one or more entities of specific interest. They are called images. For various reasons, such as noise within the sensor, errors in correcting for sun angle and other geometric factors and atmospheric interference, the signals received passively from the ground (called reflectances) have noise superimposed on spectral signatures.

For example, if the reflectances are obtained for estimating acreage of agricultural land use, the reflectance spectra differ depending upon whether the reflectances are obtained

[1]The author is grateful for NSF travel grant NSF/INT-9016175. He is also grateful to Getachew Dagne for his help with the computer simulation work in Section 3.

from vegetation, water expanses, fallow land, etc., and there is noise superimposed on these reflectance spectra. The reflectances are digitized for evaluation. Our problem is to classify the image based upon our noisy reflectance data.

To do so we set up a rectangular grid in the plane of the scene and thereby create a lattice of square picture elements called pixels. We then want to classify each pixel in the scene into one of K preassigned populations.

For example, the scene may contain a mixture of wheat and rice fields in various patterns, or it may contain a farmhouse and a river. By classifying each pixel, we are classifying the entire scene. This paper proposes a new method for classifying the pixels. It is called DNA. It is Bayesian and data analytic (a "fully" Bayesian procedure would be much too complicated to develop in this context.

We assume that there are K possible populations, or labels for each pixel. So a given pixel might really be wheat, rice, soybeans, etc., and a reflectance from a given pixel represents just one population. Signals from neighboring pixels are likely to be spatially correlated. Moreover, labels from neighboring pixels are likely to be spatially correlated. There will be a signal stochastic process and a labeling stochastic process. Each process will be indexed by a two-dimensional position vector, s (the coordinates could be latitude and longitude).

Order the pixels in the lattice in any reasonable way. Associate with pixel i an unknown label x_i, and a p-vector of observations, $\mathbf{z}_i$. Take $x_i = X(\mathbf{s}_i)$, where $X(\mathbf{s})$ denotes a stochastic labelling process, and s denotes a location defined in a two-dimensional coordinate system in the plane of the scene. We assume very little about the unknown labelling process, $X(\mathbf{s})$; only that it is ergodic.

2 DIRECTIONAL NEIGHBORHOODS APPROACH

2.1 Notation

There is a main pixel of interest that we want to classify into one of K populations. z_0 is a p-dimensional observation vector associated with the main pixel. We consider a collection of neighborhood sets, N, the ith member of which, N_i, includes r pixels defined as "neighbors" of the main pixel. The sets need not be disjoint, nor need the pixels within a neighborhood set be closest to the main pixel, in any usual sense. One of these sets will be used contextually to assist in the classification of the main pixel; we will use the data to help to select which one. The observation vectors of the neighbors (and that of the main pixel) are realizations of the spatial stochastic process $Z(\mathbf{s})$, for s defined on the plane, and they are denoted by $z_1, \ldots, z_r$; along with the observation vector of the main pixel, they will be referred to as "the data". $Z(\mathbf{s})$, given the label in a given population, will be taken to be covariance stationary and ergodic; observations from different populations are taken to be independent. The data will be augmented by prior information to determine which neighborhood set will be used in the classification of the main pixel.

2.2 Neighborhood sets

To be a "neighbor", a pixel only needs to be included in one of the neighborhood sets of the main pixel. We construct neighborhood sets according to prior beliefs about the types of regions, in a given context, that are likely to be "homogeneous", in the sense that most of the pixels within the neighborhood set have the same labels. To illustrate the procedure for constructing neighborhood sets we work with the 5×5 lattice array depicted in Fig.1. The extension of the procedure to other size neighborhoods, other size lattice arrays, and other shapes of neighborhoods is immediate. The location of each pixel in the region of interest for selecting neighborhoods is designated by one of the integers: $0, 1, \ldots, 24$, and the main pixel of interest is designated by "zero".

1	2	3	4	5
6	7	8	9	10
11	12	0	13	14
15	16	17	18	19
20	21	22	23	24

Figure 1

We first define 8 "leaf-neighborhood" sets by compass directions from pixel 0. Each neighborhood set contains 6 pixels (see Figure 2).

Set Number	Angle (in degrees)	Compass Direction	Leaf Neighborhood Set
1	45	northeast	[4,5,8,9,10,13]
2	90	east	[9,10,13,14,18,19]
3	135	southeast	[13,17,18,19,23,24] .
4	180	south	[16,17,18,21,22,23]
5	225	southwest	[12,15,16,17,20,21]
6	270	west	[6,7,11,12,15,16]
7	315	northwest	[1,2,6,7,8,12]
8	360	north	[2,3,4,7,8,9]

Figure 2

It will be seen by forming the convex hull of each neighborhood set, with the zero pixel included, that the sets so defined approximate (overlapping) leaves of a tree. The 45 degree leaf is outlined in Fig.1, for illustration. Note that leaf neighborhood sets are not symmetric about the main pixel of interest (in this case, pixel "0").

2.3 Algorithm For Contextual Classification

Let τ_β denote the predictive posterior probability that the label for the main pixel is β, conditional on the observation vectors of the main pixel, those of the pixels in a neighborhood set, and the prior information,) (we will show how to find the predictive posterior probability in Sect.2.5). That is, for $\beta = 1, 2, \ldots, K$,

$$\tau_\beta \equiv P\{X_0 = \beta | Z, \prod\} , \tag{1}$$

where X_0 denotes the label for the main pixel located at "zero", and $Z = (z_0, z_1, \ldots, z_r)$. But this probability statement leaves open the question of which neighborhood set we should take. We propose to let the data determine this issue, in a way to be described below, in this section.

Analogously, define the posterior classification probability $\tau_{i,\beta}$ that the main pixel has label β, for neighborhood set $N_i, i = 1, \ldots, L; \beta = 1, \ldots, K$; given $(Z, \prod)$. Thus,

$$\tau_{i,\beta} \equiv P\{X_0 = \beta | Z_i, \prod\} , \tag{2}$$

where Z_i denotes Z when neighborhood set Ni is used. Applying Bayes theorem in eqn. (2) gives

$$\tau_{i,\beta} \propto p_\beta p(Z_i | X_0 = \beta, \prod) , \tag{3}$$

where $\propto$ denotes proportionality, and the prior probability for label β is given by

$$p_\beta \equiv P\{X_0 = \beta | \prod\} . \tag{4}$$

Now condition on the possible labels the pixels in the neighborhood set N_i might have. The pixel labels for pixels in N_i are designated by $(X_{i1}, \ldots, X_{ir})$. This gives

$$\begin{aligned} \tau_{i,\beta} &\propto p_\beta \sum_{\rho_1=1}^{K} \cdots \sum_{\rho_r=1}^{K} p(Z_i | X_0 = \beta, X_{i1} = \rho_1, \ldots, X_{ir} = \rho_r, \prod) \\ &\times P\{X_{i1} = \rho_1, \ldots, X_{ir} = \rho_r | X_0 = \beta, \prod\}, \end{aligned} \tag{5}$$

This expression is very complicated and difficult to evaluate. We therefore seek to approximate it. We now show how a data-driven approximation can be effected.

Suppose all the pixels in a given neighborhood set had the same label as the main pixel, namely β. That is, in that region of the scene, the pixels were homogeneous. Then, we could say

$$P\{X_{i1} = \beta, \ldots, X_{ir} = \beta | X_0 = \beta\} = 1 . \tag{6}$$

Then eqn. (5) would reduce to the much simpler form,

$$\tau_{i,\beta} \propto p_\beta \cdot p(Z_i | X_0 = \beta, X_{i1} = \beta, \ldots, X_{ir} = \beta, \prod) . \tag{7}$$

This idea corresponds to local spatial continuity in the labelling process (see Switzer, 1980, and Mardia, 1984). While in some contexts (agricultural covers, for example) there might be large expanses of pixels with common labels, even in those contexts homogeneity of labelling would be unlikely to hold at or near the boundaries of regions. At a pixel near the boundary, if we took our neighborhood set symmetrically around the main pixel, it is likely (and necessary) that the neighborhood set would contain pixels with dissimilar labels. Alternatively, for the main pixel near the boundary, were we to take an asymmetric neighborhood with respect to the main pixel, and if we were to take it in the appropriate direction, using, say, one of our leaf-neighborhoods discussed in Section 2.2, we would be likely to find many pixels in the neighborhood set, if not all, having the same label as the main one. We plan to select directions for the neighborhood sets in this way regardless of whether the pixel is near a boundary or not. That's the motivation behind the approximation we propose. We implement this idea as follows.

Let $q_{i,\beta}$ denote a criterion measure of homogeneity of the labels in the neighborhood N_i relative to population β. We will explore several different potential homogeneity criteria in Section 2.4. If we evaluate $q_{i,}$ for all neighborhood sets in N, and for all β, we could choose that neighborhood set for which $q_{i,}$ is optimal. If there is no neighborhood set within N for which $q_{i,}$ is acceptable relative to some preassigned value (such as a threshold), we could augment the collection of neighborhood sets with others (preassigned neighborhood sets designed to be used for augmentation, when necessary). Once we have determined the optimal neighborhood set, say N^*, we will use that neighborhood set for all β, for that pixel. We then use N_i to evaluate eqn.(7), approximately.

We must next maximize eqn.(7) over all β. We may accomplish this by calculating all of the predictive posterior probabilities, or we may evaluate the predictive probability odds ratios. The result of this maximization yields the desired classification for the pixel. Carrying out this procedure for each pixel in the scene gives a reconstructed image of the scene.

2.4 Criterion For Homogeneity of Neighborhoods

To evaluate the degree of homogeneity of the pixel labels in a neighborhood we propose to obtain a posterior probability estimate from a preliminary classification using the classical, zero-neighbor, Bayesian classification procedure. A more sophisticated contextual classification algorithm could also be used (see, e.g., Klein and Press, 1989, 1990a, 1990b, 1993). To implement such an approach we would require some prior information, about the (unknown) population parameters of the K underlying populations represented by the labels. Such prior information is typically a sample of observation vectors taken from each of the K populations. These Bayesian classification procedures yield a posterior probability for a given pixel having the label $\beta = 1, . . , K$. Given the posterior probability estimates for the main pixel and for those in the appropriate neighborhood, we can now evaluate various criteria for degree-of-homogeneity of pixel labels based upon these posterior probability estimates. Several potential alternative criteria are enumerated below.

Once we have the posterior probability estimates for each pixel in the scene we can classify all pixels as having the label corresponding to the highest posterior probability. Let $q_{i,\beta}$ denote the proportion of pixels preclassified as having the label β in neighborhood set N_i ($q_{i,\beta}$ should include the main pixel; see explanation below).

For example, suppose there are only two possible labels: "black", and "white". We can examine the proportion of pixels that have been preclassified as having "black" labels and those that have "white" labels in each neighborhood set in N (including the main pixel to be classified currently). These proportions are the $q_{i,\beta}$ measures of homogeneity. Now we propose to select that neighborhood set that has the greatest proportion of pixels preclassified as having label β, including the main pixel. Suppose that the main pixel was preclassified as "black". Suppose further that there are six pixels in each neighborhood set (as in the case of "leaf neighborhoods"). Suppose also that the neighborhood set that has the greatest number of "black" neighbors has five "black" pixels, and including the main pixel there is therefore a directional neighborhood with six homogeneous pixels (in this case, "black"). If all other neighborhood sets have fewer than six homogeneous pixels (of either color) the algorithm would select that neighborhood that contains these five "black" pixels.

Once we have determined the optimal neighborhood set we can proceed with our contextual classification algorithm.

<u>Ties</u>

Using this approach it is almost inevitable that there will be instances in which "ties" will occur; that is, cases for which two or more neighborhoods have the same maximum proportions of pixels classified as having the same labels (for details, see Press, 1996).

2.5 Predictive Posterior Probability

We follow the predictive Bayesian approach developed by Geisser (see Geisser, 1964, 1965, 1966). We are interested in formulating a calculable expression for eqn. (1).

By Bayes theorem, eqn. (1) may be written

$$P\{X_0 = \beta | Z, \prod\} = \frac{p(Z|X_0 = \beta, \prod)P\{X_0 = \beta\}}{\sum_{\alpha=1}^{K} p(Z|X_0 = \alpha, \prod)P\{X_0 = \alpha\}} \cdot \tag{8}$$

But $p(Z|X_0 = \beta, \prod)$ may be expressed as

$$p(Z|X_0 = \beta, \prod) = \int p(Z|X_0 = \beta, \prod, \Theta_\beta) \cdot f(\Theta_\beta | \prod) d\Theta_\beta \tag{9}$$

where Θ_β denotes the set of parameters that indexes the distribution of Z_β, and $f(o)$ denotes the posterior density of these parameters given the prior information; the prior information would typically include observation vectors known to come from each of the K populations, or digitized intensity data whose origin is known. The first term in the integrand in (10) is just the likelihood function for the observational data, and the second term is readily calculated by adopting a prior on the unknown parameters, and developing the likelihood for the prior observational data whose origin is known. These calculations were carried out explicitly for the case of normally distributed data in Klein and Press, 1989, 1990a, 1990b (for different assumptions about the relative spatial relationships among the pixels corresponding to Z, Π), respectively, using the assumption of "local spatial continuity".

In the case in which the training data could be taken to be independent of the data from the pixels to be classified (the scene), the result was that the label classification probability could be taken to be proportional to a matrix T density centered at the generalized means of the training data (generalized to account for the spatial autocorrelation in the data):

$$P(X_0 = \beta | Z, \Pi) \propto \frac{P(X_0 = \beta)}{|Q + (Z - Z^*)P^{-1}(Z - Z^*)'|^{m/2}}$$

where $P\{X_0 = \beta\}$ denotes the prior classification probability, Z denotes the observation vectors for pixel 0 and for the neighborhood pizels ($Z = z_0, z_1, \ldots, z_r$), Z^* denotes a matrix whose columns are identical (because of local spatial continuity) and equal to the

"generalized mean vector" for population β in the training data; P, Q denote matrices that depend upon the "generalized sum-of-squares" matrix for the training data.

3 Simulation Study

To study the performance of DNA we carried out a variety of simulations. For all of them we chose the same scene, a symmetric pair of rectangles, as depicted in Fig. 3.

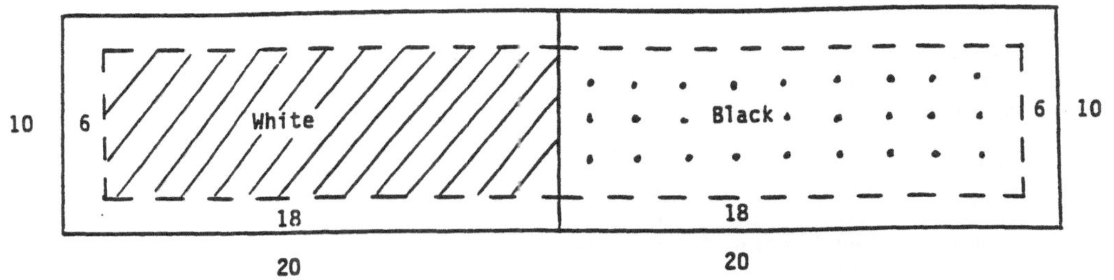

Scene Used for Simulation
Fig. 3

That is, our scene consists of a rectangle that is 10 × 40, but to avoid edge difficulties we classify only the pixels in the interior 6 × 36 pixel rectangle; and there are only two colors possible, white and black. The interior pixels divide the rectangle symmetrically into two subrectangles, on the left a rectangle composed of 6 × 18 all white pixels, and on the right, a rectangle of 6 × 18 all black pixels (shown in dotted lines). We selected a very simple scene to isolate the DNA effect from the complexity of an image.

Noisy observational data was taken to be just one-dimensional, and distributed $N(0,1)$ for the white pixels, and $N(\theta,1)$ for the black pixels, for $\theta = 0.5, 1.0, 2.0$, and 3.0.

The data vectors were permitted to be spatially correlated with structure:

$$\mathrm{corr}\{Z(0), Z(s)|Z(0) \text{ and } Z(s) \text{ have the same label}\} = e^{-C(s's)} ,$$

$$\mathrm{corr}\{Z(0), Z(s)|Z(0) \text{ and } Z(s) \text{ have different labels}\} = 0 ,$$

for some fixed, positive constant C, the same value for each label. Thus, for $C = \infty$, the observations are independent, and for $C = 0$, the observations are perfectly correlated. $\{Z(s), s \in \mathbf{R}^2\}$ is a spatial stochastic process on the plane.

For training data we used 2000 observations per population generated independently of the scene. We used "PCC", or proportion of correct classifications in a given reconstruction as the criterion of quality of classification. We carried out 300 distinct reconstructions for each case and averaged the results to obtain "mean PCC" for that case. We also calculated mean PCC for just the pixels on the left border (the right border was not necessary because of symmetry), to study the effect of DNA on border classifications.

Results were compared with those from more conventional contextual Bayesian classification (Klein and Press, 1989). The neighborhood for this solution involves using a centrally located pixel symmetrically surrounded by 8 pixels in a tick-tack-toe configuration. We refer to this reference procedure as "0".

We have used rectangular leaf neighborhoods from a 5 × 5 pixel configuration with 4 "leaves" in the 4 main compass directions, north, east, south, and west. Each leaf rectangle contained 9 pixels, including the main one.

For example, referring back to Fig. 1, the subrectangle formed by the convex hull of the pixel set: $\{8, 9, 10, 0, 13, 14, 17, 18, 19\}$ would be the easterly leaf for pixel $\{0\}$.

All pixels were taken to be "interior" pixels, in that it was assumed that every pixel was surrounded by other pixels, so every neighborhood contained 9 pixels (including the main pixel). This was done to avoid obfuscating DNA results with problems of classifying corners and edges of a scene.

We have summarized numerical results in Tables 1 and 2. There, it may be seen that the vertical dimension of the table is defined by a correlation parameter C (column 1), and a "distance between groups" that we'll call θ. We considered just 3 values for C, the cases of $C = \infty$ (independence), $C = 1$ (low correlation), and $C = 0.4$ (high correlation). We studied 4 values of θ, $\theta = 0.5, 1, 2, 3$.

The entries in Table 1 show two values in each cell, a main entry, and another in parentheses. The main entry is the mean PCC (percentage of correct classifications) overall, that is, for all 216 pixels in the scene, averaged over all 300 reconstructions. The entry in parenthesis is the mean PCC for just the left border pixels, averaged over all 300 reconstructions.

The horizontal dimension of Table 2 gives standard deviation of the simulation distributions of 300 runs for overall PCC, and for the PCC for just the left border pixels (given in parentheses).

It may be noted from Table 2 that standard deviations for the simulation distributions are small for the overall classifications (from 0.5 to about 8) and large for the border pixels. In images with a greater number of border pixels we would expect the standard deviation of the pccs of the border pixels to be much smaller.

In order to graphically show the effect of DNA at the border we have plotted the results in Figures 5A-5C for DNA compared with the reference reconstruction (Bayesian Contextual Classification using a symmetric neighborhood). Fig. 5A gives results for zero spatial correlation ($C = \infty$); Fig. 5B for "low" correlation ($C = 1$); and Fig. 5C for "high" correlation ($C = 0.4$). It may be noted that while results for high correlation are enormously better using DNA, even results for zero spatial correlation are substantially better.

Comparison of DNA Mean PCC with Mean PCC of Reference Bayesian Contextual Classification

Correlation Parameter C	Distance Between Groups	Contextual Class. (Ref.)	DNA
0.4-high corr.	0.5	60.5(40.2)	63.1(61.4)
0.4	1	70.0(33.2)	74.0(68.2)
0.4	2	85.5(20.0)	87.6(85.1)
0.4	3	93.4(10.4)	93.8(95.3)
1.0-low corr.	0.5	74.4(54.7)	78.3(62.7)
1	1	90.0(60.4)	93.4(73.6)
1	2	98.5(73.7)	99.2(89.3)
1	3	99.4(84.1)	99.8(97.1)
∞-indep.	0.5	76.4(56.3)	79.0(63.3)
	1	91.8(64.8)	93.7(73.4)
	2	98.9(81.7)	99.2(89.1)
	3	99.6(91.6)	99.8(96.8)

Entries are: mean pcc for all pixels (mean pcc for left border. pixels)
Scene is 6×36 rectangle of interior cells, left half are white, right half are black.
Noise is $N(0,1)$ vs. $N(\theta,1)$, $\theta = 0.5, 1, 2, 3$; correlation is $\exp\{-C(s's)\}$.
Each case averages 300 distinct simulations

Table 1

Standard Deviation of Overall PCC Distribution
And (In Parentheses) Standard Deviation Of The
PCC Distribution At The Left Border

Correlation Parameter C	Distance Between Groups	Contextual Class (Ref.)	DNA
0.4-high corr.	0.5	3.5(20.3)	4.3(20.6)
0.4	1	3.4(18.9)	3.7(20.4)
0.4	2	2.4(16.0)	2.9(15.9)
0.4	3	1.5(12.0)	2.0(9.2)
1.0-low corr.	0.5	5.7(27.4)	6.8(28.4)
1	1	3.5(26.4)	3.5(24.7)
1	2	0.8(23.4)	0.7(16.2)
1	3	0.6(19.4)	0.3(9.1)
∞-Indep.	0.5	5.2(28.6)	7.1(28.5)
	1	3.6(27.4)	3.7(26.8)
	2	0.8(21.8)	0.7(17.1)
	3	0.5(15.4)	0.4(9.7)

Table 2

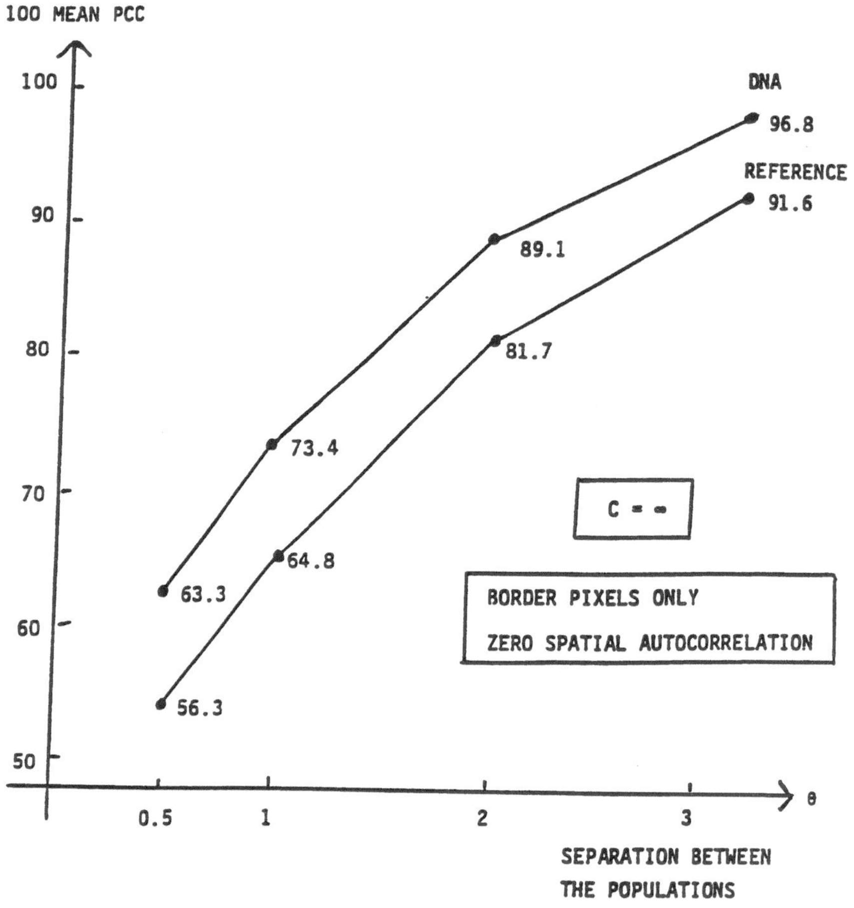

Comparison of DNA With Contexual Bayesian Classification, for Border Pixels Only – Zero Spatial correlation
Figure 5A

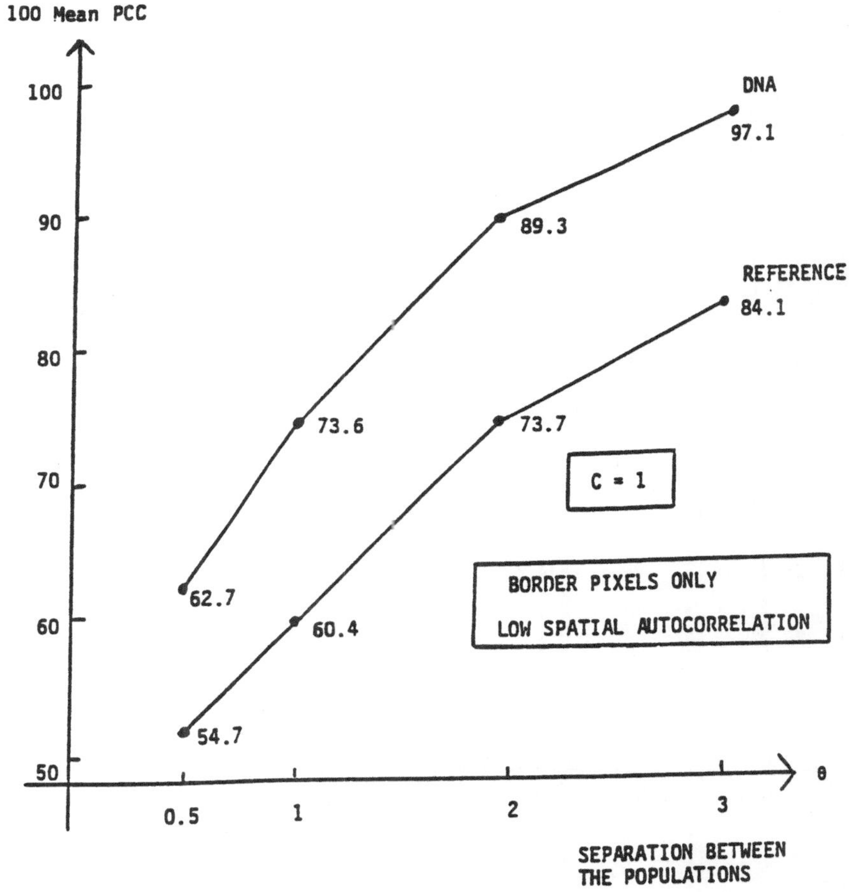

Comparison of DNA With Contextual Bayesian Classification, for Border Pixels Only-Low Spatial Correlation
Figure 5B

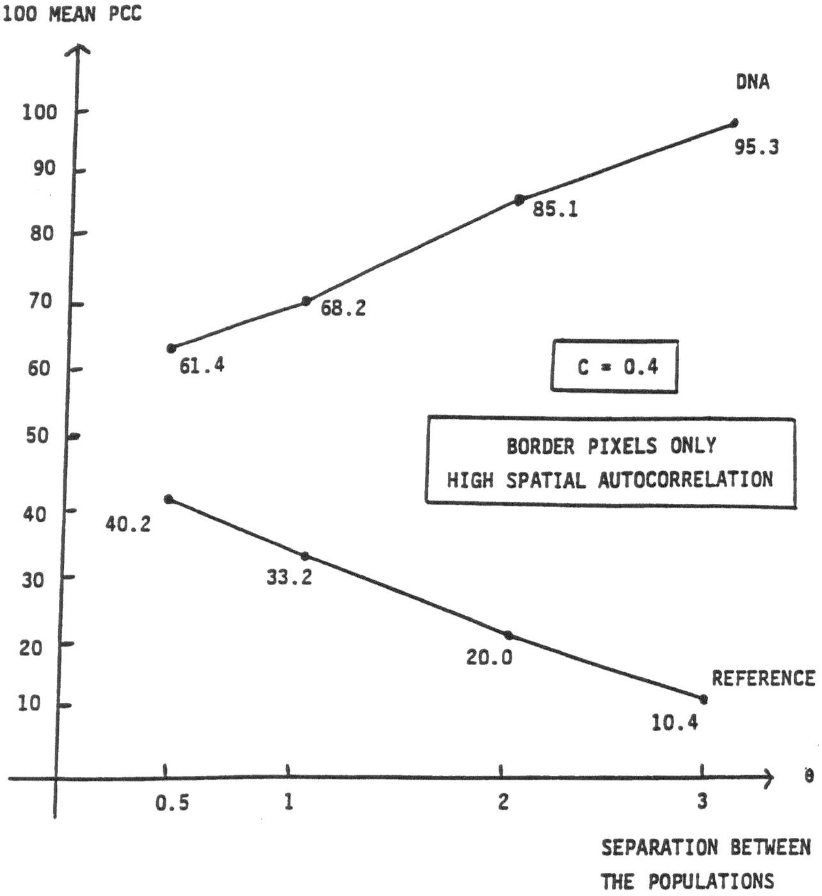

Comparison of DNA With Contextual Bayesian Classification, for Border Pixels Only-High Spatial Correlation
Figure 5C

The intuitive explanation for the poor PCC performance of the reference classification procedure at the border is the following.

The reference classification uses the assumption of "local spatial continuity". So all pixels in the neighborhood (context) are assumed to have the same label as the central one.

1. As the spatial correlation increases and becomes high, it is increasingly likely that pixels close to the central pixel will have the same label as the central pixel, which is impossible for some pixels, if the central pixel is a border pixel. This creates a conflict in the model.

   So as correlation increases, it is increasingly likely that the local spatial continuity assumption is violated, and then the probability of correct classification (PCC) declines under that model.

2. Since the variances are assumed to be the same, the posterior classification probabilities are essentially centered at the means (generalized for spatial correlation) of the two populations for classification.

   So as θ increases, the posterior classification probability for, say, "white" remains with the data centered at zero, while the posterior for, say, "black", moves the center of the data further and further away from zero.

   If we are classifying a border pixel, and we have assume that "local spatial continuity" holds, the posterior classification probability for "white" declines as θ increases because the classification point gets further and further from the mean of the "white" population, namely zero.

3. As both θ increases and spatial correlation increases, for border pixels and symmetric neighborhoods, the effect of the declining PCC intensifies.

But the directional neighborhoods approach clearly overcomes the problem.

It may be of some interest to know the time required to carry out the DNA procedure. The CPU times required for the entire scene (216 pixels) to run all 300 distinct simulations were, for example, as given in the first two numerical columns:

	CPU (seconds)	CPU (minutes)	Seconds per simulation
Reference Classification	968.32	16.14	3.23
DNA	1443.39	24.05	4.8

4 Conclusions

We conclude that at least for this very simple small scene:

(1) using DNA produces overall results at least as good (and a little better than Bayesian contextual classification;

(2) using DNA improves image reconstruction substantially at the border between populations.

(3) since DNA can be used to generate an initial solution for many image reconstruction methods whose quality depends strongly upon the quality of the initial solution, DNA should improve results for many methods, and perhaps substantially in cases involving scenes with complex borders, and perhaps when used dynamically in iterative procedures.

REFERENCES

Geisser, S. (1964), "Posterior Odds for Multivariate Normal Classification", *Journal of the Royal Statistical Soceity*, Ser. B, 26, 69-76.

Geisser, S. (1965), "Bayesian Estimation in Multivariate Analysis", *The Annals of Mathematical Statistics*, 36, 150-159.

Geisser, S. (1966), "Predictive Discrimination", in *Multivariate Analysis*, ed. P. R. Krishnaiah, New York: Academic Press, pp. 149-163.

Klein, R., and Press, S. J. (1989), "Contextual Bayesian Classification of Remotely Sensed Data", *Communications in Statistics*, Part A - Theory and Methods, 18, 3177-3202.

Klein, R., and Press, S. J. (1990a), "Bayesian Contextual Classification with Neighbors Correlated with Training Data", in *Bayesian and Likelihood Methods in Statistics and Econometrics: Essays in Honor of George A. Bernard*, eds. S. Geisser, J. Hodges, S.J. Press, and A. Zellner, New York: North-Holland pp. 337-355.

Klein, R., and Press, S. J. (1990b), "Bayesian Classification of Remotely Sensed Data When Training Data is Part of the Scene", *Revista Brasileira de Probabilidade e Estatistica*, 4, 43-67.

Klein, R., and Press, S. J. (1992), "Adaptive Bayesian Classification of Spatial Data", *Journal of the American Asssociation*, Vol. 87, No. 419, pp. 844-851.

Klein, R., and Press, S. J. (1993), "Adaptive Bayesian Classification with a Locally Proportional Prior", *Communications in Statistics - Theory and Methods*, 22(10), 2925-2940.

Mardia, K. V. (1984), "Spatial Discrimination and Classification Maps", *Communications in Statistics - Theory and Methods*, 13, 2181-2197.

Press, S. J. (1996), "The Directional Neighborhoods Approach To Contextual Classification of Images From Noisy Data", *Jour. Amer. Stat. Assn.*, in press.

Switzer, P. (1980), "Extensions of Linear Discriminant Analysis for Statistical Classification of Remotely Sensed Satellite Imagery", *Mathematical Geology*, 12, 367-376.

Case-Control Studies and Bayesian Inference

by

M. Zelen and R.A. Parker

Commentary

The Editors of this volume have kindly agreed to re-publish this paper on case-control studies in this current volume. Although published in 1986, it has not received attention from Bayesian statisticians or epidemiologists.

It is our strong view that the frequentist methods for analyzing case-control studies are applied incorrectly. The reason for the frequentist methods being incorrect in many applications is that the basis of the case-control study (equation 2 of our paper) requires that random samples be taken from populations of individuals with and without disease. This is rarely done as cases (individuals with disease) are usually compiled from available hospital records and/or disease registries. Data is drawn "conveniently". As a consequence, case control investigations on disease risk factors are often greeted with skepticism when the findings run contrary to the prevailing "wisdom". The recent article in *Science* 269: 164 (1995) discusses the limitations of epidemiologic studies, but fails to note that one of the chief tools used by epidemiologists is flawed.

The Bayesian approach to probability is appropriate for the interpreting of case control studies. Furthermore it enables one to make inferences about risk-factors without necessarily having a control group. There may be associations between disease and risk-factors which are so overwhelming that it would be ludicrous to formalize the obvious conclusions by comparison with data from a control group. Frequentist methods do not allow this flexibility.

We wish to comment on a few technical points in our paper. Our philosophy for carrying out inferences in the case control setting is to compare the ratio of the posterior distribution of the parameter relating to association of disease and a risk factor to the posterior distribution when the parameter has a value of no association. Ratios of these two posterior distributions equal to or greater than 10 are very convincing. There is little need to be more formal. We recommend determining the class of priors which results in a conclusion that a relationship exists, rather than calculating a ratio for a single specific prior probability distribution. If current views and beliefs are consistent with this group of prior probability distributions, then we would conclude that the relationship exists. This strategy has the advantage that it avoids the need for specifying a particular prior distribution to carry out a Bayesian analysis.

Our development only considered the problem when a single risk factor is to be investigated for association with disease. We did not consider the problem of several risk factors or the issue of adjusting for covariates. These are important extensions of our original theory, but are relatively straightforward.

Marvin Zelen, Dana-Farber Cancer Institute
and Harvard School of Public Health

Robert A. Parker, Beth Israel Hospital
and Harvard Medical School

August, 1995

CASE–CONTROL STUDIES AND BAYESIAN INFERENCE†

M. ZELEN AND R. A. PARKER*

Harvard School of Public Health and Dana-Farber Cancer Institute, Boston, Ma. 02115, U.S.A.

SUMMARY

We outline the methods of Bayesian inference for applications to case–control studies. These methods appear as the natural way of making inferences, since much of the controversy that surrounds a specific case–control study is subjective. We derive conjugate prior distributions of exposure, posterior distributions of the ratio of the odds of being incident with a disease both with and without exposure to a potential causal agent, and convenient approximations. In particular, we show how one may carry out 'case–control studies' without necessarily having a control group. We illustrate these ideas with the data that first showed the relationship between *in utero* exposure to diethylstilbestrol and cancer of the vagina in young girls.

KEY WORDS Case–control studies Bayesian inference Epidemiology methods Cancer of the vagina

1. INTRODUCTION

This paper applies Bayesian methods of inference to the design and analysis of case–control studies. Bayesian methods seem especially suitable since case–control studies do not represent 'experiments' which are random (or even necessarily representative) samples from a real or hypothetical population of possible experiments. In fact, cases and controls are often chosen for convenience and ease of obtaining information from individuals. It is somewhat surprising that the entire methodological literature on case–control studies is based on the frequentist interpretation of probability.[1,2]

The Bayesian philosophy seems particularly appealing and appropriate in case–control studies. Starting from a collection of cases one desires to infer if the case exposure to a causal factor is 'unusual'. One can use the information from a control group to help construct the posterior distribution of exposure. In many instances there may be so much prior information available about exposure of the population (e.g. lifestyle habits of smoking and drinking, etc.), that the limited information available from a sample of controls may generate a posterior distribution of exposure nearly the same as the prior distribution. In such situations one can carry out an analysis of the cases and their exposure without even generating data on a control group. The frequentist view of case–control studies does not permit this, which represents a serious shortcoming.

In this paper we outline the Bayesian framework for the comparison of two proportions. The formal Bayesian development is relatively straightforward. The main problem in applying these methods is determination of an appropriate prior distribution. Section 2 of this paper contains the Bayesian formulation with convenient approximations useful in applications; section 3 discusses

* Present address: Centers for Disease Control, Atlanta, GA 30333, U.S.A. Computer programs to implement the methods in this paper may be obtained by writing to Dr. Parker.

† This investigation was supported in part by grants CA-23415 and CA-23318 from the National Cancer Institute. It was presented at the Fifth International Meeting of the International Society for Clinical Biostatistics, San Marino, Sept. 1984.

0277–6715/86/030261–09$05.00

Received September 1984
Revised August 1985

methods for determining suitable prior distributions; section 4 develops methods for making inferences without necessarily having controls; and section 5 applies these methods to data relating to cancer of the vagina and diethylstilbestrol.

2. FORMULATION OF THE BAYESIAN PROBLEM

2.1 Notation and posterior distribution of the odds ratio

Let Y and Z be binary random variables defined by

$$Y = \begin{cases} 1 \text{ if individual is exposed to causal agent} \\ 0 \text{ otherwise} \end{cases}$$

$$Z = \begin{cases} 1 \text{ if individual has disease (case)} \\ 0 \text{ otherwise (control).} \end{cases}$$

The joint distribution of (Y, Z) is multinomial with four classes, denoted by $f(y, z)$. A prospective (hypothetical) study in which individuals are randomly assigned to either an exposed or non-exposed group and then followed for disease incidence would generate the distribution of disease conditional on exposure, i.e. $f(z|y)$. Alternatively, the case–control study relies on the distribution of exposure conditional on disease state, i.e. $f(y|z)$. If $f(y, z) = f_1(y)f_2(z)$, then exposure and disease status are independent.

Without loss of generality, the multinomial distribution, describing the outcome for one individual, may be written as

$$f(y, z) = e^{\alpha y + \beta yz + \gamma z}/(1 + e^{\alpha} + e^{\gamma} + e^{\alpha + \beta + \gamma}), \tag{1}$$

where (α, β, γ) are parameters. The ratio of the conditional odds is

$$e^{\beta} = \frac{f(z = 1 | y = 1)/f(z = 0 | y = 1)}{f(z = 1 | y = 0)/f(z = 0 | y = 0)} = \frac{f(y = 1 | z = 1)/f(y = 0 | z = 1)}{f(y = 1 | z = 0)/f(y = 0 | z = 0)}. \tag{2}$$

We point out that if the disease incidence is very low $f(z = 0 | y) \approx 1$, then expression (2) is, for all practical purposes, the relative risk. (These are well-known relations and we present them for completeness.) The random variables (Y, Z) are independent if and only if $\beta = 0$. Hence the inference about the association between disease and exposure depends on the value of β. Therefore, the goal of Bayesian analysis is to determine the posterior distribution of β from which one can make an inference about an association.

The case–control study generates observations from the probability distribution

$$f(y | z) = e^{(\alpha + \beta z)y}/(1 + e^{\alpha + \beta z}). \tag{3}$$

Suppose one carries out such a study. Let the data consist of the entries in the 2×2 table of Table I.

Table I

	Exposed	Non-exposed	Totals
Cases	s	$n-s$	n
Controls	r	$m-r$	m
Totals	t	$N-t$	N

The likelihood function is

$$\text{lik}(\alpha,\beta) = e^{(\alpha+\beta)s+\alpha r}/(1+e^{\alpha})^{m}(1+e^{\alpha+\beta})^{n} \tag{4}$$

and the conjugate prior distribution for (α, β) is

$$f'(\alpha,\beta) \propto e^{(\alpha+\beta)s'+\alpha r'}/(1+e^{\alpha})^{m'}(1+e^{\alpha+\beta})^{n'}, \tag{5}$$

where primes denote the parameters of the prior distribution. Therefore, the joint posterior distribution of (α, β) is

$$f''(\alpha,\beta) \propto e^{(\alpha+\beta)s''+\alpha r''}/(1+e^{\alpha})^{m''}(1+e^{\alpha+\beta})^{n''}, \tag{6}$$

where $n'' = n + n'$, $m'' = m + m'$, $s'' = s + s'$, $r'' = r + r'$.

Finally, it is necessary to integrate out α to obtain the marginal posterior distribution of β. This can be easily carried out and results in the expression

$$f''(\beta) = Ce^{\beta s''}\int_0^1 \frac{v^{t''-1}(1-v)^{N''-t''-1}}{(1-v+ve^{\beta})^{n''}}\,dv \tag{7}$$

where $t'' = r'' + s''$, $N'' = m'' + n''$ and C is a constant independent of β. The posterior distribution exists, provided $0 < t'' < N''$.

The posterior distribution of β does not exist in closed form. However, Altham[3] showed that, with appropriate choice of different prior distributions, the expression $P\{\beta < 0\}$ can bound (upper and lower bounds) the tail area probabilities associated with Fisher's exact test for the 2×2 contingency table. Latorre[4] has expressed the posterior density in terms of elementary functions.

Since interest generally focuses on $\beta = 0$, it is useful to consider $f''(\beta)/f''(0)$ in order to determine the strength of the evidence comparing any value of β to the value of $\beta = 0$. This expression is

$$f''(\beta)/f''(0) = \frac{e^{\beta s''}}{B(t'', N''-t'')}\int_0^1 \frac{v^{t''-1}(1-v)^{N''-t''-1}}{(1-v+ve^{\beta})^{n''}}\,dv. \tag{8}$$

We can readily calculate the posterior distribution using numerical integration techniques.

Note that the quantity $f''(\beta)/f''(0)$ is the ratio of two posterior probabilities. Define β_0 to be the mode of the posterior distribution of β, i.e. $f''(\beta_0) = \max_{\beta} f(\beta)$. Then if $f''(\beta_0)/f''(0)$ is close to unity it would indicate that the concentration of the area of the posterior distribution is in the neighbourhood of $\beta = 0$. Alternatively a large value of the ratio indicates that the posterior distribution is concentrated away from zero and one would infer association between disease and exposure. As an example, consider the standard normal distribution. The maximum ordinate is at the mean, which is zero. Then the ratios of the ordinate at zero to those at other selected deviates (for well-known tail areas) are as given in Table II.

Table II

Deviate	f (mean = 0)/f(deviate)	Tail area (one tail)
0	1	0.50
1.64	3.8	0.05
1.96	6.8	0.025
2.58	27.9	0.005

Thus ratios above 7 correspond to the tail areas used in practice to decide on statistical significance. We suggest that if $f''(\beta_0)/f''(0)$ exceeds 10, one should conclude that the evidence favours an association. The choice of a critical value of 10 indicates that the modal value (β_0) has ten times the posterior probability of the value $\beta = 0$. The number 10 is somewhat arbitrary and different individuals may require different critical values. In practice, the frequency interpretation of statistical evidence uses a 5 per cent or 1 per cent false positive rate. However there is no rationale for choosing these or, for that matter, any other critical values. The choice (for all practical purposes) is subjective. Hence we can adopt relatively large posterior probability ratios as evidence in favour of an association between disease and a causal factor. Alternatively one can plot $f''(\beta)/f''(0)$ or $f''(\beta)$ and judge whether the area is concentrated in a neighbourhood of $\beta = 0$.

2.2 Approximations to the posterior distribution

Evaluation of the expression for $f''(\beta)/f''(0)$ requires a computer. In this section we develop an easy to use approximation which suffices for many applications when m'' and n'' are not small.

One can obtain a normal approximation to the posterior distribution of β by use of the well-known relations between a beta random variable and two independent chi-square random variables. For this purpose, consider the transformations

$$v/(1-v) = e^{\alpha}, \; u/(1-u) = e^{\alpha+\beta}.$$

Substituting for α and β in equation (6) gives

$$f(v, u) \propto v^{r''-1}(1-v)^{m''-r''-1}u^{s''-1}(1-u)^{n''-s''-1}.$$

Thus (U, V) are independent random variables following beta distributions. Since

$$\beta = \log[u/(1-u)] - \log[v/(1-v)], \tag{9}$$

we can write (9) as

$$\beta = \log(\chi_1^2/\chi_2^2) - \log(\chi_3^2/\chi_4^2),$$

where χ_i^2 are independent random variables following a chi-square distribution with ν_i degrees of freedom. The values of the ν_i are

$$\nu_1 = 2s'', \quad \nu_2 = 2(n''-s''), \quad \nu_3 = 2r'', \quad \nu_4 = 2(m''-r'').$$

This substitution used the relationship that

$$U = \chi_1^2/(\chi_1^2 + \chi_2^2), \quad V = \chi_3^2/(\chi_3^2 + \chi_4^2).$$

Hence we can use the moments of $\log \chi^2$ to find the moments of β. Since

$$E(\log \chi^2) = \log \nu + O(\nu^{-1}),$$
$$\mathrm{Var}(\log \chi^2) = 2/\nu + O(\nu^{-2}),$$

the approximate moments of the posterior distribution of β are

$$\beta_0 = E''(\beta) \simeq \log \frac{s''/(n''-s'')}{r''/(m''-r'')},$$

$$\sigma^2 = V''(\beta) \simeq \frac{1}{s''} + \frac{1}{n''-s''} + \frac{1}{r''} + \frac{1}{n''-r''}. \tag{10}$$

Furthermore, $\log \chi^2$ is approximated by a normal distribution (for large ν) which allows one to approximate the posterior distribution of β by a normal distribution with mean and variance given by (10). This last result has been reported by Lindley[5] and Gart.[6] We point out that when $r' = s' = 1/2$ and $m' = n' = 1$, the above results correspond to the empirical logit. One may calculate closer approximations to the mean and variance with the use of additional terms in the series expansion of log χ^2.[7]

3. THE CHOICE OF THE PRIOR DISTRIBUTION

One of the main problems in applying Bayesian methods is the determination of suitable prior distributions. An advantage in using a conjugate prior distribution is that the parameters of the prior have a recognizable interpretation. For example, consider the prior distribution on exposure for controls; the quantities (r', m') are equivalent to having observed r' exposed individuals from among m' non-diseased individuals. Further, the prior expectation of exposure among the controls is

$$E'(e^{\alpha}/1+e^{\alpha}) = r'/m'.$$

Thus, with prior data on exposure available, one can use these data to construct a prior distribution.

The prior chosen may be taken to be completely non-informative, informative only on the exposure among controls or informative separately on the exposure among cases and among controls. The completely non-informative prior is to let $n' \to 0$, $m' \to 0$, which forces the quantities r', s' also to go to zero as $0 \leqslant r' \leqslant m'$, $0 \leqslant s' \leqslant n'$. In this case, $f'(\alpha, \beta) = 1$ $(-\infty < \alpha, \beta < \infty)$ and the prior will be an improper distribution.

Another option is to have a non-informative prior on β $(n' \to 0)$, but allow a prior which may be informative on α $(m' > 0)$. This results in

$$f'(\alpha, \beta) \propto e^{\alpha r'}/(1+e^{\alpha})^{m'}.$$

Simetimes it is convenient to choose the joint prior distribution so that the mean of β is zero. This can be done approximately by choosing $s' = r'/m'$ and $n' = 1$. Then the joint prior distribution is

$$f'(\alpha, \beta) \propto \exp\left\{\frac{r'}{m'}[\alpha(m'+1)+\beta]\right\} \Big/ (1+e)^{m'}(1+e^{\alpha+\beta}). \tag{11}$$

This prior distribution is useful and avoids the problem which arises if $s'' = n''$ and one is using the normal approximation discussed in section 2.

The prior distribution of exposure for the control population has two parameters (r', m') and thus requires two independent relationships in order to calculate the value of these parameters. If one has prior information on exposure based on data, then one can determine (r', m') directly. In the absence of data one can specify subjectively a range corresponding to exposure and have this range cover 95 per cent or 99 per cent of the prior probability. For example if one was asked to obtain a subjective assessment of the proportion of male smokers in the U.S. population, most people would agree that it is certainly within the limits 20–60 per cent. Use of this range to cover $\pm 1{\cdot}96\sigma$ limits and the normal approximation discussed in the previous section, results in the values $r' = 7{\cdot}7$ and $m' = 20{\cdot}3$. Figure 1 shows the prior distribution of exposure. The prior distribution has a mode (most likely value) of 38 per cent. It is unlikely that the proportion exposed was under 15 per cent or above 62 per cent because the ordinate of the distribution is less than 1/10 of the ordinate at the mode.

In some applications it may be more appropriate to specify the mean value of the exposure proportion $(\bar{P})$ and an upper limit for exposure (P_u). For example, for male smokers, one could specify that 38 per cent of males are smokers and that the subjective probability that the proportion

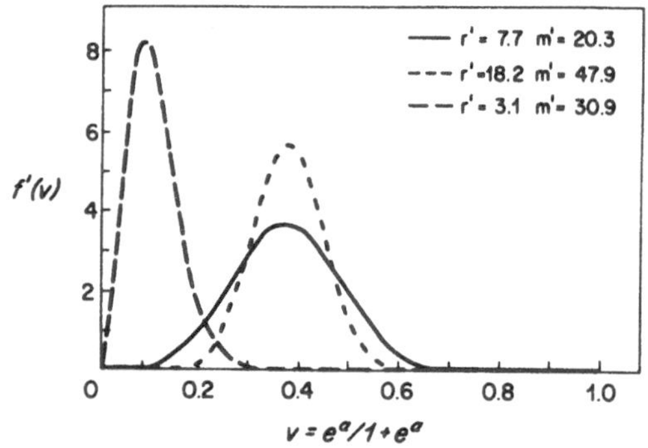

Figure 1. Density functions of prior distribution

of smokers exceeds 50 per cent is 0·05. Since

$$\bar{P} = r'/m', \log[\bar{P}/(1-\bar{P})] + 1{\cdot}64\sigma' = \log[P_u/(1-P_u)], \quad (12)$$
$$\sigma' = [1/r' + 1/(m'-r')]^{1/2},$$

we can solve for r', m'; i.e.

$$m' = [\bar{P}(1-\bar{P})(\sigma')^2]^{-1}, \quad r' = m'\bar{P}, \quad (13)$$

where

$$\sigma' = \log\left[\frac{P_u/(1-P_u)}{\bar{P}/(1-\bar{P})}\right] \Big/ 1{\cdot}64.$$

This results in $m' = 47{\cdot}9$ and $r' = 18{\cdot}2$. The exact values (without using the approximation) are $m' = 45{\cdot}6$ and $r' = 17{\cdot}3$. This prior distribution of exposure is also shown in Figure 1. Note that the mode is 37 per cent and 95 per cent of the prior probability is below 50 per cent.

4. BAYESIAN INFERENCES WITHOUT CONTROLS

A unique aspect of the use of Bayesian methods for case–control studies is that one can draw inferences about the relationship between disease and exposure without necessarily obtaining a control group. The Bayesian philosophy is that the data from a control group are used to construct a posterior distribution of exposure among non-diseased individuals. However, the prior distribution among the non-diseased population may be sufficiently informative that one can draw conclusions about the association of diseased individuals and exposure without obtaining data from a control group. This corresponds to the situation where the posterior distribution of exposure may be very similar to the prior distribution.

The posterior distribution of β enables one to determine $\max_{\beta} f''(\beta) = f''(\beta_0)$ as in equation (10). Then one decides whether a relationship exists between exposure and disease by consideration of the ratio of $f''(\beta_0)$ to $f''(0)$. If the ratio is greater than some constant K, then one would conclude that a relationship exists, i.e.

$$\frac{f''(\beta_0)}{f''(0)} \geqslant K \Rightarrow \text{relationship exists.} \quad (14)$$

Ordinarily K would be chosen as a quantity equal or greater than ten, as indicated earlier. The constant K is simply a 'critical ratio'. Assuming a non-informative prior on β and no sample information from a control group, the parameters of the posterior distribution for β are then $s'' = s$, $n'' = n$, $r'' = r'$, $m'' = m'$, $t'' = s + r'$. Similarly, with use of the prior distribution (11), $s'' = s + r'/m'$, $n'' = n + 1$, $m'' = m'$, $r'' = r'$.

One can extend these procedures without necessarily having a completely specified prior distribution on exposure. For example, suppose that (r', m') are fixed but unknown. Then the ratio $f''(\beta_0)/f''(0)$ would be a function of (r', m') which may be taken to satisfy

$$f''(\beta_0)/f''(0) = K(r', m') \geqslant K.$$

If the values of (r', m') specify families of prior distributions that are believed to be correct, then one need not go any further to conclude a relationship between disease and exposure. We illustrate these ideas in the next section.

5. EXAMPLE: ADENOCARCINOMA OF THE VAGINA AND DIETHYLSTILBESTROL (DES)

This section illustrates the methods described in the previous sections with data showing the association between exposure *in utero* to DES and adenocarcinoma of the vagina in young women. Cancer of the vagina is rare; there were only 68 case reports in over 30 years in two major hospitals.[8] Of these, approximately 90 per cent were over 40 years of age and about 90 per cent of the cases had epidermoid carcinomas. Herbst, Ulfelder and Poskanzer[9] observed a total of 8 patients, ranging in age from 14 to 22, with adenocarcinomas of the vagina. These investigators determined that 7 of the 8 cases had *in utero* exposure to diethylstilbestrol (DES), a drug which came into use during the 1940s to prevent complications of pregnancy. Hence, we will use the data $s = 7$, $n = 8$, to make an inference on disease and exposure without obtaining data from a control group. Herbst, Ulfelder and Poskanzer carried out a case–control study with four controls for each case; they concluded that a positive association existed.

In the early 1940s DES was suggested for the prevention and treatment of complications of pregnancy. The primary indications for DES use were threatened abortion (indicated by definite bleeding), or past history of spontaneous abortion (indicated by prolonged infertility or two or more previous abortions). Smith[10] reported only 632 DES treated pregnancies gathered by 117 obstetricians in approximately four years. The rate of DES treatment was (approximately) one case per obstetrician per year.

We can use other publications from that period to estimate the frequency of these complaints in the general population. Speert and Guttmacher[11] reported that only 14 cases of a series of 225 patients (6·2 per cent) had repeated bleeding/staining episodes 6 weeks or more after conception. Hertig and Livingstone,[12] in a review paper, quote a range of figures on *threatened* abortion ranging from 0·9–4·0 per cent in public hospitals and 10·9–16·0 per cent in private practice. In a major study of obstetric outcomes, Bishop[13] reported that some 10·9 per cent of women had aborted at least once in a series of 7000 pregnancies in the 1920s and 1930s, but only 0·41 per cent of the cases had aborted two or more times before carrying a pregnancy to term. Hertig and Livingstone[12] conclude that about 10 per cent of all pregnancies end in spontaneous abortions.

Since few pregnant women would have received DES prophylactically without *any* indication for its use, these studies provide estimates of the proportion of the population of pregnancies exposed to DES. Thus a prior mean exposure of 10 per cent with an upper limit of 20 per cent is likely to overestimate the true exposure in the general population. This assumes treatment of all potential patients. If we believe that the true exposed proportion in the control group has only a 1 in

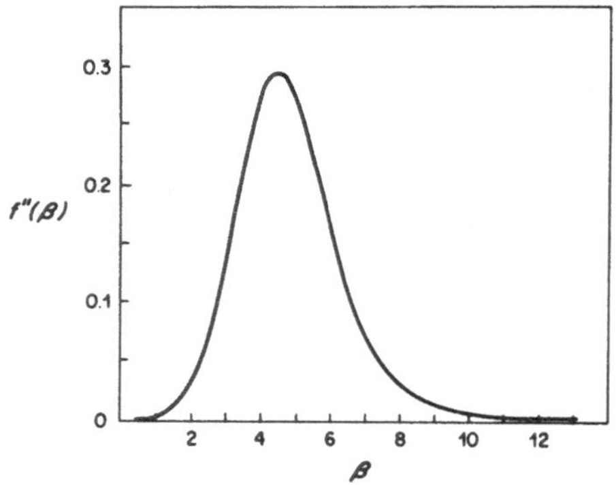

Figure 2. Posterior distribution of β (log odds-ratio) with prior distribution having parameters $r' = 3{\cdot}1$, $m' = 30{\cdot}9$, $s' = n' = 0$

Table III. Family of prior distributions on exposure which will result in critical value of 10

	Approximate results				Exact results			
$\bar{P}$	β_0	P_u	m'	r'	β_0	P_u	m'	r'
0·1	4·1	0·61	4·26	0·43	5·0	0·54	1·69	0·17
0·2	3·3	0·62	4·88	0·98	3·9	0·66	2·54	0·51
0·3	2·8	0·59	8·49	2·55	3·1	0·68	4·20	1·26
0·4	2·4	0·50	64·65	25·86	2·5	0·67	8·77	3·51
0·5	—	—	—	—	2·0	0·60	61·81	30·91
0·6	—	—	—	—	—	—	—	—

Approximate results calculated using the normal approximation (10). P_u is the 95 per cent upper limit calculated using the normal approximation.
Exact results calculated using the ratio (8). P_u is the 95 per cent upper limit calculated from the beta distribution with the specified prior parameters.

20 chance of exceeding 20 per cent, then we could obtain (r', m') by substituting in (13). Using the normal approximation with $\bar{P} = 0{\cdot}10$ and $P_u = 0{\cdot}20$ results in $m' = 45{\cdot}7$ and $r' = 4{\cdot}6$. (The exact values are $m' = 30{\cdot}9$ and $r' = 3{\cdot}1$.) The exact values for the numerical calculations will be used in the calculations described below. The exact prior distribution is shown in Figure 1. We shall take a non-informative prior distribution for the cases, e.g. $s' = n' = 0$. Therefore,

$$s'' = s + s' = 7, \quad n'' = n + n' = 8,$$
$$r'' = r' = 3{\cdot}1, \quad m'' = m' = 30{\cdot}9.$$

We use the normal approximations (10) to obtain the posterior distribution of β. Substituting in (10) results in the mean and variance of the posterior distribution, i.e. $\beta_0 = E''(\beta) = 4{\cdot}14$, $\sigma^2 = V''(\beta) = 1{\cdot}50$. The value of $\beta = 0$ corresponds to a normal deviate of $Z = \beta_0/\sigma = 3{\cdot}38$ having the ordinate 0·00132. Therefore the value of $f''(\beta_0)/f'(0)$ is $0{\cdot}399/0{\cdot}00132 = 302$. This ratio far exceeds any reasonable critical ratios and we would conclude that overwhelming evidence favoured an association between DES and cancer of the vagina. The exact result (8) leads to the posterior distribution shown in Figure 2. From this Figure, we can identify the mode β_0 as about 4·3, and the posterior distribution indicates that β less than 2 or more than 8 is unlikely, as the posterior density is less than 1/10 that of the mode.

As a further illustration, suppose we do not choose a prior distribution of exposure, but determine the class of prior distributions that result in a critical ratio $f''(\beta_0)/f''(0) = 10$. Since the ordinate at the maximum of $f''(\beta)$ is $f''(\beta_0) = 0{\cdot}40$, the ordinate at $f''(0) = 0{\cdot}04$. This corresponds to a standard normal deviate of 2·14, and thus we have the relation $\beta_0 = 2{\cdot}14\sigma''$. We can use equation (13) to solve for m' with the values $\bar{P} = r'/m'$, $s'' = 7$, and $n'' = 8$. Explicitly the result is

$$m' = \left\{\left[(\beta_0/2{\cdot}14)^2 - \left(\frac{1}{s''} + \frac{1}{n'' - s''}\right)\right]\bar{P}(1-\bar{P})\right\}^{-1}.$$

Table III summarizes the parameters of the prior distribution of exposure which results in a ratio $f'(\beta_0)/f''(0) = 10$. Note that there is no prior distribution for mean exposure of 0·60 or higher for which the density at the mode is 10 times the density at $\beta = 0$.

Thus from Table III, the data indicate the existence of a positive association between DES and cancer of the vagina if one is willing to assume that the mean exposure to DES is equal to or included in any of the priors given in Table III. For example, a prior with mean exposure of 40 per cent and an upper limit of 50 per cent results in the conclusion of an association. Priors with lower mean exposure and tighter concentration about the mean would be interpreted as having stronger evidence of exposure. The parameters of this prior distribution are well within the reports of the literature and one would conclude that such a prior is indeed reasonable.

REFERENCES

1. Breslow, N. E. and Day, N. E. *The Analysis of Case-Control Studies*, International Agency for Research on Cancer, Lyon, 1980.
2. Schlesselman, J. J. *Case-Control Studies: Design, Conduct, Analysis*, Oxford, New York, 1982.
3. Altham, P. M. E. 'Exact Bayesian analysis of a 2 × 2 contingency table, and Fisher's "Exact" significance test', *Journal of the Royal Statistical Society, Series B*, **31**, 261–269 (1969).
4. Latorre, G. 'The exact posterior distribution of the cross-ratio of a 2 × 2 contingency table', *Journal of Statistical Computation and Simulation*, **16**, 19–24 (1982).
5. Lindley, D. V. 'The Bayesian analysis of contingency tables', *Annals of Mathematical Statistics*, **35**, 1622–1643 (1964).
6. Gart, J. J. 'Alternative analyses of contingency tables', *Journal of the Royal Statistical Society, Series B*, **28**, 164–179 (1966).
7. Abramowitz, M. and Stegun, I. A. *Handbook of Mathematical Functions*, National Bureau of Standards, Applied Mathematics Series 55, U.S. Government Printing Office, 1964, p. 943.
8. Herbst, A. L., Green, T. H. and Ulfelder, H. 'Primary carcinoma of the vagina', *American Journal of Obstetrics and Gynecology*, **106**, 210–218 (1970).
9. Herbst, A. L., Ulfelder, H. and Poskanzer, D. C. 'Adenocarcinoma of the vagina', *New England Journal of Medicine*, **284**, 878–881 (1971).
10. Smith, O. W. 'Diethylstilbestrol in the prevention and treatment of complications in pregnancy', *American Journal of Obstetrics and Gynecology*, **56**, 821–834 (1948).
11. Speert, H. and Guttmacher, A. F. 'Frequency and significance of bleeding in early pregnancy', *Journal of the American Medical Association*, **155**, 712–715 (1954).
12. Hertig, A. T. and Livingstone, R. G. 'Spontaneous, threatened and habitual abortion: their pathogenesis and treatment', *New England Journal of Medicine*, **230**, 797–805 (1944).
13. Bishop, P. M. F. 'Habitual abortion: its incidence and treatment with progesterone or vitamin E', *Guy's Hospital Report*, **87**, 362–371 (1937).

Section IV

Prior Distributions and Estimation

Orthogonalizations and Prior Distributions for Orthogonalized Model Mixing

Merlise Clyde Giovanni Parmigiani
Institute of Statistics and Decision Sciences
Duke University
Durham, NC 27708-0251

Abstract

Prediction methods based on mixing over a set of plausible models can help alleviate the sensitivity of inference and decisions to modeling assumptions. One important application area is prediction in linear models. Computing techniques for model mixing in linear models include Markov chain Monte Carlo methods as well as importance sampling. Clyde, DeSimone and Parmigiani (1996) developed an importance sampling strategy based on expressing the space of predictors in terms of an orthogonal basis. This leads both to a better identified problem and to simple approximations to the posterior model probabilities. Such approximations can be used to construct efficient importance samplers. For brevity, we call this strategy orthogonalized model mixing.

Two key elements of orthogonalized model mixing are: a) the orthogonalization method and b) the prior probability distributions assigned to the models and the coefficients. In this paper we consider in further detail the specification of these two elements. In particular, after identifying the aspects of these specifications that are essential to the success of the importance sampler, we list and briefly discuss a number of different alternatives for both a) and b). We highlight the features that may make each one of the options attractive in specific situations and we illustrate some important points via a simulated data set.

1 Introduction

Prediction via model mixing is based on the marginal posterior predictive distribution obtained by integrating over a set of possible models (Geisser, 1993), and is receiving increasing attention as a way of alleviating the sensitivity of the predictions to modeling assumptions (Draper 1995, Raftery, Madigan and Volinsky 1995). Practically, the added generality of model mixing can offer more realistic uncertainty assessment, as well as ways of incorporating information from all predictors without over fitting the data. The latter is achieved by a data-based multiple shrinkage of the regression coefficients (see also George, 1986a, 1986b, George and Oman 1993).

Model mixing in large dimensional problems is computationally intensive. In most cases, the most competitive computing techniques for exploring large spaces of models are based on Markov chain Monte Carlo (MCMC) sampling. Phillips and Smith (1994), and Green (1995) discuss general algorithms. MCMC methods for variable selection are discussed by George and McCulloch (1993, 1994), Raftery, Madigan and Hoeting (1993) and Geweke (1994). In linear models, Clyde, DeSimone and Parmigiani (1996) approach model mixing by expressing the model space in terms of an orthogonal transformation of the matrix of predictors. This leads to an efficient importance sampler, based on independence of the indicators of predictor inclusion in the new basis. This strategy can provide a faster alternative to MCMC, especially when the space of models is very large, and predictors are correlated. Computational gains depend on specific problems and implementations, but can be significant because orthogonalization avoids the QR decomposition necessary during each step of the Markov chain when run with the original variables. Gains are larger for larger problems. Current applications in wavelets, where the basis is orthogonal by construction, successfully consider model spaces with 4096 linear predictors.

In addition to increased computational speed with importance sampling, orthogonalized model mixing can lead to a better identified problem. In general, the number of competing plausible models is smaller, as a result of eliminating collinearity. This approach has precedents in existing techniques for point estimation and prediction. In particular, when the goal of analysis is prediction, and the explanatory variables are highly correlated, estimators based on transforming the predictors to a new set of orthogonal variables and using a only a subset of the transformed predictors can have a smaller sum mean square error for the coefficients than the usual least squares estimates (see Jolliffe 1986, Weisberg 1985, Stone and Brooks 1990 and references therein). Examples include principal components regression, latent root regression, or partial least squares. One drawback to these methods is that they require choosing the number of orthogonal predictors to use. Using too few components can result in over shrinking. The orthogonalized model mixing can be viewed as a Bayesian generalization of these estimators that allows the data to automatically weight the number of terms that should be included in the model, and allows for multiple shrinkage.

This paper is a follow-up to Clyde, DeSimone and Parmigiani (1996). It focuses on two key implementation details for orthogonalized model mixing with importance sampling: the type of orthogonalizations and the prior distribution. In section 2, we introduce orthogonalized model mixing and review an importance sampling strategy for sampling models. In section 3, we compare alternative strategies for specifying orthogonalizations and prior distributions. Finally, in section 4 we compare two alternative orthogonalization strategies to the original model space using a simulated data set.

2 Orthogonalized Model Mixing with Importance Sampling

2.1 Orthogonal Variables

We begin by giving the basic notation, definitions, and some standard results useful in the sequel. Let Y be the $n \times 1$ vector of observed values of the response variable, and X the $n \times p$ design matrix including all candidate predictors. X can include transformations of the variables originally recorded. To simplify presentation, Y and X are taken to be centered. This restriction is not necessary to any of the methods discussed.

Model mixing is defined hierarchically on the orthogonalized variables. We begin by assuming that

$$Y|\beta, \sigma^2 \sim N_n(X\beta, \sigma^2 I_n), \tag{1}$$

where β is $p \times 1$, σ^2 is a scalar, and I_n is an $n \times n$ identity matrix. We take the prior distribution for the model parameters, when all predictors are included, to be the natural conjugate prior:

$$\begin{aligned} \beta|\sigma^2 &\sim N_p(b_0, \sigma^2 B) \\ \nu\xi/\sigma^2 &\sim \chi^2_\nu, \end{aligned}$$

where B, ν and ξ are fixed hyper-parameters. We will later refer to a factorization of the form $B = CC'$.

Consider now an orthogonalization of the design matrix, $Z = XW$. The mean space is now represented as a collection of subspaces spanned by orthogonal variables given by the columns of Z. This representation is not unique: it could be defined by principal components, latent root regression, or partial least squares, for example. We will comment on the choice of the orthogonalization method in Section 3. The linear model in (1) can be rewritten in terms of the orthogonalized variables as

$$Y = Z\alpha + e,$$

where $\alpha = W^{-1}\beta$. The transformed prior distribution on α, conditional on inclusion of all predictors is:

$$\alpha|\sigma^2 \sim N_p(a_0, \sigma^2 A) \tag{2}$$

$$\nu\xi/\sigma^2 \sim \chi^2_\nu, \tag{3}$$

where $a_0 = W^{-1}b_0$ and $A = W^{-1}B(W^{-1})'$.

Under conjugate prior distributions, computation of posterior and predictive distributions can be carried out using standard least squares regression techniques, by augmenting the Y and Z matrices as follows:

$$\tilde{Y} = \begin{bmatrix} Y \\ A^{-1/2}a_0 \end{bmatrix} \quad \text{and} \quad \tilde{Z} = \begin{bmatrix} Z \\ A^{-1/2} \end{bmatrix}$$

where $\tilde{Y}$ is $(n+p) \times 1$, $\tilde{Z}$ is $(n+p) \times p$, and $A^{-1/2}$ is a symmetric square root of A^{-1}. We can then proceed by analyzing the augmented model

$$\tilde{Y} = \tilde{Z}\alpha + e.$$

by standard least squares techniqes. In particular, the posterior mean of α is given by:

$$\tilde{\alpha} = (\tilde{Z}'\tilde{Z})^{-1}\tilde{Z}\tilde{Y} = (Z'Z + A^{-1})^{-1}(Z'Y + A^{-1}a_0).$$

Since Z is orthogonal, the matrix $\Lambda \equiv Z'Z$ is diagonal. To help in constructing efficient importance samplers, we will later assume that A is diagonal. Then the posterior mean of α can be computed coordinate-wise as

$$\tilde{\alpha}_i = \frac{\lambda_i}{\lambda_i + a_i^{-1}}\hat{\alpha}_i + \frac{a_i^{-1}}{\lambda_i + a_i^{-1}}a_{0i}$$

where $\hat{\alpha}$ is the ordinary least squares estimator of α, a_i is the i-th diagonal element of A and a_{0i} is the i-th element of a_0. Different prior specifications for the a_i control the degree of shrinkage to the prior mean. This will be discussed in more detail in Section 3.

2.2 Model Mixing

The process of selecting columns of Z for prediction is modeled via a further hierarchical level in the prior distribution. In particular, define the $p \times 1$ vector γ to be a sequence of binary random variables, each indicating whether the corresponding column of Z is included in the model. The set of all possible γ's will be referred to as the orthogonalized model space when ambiguity with the original model space may occur. The prior distribution on γ is denoted by $\pi(\gamma)$. This joint specification is equivalent to assuming that the prior distribution on α is a mixture of (2) and a point mass at zero, similar to Mitchell and Beauchamp (1988). We are assuming that the set of models considered is exhaustive. In the literature on combining prediction models, it has been found that when the set of models is exhaustive, given a squared error predictive loss function, it is always optimal to combine predictions. However, if the models did not constitute an exhaustive set, it would not necessarily be optimal to combine predictions of alternative models (see Palm and Zellner, 1992, and Bernardo and Smith, 1994 for discussion and additional references).

From conjugacy, the posterior probability of models is available in closed form up to a normalizing constant. In particular, for a given model γ, let Z_γ be the matrix obtained by selecting the columns of Z that correspond to a 1 in the vector γ and let D_γ be the diagonal matrix with diagonal elements given by γ. The residual sum of squares for model γ is then:

$$S_\gamma^2 = ||\tilde{Y} - \tilde{Z}D_\gamma\tilde{\alpha}||^2 = (Y - ZD_\gamma\tilde{\alpha})'(Y - ZD_\gamma\tilde{\alpha}) + (\tilde{\alpha} - a_0)'D_\gamma A^{-1}D_\gamma(\tilde{\alpha} - a_0).$$

Finally, defining Λ_γ and A_γ as the matrices obtained by deleting rows and columns of Λ and A for i's where $\gamma_i = 0$, we can write:

$$q_\gamma = |\Lambda_\gamma + A_\gamma^{-1}|^{-1/2}|A_\gamma|^{-1/2}(\nu\xi + S_\gamma^2)^{-\frac{(n+\nu)}{2}}\pi(\gamma) \tag{4}$$

and determine the posterior probability of model γ by

$$\pi(\gamma|Y) = \frac{p(Y|\gamma)\pi(\gamma)}{\sum_{\gamma'} p(Y|\gamma')\pi(\gamma')} = \frac{q_\gamma}{\sum_{\gamma'} q_{\gamma'}},$$

where q_γ can easily be evaluated exactly.

These are all the necessary elements for addressing predictive problems, such as finding the multivariate predictive distribution $f(\cdot|Y, X_0)$, where X_0 is a specified matrix, the mean $\hat{Y}(X_0)$ of this distribution, the expected utility U associated with a decision δ whose outcome depends on future values of Y, or other quantities of interest. Denote the quantity of interest by ϕ, possibly a vector. In many cases, computations can proceed by determining ϕ_γ (the quantity of interest conditional on γ) for each γ, and then evaluating

$$\phi = \sum \phi_\gamma \pi(\gamma|Y). \tag{5}$$

In particular, it is interesting to consider the form of the posterior mean of α_i under model mixing.

$$E(\gamma_i \tilde{\alpha}_i|Y) = \pi(\gamma_i = 1|Y)\tilde{\alpha}_i.$$

This emphasizes the multiple shrinkage nature of model mixing. Some shrinkage is incorporated in $\tilde{\alpha}_i$, which depends on A and Λ. Further shrinkage is determined by the posterior model probabilities, that depend on A, Λ and also on the model specific residual sum of squares S_γ^2 and on the prior hyperparameters ν and ξ for the error variance.

2.3 Importance Sampling

Exact evaluation of $\pi(\gamma|Y)$ involves summing over all possible models, which is computationally infeasible for relatively large p. This motivates interest for stochastic searches of the model space. One appealing feature of orthogonalized model mixing is ease of sampling. In particular, in this section we discuss a stochastic search algorithm based on an importance sampling function on the model space. Orthogonalization makes it possible to draw elements of the orthogonal basis independently, with probability that approximates very closely the actual posterior probability. This results in three main advantages over conventional Markov chains for stochastic search:

a) it is faster to sample a model; one orthogonalization is necessary overall for the entire sampler. In the original model space, sampling a new model requires updating the regression at each step of the chain;

b) convergence typically occurs earlier, especially if the model space in terms of the original variables is difficult to traverse;

c) an estimate of the total mass sampled is available for inference and possibly for convergence diagnostics.

Details of one implementation of the importance sampler are discussed in Clyde, DeSimone and Parmigiani (1996).

We make two assumptions that substantially simplify the task of constructing an accurate importance sampler: the prior covariance matrix of α, $\sigma^2 A$ is assumed to be diagonal, and the prior distribution $\pi(\gamma)$ is assumed to factor as

$$\pi(\gamma) = \prod_{i=1}^{p} \pi(\gamma_i) = \prod_{i=1}^{p} \theta_i^{\gamma_i}(1 - \theta_i)^{1-\gamma_i}. \tag{6}$$

The independence assumption implied by (6) may not always be realistic when the γ's refer to the originally recorded variables. Chipman (1996) discusses practical alternatives to independence in that context. However, when dealing with orthogonal variables that are, say, principal components of the original variables, we find it appealing to specify an independence prior on the model space. In problems where the context dictates a natural dependence structure in the orthogonalized space, one could develop importance samplers based on expansions similar to that discussed below. We do not pursue this here.

To proceed, let $\tilde{Z}_i$ be the i-th column of $\tilde{Z}$. Define SSR_i^2 as the regression sum of squares from the regression of $\tilde{Y}$ on $\tilde{Z}_i$. In particular, then $\mathrm{SSR}_i^2 = ||P_{\tilde{Z}_i}\tilde{Y}||^2$, where $P_{\tilde{Z}_i} = \tilde{Z}_i\tilde{Z}_i'/(\tilde{Z}_i'\tilde{Z}_i)$ is the projection operator on the augmented column $\tilde{Z}_i$. Using diagonality of A, we can write:

$$|\Lambda_\gamma + A_\gamma^{-1}|^{-1/2}|A_\gamma|^{-1/2} = \prod_{i=1}^{p}\left(\frac{a_i^{-1}}{\lambda_i + a_i^{-1}}\right)^{\gamma_i/2} \equiv \prod_{i=1}^{p}\zeta_i^{\gamma_i/2}$$

Rewriting (4), we have:

$$\begin{aligned}\log(q_\gamma) &= c + \sum_{i=1}^{p}\gamma_i \log\left(\frac{\theta_i}{(1-\theta_i)}\right) + \frac{1}{2}\sum_{i=1}^{p}\gamma_i \log\left(\frac{a_i^{-1}}{(\lambda_i + a_i^{-1})}\right) \\ &\quad -\frac{(n+\nu)}{2}\log\left(\nu\xi + \tilde{Y}'\tilde{Y} - \sum_{i=1}^{p}\gamma_i\,\mathrm{SSR}_i^2\right).\end{aligned} \tag{7}$$

To obtain a factorization for q_γ, we need to express the last term in (7) as a linear function of γ. We do this by using a Taylor series expansion of the last term. In particular, expanding around $\nu\xi + \tilde{Y}'\tilde{Y}$, we have

$$\log\left(\nu\xi + \tilde{Y}'\tilde{Y} - \sum_{i=1}^{p}\gamma_i\,\mathrm{SSR}_i^2\right) = \log(\nu\xi + \tilde{Y}'\tilde{Y}) - \sum_{j=1}^{k}\frac{1}{j}\left[\frac{\sum_{i=1}^{p}\gamma_i\,\mathrm{SSR}_i^2}{(\nu\xi + \tilde{Y}'\tilde{Y})}\right]^j + A_k \tag{8}$$

where A_k is a remainder term. Ignoring the cross product terms after expanding the expression in square brackets,

$$\log\left(\nu\xi + \tilde{Y}'\tilde{Y} - \sum_{i=1}^{p}\gamma_i\,\mathrm{SSR}_i^2\right) = \log(\nu\xi + \tilde{Y}'\tilde{Y}) - \sum_{j=1}^{k}\frac{\sum_{i=1}^{p}\gamma_i\,\mathrm{SSR}_i^{2j}}{j(\nu\xi + \tilde{Y}'\tilde{Y})^j} + A_k.$$

It is also helpful to calibrate the sampler so that the approximate posterior probabilities match the exact posterior probabilities for the two models including no predictors and all predictors respectively. This leads to the following approximate posterior model probability:

$$\tilde{\pi}(\gamma|Y) = \prod_{i=1}^{p} p_i^{\gamma_i}(1-p_i)^{1-\gamma_i} \tag{9}$$

where:

$$p_i = \frac{\theta_i \exp\left\{\frac{1}{2}\log\zeta_i + \frac{L_N - L_F}{\tilde{L}_N - \tilde{L}_F}\sum_{j=1}^{k}\frac{(n+\nu)\,\mathrm{SSR}_i^{2j}}{2j(\nu\xi + \tilde{Y}'\tilde{Y})^j}\right\}}{1 - \theta_i + \theta_i \exp\left\{\frac{1}{2}\log\zeta_i + \frac{L_N - L_F}{\tilde{L}_N - \tilde{L}_F}\sum_{j=1}^{k}\frac{(n+\nu)\,\mathrm{SSR}_i^{2j}}{2j(\nu\xi + \tilde{Y}'\tilde{Y})^j}\right\}}.$$

Here L_N and L_F are the calibrating constants:

$$L_N = \log(\nu\xi + \tilde{Y}'\tilde{Y} - \mathrm{SSR}_1^2)$$

$$
\begin{aligned}
L_F &= \log\left(\nu\xi + \tilde{Y}'\tilde{Y} - \sum_{i=1}^{p} \mathrm{SSR}_i^2\right) \\
\tilde{L}_N &= \log(\nu\xi + \tilde{Y}'\tilde{Y}) - \sum_{j=1}^{k} \frac{\mathrm{SSR}_1^{2j}}{j(\nu\xi + \tilde{Y}'\tilde{Y})^j} \\
\tilde{L}_F &= \log(\nu\xi + \tilde{Y}'\tilde{Y}) - \sum_{j=1}^{k} \frac{\sum_{i=1}^{p} \mathrm{SSR}_i^{2j}}{j(\nu\xi + \tilde{Y}'\tilde{Y})^j}
\end{aligned}
$$

where k is the number of terms in the series expansion and needs to be large. Further discussion is in Clyde, DeSimone and Parmigiani (1996).

Generating a sample of models from $\tilde{\pi}$ is straightforward. It can be done independently on each of the elements of the orthogonal basis by generating Bernoulli random variables with probabilities p_i and does not require Markov chain Monte Carlo methods. After the initial computation of the orthogonal basis and the SSR_i's, sampling is done directly from $\tilde{\pi}$ and is very fast. Quantities of interest can the be calculated, for example, as

$$
\hat{\phi} = \sum_{\gamma \in D} w_\gamma \hat{Y}_\gamma \tag{10}
$$

where D is the set of unique sampled models and the weights w_γ can be chosen in several ways, such as:

$$
w_\gamma = \frac{q_\gamma}{\sum_{\gamma' \in D} q_{\gamma'}}.
$$

mimicking the actual mixing within the set D of sampled models, or

$$
w_\gamma = \frac{f_\gamma q_\gamma / \tilde{\pi}(\gamma)}{\sum_{\gamma' \in D} f_{\gamma'} q_{\gamma'} / \tilde{\pi}(\gamma)}
$$

based on renormalizing the importance sampling weights where f_γ is the number of times model γ was sampled.

3 Implementing Orthogonalized Model Mixing

In model-mixing, both the choice of the orthogonalization and of the prior distribution are important in determining the nature and amount of shrinkage performed by the model. For example, priors that put large weights on models with a small number of columns encourage more shrinkage. On the other extreme, a prior concentrating on $\gamma = \underline{1}$ corresponds to the full model, which may over fit the points.

We find it natural to think about the full specification of an orthogonalized mixture model as a three stage process:

1. Choice of the orthogonalization method;

2. Choice of the prior distribution on the error variance and the regression coefficients; the latter can, in turn, be specified in terms of β or α;

3. Choice of the prior distribution on the model space.

In this section we discuss alternatives for each of these three steps in turns.

3.1 Orthogonalization

There is not just a single good way of constructing a suitable orthogonalization $Z = XW$ for model mixing. The effects of the choice of the orthogonal basis on the final results of model mixing are an open question. Different orthogonalizations may result in widely different degrees of parsimony in the representation of the mean space of Y. Thus one would like the orthogonalization to achieve closeness to "target" or "optimal" subspaces, which may be better achieved by an orthogonalization based on Y. Using an orthogonalization based on Y, however, introduces sampling uncertainty in the orthogonal basis. Indirectly, this leads to data dependence in the prior distribution on α, since the model space depends on the orthogonalization. In particular, the Sliced Inverse Regression and Partial Least Squares orthogonalizations discussed below lead to data dependent priors.

Importantly, the type of orthogonalization used does have direct implications for the allowable prior covariance matrices of either α or β, if one of the goals is to implement an importance sampler.

Generalized Principal Components (Rao 1964). In generalized principal components, $Z = XW$ and W is orthonormal with respect to the inner product determined by a $p \times p$ positive definite matrix Φ, that is: $W'\Phi W = I_p$. Choosing Φ to be the identity matrix results in standard principal components analysis, where the matrix W is given by the eigenvectors of $X'X$. More generally, letting $\Phi^{-1/2}$ be the symmetric inverse square root of Φ, we can write $W = \Phi^{-1/2}U$, where U are the eigenvectors of $\Phi^{-1/2}X'X\Phi^{-1/2}$

An important example arises with principal components formed from the correlation matrix of X, rather than the covariance matrix. Let V_X be the diagonal of $X'X$, which corresponds to the sum of squares for each X since the X's are centered. Let $\Phi = V_X$ and let R denote the correlation matrix of the X's, that is

$$R = V_X^{-1/2} X'X V_X^{-1/2}.$$

Then, using the spectral decomposition, R can be written as $U\Lambda U'$ where U is the matrix of eigenvectors and W is $V_X^{-1/2}U$. The new predictor variables Z can be viewed as the principal components formed from the standardized X's.

Generalized principal components offer a simple and relatively flexible orthogonalization strategy. They are well understood and computing routines are very readily available. The resulting orthogonal variables are invariant under reorderings of the original predictors and do not require knowledge of the response Y. This feature can be appealing in non-orthogonal designed experiments, as it frees the orthogonalization process from measurement errors.

Regression on a subset k of the principal components based on the k largest eigenvalues can lead to exclusion of important directions that have small eigenvalues but are highly correlated with Y. See Jolliffe (1982) for examples. However, model mixing does not suffer from this potential drawback, as all possible subsets of principal components are incorporated in the regression.

Partial least squares (Wold *et al.* 1984). The goal of partial least squares is to find a set of uncorrelated variables Z that have the highest possible covariances with Y. The matrix W for partial least squares can be constructed column by column sequentially

(Stone and Brook 1990). Write $X'X = U\Lambda U'$ where U and Λ are the eigenvectors and eigenvalues respectively. Initially, set $W^{(1)} = W_1$ to $X'Y/||X'Y||$. The variable $Z_1 = XW_1$ then maximizes $W'X'Y$. The next goal is to find the vector W_2 that maximizes $W_2'X'Y$, under the conditions that W_2 and W_1 are uncorrelated. The matrix W can then be constructed as follows:

1. $A^{(i)} = \Lambda U'W^{(i-1)}$
2. $d^{(i)} = U'X'Y$
3. $m^{(i)} = U(d - P_{A^{(i)}}d)$
4. $W_i = m^{(i)}/||m^{(i)}||$
5. $W^{(i)} = [W^{(i-1)}|W_i]$

Repeat until $i = p$ and take $W = W^{(p)}$. Then construct $Z = XW$.

This basis also has the advantage that the variables are invariant under reordering of the original predictors, like in the generalized principal components analysis. In addition, as the orthogonalization strategy depends on the response variable, partial least squares is more likely to produce parsimonious representations of the mean response, or "optimal" target subspaces, thus leading to model spaces that are faster to explore.

A general framework incorporating both principal components and partial least squares is continuum regression (Stone and Brook, 1990). The optimality criteria leading to principal components and partial least squares are embedded in a larger family via an exponential-like connection, governed by a parameter κ. This framework lends itself to a further layer of hierarchy in model mixing, based on averaging over values of κ, each corresponding to a different orthogonalization. This would lead to a data-based choice of the basis itself. It is arguable, however, that with the many levels of hierarchy already present in the model, the data may provide very little insight into the proper orthogonalization in complex problems. Also, continuum regression orthogonalization with arbitrary ζ requires an iterative solution, and it is a more demanding computing task than principal components or partial least squares. This could offset or greatly reduce the computational advantages of importance sampling.

Gram-Schmidt Orthogonalization. The Gram-Schmidt procedure can be used to construct an orthonormal basis directly from X. Variables are ordered and each variable is made orthogonal to the previously orthogonalized variables. Such dependence on the ordering can be useful if there is *a priori* information about the relative importance of the variables. Alternatively, the order could be based on a standard forward or backwards selection. If forward selection is used, the variable with the highest explanatory power is preserved as one of the orthogonal directions in Z. Further directions capture remaining variability, hopefully leading to a parsimonious representation of the mean space to be used in model averaging or selection.

The orthogonal predictors can be constructed by using a Cholesky decomposition of the type $(X'X)^{-1} = WW'$, where W is an upper triangular matrix such that $W'(X'X)W = I_p$. This gives the desired orthogonalization based on the ordering of the columns of X, and in addition $Z'Z = I_p$.

Sliced Inverse Regression (Li 1991). Sliced inverse regression (SIR) identifies directions $W_1, \ldots W_k$, assuming that $E(Y|X)$ depends on X only through $XW_1, \ldots, XW_k$, $Y = f(XW_1, \ldots, XW_k, \epsilon)$. The set of p directions, W can be calculated and used to obtain a new set of orthogonal variables $Z = XW$. This can then be used for subset selection or model mixing. The matrix W is calculated as follows:

1. Standardize X as $T = X(U\Gamma^{1/2}U')$ where $X'X = U\Gamma U'$ from the spectral decomposition.

2. divide the range of Y into h intervals (slices) and form the $n \times h$ matrix H where $\{H\}_{ij} = 1$ if Y_i is in the j-th slice.

3. Calculate the $h \times p$ matrix of means of the X's in each slice, $M = (H'H)^{-1}HT$.

4. Calculate the spectral decomposition of the weighted covariance matrix of the sliced means: $WGW' \equiv M'H'HM$ where W is the matrix of the eigenvectors and G are the eigenvalues.

5. Set $Z = TW$, where $Z'Z = I_p$

The appeal of using SIR in the context of model mixing comes from the hope that the subspaces spanned by the Z_i are close to the "optimal" subspace for the $E(Y|X)$. As we remarked earlier, models with high posterior model probabilities will then tend to have a small number of terms. Since there is no assumption that Y and the Z_i's are linearly related, SIR may be a better starting point than partial least squares for identifying transformations that could be added to the basis. Other variations on SIR can also be used. If the model is $Y = f(Z_1, \ldots, Z_k) + \epsilon$, with $k < p$ and unknown, using a second order Taylor expansion for f would lead to using Z and cross products in the basis. An additional orthogonalization would be needed to make the additional terms also mutually orthogonal and orthogonal to Z. The mixture model would then determine the appropriate degree of shrinkage to the target subspaces. This may prove more effective for including transformations, in particular interactions, of the original X's.

Finally, in special cases it may be recommendable to use a combination of the various alternatives above for different groups of variables. For example, in a designed experiment, columns with the main effect of one factor of interest could be included directly in Z, while the remainder of the columns, perhaps including continuous and not controlled measurements, could be orthogonalized by principal components or other methods.

3.2 Prior Distributions

In the discussion of the importance sampling algorithm of Section 2.3, three important restrictions were imposed on the prior distribution:

1. The prior on β and σ was assumed to be the natural conjugate prior. This implies a natural conjugate prior on α and vice versa;

2. The prior covariance of α, $\sigma^2 A$ was assumed to be diagonal; this assumption does not seem to be especially restrictive, as α are the coefficients of the orthogonalized variables;

3. The prior distribution on the model space was assumed to factor.

If condition 1. needs to be relaxed, analytic expressions for q_γ may not be available, leading naturally to MCMC methods. If conditions 2. and 3. are relaxed, the second term in expression (7) is not additive in the γ_i. This could be handled by applying a Taylor expansion to the second term as well, or possibly by ad hoc solutions for very special dependence structures.

In the remainder we assume that restrictions 1., 2. and 3. apply, and that model mixing is carried out by importance sampling. These restrictions still leave us with a very wide set of prior choices. We now list and briefly discuss some alternatives. Our discussion is not meant to provide a complete survey of issues in elicitation of prior distributions for linear models or model probabilities. We focus on those aspects that are most relevant to orthogonalized model mixing.

Several general goals may guide the choice of the prior distributions.

a) *Fit.* One way of interpreting the prior distribution on the parameters in prediction is as a set of tuning constants for the predictive distribution. Properties of the resulting predictions, such as shrinkage, outlier discounting and so forth, can guide the choice of the prior directly.

b) *Noninformativeness.* In some early work on Bayesian mixtures for selecting subsets of predictors, the goal is to use Bayesian techniques to model the alternative between exclusion of a variable on the one hand and inclusion with data based estimates of the coefficients on the other hand (Mitchell and Beauchamp 1988, George and McCulloch, 1993). This interpretation is appealing when the set of predictors is large and poorly understood. But this is exactly the type of scenario in which fast stochastic exploration of the model space is vital. One pragmatic implementation of this philosophy is to have vague prior distributions on the regression coefficients conditional on inclusion. Prior distributions have to be proper, but variances can be large.

c) *Speed of convergence.* This applies to sampling from the model space. Posterior model probabilities concentrating on few good models lead to faster exploration. Therefore priors leading to higher posterior concentration are more efficient. This comparison, however, should be made by keeping some measure of predictive performance fixed. To make a trivial example, a degenerate prior probability on $\gamma = 0$ leads to easily computable but perhaps worthless predictions.

d) *Relation to other methods.* Finally, priors on the coefficients might be chosen based on the equivalence, or near equivalence, between Bayesian methods and known classical methods (Foster, George and McCulloch, 1995). Ridge regression and James-Stein estimators are examples (see discussion in Weisberg 1985). This can also be useful for calibrating some of the prior hyperparameters based on equivalence with widely accepted values for the corresponding classical methods.

On a more technical terrain, one important dichotomy in the elicitation of the prior distributions is whether to specify the prior distributions in terms of the original predictors, or in terms of the orthogonalized predictors. We review some issues.

Specifying Priors Distributions on α. A straightforward way of meeting the requirement that the prior covariance of α should be diagonal is to specify it directly. This leads to $B = WAW'$. Several choices may be appropriate:

1. Arbitrary diagonal A. To evaluate the shrinkage implication of this choice, let us consider the form of the posterior mean of α.

$$\tilde{\alpha}_i = \frac{\lambda_i}{\lambda_i + a_i^{-1}}\hat{\alpha}_i + \frac{a_i^{-1}}{\lambda_i + a_i^{-1}}a_{0i} = \frac{1}{1+g_i}\hat{\alpha}_i + \frac{g_i}{1+g_i}a_{0i}.$$

where $g_i = 1/a_i\lambda_i$, so that there is differential shrinkage of each component to the prior mean. Elicitation of the a_i via the choice of the g_i can be guided by the desired amount of shrinkage in the posterior mean. Relatively noninformative choices can be achieved by setting the weights $g_i/(1+g_i)$ to be small percentages, say between 1% and 10%.

Also, the posterior model probabilities can be rewritten as:

$$\begin{aligned} \log(q_\gamma) &= c + \sum_{i=1}^{p} \gamma_i \log\left(\frac{\theta_i}{(1-\theta_i)}\right) + \frac{1}{2}\sum_{i=1}^{p}\gamma_i \log\left(\frac{g_i}{1+g_i}\right) \\ &\quad - \frac{(n+\nu)}{2}\log\left(\nu\xi + \tilde{Y}'\tilde{Y} - \sum_{i=1}^{p}\gamma_i\, \mathrm{SSR}_i^2\right). \end{aligned}$$

The g_i and θ_i have similar roles in determining model probabilities, with the difference that the g's enter into the SSR_i terms as well.

The Bayes estimator for the full model under this prior (and with the added assumption that α_0 is zero) is equivalent to a generalized ridge regression estimator that allows for differential shrinkage of the coefficients to 0. As pointed out before, model mixing allows for data based shrinkage.

2. Inverse eigenvalues. One choice is $A = g^{-1}\Lambda^{-1}$, where $\Lambda \equiv Z'Z$ and g is a scalar to be chosen. There are two important consequences of this choice: First, the shrinkage implied by the posterior mean $\tilde{\alpha}$ is the same for each of the elements of $\tilde{\alpha}$. In particular,

$$\tilde{\alpha}_i = \frac{1}{1+g}\hat{\alpha}_i + \frac{g}{1+g}a_{0i}.$$

Second, the posterior model probabilities simplify to:

$$\begin{aligned} \log(q_\gamma) &= c + \sum_{i=1}^{p} \gamma_i \log\left(\frac{\theta_i}{(1-\theta_i)}\right) + \frac{1}{2}\log\left(\frac{g}{1+g}\right)\sum_{i=1}^{p}\gamma_i \\ &\quad - \frac{(n+\nu)}{2}\log\left(\nu\xi + \tilde{Y}'\tilde{Y} - \sum_{i=1}^{p}\gamma_i\, \mathrm{SSR}_i^2\right). \end{aligned}$$

where the term $\log\left(\frac{g}{1+g}\right)\sum_{i=1}^{p}\gamma_i$ can be interpreted as a penalty for the complexity of model γ in terms of the number of predictors. A large g implies a greater shrinkage to the prior mean and a greater penalty for adding variables. In terms of the prior distribution on β, this choice of A implies $B = g^{-1}(X'X)^{-1}$ independent of the orthogonalization used.

Other approaches for choosing values of g corresponding to classical variable selection methods are discussed in Foster, George and McCulloch (1995).

3. Equal variances. A simpler alternative consists of taking $A = aI$, implying the *a priori* judgment that the variances of the coefficients α are equal. This may not be

appropriate if the columns of Z do not have a common scale. However, it can be a ways to implement pragmatic vague priors, by choosing a sufficiently large a.

If the orthogonalization is done by generalized principal components, this choice leads to $B = aWW' = a\Phi^{-1}$. In particular, standard principal components lead to $B = aI$. If the orthogonalization is done by principal components on the correlation matrix, then $B = aV_X^{-2}$

4. Training Samples. A further alternative is the elicitation method of Laud and Ibrahim (1994). This is based on expert assessment of the distribution of the response vector associated with the predictor matrix X. Model probabilities are obtained by updating suitable default prior distribution on the model parameters and the model space given the response, and by approximately averaging with respect to the elicited distribution. In our context default priors need to refer to the orthogonalized space. This approach leads to a prior that is not of the form (6). Implementation of importance sampling requires a further Taylor expansion of the type of section 2.3.

Specifying Priors on β. When the prior distribution on the coefficients is designed to incorporate some important expert opinion, or prior evidence, it may be sensible to begin by eliciting B. Elicitation of priors on coefficients of linear models are discussed by Kadane *et al.* (1980) and by Garthwaite and Dickey (1992) in the context of variable selection.

If the orthogonalization strategy is selected before elicitation, only in special cases of prior covariances for β will the orthogonalizations $Z = XW$ result in the prior covariance for α, $\sigma^2 W^{-1}B(W')^{-1}$, being diagonal. These are some of the options.

1. Arbitrary B.

A simple way of accommodating an arbitrarily specified covariance matrix $B = CC'$ is to let the orthogonalization depend on it. For example, the requirement that A is diagonal is achieved by taking Φ equal to B^{-1} in the generalized principal components. Then W is CU where U corresponds to the eigenvectors of $C'X'XC$. This amounts to first rotating X to XC and β to $C^{-1}\beta$, so that the rotated parameters are now independent, and then determining Z based on standard principal components in the rotated variables. The resulting prior covariance for α is $\sigma^2 I_p$. Constructing orthogonal variables from XC instead of X leads to $A = I_p$ when applied to other orthogonalization methods as well.

Alternatively, if the orthogonalization method of choice depends on Y, we can achieve a diagonal A by performing the orthogonalization on the augmented data,

$$\tilde{Y} = \begin{bmatrix} Y \\ C^{-1}\beta_0 \end{bmatrix} \quad \text{and} \quad \tilde{X} = \begin{bmatrix} XC \\ I_p \end{bmatrix}.$$

This strategy and orthogonalization of XC are equivalent under generalized principal components.

2. $B = g^{-1}(X'X)^{-1}$.

Here g is a known constant expressing the weight to be assigned to the prior information. This choice arises from thinking of the prior as the posterior distribution from a real or imaginary experiment using the same design X and a noninformative prior (see also Zellner 1986, Laud and Ibrahim, 1994 and references therein). This is also equivalent to $A = g^{-1}\Lambda^{-1}$, as described previously.

3. Zero Means.

Some of the mixture models for variable selection are based on $b_0 = 0$, implying also $a_0 = 0$. For example, this is the choice used by Raftery, Madigan and Hoeting (1993) and George and McCulloch (1993, 1994). This choice is usually coupled with a sufficiently large variance that the case $\gamma_i = 1$ can be interpreted, roughly, as: variable i is included and the estimated coefficient is close to the least squares estimate. Coupling $b_0 = 0$ with automatic rules for detemining A or B that lead to strongly informative prior variances on some of the coefficients of the orthogonalized variables can lead to excessive shrinkage.

Specifying Prior Distributions for the Error Variance σ. In our experience, this specification can be very important. In general, prior distributions concentrating on small values of σ encourage models that fit the data closely, while priors putting mass on larger values of σ favor more parsimonous models. The effect is overall, rather than by acting on the individual predictors, as it is the case for A. In terms of the model probabilities, we highlight the role of the term $\nu\xi$, which is interpretable as a prior sum of squares and is added to the sums of squares terms. The larger $\nu\xi$ is, the more similar the model probabilities will be, leading to more challenging posterior distributions for stochastic searches.

Hoeting, Raftery and Madigan (1995) and George and McCulloch (1994) discuss calibration of these hyperparameters. The assumption of conjugacy is often criticized for not being sufficiently flexible in representing beliefs about σ. Bernardo and Smith (1994) includes discussion and further references.

Specifying Priors on the Model Space. In orthogonalized model mixing with an independent importance sampler, the probabilities on model space are specified by assigning the prior probability θ_i that an orthogonal variable Z_i is included in the model. This is probably the most difficult part of the prior elicitation scheme. Methods for eliciting prior probabilities on model spaces expressed in terms of the original variables are available (see Laud and Ibrahim 1994). In our scheme, one cannot start with probabilities of inclusion in the original variables and map these to probabilities of inclusion in the orthogonal variables. Many of the original variables will be included in the new orthogonalized coordinates. Thus, prior probabilites have to be assessed directly on the Z_i's.

One approach is to assign equal probability, say θ, to all variables. Models with the same number of terms are then given the same probability *a priori*. Taking θ equal to .5 corresponds to uniform prior probabilities over all models. Choosing a small θ can be used to enforce parsimony. So long as $\theta < .5$, the log odds term for θ in (7) results in a constant penalty for each variable that is added to the model. On the other hand $\theta > .5$ results in a higher *a priori* weight on models with more terms. Under the independence prior with a constant θ, the number of terms in the model has a binomial distribution with mean $p\theta$, which could also be used to help choose θ by specifying the expected number of terms in the model.

If there is prior information on the inclusion probabilities of the original variables, a more informative assignment for the θ_i could be based on looking at the loadings on the original variables. Thus Z_i's that have an important original variable X_j with large coefficients $|W_{ij}|$ should have larger θ_i's.

An ad hoc choice that can substantially simplify the numerical evaluation of q_γ when model probability are extremely small or numerically unstable can be achieved by choosing θ_i in the following way: solving for

$$\left(\frac{\theta_i}{1-\theta_i}\right)^2 = \frac{g_i}{1+g_i}$$

results in model probabilities of the simplified form:

$$\log(q_\gamma) = c - \frac{(n+\nu)}{2} \log\left(\nu\xi + \tilde{Y}'\tilde{Y} - \sum_{i=1}^{p} \gamma_i \,\mathrm{SSR}_i^2\right).$$

This can be achieved either by letting

$$\theta_i = g_i\left(\sqrt{1+\frac{1}{g_i}} - 1\right)$$

when eliciting model probabilities, or by letting

$$g_i = \frac{\theta_i^2}{1-2\theta_i} \qquad \theta_i < .5$$

when eliciting A. This strategy avoids of one of the two most difficult steps in elicitation, and can improve numerical stability.

Incidentally, if a vague prior on α is desired, g_i can be chosen to be large, leading to prior model probabilities approaching $1/2$ from below and posterior model probabilities that depend on the prior specification only through the prior hyperparameters for σ^2, ν and ξ. In this case, posterior model probabilities will be ordered based on the residual sum of squares.

4 An Example

In this section we present the results of the analysis of a simulated data set. Our goal is to investigate the sensitivity of the results to the choice of the basis used for representing the model space. We focus on a relatively small dimensional problem, to have available the complete enumeration of the model space. In this way we can separate investigation of the effects of alternative orthogonalizations from the effects of speed and quality of sampling, already addressed in Clyde DeSimone and Parmigiani (1996).

We considered the following scenario: We construct an $n \times 15$ matrix X including $n = 30$ observations on 15 explanatory variable. We generated X by drawing 30 samples from a 15-dimensional normal distribution with null mean vector and correlation matrix given by $\rho_{ij} = .9^{|i-j|}$. In this way we introduce varying degrees of correlations between the predictors. We then partition the matrix X into 3 submatrices of dimension $n \times 5$, as:

$$X = \left(\; X_1 \;\vdots\; X_2 \;\vdots\; X_3 \;\right)$$

and create the two further matrices $X^T = X_1 \vdots X_2$ (for "true") and $X^A = X_1 \vdots X_3$ (for "available"). X^T is used for generating the response vector y from the model:

$$Y|\beta, \sigma^2 \sim N_n(X^T\beta, \sigma^2 I_n)$$

where β is 10×1, σ^2 is a scalar, and I_n is an $n \times n$ identity matrix. In turn, β is a 10-dimensional vector of independent normals with mean 0 and standard deviation 2, and $1.5/\sigma^2 \sim \chi^2_5$.

We assume that only X^A is available for prediction. This corresponds to the type of real life situation in which prediction by model mixing is most appropriate: Some of the variables included in the predictor set are relevant and some are not. Also, none of the prediction models that can be obtained by selecting a subset of the available predictor is the model that actually generated the data. Some of the variables in X_3 may act as proxies for the missing variables in X_2 and this may make it interesting to include them in the prediction model.

Our results concern the comparison of three prediction strategies:

1. Model mixing in the original variable space.

2. Model mixing after orthogonalization by principal components (PC).

3. Model mixing after orthogonalization by partial least squares (PLS).

The prior distributions on the regression coefficients for the orthogonalized spaces under the full model are the inverse eigenvalue prior of section 3.2. Also, priors are chosen to be the same under the full model. For all three strategies, the prior on the model space is as in Section 2.3, with $\pi(\gamma_i) = .5$ and the prior on the error variance is $3/\sigma^2 \sim \chi^2_5$.

We are displaying results for one simulated data set only, focusing on a case that seems to be highly representative. Figure 1 shows the distribution of the logarithm of the model probabilities in the three alternative representations of the model spaces. Each point corresponds to a model. Vertical bars correspond to the full model. The orthogonalizations lead to bimodal distributions, having a wider range than the distribution on the original space. Both of these features are desirable. Model spaces where all models have about the same probability are the most difficult to explore. Increasing the spread of the distribution of model probabilities corresponds to reducing model uncertainty. In this example, orthogonalization helped us divide the model space into "unlikely" models, that can be ignored in the exploration, and "likely" models, that need to be studied. It also led to models that have a higher posterior probability than any of the original ones. This has important implications for computing. The wider range and the better separation afforded by orthogonalization result in faster sampling of the high probability models, and therefore in a faster convergence of the sampling-based predictive distributions to the predictive distribution associated with exact mixing.

In practical applications with very high dimensional problems, these computational advantages can be substantial. One concern arises when orthogonalization is used as a computationally attractive substitute for model mixing in the original space. Figure 2 shows the pairwise plots of the fitted values obtained with the three methods. The agreement is very high. While this is common to most other simulated samples, no

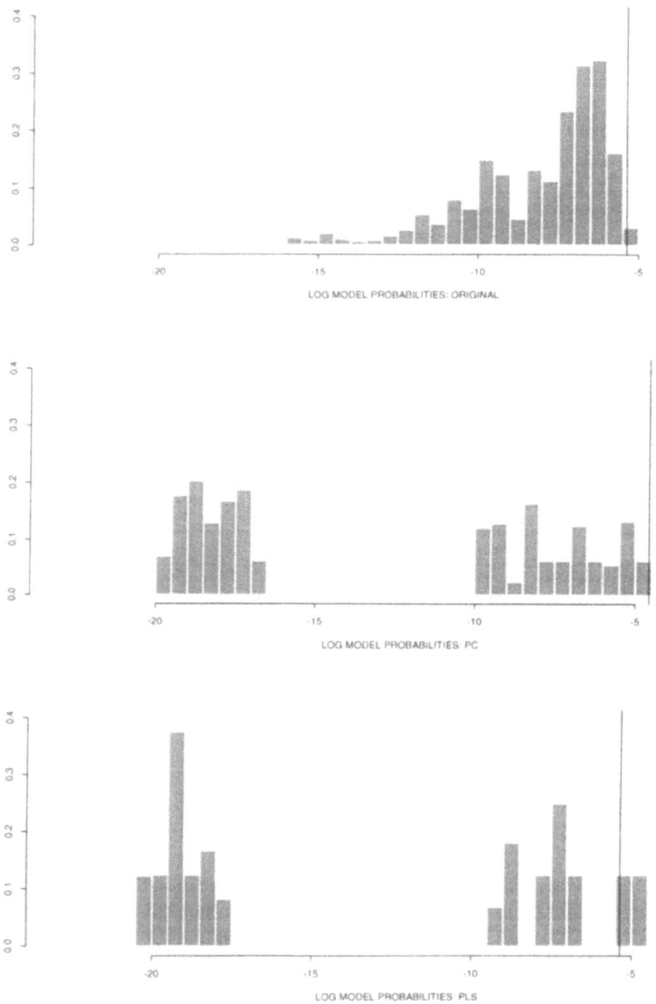

Figure 1: Distribution of posterior model probabilities under the three alternative representation of the model space. Vertical bars indicate the location of the full model.

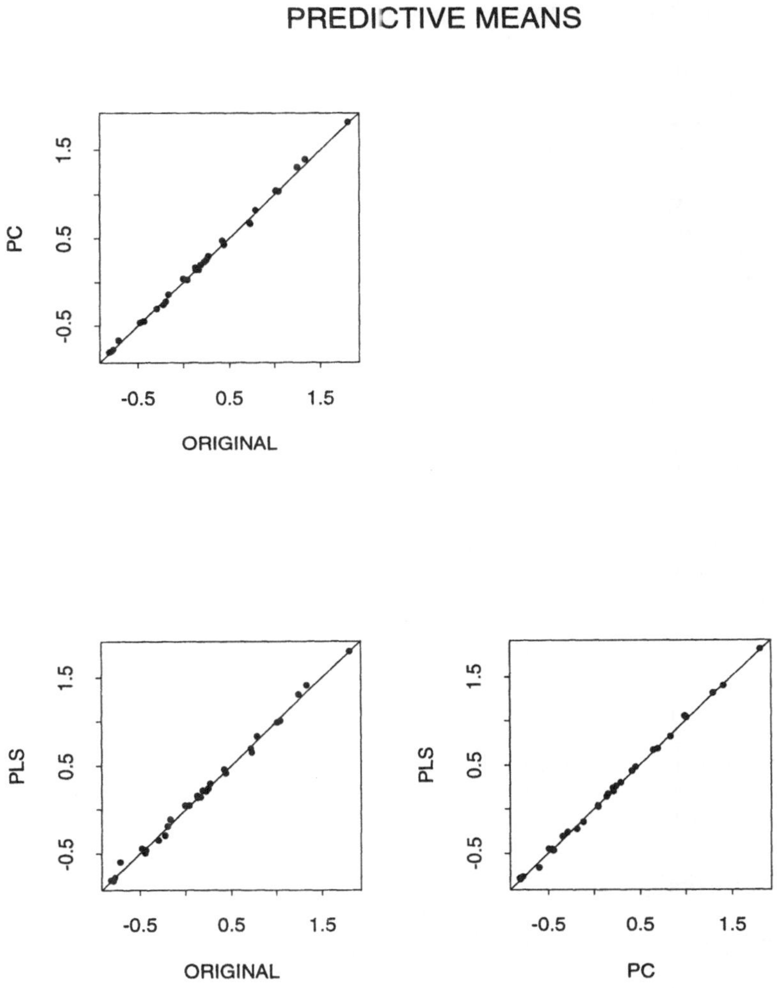

Figure 2: Pairwise plots of the fitted values obtained using model mixing in the three model spaces.

general conclusions can be drawn, as there are cases in which the fit may be strongly affected.

One interesting opportunity opened by fast enumeration of model spaces is the possibility of investigating alternative strategies for model choice. As an aside, we include Figure 3, illustrating the relationship between the posterior model probability (on the log scale) and the adjusted R^2 for the model. Agreement is in this case very high under all three strategies.

Finally, we evaluated the three strategies in term of their ability to fit the actual predictive mean for future observations, that is $X^T\beta$. We also present results obtained by conditioning on the model with highest posterior probability, in the column labeled "model selection". The table below summarizes the results.

	Model Mixing	Model Selection
Original variables	0.1003	0.1359
Principal Components	0.1119	0.1361
Partial Least Squares	0.1120	0.1362

Entries are integrated squared errors of the predictions; rows correspond to different model spaces; column correspond to the model mixing and to the model with the highest posterior probability. Mixing outperforms model selection while the various model spaces give very similar performance. This is to be expected since model selection based on a 0-1 loss function does not directly take into account the quality of prediction.

It is interesting to compare the out of sample predictive performance with a goodness of fit measure. Unadjusted values of R^2 are reported in the table below.

	Model Mixing	Model Selection
Original variables	0.999336	0.999348
Principal Components	0.998863	0.998887
Partial Least Squares	0.998872	0.998887

While all values are quite high, model selection leads to slightly higher values than model mixing. Model selection is in general more prone to over fitting the data, as it is based on more limited shrinkage.

5 Discussion

In this paper we considered implementation details and directions for future developments of orthogonalized model mixing, a method for prediction in linear models. In very large model spaces, the use of an orthogonal basis allows for more efficient sampling strategies via importance sampling. It is key to identify orthogonalizations that generate small subspaces which are more likely to contain the mean, and allow for more parsimonious models. This can also affect the efficiency of the importance sampler in traversing model space. Model averaging allows for incorporating uncertainty about the target subspaces in the posterior distribution.

We cannot recommend any specific choice as optimal at this point. We have discussed several approaches to defining orthogonal variables and some "automatic" choices for prior distributions in the context of orthogonal model mixing by importance sampling.

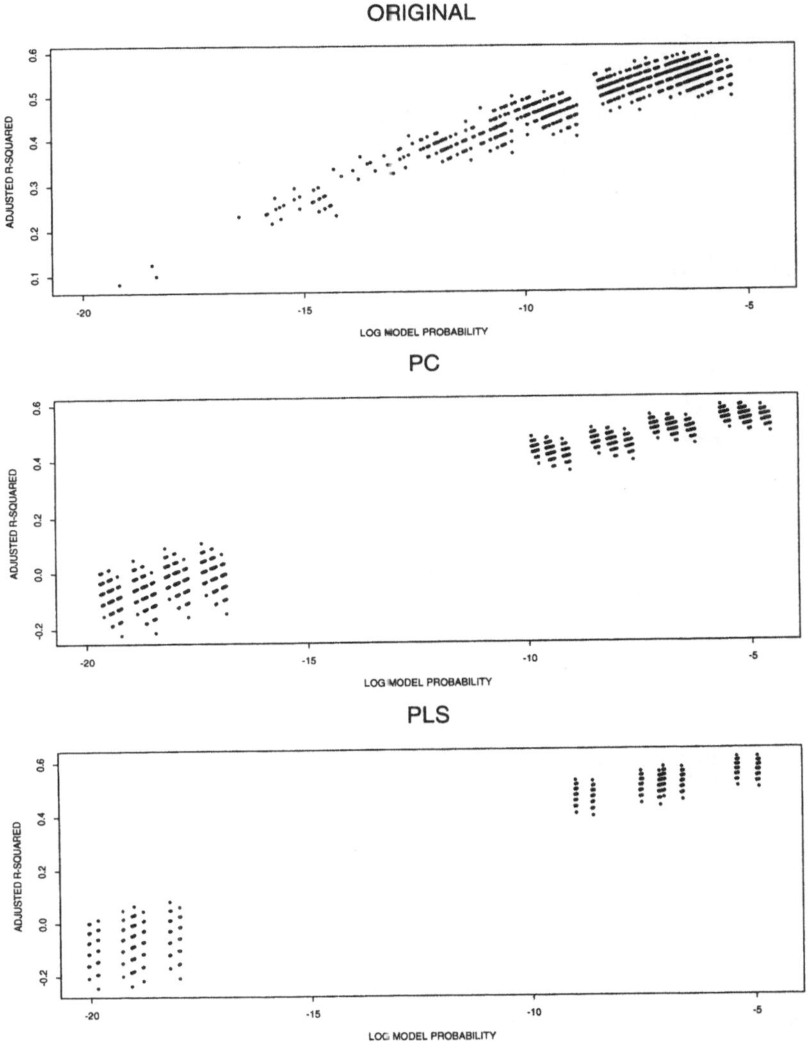

Figure 3: Posterior model probability (on the log scale) versus adjusted R^2 for the three model spaces

We attempted to highlight the features that may make each one of the options attractive in specific applications. One difficulty in elicitation is that the prior model probabilities have to be assigned to models that are defined in terms of the orthogonalized variables. We suggested that the form of the posterior model probabilities under the orthogonal variables allows one to understand the multiple shrinkage that arises under model mixing. This can aide in the elicitation of the prior distributions. Prior distributions on the coefficients can be expressed either in the original space or the orthogonalized space, with some restrictions that are pointed out in detail.

Acknowledgment

The authors thank Wesley Johnson, Arnold Zellner and an anonymous referee for useful suggestions. Work partially supported by the National Science Foundation under grant DMS-9305699.

References

Bernardo, J.M. and Smith A.F.M. (1994) *Bayesian Theory*. Wiley, N.Y.

Chipman, H. (1996). Bayesian Variable Selection with Related Predictors. *Canadian Journal of Statistics*, to appear.

Clyde, M.A., DeSimone, H., and Parmigiani, G. (1996). Prediction via Orthogonalized Model Mixing. *Journal of the American Statistical Association*, forthcoming.

Draper, D. (1995). Assessment and propagation of model uncertainty (with Discussion). *Journal of the Royal Statistical Society* 57, pp. 45–98.

Foster, D., George, E. and McCulloch, R (1995) Calibrating Bayesian variable selection procedures.

Garthwaite, P.H. and Dickey, J.M. (1992). Elicitation of prior distributions for variable-selection problems in regression. *Annals of Statistics* 20, pp. 1697-1719.

Geisser, S., Predictive inference. An introduction, Chapman & Hall, New York, 1993.

George, E.I. (1986). Minimax multiple shrinkage estimation. *Annals of Statistics* 14, pp. 188-205.

George, E.I. (1986). Combining minimax shrinkage estimators. *Journal of the American Statistical Association* 81, pp. 437-445.

George, E.I. and McCulloch, R. (1993). Variable Selection via Gibbs Sampling. *Journal of the American Statistical Association* 88, pp. 881–889.

George, E.I. and McCulloch, R. (1994). Fast Bayes Variable Selection. TR, Graduate School of Business, University of Chicago.

George, E.I. and Oman, S.D. (1993). Multiple shrinkage principal component regression. TR, University of Texas at Austin.

Geweke, J.F. (1994). Bayesian comparison of econometric models. Working Paper 532, Federal Reserve Bank of Minneapolis.

Green, P.J. (1995). Reversible Jump MCMC Computation and Bayesian Model determination. TR-94-19, University of Bristol.

Hoeting, J., Raftery, A.E., Madigan, D.M. (1995), A Method for Simultaneous Variable Selection and Outlier Identification in Linear Regression, Technical Report 95-02,

Colorado State University.

Jolliffe, I.T. (1986) Principal Component Analysis, Springer, New York.

Jolliffe, I.T. (1982). A note on the use of principal components in regression. *Applied Statistics* 31, pp. 300-303.

Kadane, J.B., Dickey, J.M., Winkler, R.L., Smith, W.S., and Peters, S.C. (1980). Interactive elicitation of opinion for a normal linear model. *Journal of the American Statistical Association* 75, pp. 845–85.

Laud, P. and Ibrahim, J.G. (1994). Predictive Specification of Prior Model Probability in Variable Selection. TR, Division of Statistics, University of Northern Illinois.

Li, K-C. (1991). Sliced inverse regression for dimension reduction. *Journal of the American Statistical Association* 86, pp. 316–327.

Mitchell, T.J. and Beauchamp, J.J. (1988). Bayesian Variable Selection in Linear Regression. *Journal of the American Statistical Association* 83 , pp. 1023–1036.

Palm F.C. and Zellner A. (1992). "To Combine or not to Combine? Issues of Combining Forecasts," *Journal of Forecasting*, 11, 687-701.

Phillips, D.B. and Smith, A.F.M. (1994). Bayesian Model Comparison via Jump Diffusions. TR-94-20, Department of Mathematics, Imperial College, U.K.

Raftery, A.E., Madigan, D.M., and Hoeting, J. (1993). Model selection and accounting for model uncertainty in linear regression models. TR 262, Department of Statistics, University of Washington.

Raftery, A.E., Madigan, D.M., and Volinski C.T. (1995). Accounting for Model Uncertainty in Survival Analysis Improves Predictive Performance (with discussion). In *Bayesian Statistics 5*, ed. J.M. Bernardo, J.O. Berger, A.P. Dawid and Smith, A.F.M.

Rao, C.R. (1964). The Use and Interpretation of Principal Components in Applied Research. *Sankhya A* 26, pp. 329–358.

Stone, M. and Brooks, R. J. (1990). Continuum regression: Cross-validated sequentially constructed prediction embracing ordinary least squares, partial least squares and principal components regression. *Journal of the Royal Statistical Society, Series B* 52, pp. 237–269.

Weisberg, S. (1985). *Applied Linear Regression. 2nd Edition.* Wiley, New York.

Wold, S., Ruhe, A., Wold, H., and Dunn, W.J.III (1984). The collinearity problem in linear regression. The partial least squares (PLS) approach to generalized inverses. *SIAM Journal on Scientific and Statistical Computing* 5, pp. 735-743.

Zellner, A., (1986). On assessing prior distribution in Bayesian regression analysis with *g*-prior distributions. In "Bayesian Inference and decision Techniques: Essays in Honor of Bruno de Finetti", 233–243, (P.K. Goel and A. Zellner eds.). North Holland, Amsterdam.

Improved Quasi-Maximum Likelihood Estimation for Stochastic Volatility Models

F. Jay Breidt and Alicia L. Carriquiry
Department of Statistics, Iowa State University, Ames, IA 50011 USA

Abstract

Jacquier, Polson and Rossi (1994, *Journal of Business and Economic Statistics*) have proposed a Bayesian hierarchical model and Markov Chain Monte Carlo methodology for parameter estimation and smoothing in a stochastic volatility model, where the logarithm of the conditional variance follows an autoregressive process. In sampling experiments, their estimators perform particularly well relative to a quasi-maximum likelihood approach, in which the nonlinear stochastic volatility model is linearized via a logarithmic transformation and the resulting linear state-space model is treated as Gaussian. In this paper, we explore a simple modification to the treatment of inlier observations which reduces the excess kurtosis in the distribution of the observation disturbances and improves the performance of the quasi-maximum likelihood procedure. The method we propose can be carried out with commercial software.

Keywords. Inliers, excess kurtosis, transformations.

1 Introduction

Financial variables such as stock returns and exchange rates are often modeled using martingale difference sequences. If a sequence of random variables observed over time is a martingale difference sequence, then both the unconditional expectation and the conditional expectation (given the past of the series) of an observation at time t are identically equal to zero. Further, the series has no serial autocorrelation. We will call a serially uncorrelated sequence with zero mean and constant unconditional variance *white noise*. An independent and identically distributed (iid) sequence with finite variance is both a martingale difference sequence and a white noise. In the iid case, the past of the series contains no information about the present or the future.

It has been shown, however, (e.g., Clark, 1973; Tauchen and Pitts, 1983; Nelson, 1988; Melino and Turnbull, 1990; Harvey, Ruiz and Shephard, 1994) that series arising in finance and econometrics cannot always be assumed to be iid. While the martingale difference property often appears plausible, the variance in a given realization seems to change over time. In fact, it is often the case that powers of the series itself exhibit serial autocorrelation, and thus it is possible to detect and model dynamics in higher order moments of the series.

Two approaches have been proposed to model time-dependent variances. The first approach, proposed by Engle (1982) and later generalized by Bollerslev (1986) and by others, uses an autoregressive conditionally heteroscedastic (ARCH, or its generalized version, GARCH) process to model the serial autocorrelation in the variances. In this approach, the variance of the series at time t is assumed to be a deterministic function of lagged values of the squared observations and of past variances. For an excellent review of this approach, see Bollerslev, Chou and Kroner (1992).

The second approach, pioneered in its earliest version in the work of Clark (1973), uses models known as stochastic volatility (SV) models. In this context, it is assumed that smooth functions of the time-dependent variances are random variables generated by an underlying stochastic process, for example an autoregressive process or a random walk. Stochastic volatility models also result from discretizing continuous-time diffusion processes such as those proposed for asset pricing (Hull and White, 1987; Harvey and Shephard, 1993). While intuitively appealing, SV models have not been popular, at least in terms of usage. The reason for the limited empirical application of these models is that, unlike the case of ARCH-type processes, the likelihood function for SV models is hard to evaluate, since it is expressed as a T-dimensional integral, where T is the number of observations.

Several methods for estimation from SV models have been proposed. A method of moments (MM) estimator, which avoids the problem of evaluating the likelihood function, was suggested by Taylor (1986), Melino and Turnbull (1990), and most recently Vetzal (1992). While easy to implement, the MM estimator was shown to be inefficient and to perform poorly over repeated sampling (Jacquier, Polson and Rossi, 1994). Nelson (1988), Harvey and Shephard (1993), Ruiz (1994) and Harvey et al. (1994), after expressing the SV model in a linear state-space form, used the usual Kalman filter recursions to produce estimators of the state vectors that are best (in the mean squared error sense) among all linear estimators. The recursions also yield the Gaussian likelihood of the data, which can be maximized to obtain parameter estimates. This approach, known as quasi-maximum likelihood (QML) is simple to implement and it is quite general. In particular, the dynamics of the stochastic process underlying the volatilities can take many different forms. Further, since QML depends only on second-order moment properties, it can be applied without modification if the distribution of the error terms in the model changes.

However, as Harvey et al. (1994) and Jacquier et al. (1994) point out (among others), the performance of QML decreases in "noisy" series. When the variance of the underlying stochastic process is small relative to the variance of the original series, QML estimators can be severely biased and have high root mean squared errors (RMSE). In addition, the transformation from a SV model to the state-space version cannot be carried out for observations with a value of zero. Indeed, the results from the transformation become suspect whenever applied to inliers, where by inliers we mean any observed value that is close to zero. Different "remedies" have been proposed to accommodate inliers, where the most widely used consists of either shifting the whole series away from zero by a small value, or just adding a small amount (for example a fraction of a percent of the series mean) to all zero observations. These inlier treatments have been criticized by, for example, Nelson (1994).

This same inlier problem applies to an approach closely related to QML (Kim and Shephard 1994, Shephard 1994, Carter 1993, Mahieu and Schotman 1994), in which the SV model is transformed into linear state-space form, and then the resulting non-Gaussian error in the observation equation is approximated by a mixture of Gaussian distributions.

Recently, Jacquier et al. (1994), using a fully Bayesian framework, derived expressions for the marginal posterior distributions of the parameters and the state vector in SV models, thereby providing a Bayesian (under the given assumptions) solution to the estimation and smoothing problems. Their approach rests upon the formulation of the nonlinear SV model as a hierarchical model, with the prior distributions for the model parameters at the top of the hierarchy. The computational problem is resolved by means of a Markov chain sampler.

The method suggested by Jacquier et al. (1994) addresses several important issues. Since the nonlinear SV model need not be expressed in linear state-space form, no transformation is necessary and the inlier problem vanishes. In addition, uncertainty about true values of the model parameters is incorporated in a natural way when computing smoothed estimates of the state vector. Furthermore, Jacquier et al. have shown, via extensive sampling experiments, that the method they propose significantly outperforms QML (and MM) both in terms of bias and RMSE, particularly in those cases where the variance of the underlying process is small.

Two drawbacks to Jacquier et al.'s procedure can, however, be pointed out. In order to derive the posterior distributions of interest, it is necessary to make strong assumptions regarding the model. Indeed, it would seem that the probabilistic model imposed on the observations, the parameters and the state vector is crucial, and that misspecification of the model would lead to estimators with poor behavior. For example, the model uses a Gaussian assumption on the disturbance in the observation equation. It is not clear that the procedure would produce reasonable results when a heavier tailed (or even skewed) distribution is, as many speculate (see, e.g., Harvey and Shephard, 1993), a more appropriate model. It is therefore questionable whether the method outlined in Jacquier et al. is robust to departures from the model's assumptions. The second drawback is one of convenience. Jacquier et al.'s procedure is not easy to implement, and in fact their algorithm requires serious modifications whenever the model is tailored to different applications.

In this paper, we propose a simple modification to the usual QML approach for estimation in SV models. In carrying out our work, our objective was to derive a method which, while inefficient in some cases, was robust to departures from model assumptions, and which was simple to implement. The modification we suggest consists in applying a linearizing transformation to shifted values of the observations, where the shift is determined by the slope of the function used for transformation at those points. This modified linearizing transformation addresses the inlier problem, and also improves the third and fourth moments of the distribution of transformed residuals in the observation equation. We show, through a sampling experiment, that our modified transformation significantly improves the performance of the usual QML estimator. Comparison of the behavior of our improved QML estimator to Jacquier et al.'s Bayes estimator is encouraging; in the

Gaussian case, and for most parameter values, our estimator performs as well as the Bayes estimator in terms of bias. The improved QML estimator has, however, a higher RMSE than the Bayes estimator, as would be expected.

In Section 2, we present the SV model and the robust linearizing transformation. The usual QML estimator is reviewed in Section 3. Results from the sampling experiment are presented in Section 4, and conclusions and directions for future work are given in Section 5.

2 Model and transformations

Consider the simple stochastic volatility model

$$y_t = \sigma_t \xi_t = \zeta \exp(\alpha_t/2)\xi_t, \quad \alpha_t = \phi\alpha_{t-1} + \eta_t \tag{1}$$

where $\{\xi_t\}$ is iid with mean zero and variance one, $\{\eta_t\}$ is an iid $(0, \sigma_\eta^2)$ sequence of random variables independent of $\{\xi_t\}$, ζ is a positive constant, and $|\phi| < 1$. This model has been considered, for example, by Harvey et al. (1994), Kim and Shephard (1994), and Jacquier et al. (1994).

Goals are usually two-fold: to estimate the parameters ζ, ϕ, and σ_η^2, and to obtain smoothed estimates of the volatilities σ_t. Though model (1) is simple, the likelihood of $(\zeta, \phi, \sigma_\eta^2)$ given $(y_1, \ldots, y_T)$ involves a T-dimensional integral and is not easy to evaluate. Jacquier et al. (1994) have developed a Markov chain simulation methodology for Bayesian inference in model (1). Their algorithm allows for numerical evaluation of all marginal posterior distributions of interest, thus avoiding the problem of directly evaluating the likelihood of $(\zeta, \phi, \sigma_\eta^2)$.

There remains some interest in simpler, albeit less efficient, estimation methods. Simple estimation methods can be used as exploratory tools in developing a model specification, which might then suggest a more sophisticated estimation procedure.

2.1 Linearizing transformation

In theory, the series (1) is simple to analyze after transformation, as suggested by Nelson (1988) and Harvey and Shephard (1993) among others. Consider the stationary process

$$\begin{aligned} x_t &= \log y_t^2 \\ &= \log \zeta^2 + \mathrm{E}\left[\log \xi_t^2\right] + \alpha_t + \left(\log \xi_t^2 - \mathrm{E}\left[\log \xi_t^2\right]\right) \\ &= \mu + \alpha_t + \epsilon_t, \end{aligned} \tag{2}$$

where $\{\epsilon_t\}$ is iid with mean zero and variance σ_ϵ^2. For example, if ξ_t is standard normal, then $\log \xi_t^2$ is distributed as the log of a χ_1^2 random variable, and the log of its moment generating function (see Bartlett and Kendall, 1946; Wishart, 1947) is

$$\log \mathrm{E}\left[\exp\left\{t \log \chi_1^2\right\}\right] = \log \mathrm{E}\left[\left(\chi_1^2\right)^t\right] = t \log 2 + \log \Gamma(1/2 + t) - \log \Gamma(1/2),$$

from which we can obtain the first two moments, $\mathrm{E}\left[\log \xi_t^2\right] = -\gamma - \log 2 \simeq -1.27$ (γ is Euler's constant) and $\sigma_\epsilon^2 = \pi^2/2$, respectively. Also, the skewness is

$$\frac{\mathrm{E}\left[(\log \xi_t^2 + \gamma + \log 2)^3\right]}{\sigma_\epsilon^3} = -14\mathcal{Z}(3)/\sigma_\epsilon^3 \simeq -1.5351$$

and the excess kurtosis is

$$\frac{\mathrm{E}\left[(\log \xi_t^2 + \gamma + \log 2)^4\right]}{\sigma_\epsilon^4} - 3 = 90\mathcal{Z}(4)/\sigma_\epsilon^4 = 4,$$

where $\mathcal{Z}(\cdot)$ is Riemann's zeta function (Abramowitz and Stegun, 1965, §6.4).

The process $\{x_t\}$ is thus a correlated signal plus an iid non-Gaussian noise, with $\mathrm{E}[x_t] = \mu$ and

$$\gamma_x(h) = \mathrm{Cov}(x_t, x_{t+h}) = \gamma(h) + \sigma_\epsilon^2 I_{\{h=0\}},$$

where $\gamma(\cdot)$ denotes the autocovariance function of $\{\alpha_t\}$ and $I_{\{h=0\}}$ is one if $h = 0$ and zero otherwise. In spite of the non-Gaussianity of $\{\epsilon_t\}$, a reasonable estimation procedure is to maximize the Gaussian likelihood of the linear state-space model (2), a procedure known as quasi-maximum likelihood estimation (QML).

2.2 Robust transformation

In practice, the series $\{y_t\}$ may contain zeroes (and other inliers), so the log transformation breaks down. Often, practitioners mean correct or work with excess returns to avoid the problem of zero observations. These procedures, which are kinds of inlier adjustments, have been criticized by Nelson (1994). An alternative inlier adjustment was presented by Fuller (forthcoming, Example 9.3.2), who suggested evaluating $\log(\cdot)$ not at the (possibly zero) measurement z, but at $z + dz$, where dz is a small increment, and then extrapolating linearly from the point $(z + dz, \log(z + dz))$ using the slope of the tangent line, $(z + dz)^{-1}$. Evaluating the extrapolation line at z, we obtain the transformation

$$\log(z + dz) - dz(z + dz)^{-1}.$$

(See Figure 1.) In the stochastic volatility context, we obtain the robustified transformation

$$\begin{aligned} x_t^* &= \log(y_t^2 + \delta\hat{\sigma}^2) - (y_t^2 + \delta\hat{\sigma}^2)^{-1}\delta\hat{\sigma}^2 \\ &= \log \sigma_t^2 + \log(\xi_t^2 + \delta\hat{\sigma}^2\sigma_t^{-2}) - (\xi_t^2 + \delta\hat{\sigma}^2\sigma_t^{-2})^{-1}\delta\hat{\sigma}^2\sigma_t^{-2} \\ &= \mu^* + \alpha_t + \epsilon_t^*, \end{aligned} \tag{3}$$

where

$$\mu^* = \log \zeta^2 + \mathrm{E}\left[\log(\xi_t^2 + \delta\hat{\sigma}^2\sigma_t^{-2}) - (\xi_t^2 + \delta\hat{\sigma}^2\sigma_t^{-2})^{-1}\delta\hat{\sigma}^2\sigma_t^{-2}\right],$$

$$\epsilon_t^* = \log(\xi_t^2 + \delta\hat{\sigma}^2\sigma_t^{-2}) - (\xi_t^2 + \delta\hat{\sigma}^2\sigma_t^{-2})^{-1}\delta\hat{\sigma}^2\sigma_t^{-2} - \mu^* + \log \zeta^2,$$

δ is some small constant and $\hat{\sigma}^2$ is the sample mean of the y_t^2. Note that x_t^* is bounded below by $\log \delta\hat{\sigma}^2 - 1$ and that the effect of the transformation is negligible for large y_t^2. That is, the transformation is flexible in that its effect depends on the degree of inlying of each observation. The transformation effect is data driven, and there is no need to decide arbitrarily which observations are to be classified as inliers.

We chose $\delta = 0.005$ as the smallest value for which excess kurtosis of the $\{\epsilon_t^*\}$ was near zero for each of the nine sets of parameter values in Table 1 of Section 4. See Figure 2(c). This choice of δ reduces the skewness of $\{\epsilon_t^*\}$ substantially, as shown in Figure 2(b). Note also that the variance of $\{\epsilon_t^*\}$ is no longer $\pi^2/2$ when ξ_t is Gaussian. We treat the variance

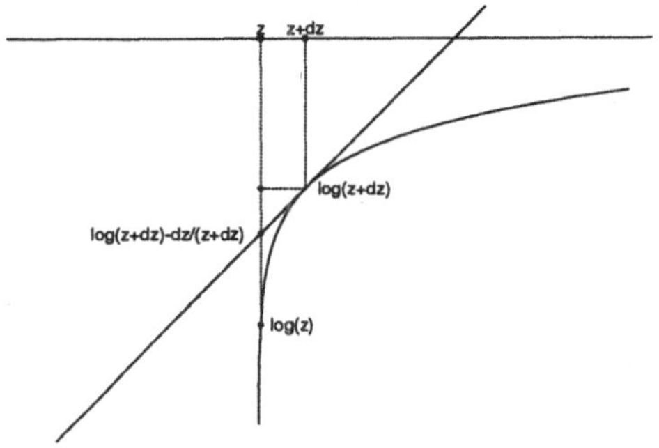

Figure 1: *Schematic diagram of the robustified transformation.*

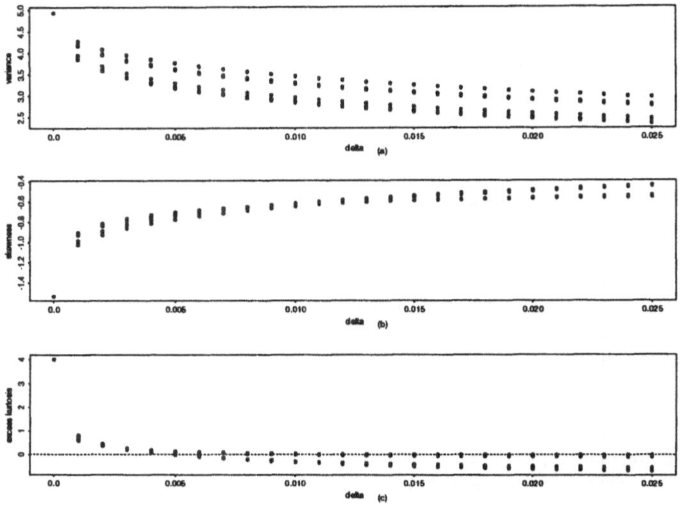

Figure 2: *Choice of* δ*: (a) variance of* $\{\epsilon_t^*\}$ *versus* δ*; (b) skewness of* $\{\epsilon_t^*\}$ *versus* δ*; and (c) excess kurtosis of* $\{\epsilon_t^*\}$ *versus* δ *for each of nine parameter settings in Table 1. At each parameter setting, statistics for each* $\delta \neq 0$ *are averages over 1000 simulated realizations of* $\{\epsilon_t^*\}$ *of length* $T = 500$*. Values at* $\delta = 0$ *are theoretical.*

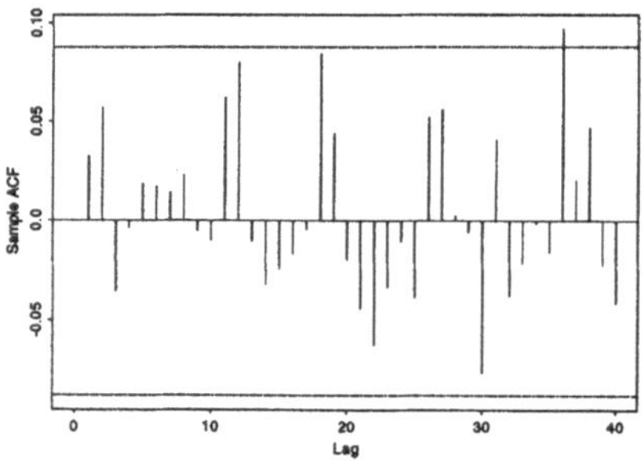

Figure 3: *Sample autocorrelation function of* $\{\epsilon_t^*\}$ *for a typical realization with* $T = 500$, $\phi = 0.95$ *and* $\sigma_\eta = 0.26$.

of $\{\epsilon_t^*\}$ as a free parameter and estimate it from the data. We have also experimented extensively with $\delta = 0.02$, with similar results.

The new errors $\{\epsilon_t^*\}$ are no longer iid, only approximately so. Nevertheless, serial correlation in the $\{\epsilon_t^*\}$ is hard to detect. Figure 3 shows the sample autocorrelation function (ACF) for a typical realization of length $T = 500$, along with the Bartlett bounds $\pm 1.96/\sqrt{T}$. For longer realizations, the dependence disappears as the sample mean of the $\{y_t^2\}$ converges to its unconditional expectation.

Figure 4(a) compares the order statistics for 50 realizations of length $T = 500$ of $\{\epsilon_t^*\}$ (with $\phi = 0.95$ and $\sigma_\eta = 0.26$) with the order statistics of $\{\epsilon_t\}$, while Figure 4(b) compares a smoothed probability density estimate for those 50 realizations with the actual probability density function (pdf) of a $\log \chi_1^2$ random variable. Note that most of the impact of the transformation is in the lower tail of the distribution; that is, on the inlying observations in the distribution of y_t.

The transformation in (3) relies on an estimator of the unconditional variance of the process. An alternative transformation would involve using estimators of the conditional variances, $\{\sigma_t^2\}$. This suggests a two-step estimation procedure: first, estimate parameters of model (1) and use these to find suitable smoothed estimates of the $\{\sigma_t^2\}$, say $\{\hat{\sigma}_t^2\}$; second, transform the original observations via

$$x_t^\dagger = \log(y_t^2 + \delta\hat{\sigma}_t^2) - (y_t^2 + \delta\hat{\sigma}_t^2)^{-1}\delta\hat{\sigma}_t^2. \tag{4}$$

Comparing the one-step robustified transformation given in (3) to the two-step estimation procedure presented in (4), it appears that the two-step estimation method would produce noticeably different parameter estimates whenever variances $\{\sigma_t^2\}$ exhibit relatively large changes over time. If the variances $\{\sigma_t^2\}$ are almost constant, then $\{\sigma_t^2\} \approx \sigma^2$ and the procedure in (4) is roughly equivalent to that given in (3).

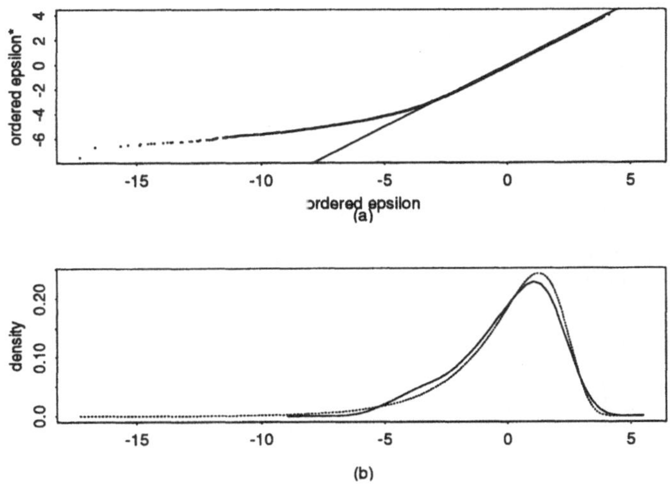

Figure 4: *Comparison of the distributions of* $\{\epsilon_t\}$ *and* $\{\epsilon_t^*\}$ *for 50 realizations with* $T = 500$, $\delta = 0.005$, $\phi = 0.95$ *and* $\sigma_\eta = 0.26$*: (a) Order statistics of* $\{\epsilon_t^*\}$ *versus order statistics of* $\{\epsilon_t\}$*, with 45° reference line; (b) Smoothed probability density estimate for* $\{\epsilon_t^*\}$ *(—) and theoretical probability density function for* $\{\epsilon_t\}$ *(···).*

3 Estimation and Smoothing

3.1 Kalman recursions and the quasi-likelihood method

Model (2) is in linear state-space form and (3) is approximately in linear state-space form. Predicted, filtered and smoothed values of the unobserved states α_t can thus be computed recursively via the Kalman recursions:

$$a_{t|t-1} = \phi a_{t-1}, \quad P_{t|t-1} = \phi^2 P_{t-1} + \sigma_\eta^2 \tag{5}$$

for the one-step-ahead predictor of α_t and its prediction error variance, and

$$a_t = a_{t|t-1} + P_{t|t-1} f_t^{-1}(x_t - a_{t|t-1}), \quad P_t = P_{t|t-1} - P_{t|t-1}^2 f_t^{-1} \tag{6}$$

for the filtered estimate of α_t and its prediction error variance, where

$$f_t = P_{t|t-1} + \sigma_\epsilon^2,$$

$a_0 = 0$ and $P_0 = \sigma_\eta^2/(1 - \phi^2)$ (e.g., Harvey, 1989, pp. 105–6). From these recursions, one can compute the innovations

$$\nu_t = x_t - a_{t|t-1}$$

and construct the Gaussian (quasi) log-likelihood,

$$\log \mathcal{L}(\psi) = -\frac{T}{2} \log 2\pi - \frac{1}{2} \sum_{t=1}^{T} \log f_t - \frac{1}{2} \sum_{t=1}^{T} \nu_t^2 / f_t. \tag{7}$$

The smoothed estimates of α_t and their variances are given by

$$a_{t|T} = a_t + P_t^*(a_{t+1|T} - \phi a_t), \quad P_{t|T} = P_t + P_t^*(P_{t+1|T} - P_{t+1|t})P_t^*, \tag{8}$$

where

$$P_t^* = \phi P_t P_{t+1|t}^{-1}$$

(Harvey, 1989, p. 154).

The state space model in equation (2) has an ARMA(1,1) reduced form,

$$(1 - \phi B)(x_t - \mu) = \eta_t + \epsilon_t - \phi \epsilon_{t-1} = z_t + \theta z_{t-1}, \tag{9}$$

where $\{z_t\}$ is a white noise (WN) sequence with mean zero and variance σ_Z^2. This implies that the following mappings are locally one-to-one:

$$\sigma_\epsilon^2 = -\theta \sigma_Z^2 \phi^{-1}, \quad \sigma_\eta^2 = (1 + \theta^2)\sigma_Z^2 + (1 + \phi^2)\theta \sigma_Z^2 \phi^{-1}. \tag{10}$$

Thus, as an alternative to maximizing (7), we can obtain QML estimates by maximizing an ARMA(1,1) likelihood. Advantages of maximizing the ARMA likelihood include readily available software and insight into the nature of the likelihood surface.

The fitting procedure we used is as follows:

- Transform $\{y_t\}$ to $\{x_t\}$ via the linearizing transformation (2) or to $\{x_t^*\}$ via the robust transformation (3). Mean-correct the transformed series to obtain $\{X_t\}$.

- Maximize the concentrated ARMA(1,1) likelihood of (ϕ, θ) given the mean-corrected transformed series, $\{X_t\}$. This likelihood (dropping irrelevant constants) is

$$\ell(\phi,\theta) = -\frac{T}{2}\log\left\{\sum_{t=1}^{T}\frac{\left(X_t - \hat{X}_t\right)^2}{r_{t-1}}\right\} - \frac{1}{2}\sum_{t=1}^{T}\log r_{t-1},$$

where

$$\hat{X}_t = \begin{cases} 0, & t=1 \\ \theta\left(X_{t-1} - \hat{X}_{t-1}\right)/r_{t-2}, & t=2,\ldots,T \end{cases}$$

and

$$r_{t-1} = \begin{cases} (1+2\theta\phi+\theta^2)/(1-\phi^2), & t=1 \\ 1+\theta^2-\theta^2/r_{t-2}, & t=2,\ldots,T \end{cases}$$

(Brockwell and Davis, 1991, §5.3).

The MLE of σ_Z^2 is then obtained as

$$\hat{\sigma}_Z^2 = \sum_{t=1}^{T}\frac{\left(X_t - \hat{X}_t\right)^2}{r_{t-1}}.$$

- Map $(\hat{\phi}, \hat{\theta}, \hat{\sigma}_Z^2)$ to $(\hat{\phi}, \hat{\sigma}_\epsilon^2, \hat{\sigma}_\eta^2)$.
- Using $(\hat{\phi}, \hat{\sigma}_\epsilon^2, \hat{\sigma}_\eta^2)$ in (5), (6) and (8), compute the smoothed estimates $\{a_{t|T}\}$.
- Estimate ζ via

$$\hat{\zeta} = \left\{T^{-1}\sum_{t=1}^{T} y_t^2 \exp(-a_{t|T})\right\}^{1/2}.$$

3.2 Two-step estimation procedure

Let $\hat{\sigma}_t^2 = \hat{\zeta}^2 \exp(a_{t|T})$ for $t = 1, \ldots, T$. If the $\hat{\sigma}_t^2$ are quite variable over the sample, a second round of calculations as in (4) can be carried out in which the transformation (3) is modified to

$$x_t^\dagger = \log(y_t^2 + \delta\hat{\sigma}_t^2) - (y_t^2 + \delta\hat{\sigma}_t^2)^{-1}\delta\hat{\sigma}_t^2,$$

where δ is a pre-specified constant, such as 0.005. Parameters can then be estimated as above, by mean-correcting the transformed series and fitting an ARMA(1,1).

4 Simulation study

4.1 Design

To assess the performance of the robust transformation in (3), and of the two-step procedure in (4), we conducted a simulation study similar to the one designed by Jacquier et al. (1994). To facilitate comparison with the results of Jacquier et al. (1994), we used for our sampling experiments the models given in Table 4 of their paper. The models are indexed by the ratio $\mathrm{Var}(\sigma_t^2)\,(\mathrm{E}[\sigma_t^2])^{-2}$ which, from Jacquier et al.'s empirical work, is expected to

		ϕ		
$\text{Var}(\sigma_t^2)\text{E}^{-2}[\sigma_t^2]$		0.90	0.95	0.98
10.0	ζ	0.01647	0.01647	0.01647
	σ_η	0.6750	0.4835	0.3082
1.0	ζ	0.02523	0.02523	0.02523
	σ_η	0.3629	0.2600	0.1657
0.1	ζ	0.02929	0.02929	0.02929
	σ_η	0.1346	0.0964	0.0614

Table 1: *Simulation experiment parameter values.*

be around 1.0. Also, as found empirically by a number of authors (e.g., Harvey and Shephard, 1993; Kim and Shephard, 1994; Jacquier et al., 1994) values of ϕ between 0.9 and 0.98 are of interest. As explained by Jacquier et al. (1994), all experiments are calibrated so that $\text{E}[\sigma_t^2] = 0.0009$, implying an approximate 20% annual standard deviation if the simulated series are thought of as weekly returns. We consider samples of size $T = 500$ and compute means and root mean squared errors over $N = 500$ simulated realizations for each of the nine parameter settings given in Table 1. The same simulated realizations are used for testing each of the estimation procedures.

We used a different parameterization than Jacquier et al. (1994). In our formulation, the scale parameter ζ can be mapped to Jacquier et al.'s autoregressive intercept parameter β via

$$\log(\zeta^2) = \frac{\beta}{1-\phi}.$$

For QML, which relies only on the second-order moment structure of the data, the cases in which the ratio of volatility variance to squared mean is 0.1 are, perhaps, unrealistically difficult without large samples or prior information. For example, a realization $\{y_t\}$ of length $T = 500$ from the model with $\phi = 0.9$ and ratio equal to 0.1 is likely to be indiscernible from a white noise series with constant conditional variance. In this case, the lag-one theoretical autocorrelations for $\{y_t^2\}$ and $\{\log y_t^2\}$ are 0.0389 and 0.0171, respectively, while the asymptotic standard error from Bartlett's formula under a white noise null hypothesis is $(500)^{-1/2} = 0.04472$. Indeed, in our 500 simulated realizations, the average lag-one sample autocorrelations for $\{y_t^2\}$ and $\{\log y_t^2\}$ are 0.033 (0.050) and 0.012 (0.046), respectively, where the simulation standard deviation appears in parentheses. Thus, practitioners carrying out exploratory analyses would be unlikely to choose a stochastic volatility model to fit to these data. Nevertheless, for the sake of completeness, we included these cases in our simulation study.

In JPR's study, MM estimation was shown to be inefficient and to perform poorly over repeated sampling. Though Andersen and Sørensen (1994) have suggested improvements in the MM methodology for SV, Andersen (1994) notes that a sample size of $T = 500$ is too low for meaningful inference by MM, and so we do not consider MM in our simulation study.

To assess the robustness of the QML approach to distributional assumptions, we repeated our simulations with some non-Gaussian disturbance terms, $\{\xi_t\}$ iid with mean

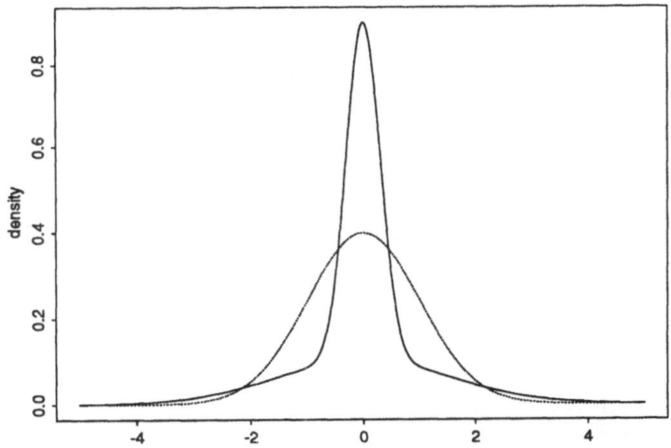

Figure 5: *Comparison of the probability density functions of contaminated normal* $(0.6N(0, 0.09) + 0.4N(0, 2.365))$ *(—) and standard normal (···).*

zero and variance one. In particular, we considered the contaminated normal distribution obtained by sampling from a $N(0, 0.09)$ distribution with probability 0.6 and from a $N(0, 2.365)$ distribution with probability 0.4. The pdf of this distribution is compared to the pdf of a standard normal in Figure 5. This is designed to be a very difficult case for QML since values near zero are selected with high probability.

4.2 Results

Results from the sampling experiments are presented in Tables 2 and 3, and are organized to be immediately comparable to those presented by Jacquier et al. (1994). The model parameters β, ϕ, and σ_η correspond to Jacquier et al.'s autoregressive intercept, autoregressive slope, and standard deviation of the underlying process, respectively. Simulation results are given for nine different true values of the model parameters. Within each cell in the table, true parameter values are displayed in the first row, Jacquier et al.'s (1994) simulation results for the Bayesian (JPR Bayes) and the usual QML (JPR QML) estimators are reproduced on the second and third rows, respectively, and in the last three rows, we show our results for the usual QML (QML0) estimator, the improved, one-step QML (QML1) estimator, and the two-step QML (QML2) estimator. Entries in the table represent average parameter values over 500 replications of the experiment, and RMSE's are given below, in parentheses.

Table 2 shows that QML1 has lower RMSE than QML0 for every parameter in every case. Reductions in RMSE range from 7% to 62%, with the greatest reduction nearly

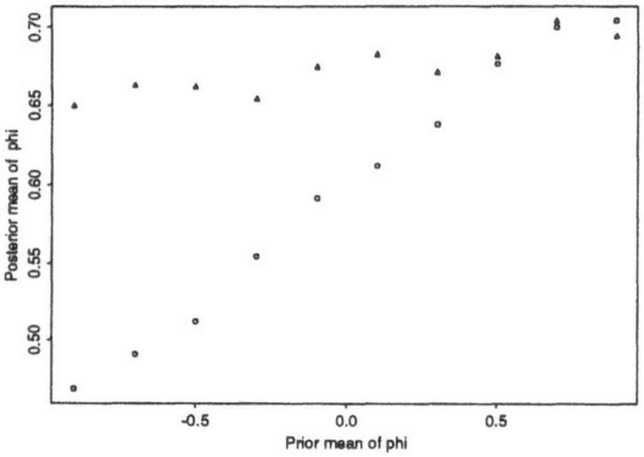

Figure 6: *Sensitivity of posterior of ϕ to prior for a realization of length $T = 500$ from a stochastic volatility model with $\phi = 0.95$ and $Var(\sigma_t^2)E^{-2}[\sigma_t^2] = 0.1$. Prior variance is one for circles, ten for triangles.*

always occuring for σ_η. In each cell, RMSE's for β and ϕ are reduced by the same amount.

As is usually the case, comparing results obtained from a fully Bayesian analysis to those obtained from a classical procedure is tricky. It is not clear how prior information should be factored into the comparison, since whenever data contain little information about the values of parameters, whatever information is incorporated through the prior becomes crucial. In the case of stochastic volatility, the likelihood function is often very "flat" over the parameter region, particularly for parameter values such as those in the bottom row of Table 2.

To assess the sensitivity of the posterior to the prior in the bottom row, we wrote a Markov chain sampler which differs from JPR's only in the way draws from

$$p(\alpha_t|\alpha_{t-1}, \alpha_{t+1}, \zeta, \phi, \sigma_\eta, \mathbf{y})$$

are obtained. We took a simulated time series for the case with $\text{Var}(\sigma_t^2)\text{E}^{-2}[\sigma_t^2] = 0.1$ and $\phi = 0.95$ and, after a very long burn-in, computed an estimate of the posterior mean of ϕ from 100,000 draws. This procedure was done for each of 20 priors for ϕ. The priors had prior means equal to $-0.9, -0.7, \ldots, 0.7$, or 0.9 and prior variances equal to one or ten. The results are given in Figure 6, which shows a clear dependence of posterior mean on prior mean for the low prior variance and some dependence of the posterior mean on the prior mean for the high variance case. (There is "jitter" in the plot in spite of the large number of draws because of the very strong correlation among the draws.)

Jacquier et al. (1994) point out that the prior distributions used in their sampling experiments were extremely diffuse. However, as is clear from their discussion in Section 1.4, prior distributions were proper, as would be required to guarantee integrability of the stationary distributions for their Markov chains. To assess further the impact of the prior, we repeated a small part of JPR's sampling experiment for the Bayes estimator, focussing on the cases in which $\phi = 0.95$ and $\mathrm{Var}(\sigma_t^2)\mathrm{E}^{-2}[\sigma_t^2] = 1.0$ or 0.1. We confirmed JPR's results for the first of these cases using four very different priors, suggesting that the likelihood was swamping the prior in this case. We found, however, that the mean and RMSE of the Bayes estimator were very sensitive to the prior variance for the case with ratio equal 0.1; for our diffuse prior choices, the mean and RMSE of our Bayes estimator were worse than QML1. It is thus possible that the prior information may have added non-negligibly to the information provided by the data for those parameter settings in the bottom row of Table 2, resulting in significantly less biased JPR Bayes estimates for all parameters.

A surprising result arises from the comparison of the performance of Jacquier et al.'s (1994) QML to the QML0 estimator we computed. Since JPR QML and QML0 are the same estimator computed from comparable samples, the difference in the performance of these two estimators was unexpected. In particular, Jacquier et al. (1994) report dismal behavior of QML for parameter values such as those in the bottom six cells of the table, whereas we found much better behavior of the usual QML for those parameter settings. Finding extreme values on a flat likelihood surface may heavily depend on numerical procedures and convergence criteria. It would seem that the difference between JPR's QML results and ours may be due to numerical accuracy.

Results obtained for the autoregressive parameter ϕ were generally good for all methods, except for those parameter values that generate sequences that are essentially indistinguishable from white noise with constant conditional variance. At least in terms of bias, all methods appear to produce comparable results. Root mean squared errors were significantly lower for the parameter estimates computed by the Jacquier et al. (1994) Bayesian method.

The performance of the estimates for both β and σ_η varies greatly from one cell of the table to the other, and from one method to another. The variance of the underlying process σ_η, is not easy to estimate using likelihood information alone, particularly when the true value of the parameter is very low. The biases in the estimates obtained from the series that were not distinguishable from white noise were very large, for all versions of the QML approach. The Bayesian method proposed by Jacquier et al. (1994) produced less biased estimates, a result that may be due to the additional information incorporated through the prior distribution for σ_η. Consider, however, the middle row of cells in Table 2, which contains parameter values similar to those that have been estimated from empirical studies (e.g., Melino and Turnbull, 1990; Vetzal, 1992; Jacquier et al., 1994). For these parameter values, both QML1 and QML2 exhibit very small biases in the estimates of ϕ and of σ_η. Again, RMSE's were higher than those for the Bayes estimator.

Bias in the estimation of the intercept parameter β was high for all parameter settings and all versions of QML (see Table 2). This is in agreement with results reported by Jacquier et al. (1994). As was mentioned in the preceding section, an alternative parameterization would include the scale parameter ζ, as given in expression (1), instead of β. Results presented in Table 3 clearly indicate that parameterizing the model in terms of ζ rather than β produces estimates for ζ that exhibit very low biases and RMSE's across all parameter settings and all versions of QML. It is worth noticing from Table 3 that the bias for ζ is nearly zero, even when data are indistinguishable from white noise with constant conditional variance.

According to Table 2, there is little difference between the one-step improved QML

(QML1) from expression (3) and the two-step (QML2) procedure outlined in (4). When the true variance of the underlying autoregressive process is relatively high, time-dependent variances in the observations change noticeably over the sample period, and we had expected use of the two-step approach to improve the performance of the QML estimator. Our simulation study did not detect important improvements of QML2 over QML1.

We present results for the non-Gaussian case in Table 4. Data for the experiment were generated from a contaminated Gaussian distribution as was described in the preceding section. Parameter settings corresponded to the middle row of cells in Table 2. We ran our experiment on those three sets of parameter values because they represent the more "realistic" scenarios. From Table 4 it seems apparent that QML1 performs significantly better than QML0 (the usual QML estimator), and that QML2 does not perform as well as the one-step QML1. In this experiment, QML1 is robust to departures from normality, as was hypothesized. The inlier treatment we propose was able to estimate the parameters with little bias. The usual QML did not perform well in the presence of inliers, as expected.

5 Discussion

The usual quasi-maximum likelihood estimator has been shown to perform poorly in stochastic volatility models, particularly when variances exhibit relatively minor changes over time (Jacquier et al., 1994; Harvey et al., 1994). Jacquier et al. (1994) proposed a method for estimation for stochastic volatility models that has been shown to behave very well, at least when the data satisfy the assumptions of the model. Their procedure, however, is very intensive from the numerical point of view, and does not easily adapt to models which depart from those assumptions.

We have proposed a simple modification to the usual QML estimator which improves its performance over repeated sampling. The modification affects the treatment of inliers rather than the estimation procedure itself. The modification reduces the skewness and the excess kurtosis of the distribution of the transformed disturbances in the observation equation.

Though our focus has been on improving the performance of QML, the transformation we propose is also applicable to the estimation technique in which the non-Gaussian error in the observation equation of the transformed SV model is approximated by a mixture of Gaussian distributions (Kim and Shephard 1994, Shephard 1994, Carter 1993, Mahieu and Schotman 1994). In fact, it may be easier to approximate the distribution of the transformed random variable ϵ_t^* by a Gaussian mixture than to approximate the distribution of ϵ_t, which has a long left tail. See Figure 4.

We have followed Jacquier et al. (1994) regarding the design of our sampling experiment, since we were interested in comparing the performance of our improved QML estimator to their Bayes estimator. We found ourselves, however, in the usual quandary: what is a fair comparison when one method uses information exogenous to the sample and the other one does not? In this particular problem, the use of prior information seems to be crucial, since for some of the parameter values considered in the experiment, the realized series are not distinguishable from white noise with constant conditional variance. It would seem reasonable to speculate that no practitioner who looks at his or her data

$\frac{Var(\sigma_t^2)}{E^2[\sigma_t^2]}$	Method	β	ϕ	σ_η	β	ϕ	σ_η	β	ϕ	σ_η
	true	-0.821	0.90	0.675	-0.4106	0.95	0.4835	-0.1642	0.98	0.308
	JPR Bayes	-0.679	0.916	0.562	-0.464	0.94	0.46	-0.19	0.98	0.35
		(0.22)	(0.026)	(0.12)	(0.16)	(0.02)	(0.055)	(0.08)	(0.01)	(0.06)
10	JPR QML	-0.99	0.88	0.70	-0.55	0.93	0.51	-0.11	0.99	0.33
		(0.48)	(0.06)	(0.16)	(0.32)	(0.04)	(0.12)	(0.09)	(0.01)	(0.07)
	QML0	-0.924	0.887	0.646	-0.480	0.941	0.402	-0.291	0.965	0.334
		(0.434)	(0.052)	(0.237)	(0.297)	(0.035)	(0.244)	(0.227)	(0.027)	(0.083)
	QML1	-0.941	0.886	0.617	-0.515	0.937	0.445	-0.284	0.966	0.303
		(0.368)	(0.044)	(0.130)	(0.255)	(0.030)	(0.093)	(0.211)	(0.025)	(0.067)
	QML2	-0.944	0.886	0.672	-0.515	0.938	0.483	-0.283	0.966	0.325
		(0.373)	(0.044)	(0.129)	(0.254)	(0.030)	(0.091)	(0.213)	(0.025)	(0.073)
	true	-0.736	0.90	0.363	-0.368	0.95	0.26	-0.1472	0.98	0.1657
	JPR Bayes	-0.87	0.88	0.35	-0.56	0.92	0.28	-0.22	0.97	0.23
		(0.34)	(0.046)	(0.067)	(0.34)	(0.046)	(0.065)	(0.14)	(0.02)	(0.08)
	JPR QML	-1.4	0.81	0.45	-1.0	0.86	0.35	-0.20	0.97	0.22
		(1.6)	(0.22)	(0.27)	(1.7)	(0.23)	(0.25)	(0.54)	(0.08)	(0.15)
1.0	QML0	-1.051	0.858	0.398	-0.610	0.918	0.279	-0.356	0.952	0.208
		(0.892)	(0.117)	(0.220)	(0.652)	(0.087)	(0.174)	(0.580)	(0.074)	(0.154)
	QML1	-0.986	0.867	0.341	-0.534	0.928	0.229	-0.336	0.955	0.191
		(0.762)	(0.099)	(0.173)	(0.430)	(0.057)	(0.135)	(0.540)	(0.068)	(0.125)
	QML2	-0.984	0.868	0.375	-0.540	0.927	0.252	-0.323	0.956	0.195
		(0.749)	(0.098)	(0.189)	(0.458)	(0.061)	(0.153)	(0.444)	(0.059)	(0.103)
	true	-0.706	0.90	0.135	-0.353	0.95	0.0964	-0.1412	0.98	0.0614
	JPR Bayes	-1.54	0.78	0.15	-1.12	0.84	0.12	-0.66	0.91	0.14
		(1.35)	(0.19)	(0.082)	(1.15)	(0.16)	(0.074)	(0.83)	(0.12)	(0.099)
	JPR QML	-5.5	0.23	0.33	-5.5	0.22	0.31	-3.5	0.49	0.35
		(5.6)	(0.79)	(0.39)	(6.0)	(0.85)	(0.41)	(4.6)	(0.67)	(0.46)
0.1	QML0	-2.478	0.658	0.341	-2.119	0.708	0.315	-1.199	0.836	0.229
		(3.100)	(0.424)	(0.414)	(2.950)	(0.403)	(0.401)	(2.001)	(0.266)	(0.405)
	QML1	-2.257	0.688	0.267	-1.822	0.749	0.244	-1.058	0.854	0.182
		(2.831)	(0.387)	(0.308)	(2.634)	(0.359)	(0.304)	(1.784)	(0.240)	(0.301)
	QML2	-2.229	0.692	0.281	-1.747	0.758	0.245	-1.002	0.862	0.181
		(2.805)	(0.384)	(0.334)	(2.535)	(0.348)	(0.301)	(1.726)	(0.231)	(0.301)

Table 2: *Mean and root mean squared error (in parentheses) for five estimation techniques: Bayesian (JPR Bayes) and quasi-maximum likelihood (JPR QML) results from Tables 7 and 6, respectively, of Jacquier, Polson and Rossi (1994); QML with conventional log-squares transformation (QML0), with robust transformation (QML1), and with two-step transformation (QML2). Statistics in this table are based on 500 simulated samples, each of length* $T = 500$.

$\frac{\mathrm{Var}(\sigma_t^2)}{\mathrm{E}^2[\sigma_t^2]}$	Method	$\phi = 0.90$	$\phi = 0.95$	$\phi = 0.98$
	true ζ	0.01647	0.01647	0.01647
	QML0	0.0179	0.0197	0.0173
		(0.0052)	(0.0079)	(0.0058)
10	QML1	0.0176	0.0168	0.0172
		(0.0049)	(0.0037)	(0.0058)
	QML2	0.0161	0.0165	0.0169
		(0.0029)	(0.0036)	(0.0056)
	true ζ	0.02523	0.02523	0.02523
	QML0	0.0256	0.0257	0.0260
		(0.0029)	(0.0037)	(0.0047)
1	QML1	0.0253	0.0257	0.0259
		(0.0028)	(0.0037)	(0.0046)
	QML2	0.0251	0.0256	0.0257
		(0.0029)	(0.0038)	(0.0045)
	true ζ	0.02929	0.02929	0.02929
	QML0	0.0283	0.0283	0.0286
		(0.0026)	(0.0027)	(0.0030)
0.1	QML1	0.0284	0.0284	0.0288
		(0.0024)	(0.0026)	(0.0027)
	QML2	0.0283	0.0284	0.0287
		(0.0025)	(0.0024)	(0.0028)

Table 3: *Mean and root mean squared error (in parentheses) for estimation of the scale parameter, ζ, via $\hat{\zeta}$ for QML with conventional log-squares transformation (QML0), with robust transformation (QML1), and with two-step transformation (QML2). Statistics in this table are based on 500 simulated samples, each of length* $T = 500$.

$\frac{\mathrm{Var}(\sigma_t^2)}{\mathrm{E}^2[\sigma_t^2]}$	Method	ζ	ϕ	σ_η	ζ	ϕ	σ_η	ζ	ϕ	σ_η
	true	0.02523	0.90	0.363	0.02523	0.95	0.26	0.02523	0.98	0.166
	QML0	0.0242	0.84	0.470	0.0247	0.91	0.331	0.0250	0.94	0.228
		(0.0034)	(0.141)	(0.317)	(0.0035)	(0.099)	(0.198)	(0.0049)	(0.090)	(0.186)
1.0	QML1	0.0242	0.85	0.392	0.0246	0.91	0.286	0.0249	0.942	0.201
		(0.0033)	(0.127)	(0.225)	(0.0036)	(0.094)	(0.182)	(0.0049)	(0.098)	(0.180)
	QML2	0.0234	0.85	0.431	0.0240	0.92	0.315	0.0245	0.944	0.217
		(0.0037)	(0.155)	(0.315)	(0.0036)	(0.086)	(0.173)	(0.0049)	(0.091)	(0.191)

Table 4: *Mean and root mean squared error (in parentheses) for contaminated Gaussian case using three estimation techniques: QML with conventional log-squares transformation (QML0), with robust transformation (QML1), and with two-step transformation (QML2). Statistics in this table are based on 500 simulated samples, each of length* $T = 500$.

prior to model fitting would even attempt to estimate stochastic volatility parameters from such data without a strong prior belief that "SV is appropriate." Our experiments show that the prior information used in the estimation procedure can also affect the performance of the estimator for these "white noise" cases. For all other parameter settings, the likelihood function is sufficiently informative to swamp the prior.

We hypothesize that even though QML is inefficient when observations are Gaussian, it may be robust when observations are not Gaussian. Our initial simulations using a contaminated Gaussian distribution seem to suggest that this is the case. We have conducted similar simulations, not presented in this paper, using a Gumbel distribution to generate the samples. Results from those experiments are also encouraging, and will be pursued.

From an operational point of view, the usual QML estimator presents obvious advantages over the estimator that was proposed by Jacquier et al. (1994). While we wrote our own software for the simulations, any commercial software (e.g., S-Plus, SAS) could be used to carry out the procedure we propose. Furthermore, the method is the same, *regardless* of the true distribution of the disturbances in the observation equation. Indeed, QML "knows" right from the start, that the distributional assumptions implicit in the use of the Kalman recursions are not correct, except in the unlikely case where the $\{\xi_t\}$ in model (1) are distributed as log-normal random variables. Finally, the QML procedure works with only minor modifications if the dynamics of $\{\alpha_t\}$ change, while the Bayesian methodology relies heavily on the special Markovian structure of the AR model.

The improved QML procedures we propose (both the one-step and the two-step methods) exhibit higher root mean squared errors than the method presented by Jacquier et al. (1994). Improvement would seem to require modification of the estimation procedure itself. We have obtained initial results that suggest that the performance of the modified QML estimator in terms of root mean squared error can be improved further, by incorporating higher-order moment information in the estimation procedure. We are currently working on this problem in collaboration with Wayne A. Fuller, and will report on it elsewhere.

Acknowledgements: The authors are grateful to Gordon C. Rausser for sparking our interest in this problem, to Wayne A. Fuller for many helpful suggestions, and to the referee for very detailed and constructive comments. We also thank Nicholas G. Polson for providing software and answering questions.

References

Abramowitz, M. and Stegun, I.A. (1965). *Handbook of Mathematical Functions with Formulas, Graphs, and Mathematical Tables*, National Bureau of Standards Applied Mathematics Series 55.

Andersen, T.G. (1994). Comment on Jacquier, Polson and Rossi's "Bayesian analysis of stochastic volatility models". *Journal of Business and Economic Statistics* **12**, 389–392.

Andersen, T.G. and Sørensen, B. (1994). Estimation of a stochastic volatility model: a Monte Carlo study. Working paper, Northwestern University, J.L. Kellogg Graduate School of Management.

Bartlett, M.S. and Kendall, D.G. (1946). The statistical analysis of variance—heterogeneity and the logarithmic transformation. Supplement to the *Journal of the Royal Statistical Society*, Vol. VIII, 128–133.

Bollerslev, T. (1986). Generalized autoregressive conditional heteroskedasticity, *Journal of Econometrics* **31**, 307–327.

Bollerslev, T., Chou, R.Y., and Kroner, K.F. (1992). ARCH modeling in finance, *Journal of Econometrics* **52**, 5–59.

Brockwell, P.J. and Davis, R.A. (1991). *Time Series: Theory and Methods,* 2nd ed., Springer-Verlag, New York.

Carter, C.K. (1993). On Markov chain Monte Carlo methods for linear state space modelling. *Unpublished Ph.D. dissertation,* University of New South Wales.

Clark, P.K. (1973). A subordinated stochastic process model with finite variances for speculative prices, *Econometrica* **41**, 135–156.

Engle, R.F. (1982). Autoregressive conditional heteroscedasticity with estimates of the variance of United Kingdom inflation. *Econometrica* **50**, 987–1007.

Fuller, W.A. (forthcoming). *Introduction to Statistical Time Series,* 2nd ed., John Wiley, New York.

Harvey, A.C. (1989). *Forecasting, Structural Time Series Models and the Kalman Filter,* Cambridge University Press, Cambridge.

Harvey, A.C., and Shephard, N. (1993). Estimation and testing of stochastic variance models. *Unpublished manuscript,* The London School of Economics.

Harvey, A.C., Ruiz, E., and Shephard, N. (1994). Multivariate stochastic variance models, *Review of Economic Studies* **61**, 247–264.

Hull, J., and White, A. (1987). The pricing of options on assets with stochastic volatilities. *Journal of Finance* **42**, 281–300.

Jacquier, E., Polson, N.G. and Rossi, P.E. (1994). Bayesian analysis of stochastic volatility models (with discussion). *Journal of Business and Economic Statistics* **12**, 371–417.

Kim, S., and Shephard, N. (1994). Stochastic volatility: likelihood inference and comparison with ARCH models. *Unpublished working paper*, Nuffield College, Oxford.

Mahieu, R. and Schotman, P. (1994). Stochastic volatility and the distribution of exchange rate news. Discussion Paper 96, Institute for Empirical Macroeconomics, Federal Reserve Bank of Minneapolis.

Melino, A. and and Turnbull, S.M. (1990). Pricing foreign currency options with stochastic volatility. *Journal of Econometrics* **45**, 239–265.

Nelson, D.B. (1988). Time series behavior of stock market volatility and returns. *Unpublished PhD dissertation,* Massachusetts Institute of Technology, Economics Department.

Nelson, D.B. (1994). Comment on Jacquier, Polson and Rossi's "Bayesian analysis of stochastic volatility models". *Journal of Business and Economic Statistics* **12**, 403–406.

Ruiz, E. (1994). Quasi-maximum likelihood estimation of stochastic volatility models. *Journal of Econometrics* **63**, 289–306.

Shephard, N. (1994). Partial non-Gaussian state space. *Biometrika* **81**, 115–131.

Tauchen, G., and Pitts, M. (1983). The price variability-volume relationship on speculative markets. *Econometrica* **51**, 485–505.

Taylor, S. (1986). *Modelling Financial Time Series,* Wiley, New York.

Vetzal, K. (1992). Stochastic short rate volatility and the pricing of bonds and bond options. *Unpublished PhD dissertation,* University of Toronto, Department of Economics.

Wishart, J. (1947). The cumulants of the z and of the logarithmic χ^2 and t distributions. *Biometrika* **34**, 170–178.

Bayesian Inference for Linear Models Subject to Linear Inequality Constraints

John F. Geweke
Research Department, Federal Reserve Bank of Minneapolis
and Department of Economics
University of Minnesota

Abstract

The normal linear model, with sign or other linear inequality constraints on its coefficients, arises very commonly in many scientific applications. Given inequality constraints Bayesian inference is much simpler than lassical inference, but standard Bayesian computational methods become impractical when the posterior probability of the inequality constraints (under a diffuse prior) is small. This paper shows how the Gibbs sampling algorithm can provide an alternative, attractive approach to inference subject to linear inequality constraints in this situation, and how the GHK probability simulator may be used to assess the posterior probability of the constraints.

1 Introduction

The normal linear regression model subject to linear inequality constraints for the coefficients arises commonly in applied econometrics as well as other scientific applications. Typically the motivating economic model restricts the signs of certain coefficients or of known linear combinations of coefficients. A well-known pedagogical example is provided by Pindyck and Rubinfeld (1981, p. 44) who take up the demand for student housing near the University of Michigan. Rent paid per person is a linear function of the number of rooms per person, with a positive coefficient, and the distance from campus, with a negative coefficient. (We return to this example below.)

The subsequent empirical work in this and other linear regression models subject to linear inequality constraints for the coefficients then focuses on two related but distinct questions. First, how plausible are the linear inequality restrictions delivered by the economic model? Second, conditional on these restrictions what is to be inferred about the coefficients of the regression model? These turn out to be nontrivial questions, and historically investigators have taken different, usually informal, approaches to these tasks. Difficulties for classical inference are discussed in Judge and Takayama (1966) and Lovell and Prescott (1970); for classical testing in Gourieroux, Holly, and Monfort (1982) and Wolak (1987).

classical inference are discussed in Judge and Takayama (1966) and Lovell and Prescott (1970); for classical testing in Gourieroux, Holly, and Monfort (1982) and Wolak (1987).

This work takes up a Bayesian approach to the problem of linear regression with linear inequality constraints on the coefficients. Extending earlier analytical work by Davis (1978), Chamberlain and Leamer (1976), and Leamer and Chamberlain (1976), it uses fast numerical methods for the determination of posterior moments and probabilities, advancing the methods reported in Geweke (1986). But whereas Geweke (1986) takes up any inequality constraints on the coefficients, this paper limits attention to inequality constraints that are linear. The more specialized algorithms provide faster, more accurate numerical approximations to posterior moments than do the more general ones. In particular, when the posterior probability of linear inequality constraints is low or the number of coefficients is large the methods in Geweke (1986) may be slow to the point of impracticality. In contrast computation time in the approach taken here does not increase systematically with the inverse of the posterior probability of the inequality constraints, and increases only linearly with the number of coefficients.

In standard notation the normal linear regression model is

$$y_{T\times 1} = X_{T\times k}\beta_{k\times 1} + \varepsilon_{T\times 1}, \quad \varepsilon \sim N(0, \sigma^2 I_T) \tag{1}$$

where y is the vector of dependent variables, X is the matrix of explanatory variables (regressors), and ε is the vector of disturbances. There are T observations and k explanatory variables. In the unconstrained case a standard diffuse reference prior for the parameter vector (β', σ) is

$$f(\beta, \sigma) \propto \sigma^{-1}. \tag{2}$$

The inequality constraints are expressed

$$a_{k\times 1} \leq D_{k\times k}\beta \leq w_{k\times 1}. \tag{3}$$

In this expression the inequalities are to read line by line: $a_i \leq \sum_{j=1}^{k} d_{ij}\beta_j \leq w_i (i = 1, ..., k)$. The matrix D is composed of real numbers and is nonsingular. The vectors a and w are composed of extended real numbers, with -∞ and $+\infty$ explicitly permitted, thus allowing single-sided inequality constraints. Since a constraint in (3) has no effect if $a_i = -\infty$ and $w_i = +\infty$, fewer than k linear inequality restrictions—perhaps only one—may be involved. Inequality constraints on more than k linear combinations are precluded by (3), a point to which the concluding section of the paper returns briefly. Extending the standard reference prior of the unconstrained model the prior distribution employed in this model is

$$\begin{aligned} f(\beta, \sigma) &\propto \sigma^{-1} \text{ if } \beta \in Q, \; f(\beta, \sigma) = 0 \text{ if } \beta \notin Q; \\ Q &\equiv \{\beta : a \leq D\beta \leq w\}. \end{aligned} \tag{4}$$

The next section takes up evaluation of the posterior probability of the hypothesis (4) relative to (2). An algorithm for the evaluation of this probability based on the Geweke-Hajivassiliou-Keane (GHK) probability simulator is constructed. (The simulator was named by Hajivassiliou, McFadden, and Ruud (1995) in a paper that first appeared in 1992. See also Keane (1990, 1993, 1994) and Geweke, Keane and Runkle (1995).) Section 3 turns to the problem of inference for β—construction of posterior means, evaluation of the posterior probabilities of regions, etc.—in the constrained model (4). A Gibbs sampling algorithm for drawing β and σ from the posterior distribution that builds on work in Geweke (1991, 1995) is presented. The algorithm described in Section 3 is superficially similar to that in Geweke (1995), but that paper treats mixed equality and inequality constraints on individual coefficients, whereas the approach here takes up inequality constraints for arbitrary sets of linear combinations of coefficients. Section 4 employs both methods in the two empirical examples used in Geweke (1986). A concluding section suggests extensions of this work.

2 Evaluating the Hypothesis of Linear Inequality Constraints

One may take two approaches to the evaluation of the inequality constraints (3) as a formal hypothesis. The first is to regard (1) and (2) as the maintained hypothesis, and evaluate the posterior probability that (3) is true. The advantage of this method is conceptual simplicity: no problems arise from the fact that (3) is not a proper prior distribution. In particular suppose that (2) is regarded as the limit of a sequence of proper prior distributions, e.g. $\beta \sim N(\beta, \alpha V)$, where β is a fixed vector, V is a fixed matrix, and $\alpha \to \infty$. Then the limiting posterior probability of the region defined by (3) is correctly given as the posterior probability of (3) when the posterior density kernel is formed as the product of (2) and the likelihood function

$$L(\beta, \sigma) = \sigma^{-T} \exp\{-[(T-k)s^2 + (\beta - b)'(X'X)(\beta - b)]/2\sigma^2\} \qquad (5)$$

corresponding to (1). (This expression employs standard notation for the least squares estimator $b \equiv (X'X)^{-1}X'y$; $s^2 \equiv (y - Xb)'(y - Xb)/(T-k)$. For details on the derivation of (5) see Zellner (1971, pp. 65-66).) Denote this probability $p_{2|1}$. The disadvantage of this approach is that (1) through (4) is frequently a hypothesis competing with (1) and (2) rather than a region of interest of the parameter space. For example, if the posterior distribution of β is centered at $\beta = 0$ in the unconstrained model, $p_{2|1}$ for $\beta \geq 0$ will generally become smaller the larger is k; yet in this case (1) through (4) is clearly competitive with (1) and (2).

The second approach is to regard (1) and (2) and (1) through (4) as competing hypotheses for which a posterior odds ratio is to be formed. Without loss of generality, assume that the prior odds ratio is 1:1. If all elements of a

and w are finite then the prior distribution (4) is proper for β and the limit of a posterior odds ratio in favor of (1) and (2) over (1) through (4) with a sequence of increasingly diffuse priors is 0. (This phenomenon is sometimes called the Lindley paradox and is well treated in the literature, e.g. Lindley (1957), Press (1989, pp. 36–37) and Bernardo and Smith (1994, p. 394).) If some elements of a are $-\infty$ and/or some elements of w are $+\infty$, then in general the limiting ratio formed from a sequence of increasingly diffuse priors depends on the particular sequence chosen. Certain specific cases are of some general interest, however. In particular if the linear inequality constraints may be brought into the form $\delta_i\beta_j \geq a_i (i = 1, ..., r)$ with $\delta_i = \pm 1$, and if the sequence of prior distributions is $\beta \sim N(a, \alpha I_k)$ for the unconstrained case and $\beta \sim N(a, \alpha I_k)$ constrained to $\delta_i\beta_j \geq a_i (i = 1, ..., r)$ for the constrained case, then the limiting posterior odds ratio (as $\alpha \to \infty$) is $p_{2:1} \equiv 2^r p_{2|1}$.

With these considerations in mind, focus on computation of the event probability $p_{2|1}$. Geweke (1986) proposed a *crude frequency simulator* as follows. In the context of (2) and (5), the marginal posterior probability distribution of σ is provided by

$$(y - Xb)'(y - Xb)/\sigma^2 \sim \chi^2(T - k). \tag{6}$$

Conditional on σ,

$$\beta \sim N[b, \sigma^2(X'X)^{-1}].$$

Hence random draws $\{\sigma^{(i)}, B^{(i)}\}_{i=1}^m$ may be taken easily. Let $d^{(i)} = 1$ if $\beta^{(i)} \in Q$, $d^{(i)} = 0$ if $B^{(i)} \notin Q$. Then $\breve{p}_{2|1} \equiv \sum_{i=1}^m d^{(i)}/m \overset{a.s}{\to} p_{2|1}$ as $m \to \infty$. Moreover, for large m the standard error of approximation of $p_{2|1}$ by $\breve{p}_{2|1}$ is given by $[p_{2|1}(1 - p_{2|1})/m]^{1/2}$. The advantage of this method is its simplicity. Its disadvantage lies in the need to make many draws, m, if $p_{2|1}$ is small; and if k is large, $p_{2:1}$may be large enough that the constraint hypothesis is competitive with the unconstrained model even though $p_{2|1}$ is small.

A more efficient method for approximating $p_{2|1}$ is based on the *GHK probability simulator,* an algorithm proposed independently by Hajivassiliou and McFadden (1990) and Keane (1990); more accessible references are Keane (1993), Keane (1994), and Geweke, Keane and Runkle (1995). Let $z \equiv D(\beta - b)$, $a^* \equiv D(a - b)$, and $w^* \equiv D(w - b)$. Since $\beta|\sigma \sim N[b, \sigma^2(X'X)^{-1}]$, $z|\sigma \sim N[0, \sigma^2 D(X'X)^{-1}D']$and $P[a \leq D\beta \leq w|\sigma] = P[a^* \leq z \leq w^*|\sigma]$. Let FF' denote the Choleski decomposition of $D(X'X)^{-1}D'$: F is the unique lower triangular matrix with positive diagonal elements such that $FF' = D(X'X)^{-1}D'$. A conventional construction for $z \mid \sigma$ is then given by

$$z|\sigma = \sigma F\varepsilon, \quad \varepsilon \sim N(0, I_k). \tag{7}$$

(Most software for generation of synthetic random multivariate normal vectors is based on this construction.) Denote a typical row of (7) $z_i = \sigma \sum_{j=1}^i f_{ij}\varepsilon_i$. The probability $P[a^* \leq z \leq w^*|\sigma]$ may be decomposed

$$P[a_1^* \leq z_1 \leq w_1^* | \sigma] \cdot P[a_2^* \leq z_2 \leq w_2^* | \sigma, a_1^* \leq z_1 \leq w_1^*] \cdot$$

$$\ldots \cdot P[a_j^* \leq z_j \leq w_j^* | \sigma, a_i^* \leq z_i \leq w_i^* (i < j)] \cdot$$

$$\ldots \cdot P[a_k^* \leq z_k \leq w_k^* | \sigma, a_i^* \leq z_i \leq w_i^* (i < k)].$$

The GHK probability simulator provides independent, unbiased simulations of each conditional probability in this product. It does so by drawing $\tilde{z}_1, \ldots, \tilde{z}_{j-1}$, from the $N(0, \sigma^2 D(X'X)^{-1} D']$ distribution subject to the constraints $a_i^* \leq z_i \leq w_i^* (i = 1, \ldots, j-1)$ and then computing $P[a_j^* \leq z_j \leq c_j^* | \tilde{z}_1, \ldots \tilde{z}_{j-1}]$. The draws are accomplished by generating

$$\begin{aligned}
\tilde{\varepsilon}_1 &\sim N(0,1) s.t. \quad a_1^*/\sigma f_{11} \leq \tilde{\varepsilon}_1 \leq w_1^*/\sigma f_{11} \\
\tilde{\varepsilon}_2 &\sim N(0,1) s.t. \quad (a_2^* - \sigma f_{21}\tilde{\varepsilon}_1)/\sigma f_{22} \leq \tilde{\varepsilon}_2 \leq (w_2^* - \sigma f_{21}\tilde{\varepsilon}_1)/\sigma f_{22} \\
&\vdots \\
\tilde{\varepsilon}_j &\sim N(0,1) s.t. \quad \left(a_j^* - \sum_{i=1}^{j-1} \sigma f_{ji}\tilde{\varepsilon}_i\right) / \sigma f_{jj} \leq \tilde{\varepsilon}_j \\
&\leq \left(w_j^* - \sum_{i=1}^{j-1} \sigma f_{ji}\tilde{\varepsilon}_i\right) / \sigma f_{jj} \qquad (8) \\
&\vdots \\
\tilde{\varepsilon}_k &\sim N(0,1) s.t. \quad \left(a_k^* - \sum_{i=1}^{k-1} \sigma f_{ki}\tilde{\varepsilon}_i\right) / \sigma f_{kk} \leq \tilde{\varepsilon}_k \\
&\leq \left(w_k^* - \sum_{i=1}^{k-1} \sigma f_{ki}\tilde{\varepsilon}_i\right) / \sigma f_{kk}
\end{aligned}$$

Let $\tilde{p}_j = P[(a_j^* - \sum_{i=1}^{j-1} \sigma f_{ji}\tilde{\varepsilon}_i)/\sigma f_{jj} \leq \tilde{\varepsilon}_j \leq (w_j^* - \sum_{i=1}^{j-1} \sigma f_{ji}\tilde{\varepsilon}_i)/\sigma f_{jj} | \tilde{\varepsilon}_1, \ldots, \tilde{\varepsilon}_{j-1}]$. Then $\tilde{p}_1 = P[a_1^* \leq z_1 \leq w_1^* | \sigma]$ and $E(\tilde{p}_j) = P[a_j^* \leq z_j \leq w_j^* | \sigma] (j > 1)$. The probabilities $\tilde{p}_j$ may be computed by direct evaluation of the univariate standard normal c.d.f. Because the $\tilde{p}_j$ are mutually independent,

$$E\left[\prod_{i=1}^{k} \tilde{p}_i\right] = P(a^* \leq z \leq w^* | \sigma).$$

(Note that in fact it is not necessary to take the last draw in (8).)

The GHK probability simulator for $p_{2|1}$ is therefore an iterative process with two steps in each iteration. In the first step, $(y - Xb)'(y - Xb)/(\sigma^{(i)})^2 \sim \chi^2(T - k)$. In the second step, one or more values of $\prod_{j=1}^{k} \tilde{p}_j$ are drawn as described in the previous paragraph, with $\sigma = \sigma^{(i)}$; let the average of these values be denoted $p^{*(i)}$. After m iterations of the two steps, $p_{2|1}$ is approximated by $\hat{p}_{2|1} \equiv \sum_{i=1}^{m} p^{*(i)}/m$. Since $E[p^{*(i)}|\sigma] = E[d^{(i)}|\sigma]$ but $0 < p^{*(i)} < 1$ whereas $d(i) = 0$ or $d^{(i)} = 1$, $var(\hat{p}_{2|1}) < var(\check{p}_{2|1})$. Hence the GHK probability simulator always provides a more accurate approximation to $p_{2|1}$ than does the crude frequency simulator, given the same number of iterations.

3 Inference Subject to Linear Constraints

A related but distinct task is to find posterior moments and probabilities corresponding to the restricted model (1) through (4). The posterior density for this model is

$$\begin{aligned} f(\beta,\sigma|y,X) &\propto \sigma^{-(T+1)}\exp\{-[(T-k)s^2 + (\beta-b)'(X'X)(B-b)]/2\sigma^2\} \\ \text{if } \beta &\in Q; \\ f(\beta,\sigma|y,X) &= 0 \quad \text{if } \beta \notin Q; \\ Q &\equiv \{\beta : a \le D\beta \le u\}. \end{aligned} \tag{9}$$

The crude frequency simulator described in the previous section may also be used to produce draws from the posterior distribution whose kernel is given by (9). One draws from the unconstrained posterior distribution, and accepts the draw if and only if $a \le D\beta \le w$. This algorithm again has the advantage of simplicity and the disadvantage of inefficiency. In this section we summarize a Gibbs sampling algorithm for generating a sequence $\{\beta^{(i)}, \sigma^{(i)}\}$ that converges in distribution to the posterior distribution whose probability density kernel is given by (9). Supporting technical details may be found in Geweke (1991).

As in the previous section define $z \equiv D(\beta - b)$, $a^* \equiv D(a-b)$, $w^* \equiv D(w-b)$. The posterior distribution of (β,σ) is then a simple transformation of the posterior distribution of (z,σ). The posterior distribution of σ conditional on z (equivalently, on $\beta = b + D^{-1}z$) is given by

$$(y - X\beta)'(y - X\beta)/\sigma^2|(\beta,y,X) \sim \chi^2(T), \tag{10}$$

which may be derived easily from (9) (Geweke 1992). The posterior distribution of z conditional on σ is

$$z|(\sigma,y,X) \sim N[0,\sigma^2 D(X'X)^{-1}D')] \ s.t.\ a^* \le z \le w^*;$$

let $R \equiv \sigma^2 D(X'X)^{-1}D'$. The distribution of

$$z_j|(\sigma, z_i (i \ne j), y, X)$$

is univariate normal, truncated below by a_j^* and above by w_j^*. Following Geweke (1991) this normal distribution has mean $b_j + \sum_{i \ne j} c_{ji} z_i$ and variance h_j^2. The vector

$$c^j \equiv (c_{j1}, ..., c_{j,j-1}, c_{j,j+1}, ..., c_{jk})'$$

is given by $c^j = -(R^{jj})^{-1}T^{j,<j}$, where R^{jj} is the element in row j and column j of R^{-1}, and $T^{j,<j}$ is row j of R^{-1} with R^{jj} deleted. The variance is $h_j^2 = (R^{jj})^{-1}$. (These expressions follow from the conventional theory for the multivariate normal distribution (Rao, 1965, p. 441) and expressions for the inverse of a partitioned symmetric matrix (Rao, 1965, p. 29). Consequently,

$$z_j|(\sigma, z_i(i \neq j), y, X) = \sum_{i \neq j} c_{ij} z_j + h_j \varepsilon_j, \tag{11}$$

$$\varepsilon_j \sim N(0,1) \quad s.t.(a_j^* - \sum_{i \neq j} c_{ij} z_j / h_j \leq \varepsilon_j \leq w_j^* - \sum_{i \neq j} c_{ij} z_j / h_j.$$

xpressions (10) and (11) provide an algorithm for producing draws for any element of (z, σ) conditional on all the other elements. Inference for the full posterior distribution whose density is given by (9) may therefore be accomplished using the Gibbs sampler described by Gelfand and Smith (1990) and Tierney (1994). Beginning with any values of σ and $z_1, ..., z_k$ in the support of (9), successively draw and replace $\sigma, z_1, ..., z_k$ using (10) and (11). Call this new draw $(z^{(1)}, \sigma^{(1)})$ and construct $\beta^{(1)} = b + D^{-1}z^{(1)}$. Then repeat the process, obtaining $\beta^{(2)}$ and $\sigma^{(2)}$, and continue on in this way. Since there is a positive probability of moving from any given $(\beta^{(j)}, \sigma^{(j)})$ to any region of the parameter space with positive posterior probability in one iteration, the sequence $\{\beta^{(j)}, \sigma^{(j)}\}$ converges in distribution to the posterior distribution whose kernel density is given by (9). If the posterior expectation $E[g(\beta, \sigma)]$ of $g(\beta, \sigma)$exists, then $g_m - m^{-1} \sum_{j=1}^{m} g(\beta^{(j)}, \sigma^{(j)}) \stackrel{a.s}{\rightarrow} E[g(\beta, \sigma)] \equiv g$. If the posterior variance $\text{var}[g(\beta, \sigma)]$ also exists, then the accuracy of the approximation of g by g_m may be assessed by computing the numerical standard error of g_m, NSE(g_m), as described in Geweke (1992). This variance provides an asymptotic (in m) approximation to the sampling standard deviation of the numerical approximation based on m iterations. Observe that the posterior mean and variance exist for $g(\beta, \sigma) = \beta_j, g(\beta, \sigma) = \sigma$, and $g(\beta, \sigma) = \sigma^2$.

4 Two Examples

Two illustrations provide an indication of the absolute and relative performance of the crude frequency simulator, GHK probability simulator, and Gibbs sampling algorithm. These examples were also used in Geweke (1986), which provides more elaborate discussion of substantive aspects.

Pindyck and Rubinfeld (1981, p. 44) provide 32 observations on rent paid, number of rooms rented, number of occupants, sex, and distance from campus in blocks for undergraduates at the University of Michigan. These data are used by the authors in developing the linear regression model at several points in their text. Denote rent paid per person by y_i, rooms per person by r_i, and distance from campus in blocks by d_i, and let s_i be a sex dummy, one for male and zero for female. The equation estimated is

$$y_i = \beta_1 + \beta_2 s_i r_i + \beta_3 (1 - s_i) r_i + \beta_4 s_i d_i + \beta_5 (1 - s_i) d_i + \varepsilon_i.$$

The inequality constraints are $\beta_2 \geq 0,\ \beta_3 \geq 0,\ \beta_4 \leq 0,\ \beta_5 \leq 0$.

Table I
Constraint Probabilities $p_{2|1}$, rent data set[a]

	Probability	Numerical standard error	Time[b]
Crude frequency simulator (1 evaluation per iteration)	.05070	.00200	8.37
GHK probability simulator (1 evaluation per iteration)	.04810	.00126	22.83
GHK probability simulator (25 evaluations per iteration)	.04921	.00020	369.88

[a]All results are based on $m = 10,000$ iterations.
[b]In seconds. All computations were carried out on a Sun Sparc IIPC 4/40, using compiled Fortran code with extensive calls to the IMSL mathematical and statistical libraries.

Alternative numerical approximations to the constraint probabilities $p_{2|1}$ are given in Table I. All approximations are based on m = 10,000 iterations for the algorithms described in Section 2. In each iteration one or more evaluations of $P(\beta \in Q|\sigma)$ may be made. In Table I results are provided for one evaluation in the crude frequency simulator, one evaluation in the GHK probability simulator, and 25 evaluations in the GHK probability simulator. Note that the numerical standard error of the GHK probability simulator with one evaluation is less than that of the crude frequency simulator (by over one-third), and the GHK probability simulator with 25 evaluations produces a probability approximation whose numerical standard error is about one-sixth of that with one evaluation. In terms of required computation time, however, the GHK probability simulator is preferred: the crude frequency simulator requires about $(8.37/369.88) \times (0.00200/0.00020)^2 = 2.26$ times as much computation time to achieve a numerical approximation of the same accuracy, while the GHK probability simulator with one evaluation per iteration requires about $(22.83/369.88) \times (0.00126/0.00020)^2 = 2.45$ times as much computation time. Finally, the substantive results show that the constrained model is competitive with the unconstrained model, using the set of reference priors described in Section 2: with a prior odds ratio of 1:1 the posterior odds ratio in favor of the constraints is $2^4 \times 0.049{:}1 = 0.78{:}1$.

Table II
Posterior moments, rent data set[a]

Parameter	Crude frequency simulator Mean	s.d.	NSE	Gibbs sampler (No skips) Mean	s.d.	NSE	Gibbs sampler (Every 10th) Mean	s.d.	NSE
β_1	37.0	35.6	.4	36.5	34.6	2.7	37.0	34.7	.7
β_2	137.6	39.3	.4	138.3	38.4	2.8	137.6	38.5	.8
β_3	124.6	40.9	.4	125.0	40.1	3.1	124.7	40.1	.7
β_4	−.917	.857	.009	−.927	.843	.011	−.922	.858	.011
β_5	−1.194	.581	.006	−1.192	.580	.010	−1.211	.578	.004
σ	40.87	5.80	.058	40.75	5.79	.098	40.79	5.84	.059
σ^2	1703.9	499.4	5.0	1693.9	497.8	8.5	1698.0	507.4	5.1
Time[b]		131.1			12.63			118.8	

[a]All results are based on m = 10,000 iterations.

[b]In seconds. All computations were carried out on a Sun Sparc I IPC 4/40, using complied Fortran code with extensive calls to the IMSL mathematical and statistical libraries.

Posterior means and standard deviations for the coefficients β_j, σ, and σ^2 are provided in Table II. Three methods were used to approximate these moments. In the first, the crude frequency simulator was run until $m = 10{,}000$ draws of (β, σ) had been made for which the value of β satisfied the constraints. This produces i.i.d. drawings from the posterior distribution, and consequently the numerical standard error is $(10{,}000)^{-1/2} = 0.01$ times the posterior standard deviation. In the second method, the Gibbs sampler was used to make 10,000 successive draws as described in Section 3. In general the Gibbs sampler produces a serially correlated sequence, and that is evident here in the numerical standard errors that are substantially higher than are obtained from a sequence of i.i.d. drawings from the posterior distribution, for many coefficients. This problem can be alleviated by increasing the number of iterations, or by recording only every nth draw: as $n \to \infty$ the Gibbs sampled parameters become serially uncorrelated and the numerical standard error approaches that of an i.i.d. sequence. The last panel of Table II shows the results corresponding to this procedure with $n = 10$. At a cost of a 10-fold increase in computation time numerical standard errors are reduced to values closer to those of the crude frequency simulator than to those of the Gibbs sampler with no skips. Execution time for this modified Gibbs sampler is still less than that for the crude frequency simulator. (Note that the crude frequency simulator stochastically records about every 20th draw, whereas the Gibbs sampler for the third panel of Table II records every 10th draw.)

The second example is taken from Bails and Peppers (1982). In Appendix G they provide 60 quarterly observations on unit sales of automobiles in the U.S., and 10 explanatory variables. They consider the normal linear regression model $y_t = \sum_{j=1}^{11} \beta_j x_{tj} + \varepsilon_t$, with y_t denoting unit sales of automobiles at time t; x_{1t}, an intercept term; x_{2t}, personal income less transfer payments; x_{3t}, index of consumer sentiment; x_{4t}, unemployment rate; x_{5t}, index of cost of car ownership; x_{6t}, average miles per gallon of current model-year cars; x_{7t}, dummy variable for automobile strikes; x_{8t}, depreciation rate of the stock of cars; x_{9t}, average price of a new car; x_{10t}, stock of automobiles; and x_{11t}, interest rate on automobile loans. (A discussion of these variables is provided by Bails and Peppers on pp. 246–247.) We use the data exactly as presented, except that x_{7t} is scaled by 10^3. The coefficients β_2, β_3, β_6, and β_8 are anticipated to be nonnegative; all the rest except the intercept are anticipated to be nonpositive. In the Bails and Peppers text the model and data are used as an instructive example of how algorithmic addition and deletion of variables can be used in conjunction with the informal imposition of sign constraints to yield a satisfactory final equation. The numbers of observations and regressors here seem to be typical of the fairly common situation in which sign constraints are imposed in an informal, descriptive regression equation.

Table III
Constraint Probabilities $p_{2|1}$, auto sales data set[a]

	Probability	Numerical standard error	Time[b]
Crude frequency simulator (1 evaluation per iteration)	2.0×10^{-4}	1.3×10^{-4}	15.96
GHK probability simulator (1 evaluation per iteration)	8.02×10^{-5}	$.38 \times 10^{-5}$	70.52
GHK probability simulator (25 evaluations per iteration)	9.06×10^{-5}	$.15 \times 10^{-5}$	1379.93

[a]All results are based on $m = 10,000$ iterations.
[b]In seconds. All computations were carried out on a Sun Sparc IIPC 4/40, using compiled Fortran code with extensive calls to the IMSL mathematical and statistical libraries.

Table III provides alternative numerical approximations to the constraint probabilities in the same style of presentation as Table I. Because the constraint probability $p_{2|1}$ is so small, the crude frequency simulator with $m =$ 10,000 iterations provides an unsatisfactory approximation. The GHK probability simulator provides this small probability to one significant figure and clearly could provide it for two figures. The crude frequency simulator requires about $[(1.3\text{x}10^{-4})/(0.38\text{x}10^{-5})]^2 \times (15.96/70.53) = 265$ times as much computation time to achieve the same numerical standard error as the GHK probability simulator. Increasing the number of evaluations by a factor of 25 leads to less than a five-fold reduction in numerical standard error. This indicates that variation in σ plays an important role in evaluation of the constraint probability. Employing the reference prior described in Section 2, the posterior odds ratio in favor of the constrained model is about $2^{10} \times 9 \times 10^{-5}$:1 = 0.092:1 when the prior odds ratio is 1:1.

Table IV
Posterior moments, auto sales data set[a]

	Crude frequency simulator			Gibbs sampler (No skips)			Gibbs sampler (Every 10th)		
Parameter	Mean	s.d.	NSE	Mean	s.d.	NSE	Mean	s.d.	NSE
β_1	−7.59	2.25	.06	−7.75	2.17	.25	−7.70	2.23	.10
β_2	.0242	.0025	.0001	.0240	.0023	.0003	.0242	.0024	.0001
β_3	.0405	.0125	.0003	.0403	.0125	.0008	.0410	.0120	.0003
β_4	−.0232	.0212	.0005	−.0236	.0221	.0004	−.0237	.0227	.0002
β_5	−3.212	1.110	.042	−3.458	1.093	.080	−3.190	1.122	.024
β_6	.1281	.0928	.0027	.1375	.1036	.0099	.1297	.0969	.0030
β_7	−.1282	.0296	.0009	−.1277	.0306	.0004	−.1287	.0307	.0002
β_8	33.04	26.08	.90	32.36	26.30	1.39	32.45	25.98	.53
β_9	−.421	.322	.012	−.372	.287	.027	−.432	.329	.012
β_{10}	−.0160	.0143	.0004	−.0146	.0128	.0010	−.0154	.0140	.0003
β_{11}	−.1285	.0953	.0020	−.1175	.0906	.0064	−.1270	.0957	.0020
σ	.5439	.0537	.0015	.5431	.0557	.0016	.5454	.0562	.0005
σ^2	.2987	.0599	.0016	.2981	.0621	.0017	.3006	.0629	.0006
Time[b]		13846.24			300.15			2939.66	

[a] All results are based on m = 10,000 iterations.

[b] In seconds. All computations were carried out on a Sun Sparc I IPC 4/40, using complied Fortran code with extensive calls to the IMSL mathematical and statist cal libraries.

Posterior means and standard deviations for the model parameters are provided in Table IV. Results for the crude frequency simulator are based on $m =$ 1,000 draws of (β, σ) from the support of the constrained posterior distribution. This required over four hours of execution time. The Gibbs sampler produces drawings with very high positive serial correlation for most parameters, and results for $m = 10{,}000$ with no skips yield very poor approximations and these results are not shown in Table IV. When every 10th draw is used until $m =$ 10,000 draws have been obtained, numerical standard errors are as high as 10 percent of the posterior standard deviation for several coefficients. Increasing to every 100th draw lowers this error to about 4 percent. The Gibbs sampler requires about 49 minutes to achieve this accuracy, whereas the crude frequency simulator demands almost four hours. For this problem, the Gibbs sampler is clearly the method of choice.

5 Extensions

The treatment of linear inequality constraints developed in this paper may be extended and applied in models other than the normal linear regression model. The presentation here has limited treatment to inequality constraints on no more than k linear combinations of the k coefficients. More than k linear combinations can be handled by combining the GHK probability simulator and Gibbs sampler with an accept-reject algorithm. Given $s > k$ constraints, apply the Gibbs sampler using k of the linear constraints, but accept only those draws that satisfy the other $s - k$ constraints, and retain the acceptance rate. The retained sample converges in distribution to the posterior distribution. Use the GHK probability simulator to approximate the probability of the same k constraints, and then scale it by the acceptance rate from the application of the Gibbs sampler to obtain an approximation to $p_{2|1}$.

The semi-informative prior distribution with kernel density $\sigma^{-1}\exp\{-R\beta - r)'\Psi^{-1}(R\beta - r)\}$ can be used in place of (4). Conditional on σ the distribution of β is still truncated multivariate normal, and consequently the GHK probability simulator and Gibbs sampler may be applied in the same way. The only modification of substance is that since the variance of this multivariate normal distribution is $[\sigma^{-2}X'X + R'\Psi^{-1}R]^{-1}$, the Choleski decomposition of the $k \times k$ variance matrix must be recomputed each iteration in both the crude frequency simulator and the GHK probability simulator. In the Gibbs sampler the vectors c^j must be recomputed, requiring the inversion of a $k \times k$ matrix, each iteration. The times required to compute a Choleski decomposition, and to invert a matrix, are similar and proportional to the cube of the dimension of the matrix. Consequently when the number of regressors, k, is large (say, $k \geq 8$) computation time with a semi-informative prior will be roughly proportional to the number of iterations. In the automobile sales example, the crude frequency simulator requires more than 100 times as many iterations as the Gibbs sampler. For this case, the Gibbs sampler would clearly have been the method of choice had a semi-informative prior been employed.

The GHK probability simulator and the Gibbs sampler can generally be employed in any Gibbs sampling algorithm in which the linear inequality restrictions apply to elements of a vector whose conditional distribution is normal. This extension applies to a wide range of models, for example the seemingly unrelated regressions model, censored regression model, and the multinomial probit model. A particularly interesting class of cases is formed by scale-mixture normal models. For example, the models

$$y_t = x_t'\beta + \varepsilon_t, \quad \varepsilon_t \overset{IID}{\sim} t(0, \sigma^2; \nu)$$

and

$$y_t = x_t'\beta + \varepsilon_t, \quad \varepsilon_t \overset{ID}{\sim} N(0, \sigma^2 v_t)$$

are equivalent if in the prior distribution the v_t are mutually independent, $\nu v_t^{-1} \sim \chi^2(v)$ (Geweke 1993). Conditional on σ and the v_t, the distribution of β is truncated normal, and the GHK probability simulator and Gibbs sampler may be applied to treat linear inequality constraints in the way described in this paper. Other prior distributions for the v_t produce other unconditional distributions for the disturbance term. The methods for treatment of linear inequality constraints described in this paper may therefore be extended to a variety of nonnormal linear regression models.

6 Acknowledgements

The editor and a referee, who bear no responsibility for any errors or omissions, have provided suggestions that improved the paper. This work was supported in part by National Science Foundation Grant SES-9210070. The views expressed in this paper are those of the author and not necessarily those of the Federal Reserve Bank of Minneapolis or the Federal Reserve System.

References

Bails, D. G., and L. C. Peppers (1982), Business fluctuations. Englewood Cliffs: Prentice-Hall.

Bernardo, J. M., and A. F. M. Smith (1994), Bayesian Theory. New York: Wiley.

Chamberlain, G., and E. Leamer (1976), Matrix weighted averages and posterior bounds, Journal of the Royal Statistical Society, Series B, 38, 73–84.

Davis, W. W. (1978), Bayesian analysis of the linear model subject to linear inequality constraints, Journal of the American Statistical Association 73, 573–79.

Gelfand, A. E., and A. F. M. Smith (1990), Sampling based approaches to calculating marginal densities, Journal of the American Statistical Association 85, 398–409.

Geweke, J. (1986), Exact inference in the inequality constrained normal linear regulation model, Journal of Applied Econometrics 1, 127–41.

Geweke, J. (1991), Efficient simulation from the multivariate normal and student-t distributions subject to linear constraints. In E. M. Keramidas (ed.), Computing Science and Statistics: Proceedings of the 23rd Symposium on the Interface, pp. 571–78. Fairfax, VA: Interface Foundation of North America.

Geweke, J. (1992), Evaluating the accuracy of sampling-based approaches to the calculation of posterior moments. In J. O. Berger et al. (eds.), Proceedings of the Fourth Valencia International Meeting on Bayesian Statistics. Oxford: Oxford University Press.

Geweke, J. (1993), Bayesian treatment of the student-t linear model, Journal of Applied Econometrics 8, S19–S40.

Geweke, J. (1995), Variable selection and model comparison in regression. Forthcoming in J. O. Berger, J. M. Bernardo, A. P. Dawid, and A. F. M. Smith (eds.), Proceedings of the Fifth Valencia International Meeting on Bayesian Statistics. Also, Research Department Working Paper 539, Federal Reserve Bank of Minneapolis, (1994).

Geweke, J., M. Keane, and D. Runkle (1995), Recursively simulating multinomial multiperiod probit probabilities, American Statistical Association 1994 Proceedings of the Business and Economic Statistics Section, forthcoming.

Gourieroux, C., A. Holly, and A. Monfort (1982), Likelihood ratio test, Wald test, and Kuhn-Tucker test in linear models with inequality constraints on the regression parameters, Econometrica 50, 63–80.

Hajivassiliou, V., and D. McFadden (1990), The method of simulated scores for the estimation of LDV models with an application to external debt crises. Cowles Foundation Discussion Paper 967, Yale University.

Hajivassiliou, V., D. McFadden, and P. Ruud (1995), Simulation of multivariate normal orthant probabilities: Methods and programs, Journal of Econometrics, forthcoming.

Judge, G. G., and T. Takayama (1966), Inequality restrictions in regression analysis, Journal of the American Statistical Association 61, 166–81.

Keane, M. (1990), Four essays in empirical macro and labor economics. Ph.D. dissertation, Brown University.

Keane, M. (1993), A computationally practical simulation estimator for panel data, Econometrica 62, 95–116.

Keane, M. (1994), Simulation estimation for panel data models with limited dependent variables. In G. S. Maddala, C. R. Rao, and H. D. Vinod (eds.), Handbook of Statistics vol. 11, pp. 545–71. Amsterdam: Elsevier Science Publishers.

Leamer, E., and G. Chamberlain (1976), A Bayesian interpretation of pretesting, Journal of the Royal Statistical Society, Series B, 38, 85–94.

Lindley, D. V. (1957), A statistical paradox, Biometrika 44, 187–92.

Lovell, M. C., and E. Prescott (1970), Multiple regression with inequality constraints: pretesting bias, hypothesis testing, and efficiency, Journal of the American Statistical Association 65, 913–25.

Pindyck, R. S., and D. L. Rubinfeld (1981), Econometric models and economic forecasts. New York: McGraw-Hill.

Press, S. J. (1989), Bayesian statistics: Principles, models, and applications. New York: John Wiley.

Rao, C.R. (1965), Linear statistical inference and its applications. New York: Wiley.

Tierney, L., (1994), Markov chains for exploring posterior distributions, (with discussion and rejoinder), Annals of Statistics, **22**: 1701–1762. (Also, School of Statistics Technical Report 560, University of Minnesota, (1991).

Wolak, F. A. (1987), An exact test for multiple inequality and equality constraints in the linear regression model, Journal of the American Statistical Association 82, 782–93.

Zellner, A. (1971), An introduction to Bayesian analysis in econometrics. New York: Wiley.

The Joint Asymptotic Distribution of the Maximum Likelihood and Mantel-Haenszel Estimators of the Common Odds Ratio in k 2×2 Tables

Samuel W. Greenhouse and Joseph L. Gastwirth
The George Washington University
Department of Statistics
Washington, DC 20052

Abstract

The Mantel-Haenszel (MHE) and Maximum Likelihood (MLE) estimators of an assumed common odds ratio in the analysis of several (k) 2×2 tables are usually found to be quite close. This suggests their joint asymptotic distribution should have a high correlation. Since the MLE cannot be obtained as an explicit function of the observations, we instead utilize an asymptotically equivalent surrogate for the MLE enabling us to find a large sample representation for the two estimators from which we obtain their joint asymptotic distribution. This is found for both the uncondtional and conditional likelihoods, under the assumption that k is fixed and the sample sizes within each table approach infinity.

1. Introduction

Since it has long been observed that the Mantel-Haenszel (MHE) and maximum likelihood (MLE) estimators of the common odds ratio in several (k) 2×2 tables are usually quite close (see, for example, Breslow and Liang 1982), their joint distribution should have a high correlation. The purpose of this paper is to present an asymptotic representation of the two estimators from which we obtain their asymptotic bivariate normal distirbution where k is fixed. Since the MLE cannot be obtained in closed form, our approach is to utilize an asymptotically equivalent surrogate for the MLE (Gart, 1962) which enables us to apply large sample methodology.

In section 2, we present two sets of data with the MH and ML estimators in each illustrating their closeness. Both data sets are examples of the conditional model in which all marginals are assumed to be fixed. After reviewing some background material in section 3, we present in section 4 the unconditional case (in which the sampling numbers only are held fixed) and in section 5 the conditional case.

Although not directly related to the main thesis of this paper, in section 6 we present some interesting properties in the unconditional case of the ML estimates of the "success" probabilities P_{1i}, the probability of the event of interest in the i-th table. Because the odds ratio Ψ is common to all k tables the ML estimate of P_{1i}, $\hat{P}_{1i}$, and its variance are no longer the ordinary binomial values. Furthermore, $\hat{P}_{1i}$, and $\hat{P}_{1j}$ have a non-zero covariance despite the fact that the k tables are independent sampling-wise. In section 7 we present the analyses of the two examples in which the asymptotic orrelation between the MHE and the MLE of the common Ψ is quite large ($> .99$) even though Ψ's differ from one (the case where the MHE and the MLE are equally efficient).

2. Motivating Examples

In this section we present two data sets from different areas of application in which a summary odds ratio can be used to assess the meaningfulness of the effect under investigation. Table 1 comes from a law case, Agerwal and McKee (Gastwirth, 1984; 1988) concerning the fairness of a firm's promotion process. Note that the promotions (x) to a higher pay level are selected from those at the previous level so the conditional model is appropriate.

Salary Level (i)	Majority x_{1i}	Majority n_{1i}	Majority p_{1i}	Minority x_{2j}	Minority n_{2i}	Minority p_{2i}
1	35	238	.147	3	19	.158
2	45	147	.306	7	39	.179
3	54	235	.230	17	87	.195
4	77	242	.318	34	143	.238

Table 1: Promotion data from Agarwal vs. McKee for the period 1970-74

The MH and ML estimators are 1.432 and 1.451 respectively.

Similar data arise in equal employment cases concerned with the disparate impact of a test or employment criteria on a minority group. In this application, the data may be stratified by job type and the proportions of each group passing the test are of interest. Since these proportions of persons passing the test are not fixed in advance, the data would be consistent with the unconditional model.

The second data set is from the first case-control study (Wertheimer and Leeper, 1979) exploring the possible relationship between electromagnetic fields (EMF) and the development of childhood cancers. Cases of cancer and controls were classified as being exposed to a high (HCC) or low (LCC) level of EMF. As prior studies had suggested that living in a heavy traffic area was also related to childhood cancer, the data in Table 2 are stratified by heavy or light traffic levels. (These data are taken from Table 9 of the cited reference).

Traffic Level(i)	(j)	HCC x_{ji}	HCC p_{ji}	LCC $n_{ji} - x_{ji}$	LCC q_{ji}	n_{ji}
Heavy (1)	Cases (1)	32	.432	42	.568	74
	Controls (2)	9	.188	39	.812	48
						$n_1 = 122$
Light(2)	Cases (1)	84	.201	333	.799	417
	Controls (2)	53	.125	371	.875	424
						$n_2 = 841$

Table 2: Cases of Cancer and Controls Exposed to high (HCC) and Low (LCC) Wiring Configurations Stratified by Traffic Level

Here x_{ji} is the number of highly exposed subjects in the j-th sampling group ($j = 1$ is a case, $j = 2$ is a control) and in the i-th traffic stratum ($i = 1$ is exposed to heavy traffic, $i = 2$ to light traffic). In this example, the MH and ML estimators are 1.97 and 2.10 respectively.

In a case-control study, an estimate of relative risk is not obtainable from the oberved data alone. Under certain conditions (Cornfield, 1951) the odds ratio (OR) is a good approximation to this relative risk. Of course, the OR is a well known measure of association (Mosteller, 1968). It is the parameter in the non-central hypergeometric distribution that arises when analyzing the 2×2 table by conditioning on the four marginal totals.

3. Notation and Known Results

We consider k pairs of independent binomial variates x_{1i} and x_{2i}, with parameters (n_{1i}, P_{1i}) and (n_{2i}, P_{2i}) respectively, $i = 1, 2, \ldots k$. The common odds ratio is

$$\Psi = \frac{P_{1i}Q_{2i}}{Q_{1i}P_{2i}}, \quad i = 1, 2, \ldots k \tag{1}$$

where $Q_{ji} = 1 - P_{ji}, 0 < P_{ji} < 1$. The sample odds ratio in the ith table is given by

$$\hat{\Psi}_i = \frac{p_{1i}q_{2i}}{q_{1i}p_{2i}} \quad i = 1, 2, \ldots k \tag{2}$$

where $p_{ji} = \frac{x_{ji}}{n_{ji}}$, $q_{ji} = 1 - p_{ji}$, $0 < p_{ji} < 1$, $j = 1, 2$.

We define the MHE of Ψ, (Mantel-Haenszel, 1959), as

$$\Psi_{MH} = \frac{\Sigma x_{1i}(n_{2i} - m_i + x_{1i})/n_i}{\Sigma(n_{1i} - x_{1i})(m_i - x_{1i})/n_i} \tag{3}$$

where $m_i = x_{1i} + x_{2i}$ and $n_i = n_{1i} + n_{2i}$. The form of the MHE is the same in both the unconditional and conditional settings. Equation (3) can be expressed as

$$\Psi_{MH} = \Sigma \frac{a_i q_{1i} p_{2i}}{\Sigma a_i q_{1i} p_{2i}} \hat{\Psi}_i = \Sigma \hat{c}_i \hat{\Psi}_i \tag{4}$$

where $a_i = \frac{n_{1i}n_{2i}}{n_i}$ and $\hat{c}_i = \frac{a_i q_{1i} p_{2i}}{\Sigma a_i q_{1i} p_{2i}}$.

4. The Unconditional Case or Product Binomial Sampling

In this case, only the marginal totals representing sample sizes n_{1i} and n_{2i} are fixed in each table. Hence x_{ji}, $j = 1, 2$, are independent and can range between 0 and n_{ji}, respectively. The likelihood of the ith table is

$$L_i = \binom{n_{1i}}{x_{1i}} \binom{n_{2i}}{x_{2i}} P_{1i}^{x_{1i}} Q_{1i}^{n_{1i} - x_{1i}} P_{2i}^{x_{2i}} Q_{2i}^{n_{2i} - x_{2i}} \tag{5}$$

After substituting

$$P_{2i} = \frac{P_{1i}}{(P_{1i} + Q_{1i}\Psi)}, \quad Q_{2i} = \frac{Q_{1i}\Psi}{(P_{1i} + Q_{1i}\Psi)}$$

(5) becomes

$$L_i \propto \Psi^{n_{2i}-x_{2i}} P_{1i}^{m_i} Q_{1i}^{n_i-m_i} (P_{1i}+Q_{1i}\Psi)^{-n_{2i}} \tag{6}$$

The ML estimates of Ψ and P_{1i} from the i-th table respectively are

$$\hat{\Psi}_i = \frac{x_{1i}(n_{2i}-x_{2i})}{x_{2i}(n_{1i}-x_{1i})} = \frac{p_{1i}q_{2i}}{q_{1i}P_{2i}},\ \hat{P}_i = p_{1j} = \frac{x_{1j}}{n_{1j}}. \tag{7}$$

From the inverse of the information matrix based on L_i we find,

$$\begin{aligned} \text{var}\ (\hat{\Psi}_i) &\doteq \Psi^2(v_{1i}^{-1}+v_{2i}^{-1}) = \Psi^2 v_i \\ \text{var}\ (\hat{P}_{1i}) &\doteq \tfrac{P_{1i}Q_{1i}}{n_{1i}} \\ \text{cov}\ (\hat{\Psi}_i, \hat{P}_{1i}) &\doteq \tfrac{\Psi}{n_{1i}} \end{aligned} \tag{8}$$

where, $v_{ji} = n_{ji}P_{ji}Q_{ji}$, $v_i = v_{1i}^{-1} + v_{2i}^{-1}$.
For all k tables, the likelihood is

$$L = \prod^{k} L_i \propto \prod_{i=1}^{k} \Psi^{n_{2i}-x_{2i}} P_{1i}^{m_i} Q_{1i}^{n_i-m_i}(P_{1i}+Q_{1i}\Psi)^{-n_{2i}}. \tag{9}$$

The MLE, $\hat{\Psi}_{\text{ML}}$, is not an explicit expression of the observations but must satisfy

$$\sum x_{1i} = \sum E(x_{1i} \mid P_{1i}, \Psi).$$

The asymptotic variance of $\hat{\Psi}_{\text{ML}}$, however, is attainable in closed form in the usual way from the inverse of the information matrix, namely,

$$\text{var}\ (\hat{\Psi}_{\text{ML}}) \doteq \frac{\Psi^2}{\Sigma(v_{1i}^{-1}+v_{2i}^{-1})^{-1}} = \frac{\Psi^2}{\Sigma v_i^{-1}}. \tag{10}$$

(Maximum likelihood estimates of P_{1i} and their variances and covariances will be discussed in section 6). Throughout this section we assume that as $n = \Sigma n_i$ goes to infinity $n_i/n \to \xi_i$, $0 < \xi_i < 1$, and in each table $n_{1i}/n_i \to \lambda_i$, $0 < \lambda_i < 1$.

We now define a closed form estimate of Ψ which is asymptotically efficient (Gart 1962) namely,

$$\hat{\Psi}_{\text{MLS}} = \Sigma \hat{w}_i \hat{\Psi}_i, \tag{11}$$

where $\hat{w}_i = \hat{v}_i^{-1}/\Sigma\hat{v}_i^{-1}$ and $\hat{v}_i = \hat{v}_{1i}^{-1} + \hat{v}_{2i}^{-1}$, $\hat{v}_{ji} = n_{ji}p_{ji}q_{ji}$. The choice of $\hat{w}_i$ is obtained from (8) as each $\hat{\Psi}_i$ is weighted by the inverse of its variance.

Let $\hat{g}_{\text{MLS}}(\{p_{1i},p_{2i}\}_i) = \Sigma\hat{w}_i\hat{\Psi}_i = \hat{\Psi}_{\text{MLS}}$ and $\hat{g}_{MH}(\{p_{1i},p_{2i}\}_i) = \Sigma\hat{c}_i\hat{\Psi}_i = \hat{\Psi}_{MH}$ where $\hat{w}_i, \hat{c}_i, \hat{\Psi}_i$ are as defined previously and $\{.,.\}_i$ denotes the collection over all i. We note that $g_{MLS}\{P_{1i},P_{2i}\} = g_{MH}\{P_{1i},P_{2i}\} = \Psi$ so that a Taylor series expansion for either function is

$$\hat{g}\{p_{1i},p_{2i}\} = \Psi + \sum\left[(p_{1i}-P_{1i})\hat{g}'_{p_{1i}}\{P_{1i},P_{2i}\} + (p_{2i}-P_{2i})\hat{g}'_{p_{2i}}\{P_{1i},P_{2i}\}\right] + o_p(n^{-1/2}). \tag{12}$$

The partial derivatives, $\hat{g}'_{p_{ji}}(P_{1i},P_{2i})$, of $\hat{g}$ with respect to p_{ji} evaluated at $p_{ji} = P_{ji}$, $j = 1,2$, are considerably simplified by noting that

$$\begin{aligned}\frac{\partial \hat{g}_{MLS}\{p_{1i},p_{2i}\}}{\partial p_{ji}} &= \frac{\partial}{\partial p_{ji}}\frac{\sum \hat{v}_i^{-1}\hat{\Psi}_i}{\sum \hat{v}_i^{-1}}\Big|_{(p_{ji}=P_{ji})} \\ &= \frac{v_i^{-1}}{\sum v_i^{-1}}\frac{\partial \hat{\Psi}_i}{\partial p_{ji}} + \frac{\Psi}{\sum v_i^{-1}}\frac{\partial \hat{v}_i^{-1}}{\partial p_{ji}} - \Psi\frac{1}{\sum v^{-1}}\frac{\partial \hat{v}_i^{-1}}{\partial p_{ji}} \\ &= w_i\frac{\partial \hat{\Psi}_i}{\partial p_{ji}}\end{aligned} \tag{13}$$

and

$$\begin{aligned}\frac{\partial \hat{g}_{MH}\{p_{1i},p_{2i}\}}{\partial p_{ji}} &= \frac{\partial}{\partial p_{ji}}\frac{\sum a_i q_{1i}p_{2i}\hat{\Psi}_i}{\sum a_i q_{1i}p_{2i}}\Big|_{p_{ji}=P_{ji}} \\ &= \frac{1}{\sum a_i Q_{1i}P_{2i}}\frac{\partial \sum a_i q_{1i}p_{2i}\hat{\Psi}_i}{\partial p_{ji}} - \frac{\Psi}{\sum a_i Q_{1i}P_{2i}}\frac{\partial \sum a_i q_{1i}p_{2i}}{\partial p_{ji}} \\ &= \frac{a_i Q_{1i}P_{2i}}{\sum a_i Q_{1i}P_{2i}}\frac{\partial \hat{\Psi}_i}{\partial p_{ji}} = c_i\frac{\partial \hat{\Psi}_i}{\partial p_{ji}}.\end{aligned} \tag{14}$$

We note that

$$\begin{aligned}\frac{\partial \hat{\Psi}_i}{\partial p_{1i}} &= \frac{\Psi}{P_{1i}Q_{1i}} \\ \frac{\partial \hat{\Psi}_i}{\partial p_{2i}} &= \frac{-\Psi}{P_{2i}Q_{2i}}.\end{aligned} \tag{15}$$

After substituting (15) into (13) and (14) and inserting the latter into the Taylor expansion (12), we obtain the asymptotic representations for the two estimators:

$$\begin{aligned}\hat{\Psi}_{MLS} &= \Psi + \Psi\sum w_i\left[\frac{p_{1i}-P_{1i}}{P_{1i}Q_{1i}} - \frac{p_{2i}-P_{2i}}{P_{2i}Q_{2i}}\right] + o_p(n^{-1/2}) \\ \hat{\Psi}_{MH} &= \Psi + \Psi\sum c_i\left[\frac{p_{1i}-P_{1i}}{P_{1i}Q_{1i}} - \frac{p_{2i}-P_{2i}}{P_{2i}Q_{2i}}\right] + o_p(n^{-1/2}).\end{aligned} \tag{16}$$

Since p_{1i} and p_{2i} are independent in the unconditional case the large sample variance of $\hat{\Psi}_{MLS}$ from (16) is:

$$\begin{aligned}\text{var}\,(\hat{\Psi}_{MLS}) &= \Psi^2\sum w_i^2\left(\frac{1}{n_{1i}P_{1i}Q_{1i}} + \frac{1}{n_{2i}P_{2i}Q_{2i}}\right) \\ &= \Psi^2\frac{\sum v_i^{-2}(v_{1i}^{-1}+v_{2i}^{-1})}{(\sum v_i^{-1})^2} \\ &= \Psi^2(\sum v_i^{-1})^{-1}.\end{aligned}$$

Since the variance of $\hat{\Psi}_{MLS}$ equals the variance of the MLE, (eq. 10), $\hat{\Psi}_{MLS}$ is asymptotically efficient.

From equations (16) and the fact that $\sqrt{n_i}(p_{ji} - P_{ji})$ are jointly asymptotically normally distributed, we have that $\hat{\Psi}_{MH}$ and $\hat{\Psi}_{ML}$ are jointly distributed asymptotically as a bivariate normal distribution with common mean Ψ and

$$\begin{aligned} &\text{var }(\hat{\Psi}_{MLS}) = \text{ var }(\hat{\Psi}_{ML}) = \Psi^2 \sum w_i^2 v_i, \quad \text{var }(\hat{\Psi}_{MH}) = \Psi^2 \sum c_i^2 v_i \\ &\text{Covariance }(\hat{\Psi}_{MLS}, \hat{\Psi}_{MH}) = \Psi^2 \sum w_i c_i v_i. \end{aligned}$$

It follows from the asymptotic equivalence of $\hat{\Psi}_{MLS}$ and $\hat{\Psi}_{ML}$ that the asymptotic correlation between Ψ_{ML} and Ψ_{MH} is

$$\rho = \frac{\sum w_i c_i v_i}{[\sum u^2 v_i \cdot \sum c_i^2 v_i]^{1/2}} \tag{17}$$

We note that the asymptotic variance of $\hat{\Psi}_{MH}$ agrees with that given by Guilbaud (1983).

5. The Conditional Case

In this section, we constrain the variables $x_{1i} + x_{2i} = m_i$, where the integers m_i are fixed in each of the k tables. The conditional likelihood of the ith table is

$$L_i = A \begin{pmatrix} n_{1i} \\ x_{1i} \end{pmatrix} \begin{pmatrix} n_{2i} \\ m_i - x_{1i} \end{pmatrix} \Psi^{x_{1i}}, x_{1i} = 0, 1, 2 \ldots, s_i \tag{18}$$

and $A^{-1} = \sum_{x_{1i}=0}^{s_i} \begin{pmatrix} n_{1i} \\ x_{1i} \end{pmatrix} \begin{pmatrix} n_{2i} \\ m_i - x_{1i} \end{pmatrix} \Psi^{x_{1i}}$, where $s_i = \min[m_i, n_{1i}]$. The MLE of Ψ from the i-th table is no longer the sample estimate, $\Psi_i = p_{1i}q_{2i}/q_{1i}p_{2i}$, as it was in the unconditional case. We denote the MLE by Ψ_i(ML) and note that although Ψ_i(ML) is not obtainable in closed form, its large sample variance is simply $\frac{\Psi^2}{\text{var }(x_{1i})}$.

The likelihood of Ψ from all k tables is $L = \prod L_i$. The MLE of Ψ, $\hat{\Psi}_{ML}$, is not an explicit function of the observations and satisfies

$$\sum x_{1i} = \sum E(x_{1i} \mid m_i, n_{1i}, n_{2i}, \Psi).$$

The large sample variance of $\hat{\Psi}_{ML}$ is

$$\text{var } \hat{\Psi}_{ML} \doteq \left[E\left(-\frac{\partial^2 \log L}{\partial \Psi^2}\right)\right]^{-1} = \frac{\Psi^2}{\sum \text{ var }(x_{1i})}. \tag{19}$$

For both var $\hat{\Psi}_i$(ML) and var $\hat{\Psi}_{ML}$ we need the first two moments of the non-central hypergeometric probability function As is well known the exact moments are difficult to find even for moderate m_i and n_{1i}. We therefore use the mean and variance of the approximating normal density to L_i (Stevens, 1951; Cornfield, 1956; Hannan and Harkness, 1963), namely

$$x_{1i} \sim N(n_{1i}P_{1i}, \text{ var }(x_{1i})),$$

where

$$\text{var } (x_{1i}) = [(n_{1i}P_{1i}Q_{1i})^{-1} + (n_{2i}P_{2i}Q_{2i})^{-1}]^{-1} = v_i^{-1}. \tag{20}$$

The P_{ji} are the unique real numbers satisfying $0 < P_{ji} < 1$, $P_{1i}Q_{2i}/Q_{1i}P_{2i} = \Psi$ and $n_{1i}P_{1i} + n_{2i}P_{2i} = m_i$. If we let $\tilde{x}_{1i}$ be the mean, $E(x_{1i} \mid m_i, n_{i1}, n_{2i}, \Psi)$, of the non-central hypergeometric function with parameter $\hat{\Psi}$, then the above assumptions imply $P_{1i} = \tilde{x}_{1i}/n_{1i}$. Hence, the large sample variance of the MLE obtained from (18) is

$$\text{var } \hat{\Psi}_i(ML) = \frac{\Psi^2}{\text{var } x_{1i}} = \Psi^2 v_i^{-1}. \tag{21}$$

$$\text{var } \hat{\Psi}_{ML} = \frac{\Psi^2}{\sum v_i^{-1}}. \tag{22}$$

Using values of $\hat{w}_i, \hat{v}_i, \hat{c}_i, a_i$ and p_{ji} as defined previously, we take the same surrogate estimate of the MLE, $\hat{\Psi}_{MLS} = \sum \hat{w}_i \hat{\Psi}_i$ as in the unconditional case (11), and the same estimate of the MHE, $\hat{\Psi}_{MH} = \sum \hat{c}_i \hat{\Psi}_i$, as defined for the unconditional case (4) (Guilbaud, 1983 and Harkness 1965). Because of the constraints $x_{1i} + x_{2i} = m_i$, m_i fixed, the estimators Ψ_{MLS} and Ψ_{MH} are each functions of only one variable x_{1i}, from each table.

We now assume that as $n = \sum n_i \to \infty, m_i$ and n_{1i} increase so that $m_i/n_i \to \lambda_i$ and $n_{1i}/n_i \to \xi_i$ for some λ_i and ξ_i satisfying $0 < \lambda_i < 1, 0 < \xi_i < 1$, $i = 1, 2, \dots k$. It then follows (Hannan and Harkness 1963 and Harkness 1965) that the limiting distribution of x_{1i} is normal with mean $\tilde{x}_{1i}$ and variance v_i^{-1}. Let $\hat{\Psi}_{MLS} = \hat{g}_{MLS}(x_{1i})$ and $\hat{\Psi}_{MH} = \hat{g}_{MH}(x_{1i})$. Then note that $\hat{g}_{MLS}(\tilde{x}_{1i}) = \hat{g}_{MH}(\tilde{x}_{1i}) = \Psi$. A first order Taylor series expansion for either $\hat{g}$ function is

$$\hat{g}(x_{1i}) = \Psi + \sum (x_{1i} - \tilde{x}_{1i}) \frac{\partial \hat{g}}{\partial x_{1i}} \Big|_{x_{1i} = \tilde{x}_{1i}} + o_p(n^{-1/2}) \tag{23}$$

It is fairly easy to show that

$$\frac{\partial \hat{g}_{MLS}}{\partial x_{1i}} = w_i \frac{d\hat{\Psi}_i}{dx_{1i}}, \quad \frac{d\hat{g}_{MH}}{dx_{1i}} = c_i \frac{d\hat{\Psi}_i}{dx_{1i}} \tag{24}$$

and

$$\frac{d\hat{\Psi}_i}{dx_{1i}} = \Psi(v_{1i}^{-1} + v_{2i}^{-1}) = \Psi v_i. \tag{25}$$

Substituting (24) and (25) into (23) yields the asymptotic representations:

$$\begin{aligned} \hat{\Psi}_{MLS} &= \Psi + \Psi \sum w_i v_i (x_{1i} - \tilde{x}_{1i}) + o_p(n^{-1/2}) \\ \hat{\Psi}_{MH} &= \Psi + \Psi \sum c_i v_i (x_{1i} - \tilde{x}_{1i}) + o_p(n^{-1/2}). \end{aligned} \tag{26}$$

Therefore $\hat{\Psi}_{MLS}$ and $\hat{\Psi}_{MH}$ are asymptotically bivariate normal with common mean Ψ and with the same asymptotic covariance matrix that was obtained in the unconditional case and consequently the same asymptotic correlation cf. (17). The covariance follows since covar $(x_{1i} - \hat{x}_{1i}, x_{1j} - \hat{x}_{1j}) = 0$ for $i \neq j$.

Note that var $\hat{\Psi}_{MLS} = \Psi^2 \sum w_i^2 v_i = \Psi^2 / \sum v_i^{-1}$ which is the same as var $\hat{\Psi}_{ML}$ (22). Therefore, Ψ_{MLS} is asymptotically fully efficient. We remark that the var Ψ_{MH} agrees with Guilbaud (1983).

6. Maximum Likelihood Estimates of P_{1i} in the Unconditional Case.

We return to the unconditional case to consider the nature of the maximum likelihood estimates of the P_{1i}. From the likelihood (9),
$L(\Psi, P_{11}, P_{12}, \ldots P_{1k} \mid (x_{11}x_{21}), \ldots, (x_{1k}x_{2k}))$, we have ($\ell = \log L$),

$$\begin{aligned} \frac{\partial \ell}{\partial \Psi} &= \frac{\sum(n_{2i} - x_{2i})}{\Psi} - \sum \frac{n_{2i} - Q_{1i}}{P_{1i} + Q_{1i}\Psi} \\ \frac{\partial \ell}{\partial P_{1i}} &= \frac{m_i}{P_{1i}} - \frac{n - m}{Q_{1i}} - \frac{n_{2i}(1 - \Psi)}{P_{1i} + Q_{1i}\Psi}, i = 1, 2, \ldots k, \end{aligned} \tag{27}$$

and (Gart 1962),

$$\begin{aligned} E\left(-\frac{\partial^2 \ell}{\partial \Psi^2}\right) &= \left(\frac{1}{\Psi^2}\right) \sum v_{2i} \\ E\left(-\frac{\partial^2 \ell}{\partial P_{1i} \partial \Psi}\right) &= \left(-\frac{1}{\Psi}\right) \frac{n_{1i} v_{2i}}{v_{1i}}, i = 1, 2, \ldots, k \\ E\left(-\frac{\partial^2 \ell}{\partial P_{1i}^2}\right) &= \frac{n_{1i}^2 (v_{1i} + v_{2i})}{v_{1i}^2}, i = 1, 2, \ldots, k \\ E\left(-\frac{\partial^2 \ell}{\partial P_{1i} \partial P_{1j}}\right) &= 0,\ i \neq j. \end{aligned} \tag{28}$$

Setting eqs. (27) equal to zero does not generally lead to $\hat{P}_{1i} = x_{1i}/n_{1i}$.

Relations (28) are the elements of the $(k+1) \times (k+1)$ Information Matrix. Taking advantage of the fact that the lower $k \times k$ matrix is diagonal we most conveniently obtain the inverse by using the formulae given by Anderson (1984, theorem A.3.3, p. 594). Recalling that $v_{ji} = n_{ji} P_{ji} Q_{ji}$, we obtain the following properties of the $\hat{P}_{1i}$ and $\hat{\Psi}$, the maximum likelihood estimators of P_{1i} and Ψ.

$$\begin{aligned} \text{var } \hat{\Psi} &= \frac{\Psi^2}{\sum v_i^{-1}} \\ \text{cov } (\hat{\Psi}, \hat{P}_{1i}) &= \frac{\Psi}{n_{1i}} \frac{v_i^{-1}}{\sum v_i^{-1}} = \frac{\Psi}{n_{1i}} w_i \\ \text{var } \hat{P}_{1i} &= \frac{P_{1i} Q_{1i}}{n_{1i}} \frac{(v_{1i} + v_{2i} w_i)}{(v_{1i} + v_{2i})} \\ \text{cov } (\hat{P}_{1i}, \hat{P}_{1j}) &= \frac{1}{n_{1i} n_{1j}} w_i w_j \sum v_i^{-1}, \end{aligned} \tag{29}$$

where again $v_i = v_{1i}^{-1} + v_{2i}^{-1}$ and $w_i = v_i^{-1} / \sum v_i^{-1}$. Thus, in the unconditional analysis where Ψ is common to all k 2×2 tables, there is information in all $(k-1) p_{1j}$'s in estimating P_{1i} (see eq. 27) and the variance of the $\hat{P}_{1i}$ (see eq. 29) as a result of their contributions to the estimate of the common Ψ.

7. Examples

In this section, we analyze the two data sets described in section 2. Before proceeding to compute the estimates of the common odds ratio for the data sets we tested this assumption. The P-values of the Breslow-Day test for homogeneity in the Agarwal data was about .76 and for the EMF data about .18. Thus both data sets are consistent with the assumption of a common odds ratio.

The values of $\hat{\Psi}_i, \hat{c}_i, \hat{w}_i$ for the $i = 1, \ldots, 4$ levels in Table 1 are (.920, .0782, .0670), (2.017, .1262, .1390), (1.228, .3144, .2955) and (1.496, .4812, .4986) respectively. We then find $\hat{\Psi}_{MH} = \sum \hat{c}_i \hat{\Psi} = 1.4324$, $\hat{\Psi}_{MLS} = \sum \hat{w}_i \hat{\Psi}_i = 1.4508$, var $\hat{\Psi}_{MH} = \hat{\Psi}^2 \sum \hat{c}_i^2 \hat{v}_i = .06074$, var $\hat{\Psi}_{MLS} = \hat{\Psi}^2 \sum \hat{w}_i^2 \hat{v}_i = .06046$ and covar $(\hat{\Psi}_{MH}, \hat{\Psi}_{MLS}) = \hat{\Psi}^2 \sum \hat{c}_i \hat{w}_i \hat{v}_i = .06045$. The estimated asymptotic correlation between $\hat{\Psi}_{MH}$ and $\hat{\Psi}_{MLS}$ is .9975.

For the EMF data the values of $\hat{\Psi}_i, \hat{c}_i, \hat{w}_i$ for the $i = 1, 2$ levels in Table 2 are (3.30, .129, .160) and (1.77, .871, .840) respectively. We then find $\hat{\Psi}_{MH} = \sum \hat{c}_i \hat{\Psi}_i = 1.97$, $\hat{\Psi}_{MLS} = \sum \hat{w}_i \hat{\Psi}_i = 2.01$, var $\hat{\Psi}_{MH} = \hat{\Psi}^2 \sum \hat{c}_i \hat{v}_i = .1234$, var $\hat{\Psi}_{MLS} = \hat{\Psi}^2 \sum \hat{w}_i^2 \hat{v}_i = .1226$, and covariance $(\hat{\Psi}_{MH}, \hat{\Psi}_{MLS}) = \hat{\Psi}^2 \sum \hat{c}_i \hat{w}_i \hat{v}_i = .1226$. Therefore, the estimated asymptotic correlation is .997.

We obtained the exact conditional likelihood estimate of Ψ using the program of Thomas (1975) for each of the examples. In the first example $\Psi_{ML} = 1.435$ and in the second $\Psi_{ML} = 1.962$. We also analyzed an interesting set of data of 10 case control studies relating the exposure of cigarette smoking to the incidence of lung cancer. The data were presented by Cornfield (1956) and used by Gart (1962) to illustrate the application of three asymptotically efficient estimators. The estimate of the asymptotic correlation coefficient is .9902. All three examples are consistent with the findings, obtained by simulation, of Donner and Hauck (1986).

Acknowledgment

This work was supported by a grant from the National Science Foundation. We are very grateful to Beth Saunders for computational assistance.

References

Agarwal v. McKee (1977). 19 F.E.P. Cases 503 (N.D. Cal 1977)

Anderson, T.W. (1984). *An Introduction to Multivariate Statistical Analysis.* 2nd Ed. New York: Wiley.

Breslow, N.E. and Day, N.E. (1980). *Statistical Methods in Cancer Research.* Vol. 1, IARC, Lyon.

Breslow, N.W. and Liang, K.Y. (1982). The variance of the Mantel-Haenszel estimator, *Biometrics*, 38, 943-952.

Cornfield, J. (1951). A method for estimating comparative rates from clinical data. *Journal of the National Cancer Institute*, 11, 1269-1275.

Cornfield, J. (1956). A statistical problem arising from retrospective studies. In Neyman, J. Ed., *Proceedings of the Third Berkeley Symposium IV*, Berkeley, University of California Press, 133-148.

Donner, A. and Hauck, W.W. (1986). The large-sample relative efficiency of the Mantel-Haenszel estimator in the fixed-strata case. *Biometrics*, 42, 537-545.

Gart, J.J. (1962). On the combination of relative risks. *Biometrics*, 18, 601-610.

Gastwirth, J.L. (1984). Statistical methods for analyzing claims of employment discrimination. *Industrial and Labor Relations Review*, 38, 75-86.

Gastwirth, J.L. (1988). *Statistical Reasoning in Law and Public Policy*, Vol. 1, New York, Academic Press.

Guilbaud, O. (1983). On the large sample distribution of the Mantel-Haenszel odds-ratio estimator. *Biometrics*, 39, 523-525.

Hannan, J. and Harkness, W.L. (1963). Normal approximation to the distribution of two independent binomials, conditional on fixed sum. *The Annals of Mathematical Statistics*, 34, 1593-1595.

Harkness, W.L. (1965). Properties of the extended hypergeometric distribution. *The Annals of Mathematical Statistics*, 36, 938-945.

Mantel, N. and Haenszel, W. (1959). Statistical aspects of the analysis of data from retrospective studies of disease. *Journal of the National Cancer Institute*, 22, 719-748.

Mosteller, F. (1968). Association and estimation in contingency tables. *Journal of the American Statistical Association*, 63, 1-28.

Stevens, W.L. (1951). Mean and variance of an entry in a contingency table. *Biometrika*, 38, 468-470.

Thomas, D.G. (1975). Exact and asymptotic methods for the combination of 2 × 2 tables. *Computers and Biomedical Research*, 8, 423-446.

Wertheimer, N. and Leeper, E. (1979). Electrical wiring configurations and childhood cancer. *American Journal of Epidemiology*, 109, 273-384.

Section V

Posterior Odds, Testing and Model Selection

ON THE JUSTIFICATION OF DEFAULT AND INTRINSIC BAYES FACTORS*

James O. Berger
Department of Statistics
Purdue University
West Lafayette, IN 47907-1399 USA

and

Luis R. Pericchi
CESMa and
Departamento de Matemáticas
Universidad Simón Bolívar
Apartado 8900
Caracas 1080A, VENEZUELA

ABSTRACT:

In Bayesian model selection or hypothesis testing, it is difficult to develop default Bayes factors, since (improper) noninformative priors cannot typically be used. In developing such default Bayes factors, we feel that it is important to keep several principles in mind. The first is that the default Bayes factor should correspond, in some sense, to an actual Bayes factor with a (sensible) prior, which we call an intrinsic prior. The second principle is that such priors should be properly calibrated across models, in the sense of being "predictively matched." These notions will be described and illustrated, primarily using examples involving the intrinsic Bayes factor, a recently proposed default Bayes factor. It will be seen that intrinsic Bayes factors seem to correspond to actual Bayes factors with proper priors, at least for nested model scenarios. The corresponding intrinsic priors are specifically given for the normal linear model.

1. INTRODUCTION

There are a number of compelling reasons to consider use of Bayes factors in model selection and hypothesis testing. There are also a number of compelling reasons for development of 'default' or 'automatic' Bayes factors, especially in the preliminary stages of modelling when careful specification of subjective priors for all models under consideration is typically not feasible. For discussion of these issues, see Jeffreys (1961), Edwards, Lindman, and Savage (1963), Berger and Sellke (1987), Berger and Delampady (1987), Draper (1995), Kass and Raftery (1995), Madigan and Raftery (1995), and Berger and Pericchi (1993).

---

* This research was supported by the National Science Foundation, Grants DMS-8923071 and DMS-9303556, and by BID-CONICIT (Venezuela).

There are two main difficulties with the development of default Bayes factors. The first is the well-known difficulty that, when the models or hypotheses have parameter spaces of differing dimension, one cannot use only (improper) noninformative priors for computing the Bayes factors; improper priors are unaffected by multiplication by an arbitrary positive constant, but such arbitrary constants directly affect Bayes factors. The second difficulty in developing default (or even subjective) Bayes factors is that parameters do not typically have meaning independent of the model. Although this difficulty is also well-known, it is less often discussed, and is of enough importance to deserve emphasis through an example.

Example. We wish to predict automotive fuel consumption, Y, from the weight, X_1, and engine size, X_2, of a vehicle. Two models are entertained:

$$M_1 : Y = X_1\beta_1 + \varepsilon_1, \quad \varepsilon_1 \sim \mathcal{N}(0, \sigma_1^2)$$
$$M_2 : Y = X_1\beta_1 + X_2\beta_2 + \varepsilon_2, \quad \varepsilon_2 \sim \mathcal{N}(0, \sigma_2^2).$$

Thinking, first, about M_2, suppose the elicited prior density is of the form $\pi_2(\beta_1, \beta_2, \sigma_2) = \pi_{21}(\beta_1).\pi_{22}(\beta_2) \cdot \pi_{23}(\sigma_2)$. It is then quite common to choose, as the M_1 prior, $\pi_1(\beta_1, \sigma_1) = \pi_{21}(\beta_1) \cdot \pi_{12}(\sigma_1)$, i.e., to use the same prior for β_1 as in Model 1. The problem, of course, is that β_1 has a different meaning (and value) under M_1 than under M_2. For instance, regressing fuel consumption on weight alone will yield a larger coefficient than regressing on both weight and engine size, because of the considerable positive correlation between weight and engine size. Even worse, conceptually, would be to equate σ_1 and σ_2 and give them the same prior; clearly σ_1 will typically be larger than σ_2.

The first approach to overcoming these difficulties was that proposed by Jeffreys (1961). He proposed the use of orthogonal parameters (i.e., parameters for which the corresponding expected Fisher information matrix is diagonal, or block diagonal if the parameters are to be handled in blocks), presumably in an effort to overcome the type of difficulty illustrated in the above example. That use of orthogonal parameters overcomes this difficulty is a belief in the statistical folklore and is undoubtedly true in certain asymptotic senses, but we have not seen a clear Bayesian argument as to why this should be so. The other problems with orthogonalization are (i) it is frequently extremely difficult or impossible to find orthogonal parameters, and (ii) orthogonal parameters typically have no intuitive meaning, and so models expressed in terms of subsets of orthogonal parameters often have no meaning. Nevertheless, the use of orthogonal parameters, when possible, appears to be a quite effective tool. Jeffreys (1961) provides a number of convincing examples. For a modern successful use of the idea, see Clyde and Parmigiani (1995).

Jeffreys (1961) dealt with the issue of indeterminacy of noninformative priors by (i) only using noninformative priors for common (orthogonal) parameters in the models, so that the arbitrary multiplicative constant for the priors would cancel in all Bayes factors, and (ii) using default proper priors for parameters that would occur in one model but not the other. He presented arguments for appropriate default proper priors, but mostly on a case-by-case basis. This line of development has been successfully followed by several others, for instance by Zellner and Siow (1980).

Although use of particular default proper priors can be criticized for being somewhat arbitrary, one cannot be too demanding here. Any automatic procedure is going to contain some quite arbitrary features, and we feel that the Jeffreys approach is among those with the least objectionable arbitrary feartures. Indeed, we feel that any default Bayes factor should correspond (in some sense, perhaps asymptotic) to use of an actual Bayes factor with some proper prior distribution; if not, the Bayes factor is not compatible with Bayesian reasoning, and we feel that it is then probably uninterpretable. Furthermore, we feel that the best method of evaluating such 'good' default Bayes factors is to find the prior distribution to which they correspond, which we call the intrinsic prior, and to determine whether or not this distribution is sensible. In Section 3 we carry out this program for the intrinsic Bayes factor, a default Bayes factor that was proposed in Berger and Pericchi (1993, 1995).

Another general approach to overcoming the difficulties discussed above is the idea of selecting prior distributions that are somehow "matched" across models. Suzuki (1983) proposed matching the entropies of the prior distributions, or perhaps matching the entropies of an intermediate sequence of priors that converge to noninformative priors upon renormalization.

Perhaps more natural is to attempt to choose priors to match predictives. The underlying motivation is the foundational Bayesian view that one should concentrate on predictive distributions of observables; models and priors are, at best, convenient abstractions. According to this perspective, it is a predictive distribution $m(\mathbf{y})$ that describes reality, where $\mathbf{y}$ is a variable of predictive interest. We can choose to represent $m(\mathbf{y})$ as $m_i(\mathbf{y}) = \int f_i(\mathbf{y} \mid \theta_i)\pi_i(\theta_i)d\theta_i$, where f_i is a model and π_i a prior, but these are merely a convenient abstraction.

From this perspective, if one is comparing models $M_1 : f_1$ versus $M_2 : f_2$, then the priors π_1 and π_2 should be chosen so that $m_1(\mathbf{y})$ and $m_2(\mathbf{y})$ are as close as possible. Thus we think of π_1 and π_2 as being properly calibrated if, when filtered through the models M_1 and M_2, they yield similar predictives. This could be assessed by defining some distance measure, $d(m_1, m_2)$, and calling π_1 and π_2 calibrated if $d(m_1, m_2)$ is small. We explore this formal approach elsewhere, here being content simply with showing that intrinsic priors which arise from intrinsic Bayes factors seem to be well-calibrated.

One key issue in operationalizing this idea is that of choosing the variable $\mathbf{y}$ at which a predictive match is desired. It seems natural, in the exchangeable case, to choose $\mathbf{y}$ to be an "imaginary" minimal training sample, which is typically the smallest set of observations for which the various model parameters are identifiable.

The ideas here are related to ideas of elicitation through predictives (cf, Kadane, et.al., 1980). Also, a similar use of predictive matching to define priors for model selection can be found in Laud and Ibrahim (1993) and Ibrahim and Laud (1994).

2. THE INTRINSIC BAYES FACTOR

2.1 Definition of IBF's

Suppose that we are comparing q models for the data $\mathbf{x}$,

$$M_i\colon \mathbf{X} \textit{ has density } f_i(\mathbf{x}|\theta_i), \quad i = 1, \ldots, q,$$

and that we only have available default priors $\pi_i^N(\theta_i)$, $i = 1, \ldots, q$. The general strategy for defining IBF's starts with the definition of a proper and minimal training sample. The entire sample $\mathbf{x}$ is divided into two subsamples: $\mathbf{x}(l)$, which is the training sample, and $\mathbf{x}(-l)$ the remaining observations used for discrimination. Define the marginal or predictive densities of $\mathbf{X}$,

$$m_i^N(\mathbf{x}) = \int f_i(\mathbf{x}|\theta_i)\pi_i^N(\theta_i)d\theta_i.$$

Definition. A training sample, $\mathbf{x}(l)$, is called *proper* if $0 < m_i^N(\mathbf{x}(l)) < \infty$ for all M_i, and *minimal* if it is proper and no subset is proper. (Note that, if $\mathbf{x}(l)$ is proper, then all posteriors, $\pi_i^N(\theta_i|\mathbf{x}(l))$, are proper.)

The "standard" use of a training sample to define a Bayes factor is based on using $\mathbf{x}(l)$ to "convert" the improper $\pi_i^N(\theta_i)$ to proper posteriors, $\pi_i^N(\theta_i|\mathbf{x}(l))$, and using the latter to define a Bayes factor for the remaining data $\mathbf{x}(-l)$. The result, for comparing M_j to M_i, is (with obvious notation)

$$\begin{aligned} B_{ji}(l) =& \frac{\int f_j(\mathbf{x}(-l)|\theta_j, \mathbf{x}(l))\pi_i^N(\theta_j|\mathbf{x}(l))d\theta_j}{\int f_i(\mathbf{x}(-l)|\theta_i, \mathbf{x}(l))\pi_i^N(\theta_i|\mathbf{x}(l))d\theta_i} \\ =& B_{ji}^N \cdot B_{ij}^N(l), \end{aligned} \tag{1}$$

where

$$B_{ji}^N = \frac{m_j^N(\mathbf{x})}{m_i^N(\mathbf{x})} \quad \text{and} \quad B_{ij}^N(l) = \frac{m_i^N(\mathbf{x}(l))}{m_j^N(\mathbf{x}(l))} \tag{2}$$

are the Bayes factors that would be obtained for the full data $\mathbf{x}$ and training sample $\mathbf{x}(l)$, respectively, if one were to blindly use π_i^N and π_j^N.

While $B_{ji}(l)$ no longer depends on the scales of π_j^N and π_i^N, it does depend on the arbitrary choice of the (minimal) training sample $\mathbf{x}(l)$. To eliminate this dependence and to increase stability, a natural idea is to average the $B_{ji}(l)$ over all possible training samples $\mathbf{x}(l)$, $l = 1, \ldots, L$. Thus, in Berger and Pericchi (1993), we defined the *arithmetic IBF* (AIBF) and *geometric IBF* (GIBF) as, respectively,

$$B_{ji}^{AI} = \frac{1}{L}\sum_{l=1}^{L} B_{ji}(l) = B_{ji}^N \cdot \frac{1}{L}\sum_{l=1}^{L} B_{ij}^N(l), \tag{3}$$

$$B_{ji}^{GI} = \left(\prod_{l=1}^{L} B_{ji}(l)\right)^{1/L} = B_{ji}^N \cdot \left(\prod_{l=1}^{L} B_{ij}^N(l)\right)^{1/L}. \tag{4}$$

An important point, observed in Berger and Pericchi (1993), is that the average of the correction factors, $B_{ij}^N(l)$, must converge (for large samples) in order for B_{ji}^{AI} to correspond to a proper Bayes factor. To this end, it is typically necessary to place the more "complex" model in the numerator of the AIBF, i.e., to let M_j be the more complex model. We then *define* B_{ij}^{AI} by

$$B_{ij}^{AI} = 1/B_{ji}^{AI}. \tag{5}$$

2.2 The IBF for Two Non-Nested Examples

The following two scenarios will be used to illustrate several of the issues raised in the Introduction.

The IBF for Fixed Design Linear Models

Assume that we are considering the Linear Models

$$M_j : \mathbf{Y} = \mathbf{X}\boldsymbol{\beta}_j + \sigma_j \boldsymbol{\varepsilon}_j, \tag{6}$$

for $j = 1, \ldots, q$ alternative error models $\boldsymbol{\varepsilon}_j \sim g_j$; here $\mathbf{Y}$ is $n \times 1$, $\mathbf{X}$ is $n \times k, \boldsymbol{\beta}_j \epsilon R^k$ is $k \times 1, \sigma_j > 0$, and $\boldsymbol{\varepsilon}_j$ is $n \times 1$. Note that the design matrix, $\mathbf{X}$, is assumed to be fixed across models. We label the unknown $\boldsymbol{\beta}$ and σ by j, so as to emphasize that parameters can have different meanings within different models. We will use reference default priors, $\pi_j^N(\boldsymbol{\beta}_j, \sigma_j) = 1/\sigma_j$. A minimal training sample can be seen to be any $(k+1)$-vector $\mathbf{y}(l)$ with corresponding sub-matrix $\mathbf{X}(l)$, of $\mathbf{X}$, such that $\mathbf{X}^t(l)\mathbf{X}(l)$ is nonsingular. Let $|\mathbf{A}|$ denote the determinant of a matrix $\mathbf{A}$.

Lemma 1. *In the above situation, if $g_j(\mathbf{v}) = g_j(-\mathbf{v})$, then the marginal density of the minimal training sample $\mathbf{y}(l)$ is*

$$m_j^N(\mathbf{y}(l)) = [2|\mathbf{X}^t(l)\mathbf{X}(l)|^{1/2}|\mathbf{y}(l) - \mathbf{X}(l)(\mathbf{X}^t(l)\mathbf{X}(l))^{-1}\mathbf{X}^t(l)\mathbf{y}(l)|]^{-1}. \tag{7}$$

Lemma 1 is established in Berger, Pericchi and Varshavsky (1994). It is a quite surprising result because $m_j(\mathbf{y}(l))$ does not depend in any way on g_j. For instance, it holds when g_j is any $\mathcal{N}_n(0, \Sigma_j)$ distribution, regardless of Σ_j. It also holds for nonnormal distributions.

This provides our first illustration of the "predictive matching" idea described in the Introduction. Indeed, Lemma 1 suggests that the reference prior is properly calibrated for comparison of *any* models of the form (6), in that the predictives for a minimal sample are then identical. (Note that this will not be the case if other noninformative priors, e.g., the Jeffreys prior, are used.) This result greatly simplifies the model elaboration for linear models, since then all $B_{ij}^N(l)$ clearly equal one (see (2)) and hence (from (3) and (4))

$$B_{ji}^{AI} = B_{ji}^{GI} = B_{ji}^N \tag{8}$$

Comparison of Exponential and Lognormal Models

Suppose $X_1, \ldots, X_n$ are i.i.d. according to one of the following models:

$$M_1\colon f_1(x_i|\theta_1) = \theta_1^{-1}\exp\{-x_i/\theta_1\} \quad (\text{Exponential}(\theta_1)),$$

$$M_2\colon f_2(x_i|\mu, \sigma) = \frac{\exp\{-(\log x_i - \mu)^2/(2\sigma^2)\}}{\sqrt{2\pi}\sigma x_i} \quad (\text{Lognormal}(\mu, \sigma)).$$

For M_1 and M_2, the standard noninformative priors are $\pi_1^N(\theta_1) = 1/\theta_1$ and $\pi_2^N(\mu, \sigma) = 1/\sigma$. Calculation yields, for $\mathbf{x} = (x_1, \ldots, x_n)$,

$$m_1^N(\mathbf{x}) = \frac{\Gamma(n)}{(\Sigma x_i)^n}, \qquad m_2^N(\mathbf{x}) = \frac{\Gamma((n-1)/2)}{(\prod_{i=1}^n x_i)\pi^{(n-1)/2}2\sqrt{n}S_y^{(n-1)}},$$

where $S_y^2 = \sum_{i=1}^{n}(y_i - \bar{y})^2, y_i = \log x_i$. It is easy to see that minimal training samples are of the form $\mathbf{x}(l) = (x_i, x_j), x_i \neq x_j$, so that

$$m_1^N(\mathbf{x}(l)) = \frac{1}{(x_i + x_j)^2}, \qquad m_2^N(\mathbf{x}(l)) = \frac{1}{2x_i x_j |\log(x_i/x_j)|}.$$

The IBF can thus be computed using (2) and (3) or (4). We defer discussion of this example to Section 3.3.

2.3 The IBF for the Normal Linear Model

Suppose, for $j = 1, \ldots, q$, that model M_j for the data $\mathbf{Y}(n \times 1)$ is the linear model

$$M_j : \mathbf{Y} = \mathbf{X_j}\boldsymbol{\beta}_\mathbf{j} + \boldsymbol{\varepsilon}_\mathbf{j}, \qquad \boldsymbol{\varepsilon}_\mathbf{j} \sim \mathcal{N}_\mathbf{n}(\mathbf{0}, \sigma_\mathbf{j}^2 \mathbf{I_n}),$$

where σ_j^2 and $\boldsymbol{\beta}_j = (\beta_{j1}, \beta_{j2}, \ldots, \beta_{jk_j})^t$ are unknown, and $\mathbf{X_j}$ is an $(n \times k_j)$ given design matrix of rank $k_j < n$. Let

$$\hat{\boldsymbol{\beta}}_j = (\mathbf{X}_j^t \mathbf{X}_j)^{-1} \mathbf{X}_j^t \mathbf{y} \textit{ and } R_j = |\mathbf{y} - \mathbf{X_j}\hat{\boldsymbol{\beta}}_\mathbf{j}|^2$$

denote the least squares estimator for $\boldsymbol{\beta}_j$ and residual sum of squares, respectively.

We will consider default priors of the form

$$\pi_j^N(\boldsymbol{\beta}_j, \sigma_j) = \sigma_j^{-(1+q_j)}, \quad q_j > -1. \tag{9}$$

Common choices of q_j are $q_j = 0$ (the reference prior; cf., Bernardo, 1979, and Berger and Bernardo, 1992) or $q_j = k_j$ (the Jeffreys prior). When comparing model M_i nested in M_j, we will also consider a *modified Jeffreys prior*, having $q_i = 0$ and $q_j = k_j - k_i$. This is intermediate between the reference and Jeffreys priors.

It is easy to show, for these priors, that a minimal training sample $\mathbf{y}(l)$, with corresponding design matrices $\mathbf{X}_j(l)$ (under the M_j), is a sample of size $m = \max\{k_j\} + 1$ such that all $(\mathbf{X}_j^t(l)\mathbf{X}_j(l))$ are nonsingular. (Note that if $q_j = -1$, i.e., constant noninformative priors are used, then one would instead need $m = \max\{k_j\} + 2$.)

Computation yields that

$$B_{ji}^N = \frac{\pi^{(k_j - k_i)/2}}{2^{(q_i - q_j)/2}} \cdot \frac{\Gamma((n - k_j + q_j)/2)}{\Gamma((n - k_i + q_i)/2)} \cdot \frac{|\mathbf{X}_i^t \mathbf{X}_i|^{1/2}}{|\mathbf{X}_j^t \mathbf{X}_j|^{1/2}} \cdot \frac{R_i^{(n-k_i+q_i)/2}}{R_j^{(n-k_j+q_j)/2}}, \tag{10}$$

and that $B_{ij}^N(l)$ is given by the inverse of this expression with n, $\mathbf{X}_i$, $\mathbf{X}_j$, R_i, and R_j replaced by m, $\mathbf{X}_i(l)$, $\mathbf{X}_j(l)$, $R_i(l)$, and $R_j(l)$, respectively; here $R_i(l)$ and $R_j(l)$ are the residual sums of squares corresponding to the training sample $\mathbf{y}(l)$, i.e.,

$$R_j = |\mathbf{y}(l) - \mathbf{X}_j(l)\hat{\boldsymbol{\beta}}_j(l)|^2, \qquad \hat{\boldsymbol{\beta}}_j(l) = (\mathbf{X}_j^t(l)\mathbf{X}_j(l))^{-1}\mathbf{X}_j^t(l)\mathbf{y}(l). \tag{11}$$

Inserting these expressions in (1) results in the following arithmetic IBF's in (3) for the three default priors being considered. (For the corresponding geometric IBF's, simply replace the arithmetic averages by geometric averages.)

Using the Jeffreys prior:

$$B_{ji}^{AI} = \frac{|\mathbf{X}_i^t\mathbf{X}_i|^{1/2}}{|\mathbf{X}_j^t\mathbf{X}_j|^{1/2}} \cdot \left(\frac{R_i}{R_j}\right)^{n/2} \cdot \frac{1}{L}\sum_{l=1}^{L} \frac{|\mathbf{X}_j^t(l)\mathbf{X}_j(l)|^{1/2}}{|\mathbf{X}_i^t(l)\mathbf{X}_i(l)|^{1/2}} \cdot \left(\frac{R_j(l)}{R_i(l)}\right)^{m/2}. \tag{12}$$

Using the Modified Jeffreys prior: Defining $p = k_j - k_i$,

$$B_{ji}^{AI} = \frac{|\mathbf{X}_i^t\mathbf{X}_i|^{1/2}}{|\mathbf{X}_j^t\mathbf{X}_j|^{1/2}} \cdot \left(\frac{R_i}{R_j}\right)^{(n-k_i)/2} \cdot \frac{1}{L}\sum_{l=1}^{L} \frac{|\mathbf{X}_j^t(l)\mathbf{X}_j(l)|^{1/2}}{|\mathbf{X}_i^t(l)\mathbf{X}_i(l)|^{1/2}} \cdot \left(\frac{R_j(l)}{R_i(l)}\right)^{(p+1)/2}. \tag{13}$$

Using the Reference prior: Defining $p = k_j - k_i$ and

$$C = \frac{\Gamma((n-k_j)/2)\Gamma((k+1)/2)}{\Gamma((n-k_i)/2)\Gamma(1/2)}, \tag{14}$$

$$B_{ji}^{AI} = \frac{|\mathbf{X}_i^t\mathbf{X}_i|^{1/2}}{|\mathbf{X}_j^t\mathbf{X}_j|^{1/2}} \cdot \frac{R_i^{(n-k_i)/2}}{R_j^{(n-k_j)/2}} \cdot \frac{C}{L}\sum_{l=1}^{L} \frac{|\mathbf{X}_j^t(l)\mathbf{X}_j(l)|^{1/2}}{|\mathbf{X}_i^t(l)\mathbf{X}_i(l)|^{1/2}} \cdot \frac{(R_j(l))^{1/2}}{(R_i(l))^{(p+1)/2}}. \tag{15}$$

For Known σ^2: If the σ_j^2 are known and equal σ^2, and the $\pi_j^N(\boldsymbol{\beta}_j) = 1$, then

$$B_{ji}^{N} = (2\pi\sigma^2)^{(k_j-k_i)/2} \cdot \frac{|\mathbf{X}_i^t\mathbf{X}_i|^{1/2}}{|\mathbf{X}_j^t\mathbf{X}_j|^{1/2}} \cdot \exp\left\{-\frac{1}{2\sigma^2}(R_j - R_i)\right\}. \tag{16}$$

Here, a minimal training sample is a sample of size $m = \max\{k_j\}$ such that all $(\mathbf{X}_j^t(l)\mathbf{X}_j(l))$ are nonsingular, and $B_{ji}^N(l)$ is as in (16) with $\mathbf{X}_i$, $\mathbf{X}_j$, R_i, and R_j replaced by $\mathbf{X}_i(l)$, $\mathbf{X}_j(l)$, $R_i(l)$, and $R_j(l)$. Thus the arithmetic intrinsic Bayes factor from (1) and (3) is

$$B_{ji}^{AI} = \frac{|\mathbf{X}_i^t\mathbf{X}_i|^{1/2}}{|\mathbf{X}_j^t\mathbf{X}_j|^{1/2}} \cdot \exp\left\{-\frac{1}{2\sigma^2}(R_j - R_i)\right\}$$
$$\times \frac{1}{L}\sum_{l=1}^{L} \frac{|\mathbf{X}_j^t(l)\mathbf{X}_j(l)|^{1/2}}{|\mathbf{X}_i^t(l)\mathbf{X}_i(l)|^{1/2}} \cdot \exp\left\{-\frac{1}{2\sigma^2}(R_j(l) - R_i(l))\right\}. \tag{17}$$

We discuss intrinsic priors and predictive matching for IBF's in Section 3.2.

3. INTRINSIC PRIORS FOR IBF's

3.1 Definition and Motivation.

Our major goal is to show that arithmetic IBF's correspond to actual Bayes factors with respect to what we call an *intrinsic prior*. We view the fact that IBFs tend to correspond to actual Bayes factors w.r.t. (sensible) intrinsic priors to be their strongest justification. Hence, determination of the intrinsic priors is of inherent theoretical interest, as well as providing the best insight into the behavior of IBFs.

There are also potential practical benefits in determining intrinsic priors. One obvious benefit is that the intrinsic priors could themselves be used, in place of the π_i^N, to compute actual Bayes factors. This would eliminate the need for training sample computations and eliminate concerns about stability of the IBFs. Indeed, one could alternatively view the IBF procedure as a method to apply to "imaginary training samples," so as to determine actual conventional priors to be used for model selection and hypothesis testing. This could be viewed as the complement to, say, the reference prior theory (Bernardo, 1979; Berger and Bernardo, 1992), which also uses imaginary samples to develop conventional priors for estimation and related problems.

While this latter view of the IBF methodology has considerable philosophical appeal, there are pragmatic arguments against actually operating in this fashion. Foremost among these arguments is that it is often very difficult to determine intrinsic priors. In contrast, IBFs are typically extremely easy to determine.

The formal definition of an intrinsic prior, given in Berger and Pericchi (1993), was based on an asymptotic analysis, utilizing the following approximation to a Bayes factor:

$$B_{ji} = B_{ji}^N \cdot \frac{\pi_j(\hat{\theta}_j)\pi_i^N(\hat{\theta}_i)}{\pi_j^N(\hat{\theta}_j)\pi_i(\hat{\theta}_i)}(1 + o(1)); \tag{18}$$

here B_{ji} denotes the Bayes factor associated with priors π_j and π_i, π_i^N and π_j^N are the noninformative priors used to compute B_{ji}^N, and $\hat{\theta}_i$ and $\hat{\theta}_j$ are the MLEs under M_i and M_j. (The approximation in (18) holds more generally than the more standard Schwarz approximation that is discussed, for instance in Schwarz, 1978, Gelfand and Dey, 1994, and Kass and Raftery, 1995.)

To define intrinsic priors, equate (18) with (3) or (4), yielding

$$\frac{\pi_j(\hat{\theta}_j)\pi_i^N(\hat{\theta}_i)}{\pi_j^N(\hat{\theta}_j)\pi_i(\hat{\theta}_i)}(1 + o(1)) = \tilde{B}_{ij}^N, \tag{19}$$

where we define $\tilde{B}_{ij}^N$ to be either the arithmetic or geometric average of the $B_{ij}^N(l)$. We next need to make some assumptions about the limiting behavior of the quantities in (19). The following are typically satisfied, and will be assumed to hold as the sample size grows to infinity:

(i) Under M_j, $\hat{\theta}_j \to \theta_j$, $\hat{\theta}_i \to \psi_i(\theta_j)$, and $\tilde{B}_{ij}^N \to B_j^*(\theta_j)$.
(ii) Under M_i, $\hat{\theta}_i \to \theta_i$, $\hat{\theta}_j \to \psi_j(\theta_i)$, and $\tilde{B}_{ij}^N \to B_i^*(\theta_i)$. (20)
(iii) For $k = i$ or $k = j$, the following limits exist:

$$B_k^*(\theta_k) = \begin{cases} \lim\limits_{L\to\infty} \quad E_{\theta_k}^{M_k}\left[\frac{1}{L}\sum\limits_{l=1}^{L} B_{ij}^N(l)\right] & \text{arithmetic case} \\ \lim\limits_{L\to\infty} \quad \exp\left\{E_{\theta_k}^{M_k}\left[\frac{1}{L}\sum\limits_{l=1}^{L} \log B_{ij}^N(l)\right]\right\} & \text{geometric case;} \end{cases} \tag{21}$$

if the $\mathbf{X}(l)$ are exchangeable, then the limits and averages over L can be removed.

Passing to the limit in (19), first under M_j and then under M_i, results in the following two equations which define the *intrinsic prior* (π_j^I, π_i^I)

$$\frac{\pi_j^I(\theta_j)\pi_i^N(\psi_i(\theta_j))}{\pi_j^N(\theta_j)\pi_i^I(\psi_i(\theta_j))} = B_j^*(\theta_j), \tag{22}$$

$$\frac{\pi_j^I(\psi_j(\theta_i))\pi_i^N(\theta_i)}{\pi_j^N(\psi_j(\theta_i))\pi_i^I(\theta_i)} = B_i^*(\theta_i). \tag{23}$$

The motivation, again, is that priors which satisfy (22) and (23) would yield answers which are asymptotically equivalent to use of the intrinsic Bayes factors. We note that solutions are not necessarily unique, do not necessarily exist, and are not necessarily proper (cf, Dmochowski, 1994).

As a simple example of the above ideas, consider the fixed design linear model from Section 2.2. It is clear that $B_j^*(\theta_j) = B_i^*(\theta_i) = 1$; it follows trivially that solutions to (22) and (23) are given by

$$\pi_k^I(\theta_k) = \pi_k^N(\theta_k), \quad k = i, j.$$

Thus the intrinsic priors are merely the original noninformative priors. (Note, however, that this happens only because we used the reference noninformative priors; it would not happen, for instance, had the Jeffreys noninformative prior been used.)

3.2 Intrinsic Priors for Arithmetic IBF's in Nested Linear Models

Here we consider the normal linear model situation of Section 2.3. For the nested situation and use of arithmetic IBF's, it will be shown that proper intrinsic priors exist. Model M_i will be said to be nested in M_j if $\mathbf{X}_i$ consists of a subset of the columns of $\mathbf{X}_j$. (More general types of nesting can be reduced to this by transformation.) In fact, we will assume that the covariates have been ordered so that $\mathbf{X}_j = (\mathbf{X}_i\ \mathbf{X}^*)$ (the concatenation of the two matrices, not the product), and that the parameterization has been chosen so that $X_i^t X^* = 0$. Writing $\boldsymbol{\beta}_j^t = (\boldsymbol{\beta}_0^t, \boldsymbol{\beta}^{*t})$, it is convenient to write $\pi_j(\theta_j) = \pi_j(\boldsymbol{\beta}_j, \sigma_j)$ as

$$\pi_j(\boldsymbol{\beta}_j, \sigma_j) = \pi_j^1(\boldsymbol{\beta}^*|\boldsymbol{\beta}_0, \sigma_j) \cdot \pi_j^2(\boldsymbol{\beta}_0, \sigma_j). \tag{24}$$

Note that $(\boldsymbol{\beta}_0, \sigma_j)$ is the analogue, under M_j, of $(\boldsymbol{\beta}_i, \sigma_i)$ under M_i. As discussed in the Introduction, we do not make the common mistake of identifying these parameters as being equal but, as they are related "nuisance" location-scale parameters, it is natural to assign them the same noninformative prior. We in fact will *choose* this common prior to be the same as $\pi_i^N(\boldsymbol{\beta}_i, \sigma_i) = \sigma_i^{-(1+q_i)}$, so that

$$\pi_i(\boldsymbol{\beta}_i, \sigma_i) = \sigma_i^{-(1+q_i)}, \quad \pi_j^2(\boldsymbol{\beta}_0, \sigma_j) = \sigma_j^{-(1+q_i)}. \tag{25}$$

If $(\boldsymbol{\beta}_i, \sigma_i)$ and $(\boldsymbol{\beta}_0, \sigma_j)$ really were the same parameters, this choice would be noncontroversial. As they are not necessarily the same parameters, however, it could be argued that π_i and π_j^2 may not be properly "calibrated." If, however, $q_i = 0$ (i.e., the original π_i^N is the reference prior), then π_i and π_j^2 are themselves the reference priors, and we saw in Section 2.2 that this seems to provide a type of predictive "calibration" for *any* location-scale models. Thus our argument that IBF's correspond to sensible real Bayes factors is strongest if the IBF is defined for reference π_i^N, which occurs in either the "reference prior case" or the "modified Jeffreys prior case." (In fact, we will see that an "adjustment" of $\pi_j^2(\boldsymbol{\beta}_0, \sigma_j)$ is needed for the Jeffreys prior case.)

For this situation, the conditions in (20) can be shown to hold, with $\psi_i(\theta_j) = (\boldsymbol{\beta}_0, \sigma_j)$ and $\psi_j(\theta_i) = \theta_i = (\boldsymbol{\beta}_i, \sigma_i)$, providing the limits in (21) exist. There can be a certain ambiguity in defining this limit when the design matrix is unpatterned; we will thus assume that, as $n \to \infty$, the design matrix is patterned or replicated in such a way that the limits in (21) exist.

Next, observe that expectation in (21) under M_i is equivalent to expectation under M_j with $\theta_j = ((\boldsymbol{\beta}_0, 0), \sigma_j)$. It is then straightforward to show that (22) and (23) are both equivalent to the single equation.

$$\pi_j^1(\boldsymbol{\beta}^*|\boldsymbol{\beta}_0, \sigma_j) = \sigma_j^{(q_i - q_j)} \cdot B_j^*(\theta_j)$$
$$= \sigma_j^{(q_i - q_j)} \cdot \frac{1}{L} \sum_{l=1}^{L} E^{M_2}_{(\boldsymbol{\beta}_j, \sigma_j)}[B_{ij}^N(l)]. \tag{26}$$

Interestingly, the expectations in (26) can be computed in closed form; see the Appendix. Using these expressions, the "intrinsic priors" in (26) can be written as follows. (We also include the result for the known variance case; the analogue of (26) for this case is easy to derive using Fact(v) from the Appendix.)

Unknown σ_i^2 and σ_j^2:

$$\pi_j^1(\boldsymbol{\beta}^*|\boldsymbol{\beta}_0, \sigma_j) = \frac{\sigma_j^{(q_i - q_j)} C^*}{L} \cdot \sum_{l=1}^{L} \frac{|\mathbf{X}_j^t(l)\mathbf{X}_j(l)|^{1/2}}{|\mathbf{X}_i^t(l)\mathbf{X}_i(l)|^{1/2}} \cdot \psi(\lambda(l), \sigma_j), \tag{27}$$

where C^* is defined in (A2) of the Appendix, $\psi(\lambda(l), \sigma_j)$ is either (A3), (A4), or (A5), depending on the default prior used, and

$$\lambda(l) = \sigma_j^{-2} \boldsymbol{\beta}^{*t} \mathbf{X}^{*t}(l)(\mathbf{I} - \mathbf{X}_i(l)[\mathbf{X}_i^t(l)\mathbf{X}_i(l)]^{-1}\mathbf{X}_i^t(l))\mathbf{X}^*(l)\boldsymbol{\beta}^*. \tag{28}$$

Known $\sigma_i^2 = \sigma_j^2 = \sigma^2$: Defining $p = k_j - k_i$ (recall that k_j is the dimension of $\boldsymbol{\beta}_j$)

$$\pi_j^1(\boldsymbol{\beta}^*|\boldsymbol{\beta}_0) = \frac{1}{(4\pi\sigma^2)^{p/2}} \cdot \frac{1}{L} \sum_{l=1}^{L} \frac{|\mathbf{X}_j^t(l)\mathbf{X}_j(l)|^{1/2}}{|\mathbf{X}_i^t(l)\mathbf{X}_i(l)|^{1/2}} \cdot \exp\{-\lambda(l)/4\}. \tag{29}$$

Of course, we have not yet answered the big question: is $\pi_j^1(\boldsymbol{\beta}^*|\boldsymbol{\beta}_0, \sigma_j)$ a proper distribution? If so, we have established the Bayesian correspondence of IBF's.

Consider, first, the case of known $\sigma_i^2 = \sigma_j^2 = \sigma^2$. It is straightforward to show that

$$\boldsymbol{\Sigma}(l) \equiv (\mathbf{X}^*(l)^t(\mathbf{I} - \mathbf{X}_i(l)[\mathbf{X}_i^t(l)\mathbf{X}_i(l)]^{-1}\mathbf{X}_i^t(l))\mathbf{X}^*(l))^{-1}\sigma^2 \tag{30}$$

has determinant

$$|\boldsymbol{\Sigma}(l)| = \sigma^{2p}|\mathbf{X}_i^t(l)\mathbf{X}_i(l)|/|\mathbf{X}_j^t(l)\mathbf{X}_j(l)|.$$

Hence (29) can be written

$$\pi_j^1(\boldsymbol{\beta}^*|\boldsymbol{\beta}_0) = \frac{1}{L}\sum_{l=1}^{L}\pi_l(\boldsymbol{\beta}^*), \tag{31}$$

where the π_l are $\mathcal{N}_p(0, \frac{1}{2}\boldsymbol{\Sigma}(l))$ distributions. Thus π_j^1 is a mixture of normals, and is trivially a proper distribution. The following theorem deals with the unknown variance case.

Theorem 1. *For the reference prior and modified Jeffreys prior cases, $\pi_j^1(\boldsymbol{\beta}^*|\boldsymbol{\beta}_0, \sigma_j)$ in (27) is a proper density. For the Jeffreys prior case,*

$$\int \pi_j^1(\boldsymbol{\beta}^*|\boldsymbol{\beta}_0, \sigma_j)d\boldsymbol{\beta}^* = C_0 = \frac{\Gamma((k_i+1)/2)\Gamma((p+1)/2)}{\Gamma((k_j+1)/2)\Gamma(1/2)}.$$

Proof. We freely use notation and facts from the Appendix. For the Jeffreys prior case,

$$\pi_j^1(\boldsymbol{\beta}^*|\boldsymbol{\beta}_0, \sigma_j) = \frac{1}{L}\sum_{l=1}^{L} g_l(\boldsymbol{\beta}^*),$$

$$g_l(\boldsymbol{\beta}^*) = \frac{C^{**}}{(2\pi\sigma_j^2)^{p/2}} \cdot \frac{|\mathbf{X}_j^t(l)\mathbf{X}_j(l)|^{1/2}}{|\mathbf{X}_i^t(l)\mathbf{X}_i(l)|^{1/2}} \cdot e^{-\lambda(l)/2} \cdot M\left(\frac{p+1}{2}, \frac{p+k_j+2}{2}, \frac{\lambda(l)}{2}\right).$$

The transformation $\boldsymbol{\beta}^* \to \lambda(l)$ has Jacobian

$$\frac{|\mathbf{X}_i^t(l)\mathbf{X}_i(l)|^{1/2}}{|\mathbf{X}_j^t(l)\mathbf{X}_j(l)|^{1/2}} \cdot \frac{(\pi\sigma_j^2)^{p/2}}{\Gamma(p/2)} \cdot \lambda(l)^{(p-2)/2},$$

so that (writing $\lambda = \lambda(l)$)

$$\int g_l(\boldsymbol{\beta}^*)d\boldsymbol{\beta}^* = \frac{C^{**}}{2^{p/2}\Gamma(p/2)}\int_0^\infty \lambda^{(p-2)/2}e^{-\lambda/2}M(\frac{p+1}{2}, \frac{p+k_j+2}{2}, \frac{\lambda}{2})d\lambda.$$

Using Fact (ii), and integrating term by term yields

$$\begin{aligned}
&\int_0^\infty \lambda^{(p-2)/2}e^{-\lambda/2}M(\frac{p+1}{2}, \frac{p+k_j+2}{2}, \frac{\lambda}{2})d\lambda \\
&= \frac{\Gamma((p+k_j+2)/2)}{\Gamma((p+1)/2)}\sum_{j=0}^{\infty}\frac{\Gamma(j+(p+1)/2)}{\Gamma(j+(p+k_j+2)/2)(j!)2^j} \cdot \int_0^\infty \lambda^{(j-1+p/2)}e^{-\lambda/2}d\lambda \\
&= \frac{\Gamma((p+k_j+2)/2)}{\Gamma((p+1)/2)}\sum_{j=0}^{\infty}\frac{\Gamma(j+(p+1)/2)\Gamma(j+p/2)2^{p/2}}{\Gamma(j+(p+k_j+2)/2)(j!)} \\
&= 2^{p/2}\Gamma(p/2)F(\frac{p+1}{2}, \frac{p}{2}, \frac{p+k_j+2}{2}, 1) \\
&= \frac{2^{p/2}\Gamma(p/2)\Gamma((p+k_j+2)/2)\Gamma((k_i+1)/2)}{\Gamma((k_j+1)/2)\Gamma((k_j+2)/2)},
\end{aligned}$$

where F is the hypergeometric function and we have used 15.1.20 of Abramowitz and Stegun (1970). Combining terms and simplifying yields C_0.

The identical argument works for the reference and modified Jeffreys prior cases, but now the integral equals 1.

It is interesting that $\pi_j^1(\boldsymbol{\beta}^*|\boldsymbol{\beta}_0,\sigma_j)$ is proper for the reference prior and modified Jeffreys prior cases, but is *not* for the Jeffreys prior case. This suggests that our choice of $\pi_i(\boldsymbol{\beta}_i,\sigma_i) = \sigma_i^{-(1+q_i)}$ and $\pi_j^2(\boldsymbol{\beta}_0,\sigma_j) = \sigma_j^{-(1+q_i)}$ for the Jeffreys prior IBF are not properly "calibrated"; choosing $\pi_j^2(\boldsymbol{\beta}_0,\sigma_j) = C_0^{-1}\sigma_j^{-(1+q_i)}$ would ensure that $\pi_j^1(\boldsymbol{\beta}^*|\boldsymbol{\beta}_0,\sigma_j)$ is then proper, and is hence perhaps the correct calibration of π_j^2.

The nature of $\pi_j^1(\boldsymbol{\beta}^*|\boldsymbol{\beta}_0,\sigma_j)$ is of considerable interest in providing insight into the behavior of the associated IBF's. In the known variance case, $\pi_j^1(\boldsymbol{\beta}^*|\boldsymbol{\beta}_0)$ is rather simple, and clearly has mean 0 and covariance

$$\boldsymbol{\Sigma}^* = \frac{1}{2L}\sum_{l=1}^{L}\boldsymbol{\Sigma}(l). \tag{32}$$

Note that, in balanced cases where the $\boldsymbol{\Sigma}(l)$ are equal, $\pi_j^1(\boldsymbol{\beta}^*|\boldsymbol{\beta}_0)$ is just a single normal prior, and is similar to the prior used for model comparison by Zellner and Siow (1980). Seeing how $\boldsymbol{\Sigma}^*$ differs from the Zellner and Siow covariance matrix in unbalanced cases would be of considerable interest.

The behavior of $\pi_j^1(\boldsymbol{\beta}^*|\boldsymbol{\beta}_0,\sigma_j)$ in the unknown variance case is more difficult to ascertain. For the modified Jeffreys prior case and $p = (k_j - k_i)$ an odd integer, simple closed form expressions are available, as shown following (A4). For instance, when $p = 1$, using (27) and (28) yields

$$\begin{aligned}\pi_j^1(\boldsymbol{\beta}^*|\boldsymbol{\beta}_0,\sigma_j) &= \frac{1}{\sqrt{2\pi\sigma_j^2}}\cdot\frac{1}{L}\sum_{l=1}^{L}\frac{|\mathbf{X}_j^t(l)\mathbf{X}_j(l)|^{1/2}}{|\mathbf{X}_i^t(l)\mathbf{X}_i(l)|^{1/2}}\cdot\frac{1}{\lambda(l)}(1-e^{-\lambda(l)/2})\\ &= \frac{1}{L}\sum_{l=1}^{L}\frac{1}{2\sqrt{\pi V(l)}}\cdot\frac{1}{(\beta^{*2}/V(l))}\cdot(1-e^{-\beta^{*2}/V(l)}),\end{aligned} \tag{33}$$

where

$$V(l) = 2\sigma_j^2/[\mathbf{X}^*(l)^t(\mathbf{I}-\mathbf{X}_i(l)(\mathbf{X}_i^t(l)\mathbf{X}_i(l))^{-1}\mathbf{X}_i^t(l))\mathbf{X}^*(l)].$$

Each of the densities in this mixture is very similar to a Cauchy $(0, \sqrt{V(l)})$ density (never differing by more than 15%). This Cauchy density is similar to that recommended by Jeffreys (1961) or Zellner and Siow (1980).

In general, it can be shown (for the reference and modified Jeffreys cases) that $\pi_j^1(\boldsymbol{\beta}^*|\boldsymbol{\beta}_0,\sigma_j)$ is a mixture of densities that behave like $\mathcal{T}_p(1,0,\boldsymbol{\Sigma}^*(l))$ densities: p-variate t-densities with 1 degree of freedom, location 0, and scale matrix

$$\boldsymbol{\Sigma}^*(l) = 2\sigma_j^2[\mathbf{X}^*(l)^t(\mathbf{I}-\mathbf{X}_i(l)(\mathbf{X}_i^t(l)\mathbf{X}_i(l))^{-1}\mathbf{X}_i^t(l))\mathbf{X}^*(l)]^{-1}.$$

The fact that the degree of freedom here is minimal, seems related to the fact that minimal training samples were used.

As a final comment, note that an analogous derivation of intrinsic priors for geometric IBF's can be performed. However, the analogous expressions for $\pi_j^1(\boldsymbol{\beta}^*|\boldsymbol{\beta}_0,\sigma_j)$ are considerably more involved, and also do not appear to be proper distributions.

3.3 Intrinsic Priors in Nonnested Models: An Example

For nonnested models, finding a solution to (22) and (23) is often more difficult. Consider comparison of the nonnested models M_1: Exponential (θ_1) and M_2: Lognormal (μ,σ), introduced in Section 2.2. Assumption (20) can be shown to be satisfied, since

$$\begin{aligned}\text{under } M_1, \quad \hat{\theta}_2 = (\hat{\mu},\hat{\sigma}) &= (\overline{y},(S_y^2/n)^{1/2})\\ &\overset{(n\to\infty)}{\longrightarrow}(E_{\theta_1}^{M_1}[\overline{Y}],(\frac{1}{n}E_{\theta_1}^{M_1}[S_y^2])^{1/2})\\ &\overset{(n\to\infty)}{\longrightarrow}\psi_2(\theta_1)\equiv(\log\theta_1-0.5772,1.2825); \quad (34)\\ \text{under } M_2, \quad \hat{\theta}_1 = \overline{x} \to \psi_1(\mu,\sigma) &= E_{(\mu,\sigma)}^{M_2}[\overline{X}] = \exp\{\mu+\tfrac{1}{2}\sigma^2\}. \quad (35)\end{aligned}$$

Also, (21) becomes

$$B_1^* = \begin{cases} E_{\theta_1}^{M_1}\left[\frac{2X_iX_j|\log(X_i/X_j)|}{(X_i+X_j)^2}\right] & \text{arithmetic case}\\ \exp\left\{E_{\theta_1}^{M_1}\left[\log\left(\frac{2X_iX_j|\log(X_i/X_j)|}{(X_i+X_j)^2}\right)\right]\right\} & \text{geometric case}\end{cases}$$
$$= \begin{cases} 0.2954 & \text{arithmetic case}\\ 0.2383 & \text{geometric case;}\end{cases} \tag{36}$$

$$B_2^* = \begin{cases} E_{(\mu,\sigma)}^{M_2}\left[\frac{2X_iX_j|\log(X_i/X_j)|}{(X_i+X_j)^2}\right] & \text{arithmetic case}\\ \exp\left\{E_{(\mu,\sigma)}^{M_2}\left[\log\left(\frac{2X_iX_j|\log(X_i/X_j)|}{(X_i+X_j)^2}\right)\right]\right\} & \text{geometric case}\end{cases}$$
$$= \begin{cases} H^A(\sigma)\equiv E^Z\left[\frac{\sqrt{2}\sigma|Z|}{1+\cosh(\sqrt{2}\sigma Z)}\right] & \text{arithmetic case}\\ H^G(\sigma)\equiv\frac{3\sigma}{2}\cdot\exp\left\{-2E^Z\left[\log\left(1+e^{\sqrt{2}\sigma Z}\right)\right]\right\} & \text{geometric case,}\end{cases} \tag{37}$$

where $Z\sim\mathcal{N}(0,1)$. (The derivations above are straightforward.)

For the arithmetic case, equations (22) and (23) thus become

$$\frac{\pi_2^I(\mu,\sigma)(1/\exp\{\mu+\frac{1}{2}\sigma^2\})}{(1/\sigma)\pi_1^I(\exp\{\mu+\frac{1}{2}\sigma^2\})} = H^A(\sigma), \tag{38}$$

$$\frac{\pi_2^I(\log\theta_1-0.5772,1.2825)(1/\theta_1)}{(1/1.2825)\pi_1^I(\theta_1)} = (0.2954). \tag{39}$$

We have not attempted to characterize the solutions to (38) and (39) in general. The equations are fairly easy to solve, however, if one assumes that

$$\pi_2^I(\mu,\sigma) = \pi_{21}^I(\mu)\pi_{22}^I(\sigma). \tag{40}$$

Indeed, the solutions are then given (up to multiplication of π_1^I and division of π_2^I by an arbitrary positive constant) by

$$\pi_1^I(\theta_1) = 2/\theta_1^c$$
$$\pi_2^I(\mu,\sigma) = \frac{1}{2\sigma} H^A(\sigma)\exp\{(1-c)(\mu + \frac{1}{2}\sigma^2)\}, \tag{41}$$

where $c = 1.1291$. A similar analysis for the geometric IBF yields, as the intrinsic priors, the expressions in (41) with H^A replaced by H^G and $c = 1.2602$.

To obtain some insight into the behavior of these priors, it is useful to reparameterize M_2 by (ν,σ), where $\nu = \exp\{\mu + \sigma^2/2\}$ is the lognormal mean. Then

$$\pi_2^I(\mu,\sigma) \longrightarrow \frac{2}{\nu^c} \cdot \frac{H(\sigma)}{2\sigma},$$

where H is either H^A or H^G. The point of this transformation is that θ_1 and ν are then both the mean parameters of their respective distributions, and are given the same improper prior. Curiously, however, it is not the usual inverse noninformative prior. We speculate that this noninformative prior might prove to provide a better predictive match for these "common" mean parameters.

The "nuisance" parameter, σ, receives the prior $\pi_{22}^I(\sigma) = H(\sigma)/(2\sigma)$. It is easy to show that $\pi_{22}^I(\sigma)$ is monotonically decreasing, with the following limiting behavior:

$$\text{as } \sigma \to 0, \quad \pi_{22}^I(\sigma) \cong \begin{cases} 1/(2\sqrt{\pi}) & \text{arithmetic case,} \\ 3/16 & \text{exponential case,} \end{cases}$$

$$\text{as } \sigma \to \infty, \quad \pi_{22}^I(\sigma) \cong \begin{cases} 1/(\sqrt{\pi}\sigma^2) & \text{arithmetic case,} \\ \frac{3}{4}\exp(-2\sigma/\sqrt{\pi}) & \text{exponential case.} \end{cases}$$

It is thus clear that $\pi_{22}^I(\sigma)$ is integrable; indeed, we have normalized (41) so that, in the arithmetic case, $\pi_{22}^I(\sigma)$ is a proper density.

The pattern we have observed thus seems to be holding: for parameters that are in some sense "common," the intrinsic priors are the same and are of a noninformative type, while parameters that exist only in one of the models receive proper intrinsic priors. For a variety of other examples and characterizations of intrinsic priors, see Dmochowski (1994).

APPENDIX

Proof of Equation (27): Defining $p = k_j - k_i$, note that

$$\frac{1}{L}\sum_{l=1}^{L} E^{M_j}_{\beta_j,\sigma_j}[B^N_{ij}(l)] = \frac{C^*}{L}\sum_{l=1}^{L} \frac{|\mathbf{X}_j^t(l)\mathbf{X}_j(l)|^{1/2}}{|\mathbf{X}_i^t(l)\mathbf{X}_i(l)|^{1/2}} E^{M_j}_{\beta_j,\sigma_j}\left[\frac{(R_j(l))^{(q_j+1)/2}}{(R_i(l))^{(q_i+p+1)/2}}\right], \tag{A1}$$

where

$$C^* = \frac{\pi^{-p/2}}{2^{(q_j-q_i)/2}} \cdot \frac{\Gamma((q_i+p+1)/2)}{\Gamma((q_j+1)/2)}. \tag{A2}$$

The expectation in (A1) can be evaluated in closed form for the default priors we consider. The answers are in terms of Kummer's function, $M(a,b,c)$ (see Abramowitz and Stegun, 1970, Chapter 13).

In the proofs, the following standard facts will be repeatedly used ; all notation is taken from Sections 2.3 and 3.2.

(i) Under M_j,

$$W = \frac{R_j(l)}{\sigma_j^2} \sim \chi_1^2,$$
$$V = \frac{R_i(l) - R_j(l)}{\sigma_j^2} \sim \chi_p^2(\lambda(l)),$$

where χ_ν^2 denotes the central chi-square distribution with ν degrees of freedom, and $\chi_p^2(\lambda(l))$ is the noncentral chi-square distribution with $p = k_j - k_i$ degrees of freedom and noncentrality parameter $\lambda(l)$ which, in the nested case, is given by (28). Also, W and V are independent.

(ii)

$$M(a,b,z) = \frac{\Gamma(b)}{\Gamma(a)} \sum_{j=0}^{\infty} \frac{\Gamma(a+j)}{\Gamma(b+j)} \cdot \frac{z^j}{j!}.$$

(iii)

$$E[h(\chi_\nu^2(\lambda))] = \sum_{j=0}^{\infty} \frac{(\lambda/2)^j \exp\{-\lambda/2\}}{j!} \cdot E[h(\chi_{\nu+2j}^2)].$$

(iv) With obvious abuse of notation,

$$E\left[\left(\frac{\chi_1^2}{\chi_1^2+\chi_\nu^2}\right)^s\right] = \frac{\Gamma(s+1/2)\Gamma((\nu+1)/2)}{\Gamma(1/2)\Gamma(s+(\nu+1)/2)},$$

providing χ_1^2 and χ_ν^2 are independent.

(v)

$$E[\exp\{-\frac{1}{2}\chi_p^2(\lambda)\}] = 2^{-p/2}e^{-\lambda/4}.$$

(vi)

$$\frac{\Gamma(1+p/2)\Gamma((p+1)/2)}{\Gamma(1/2)\Gamma(p+1)} = 2^{-p}.$$

Lemma 2. For the various noninformative priors, the expectations in (A1) are given in the following expressions:

Using the Jeffreys prior: Here $q_i = k_i$ and $q_j = k_j$, and the expectation in (A1) becomes

$$E_{\beta_j,\sigma_j}^{M_2}\left[\left(\frac{R_j(l)}{R_i(l)}\right)^{(k_j+1)/2}\right] = C^{**}e^{-\lambda(l)/2}M\left(\frac{p+1}{2}, \frac{p+k_j+2}{2}, \frac{\lambda(l)}{2}\right), \qquad \text{(A3)}$$

where

$$C^{**} = \frac{\Gamma((k_j+2)/2)\Gamma((p+1)/2)}{\Gamma((k_j+p+2)/2)\Gamma(1/2)}.$$

Using the Modified Jeffreys prior: Here $q_i = 0$ and $q_j = k_j - k_i = p$, and the expectation in (A1) becomes

$$E\left[\left(\frac{R_j(l)}{R_i(l)}\right)^{(p+1)/2}\right] = 2^{-p}e^{-\lambda(l)/2}M(\frac{p+1}{2},\ p+1,\ \frac{\lambda(l)}{2}) \tag{A4}$$

$$= \begin{cases} \frac{1}{\lambda(l)}[1-e^{-\lambda(l)/2}] & \text{if } p=1 \\ \frac{3}{\lambda(l)^2}[(1-\frac{4}{\lambda(l)})+(1+\frac{4}{\lambda(l)})e^{-\lambda(l)/2}] & \text{if } p=3 \\ \frac{15}{\lambda(l)^3}[(1-\frac{12}{\lambda(l)}+\frac{48}{\lambda(l)^2})-(1+\frac{12}{\lambda(l)}+\frac{48}{\lambda(l)^2})e^{-\lambda(l)/2}] & \text{if } p=5. \end{cases}$$

Using the Reference prior: Here $q_i = q_j = 0$, and the expectation in (A1) becomes

$$E\left[\frac{(R_j(l))^{1/2}}{(R_i(l))^{(p+1)/2}}\right] = \frac{\exp\{-\lambda(l)/2\}}{\sigma_j 2^{p/2}\Gamma((p+2)/2)}M(\frac{1}{2}, \frac{p+2}{2}, \frac{\lambda(l)}{2}). \tag{A5}$$

Proof of Lemma 2: Using, in order, Facts (i), (iii), and (iv), we obtain

$$E^{M_2}\left[\left(\frac{R_j(l)}{R_i(l)}\right)^{(k_j+1)/2}\right] = E\left[\left(\frac{W}{W+V}\right)^{(k_j+1)/2}\right]$$

$$= \sum_{j=0}^{\infty}\frac{(\lambda(l)/2)^j\exp\{-\lambda(l)/2\}}{j!}\cdot E\left[\left(\frac{\chi_1^2}{\chi_1^2+\chi_{p+2j}^2}\right)^{(k_j+1)/2}\right]$$

$$= e^{-\lambda(l)/2}\sum_{j=0}^{\infty}\frac{(\lambda(l)/2)^j}{j!}\cdot\frac{\Gamma((k_j+2)/2)\Gamma((p+2j+1)/2)}{\Gamma(1/2)\Gamma((p+k_j+2j+2)/2)}.$$

Using Fact (ii), (A3) follows immediately. The proof of (A4) is almost identical, but also uses Fact (vi). The explicit forms given for $p = 1, 3, 5$ follow from representations of M.

To prove (A5) use, in order, Facts (i) and (iii) to obtain

$$E\left[\frac{(R_j(l))^{1/2}}{(R_i(l))^{(p+1)/2}}\right] = \sum_{j=0}^{\infty}\frac{(\lambda(l)/2)^j\exp\{-\lambda(l)/2\}}{j!}\cdot E\left[\frac{\sqrt{\chi_1^2}}{(\chi_1^2+\chi_{p+2j}^2)^{(p+1)/2}}\right].$$

Defining $c_j^{-1} = 2^{(p+2j+1)/2}\Gamma(1/2)\Gamma((p+2j)/2)$, it is clear that

$$E\left[\frac{\sqrt{\chi_1^2}}{(\chi_1^2+\chi_{p+2j}^2)^{(p+1)/2}}\right] = \int_0^\infty\int_0^\infty\frac{c_j y^{(j-1+p/2)}e^{-(x+y)/2}}{(x+y)^{(p+1)/2}}dxdy$$

$$= c_j 2^{(j+1/2)}\Gamma(j+1/2)/(j+p/2).$$

Algebra, together with Fact (ii), yields the result.

References

Abramowitz, M. and Stegun, I. (1970), *Handbook of Mathematical Functions*, National Bureau of Standards Applied Mathematics Series 55.

Berger, J. and Bernardo, J. M. (1992), "On the Development of the Reference Prior Method," in *Bayesian Statistics IV*, eds. J. M. Bernardo, et. al., London: Oxford University Press, pp. 35–60.

Berger, J. and Delampady, M. (1987), "Testing Precise Hypotheses," *Statistical Science*, 3, 317–352.

Berger, J. and Pericchi, L. (1993), "The Intrinsic Bayes Factor for Model Selection and Prediction," Technical Report 93-43C, Purdue University, Department of Statistics.

Berger, J. and Pericchi, L. (1995), "The Intrinsic Bayes Factor for Linear Models," in *Bayesian Statistics* V, eds. J. M. Bernardo, et. al., London: Oxford University Press, pp. 23–42.

Berger, J., Pericchi, L., and Varshavsky, J. (1995), "An Identity for Linear and Invariant Models, with Application to Non-Gaussian Model Selection," Technical Report 95-7C, Purdue University, Department of Statistics.

Berger, J. and Sellke, T. (1987), "Testing a Point Null Hypothesis: the Irreconcilability of P-Values and Evidence," *Journal of the American Statististical Association*, 82, 112–122.

Bernardo, J. M. (1979), "Reference Posterior Distributions for Bayesian Inference," *Journal of the Royal Statistical Society*, Ser. B, 41, 113–147.

Clyde, M. and Parmigiani, G. (1995), "Orthogonalizations and Prior Distributions for Orthogonalized Model Mixing," ISDS Discussion Paper 95-07, Duke University.

de Vos, A. F. (1993), "A Fair Comparison Between Regression Models of Different Dimension," Technical Report, The Free University, Amsterdam.

Dmochowski, J. (1994), "Intrinsic Priors Via Kullback-Liebler Geometry," Technical Report 94-15, Purdue University, Department of Statistics.

Draper, D. (1995), "Assessment and Propogation of Model Uncertainty," *Journal of the Royal Statistical Society*, Ser. B, 57, 45–98.

Edwards, W., Lindman, H. and Savage, L. J. (1963), "Bayesian Statistical Inference for Psychological Research," *Psychological Review*, 70, 193–242.

Gelfand, A. E. and Dey, D. K. (1994), "Bayesian Model Choice: Asymptotics and Exact Calculations," *Journal of the Royal Statistical Society*, Ser. B, 56, 501–514.

Gelfand, A. E., Dey, D. K. and Chang, H. (1992), "Model Determination Using Predictive Distributions with Implementations Via Sampling-Based Methods," in *Bayesian Statistics 4*, eds. J. M. Bernardo, et. al., London: Oxford University Press, pp. 147–167.

Ibrahim, J. and Laud, P. (1994), "A Predictive Approach to the Analysis of Designed Experiments," *Journal of the American Statistical Association*, 89, 309-319.

Iwaki, K. (1995), "Posterior Expected Marginal Likelihood for Comparison of Hypotheses," Technical Report, Purdue University.

Jeffreys, H. (1961), *Theory of Probability*, London: Oxford University Press.

Kadane, J.B., Dickey, J., Winkler, R., Smith, W., and Peters, S. (1980), "Interactive Elicitation of Opinion for a Normal Linear Model," *Journal of the American Statistical Association*, 75, 845-854.

Kass, R. E. and Raftery, A. (1995), "Bayes Factors," to appear in the *Journal of the American Statistical Association*.

Laud, P.W. and Ibrahim, J. (1993), "Predictive Model Selection," Technical Report 93-01, Northern Illinois University.

Madigan, D. and Raftery, A. E. (1995), "Model Selection and Accounting for Model Uncertainty In Graphical Models Using Occam's Window," to appear in the *Journal of the American Statistical Association*.

O'Hagan, A. (1995), "Fractional Bayes Factors for Model Comparisons," *Journal of the Royal Statistical Society*, Ser. B, 57, 99–138.

Pericchi, L. R. and Pérez, M. E. (1994), "Posterior Robustness with More than One Sampling Model," *Journal of Statistical Planning and Inference*, 40, 279–294.

Poirier, D. J. (1985), "Bayesian Hypothesis Testing in Linear Models with Continuously Induced Conjugate Priors Across Hypotheses," in *Bayesian Statistics 2*, eds. J. M. Bernardo, et. al., New York: Elsevier, pp. 711–722.

Suzuki, Y. (1983), "On Bayesian Approach to Model Selection," in Proceedings of the International Statistical Institute, Madrid Vol.1, 288-291.

Schwarz, G. (1978), "Estimating the Dimension of a Model," *Annals of Statistics*, 6, 461–464.

Zellner, A. and Siow (1980), "Posterior Odds for Selected Regression Hypotheses," in *Bayesian Statistics 1*, eds. J. M. Bernardo, et. al., Valencia: Valencia University Press, pp. 585–603.

Tests on Independence of VNTR Alleles in the Chinese Population in Hong Kong

Wing K. FUNG
Department of Statistics,
The University of Hong Kong
Pokfulam Road, Hong Kong

Abstract

The issue of independence of fragment lengths within and between VNTR loci is a controversial topic in DNA profiling (fingerprinting). This paper discusses a battery of tests and applies them for testing independence of VNTR data for the Chinese population in Hong Kong. The independence assumption seems to be satisfied generally. This finding is very different from those obtained by Geisser and Johnson (1993) that the assumption is largely invalid for Blacks, Caucasians and Hispanics. Some discussion on the validity of the tests is also given.

1. Introduction

DNA profiling has been considered as a very powerful but controversial tool for forensic identification since its introduction (Jeffreys et al., 1985). The restriction fragment length polymorphism (RFLP) analysis of variable number tandem repeats (VNTR) loci is often used. Large numbers of alleles are found for these loci. A restriction enzyme cuts a sample of DNA into fragments of various lengths at these loci.

DNA profiling may be regarded as one of the greatest discoveries in forensic science since the introduction of dermal fingerprinting. Unique identification using DNA profiling is possible in principle since no two persons, except for identical twins, have the same DNA sequence. However, in practice, only a few sites (loci) are examined. Most laboratories nowadays use 3-5 loci where each locus usually contains (at most) two fragments. The DNA fragments at these loci extracted from the blood specimens found at the crime scene are compared with those obtained from the suspect for identification. A match between two DNA patterns may already be considered strong evidence that the two samples came from the same source. The DNA experts often calculate and report the probability that the DNA profiles of a randomly selected person will match with those found in the crime scene. These random match probabilities are often as small as one in a million as found in the forensic literature or in some court cases (see examples in Berry, 1990 say).

One major controversy in DNA profiling is about the validity of the assumption employed for calculating the random match probabilities. They are derived via the multiplication of marginal probabilities of occurrence for individual fragments. The multiplication rule (National Research Council Report, 1992) is of course based on the assumption that the fragments within and between loci are statistically independent (They are called Hardy-Weinberg and linkage equilibrium respectively in genetics). Weir (1992) found that the assumption is generally valid for the ethnic groups including Blacks, Caucasians and Hispanics. However, Geisser and Johnson (1992, 1993) proposed simple tests and had the opposite findings. Devlin and Risch (1993), and Geisser and Johnson (1993) discussed the appropriateness of the Geisser-Johnson tests for independence of VNTR fragments. There are still debates over this controversial issue, and interested readers may refer to Geisser and Johnson (1995), Roeder (1994) and the discussion therein, and Weir (1993). Some additional references in the statistical literature include Berry et al. (1992), and Devlin et al. (1992).

The independence issue has been well discussed for ethnic groups in Western countries, but not for people in Asian countries. The paper examines the independence assumption — both the within and between loci issues — for the Hong Kong Chinese (Sections 3 and 4). The DNA profiling technique has been used in Hong Kong for a few years. It has been used in courts only since early 1994. The Hong Kong Chinese population database was collected by the Hong Kong Government Laboratory (Tsui and Wong, 1995) using the HAE III enzyme at loci D1S7, D2S44, D4S139 and D10S28. The same enzyme is also used by the FBI (Budowle et al., 1991) and the Royal Canadian Mounted Police (Waye, 1989) in North America, and Singapore (Chow et al., 1993) in Asia. The Hong Kong database is of size $208(n)$ persons and is discussed in Section 2. The rarity of a DNA match is shown using the pairwise comparison within the database (Section 2). Some concluding remarks are given in Section 5.

2. Occurrence of One or More Loci Match

Due to the limitation of the RFLP technology, the fragments are measured with errors. Let z_1 and z_2 be the measured fragment lengths of alleles of two individuals. The fragments are called a match if

$$\frac{|z_1 - z_2|}{0.5(z_1 + z_2)} < 0.05. \tag{1}$$

As an individual has two fragments for each locus (one from each parent), a single-locus match would be declared if both fragments for two individuals satisfy the criterion. The 5% cut-off is used by the Hong Kong Government Laboratory and the FBI. Other laboratories may use a different cut-off such as 4% in Singapore (Chow et al., 1994). Since population reference databases usually contain fresh blood DNA samples with little contamination and less measurement error variability, we also tried the 2.4% cut-off as suggested by Risch and Devlin (1992). For brevity, they are omitted.

Table 1 shows the single-locus matching results. The total number of between-person comparisons is 21528 (208 × 207/2) for a database of size 208 persons. The observed (average) probabilities of single-locus matches can then be estimated. They are all rather small with the highest probability being only 2.2% (D2S44), thus showing the rarity of a random match in DNA profiling. The loci D2S44 and D4S139 are relatively less discriminating than the other two loci.

The observed number of two-locus matches $O(M)$ can be obtained in a similar way for every possible pair of loci. The results are given in column 2 of Table 2. The observed numbers of two-locus matches are much smaller than those of single-locus matches. Only 19 matches out of 129168 [$= 6 \times (208 choose 2)$] comparisons are found. The largest observed number of matches is only 9 for the pair of loci D2S44 and D4S139. No two-locus match is found for the more discriminating pair D1S7 and D10S28. The expected numbers of matches $E(M)$ assuming independence of loci are estimated and they are given in the third column of Table 2. They are very close to the corresponding observed numbers.

Table 1. *The observed number of matches $O(M)$ and the estimated probability of a random match $P(M)$ for single locus.*

Locus	D1S7	D2S44	D4S139	D10S28
$O(M)$	132	475	391	174
$P(M)$	.0061	.0221	.0182	.0081

Since the probabilities of single-locus random matches at each locus are small (Table 1), it suggests that a random three-locus or four-locus match in the database would be very infrequent. This expectation is met since no three-locus or four-locus matches are found.

Assuming that the match probabilities across loci are independent, the (average) probability of a random four-locus match, evaluated as the product of the four single-locus probabilities (Table 1), is given as 1.99×10^{-8}. The random match probability estimate for individual case of course would be different. The next section examines the validity of the independence assumption which is the basis of the multiplication rule for probability assessment.

Table 2. *The observed and expected numbers (assuming independence) of matches, $O(M)$ and $E(M)$, for pairs of loci. The statistic T_s is used for testing independence of loci, with p-value P_1 obtained from a bootstrap simulation of 1000 replicates.*

Loci	$O(M)$	$E(M)$	T_s	P_1
D1, D2	1	2.9	1.250	.30
D1, D4	2	2.4	.065	.83
D1, D10	0	1.1	1.080	.33
D2, D4	9	8.5	.026	.86
D2, D10	4	3.8	.010	.93
D4, D10	3	3.2	.008	.93

3. Tests for Independence or Zero Correlation

3.1 Analysis of independence between loci

Using the numbers of matches in the previous section, the following 2×2 contingency table is constructed for each combination of loci.

		Locus 2		Total
		Match	No match	
Locus 1	Match	a	b	A_1
	No match	c	d	A_2
	Total	B_1	B_2	M

The usual test for independence $T_s = M(ad - bc)^2/(A_1 A_2 B_1 B_2)$ is constructed, with $M = 21528$ being the number of comparisons. The results are given in the fourth column of Table 2. Since the comparisons are between individuals in the database, they are not independent observations. The usual asymptotic χ_1^2 null distribution for the test is inappropriate, and so we approximate the distribution using a bootstrap simulation (Risch and Devlin, 1992).

A random sample of 208 cases (fragment pairs) is drawn independently with replacement for each locus of the loci pair from the database. The two random samples are then randomly paired, and a table is constructed as above. The test statistic T_s is then calculated for this bootstrap sample of fragment lengths which is simulated under the null hypothesis of independent fragments between loci. The procedure is repeated 1000 times for approximating the null distribution of T_s. The approximate p-value P_1 of the observed T_s for each loci pair is given in the fifth column of Table 2. We notice that the p-values are all above 0.3, indicating that the hypothesis of independent fragment lengths between loci is not rejected.

3.2 The correlations of fragment lengths

At each DNA locus, two VNTR fragments, one from either parent, are observed for an individual. Let x_{ij}, $j = 1, 2$ be the measured fragment lengths for individual i at a certain locus. One common way for studying independence is to use correlation analysis. The Hardy-Weinberg law suggests the independence of fragment lengths within loci which gives a zero intraclass correlation. In DNA profiling, the source of the two fragments is unidentifiable because it is generally not possible to distinguish whether a fragment came from the mother or the father. Thus, the usual Pearson correlation coefficient would not be appropriate for estimating the intraclass correlation. This identifiability problem is also found in other circumstances such as in that discussed by Fisher (1958) about the estimation of correlation between the heights of two brothers, without knowledge about which brother was older.

Let ρ_x be the intraclass correlation between the two fragment lengths, and σ_x^2 be the variance of the fragment length. An analysis of variance table for fragment lengths can be constructed (Weir, 1992) and it is given in Table 3. Thus the intraclass correlation is estimated as,

$$\hat{\rho}_x = \frac{MSB_x - MSW_x}{MSB_x + MSW_x}. \tag{2}$$

The estimates for the Hong Kong Chinese database are given in Table 4. They are all very close to zero except $\hat{\rho}_x$ for D1S7 which is 0.16.

Table 3. *Analysis of variance for fragment lengths within loci*

Source	d.f.	Mean square (MS)	Expected MS
Between individuals	$n-1$	$MSB_x = \frac{1}{2(n-1)}\left(\sum_i x_{i\cdot}^2 - \frac{1}{n}x_{\cdot\cdot}^2\right)$	$(1+\rho_x)\sigma_x^2$
Within individuals	n	$MSW_x = \frac{1}{n}\left(\sum_i\sum_j x_{ij}^2 - \frac{1}{2}\sum_i x_{i\cdot}^2\right)$	$(1-\rho_x)\sigma_x^2$

For two fragment lengths x_{ij} in each of n individuals $(i = 1, \cdots, n,\ j = 1, 2)$.
$x_{i\cdot} = \sum_j x_{ij},\ x_{\cdot\cdot} = \sum_i x_{i\cdot}$.

The significance of $\hat{\rho}_x$ under the null hypothesis of zero correlation is obtained from bootstrap simulation. The two fragments of each case are independently selected with replacement from the database of size 416 ($2n$) for each locus. A sample of 208 independent cases is randomly selected with the intraclass correlation estimated using equation (2). The procedure is repeated 1000 times to obtain the (approximate) null distribution of $\hat{\rho}_x$. It is found that only the $\hat{\rho}_x$ for locus D1S7 is significant at the 5% (but not at 1%) level. Thus, the correlations between fragment lengths within loci seem rather small. Intraclass correlation being zero of course does not imply independence because the distribution of the fragment lengths is highly non-normal (Tsui and Wong, 1995).

Table 4. *Estimates of within locus correlations*

Locus	D1S7	D2S44	D4S139	D10S28
Estimated Correlation	.16*	.01	−.02	−.02

* significantly different from zero at 5% (but not 1%) level.

The correlations between fragment lengths at different loci can be studied in the following way. Let x_{ij} and y_{ij}, $j = 1, 2$ be the fragment lengths for individual i at two loci, let ρ_{xy_1} be the correlation between two fragments at these loci when the two fragments came from the same parent, and ρ_{xy_2} be that when they came from different parents. Since it is impossible to distinguish whether the two fragments came from the same parent or different parents, only the average of the two correlations can be estimated. Since $E(MPB_{xy}) = \sigma_x\sigma_y(\rho_{xy_1} + \rho_{xy_2})/2$, it follows that a moment estimate for the average correlation is given as (Weir, 1992),

$$\frac{1}{2}\widehat{(\rho_{xy_1} + \rho_{xy_2})} = \frac{2(MPB_{xy})}{\sqrt{(MSB_x + MSW_x)(MSB_y + MSW_y)}}, \tag{3}$$

where

$$MPB_{xy} = \frac{1}{n-1}\left[\sum_{i=1}^{n} X_iY_i - \frac{1}{n}\sum_{i=1}^{n} X_i \sum_{i=1}^{n} Y_i\right], \tag{4}$$

with $X_i = (x_{i1} + x_{i2})/2$, $Y_i = (y_{i1} + y_{i2})/2$, MSB_x and MSW_x defined in Table 3 (MSB_y and MSW_y are defined accordingly for y_{ij}).

Table 5 gives the estimates of correlation coefficients of fragment lengths for each pair of loci. The magnitudes are all below 0.1. A bootstrap simulation with 1000 replicates is conducted for determining the significance of each estimate under the null hypothesis of zero correlation. The estimates are all insignificant at the 5% level. It seems that the assumption of independent fragment lengths between loci may not be unreasonable.

Table 5. *Estimates of between loci correlations*

Locus pair	D1, D2	D1, D4	D1, D10	D2, D4	D2, D10	D4, D10
Estimated Correlation	−.08	−.03	−.04	−.03	.01	.04

4 Geisser and Johnson Tests for Independence

4.1 Within loci

Recently Geisser and Johnson (1992, 1993, 1995) suggested the following tests for testing independence within loci. Let the $2n$ observations of each locus be divided into q ordered quantiles, $Q_1, \cdots, Q_q$ where $q \geq 2$. We count, for each locus, the number of pairs n^*_{ij} such that one of the pair is in Q_i and the other in Q_j, $i \neq j$, and the number n_{ii} for both in Q_i. Under the null hypothesis of independent fragments within loci, it can be shown that approximately $E(n^*_{ij}) = 2n/q^2$ and $E(n_{ii}) = n/q^2$. Thus, Geisser and Johnson (1992) suggested the Pearson χ^2 test

$$X = \frac{q^2}{n}\sum_{i=1}^{q}\left(n_{ii} - \frac{n}{q^2}\right)^2 + \frac{q^2}{2n}\sum\sum_{i<j}\left(n^*_{ij} - \frac{2n}{q^2}\right)^2 \tag{5}$$

that is asymptotically chi-squared distributed with $q(q-1)/2$ df under the null hypothesis. Moreover, it is often argued that the data would exhibit a positive dependence because of substructuring in the population under study. One would then expect the sum of the diagonal entries to be greater than those under independence. Since under independence, approximately $E(\sum n_{ii}/n) = 1/q$ and $\text{var}(\sum n_{ii}/n) = (1 - 1/q)/(qn)$, Geisser and Johnson (1992) suggested using

$$Z = \left(q\sum_{i=1}^{q} n_{ii} - n\right) \Big/ \sqrt{n(q-1)} \tag{6}$$

which is asymptotically distributed as standard normal under the null. This test is more sensitive to the positive dependence alternative. They found a very strong departure from independence in the FBI's Black, Caucasian and Hispanic databases, which was contrary to Weir's (1992) findings.

We apply the tests to our data set using $q = 2, \cdots, 6$ as in Geisser and Johnson (1993). The results are listed in Table 6. The same locus D1S7 shows some violation of independence for $q = 3$ and 4, and the tests also indicate that D4S139 has a positive dependence structure. We however do not have the alarming results as in Geisser and Johnson's (1993) analysis on the FBI's Black, Caucasian and Hispanic databases. If we investigate the data of D4S139 carefully, we realize that the findings may not be too surprising. This locus has less variability than the other three loci. Locus D4S139 occupies only 16 bins out of a total of 31 fixed bins (The fixed bin approach assigns fragments into 31 pre-defined fixed bins that are of similar sizes. The approach is commonly adopted in Hong Kong, Singapore and by the FBI), whereas loci D1S7, D2S44 and D10S28 are spread over 28, 20 and 23 bins respectively (Tsui and Wong, 1995). Moreover, over 60% of the D4S139 fragments fall within four bins only.

Another special feature of D4S139 is that it has the largest number of single fragments. It has 27 single fragments compared with 5, 16 and 8 for D1S7, D4S139 and D10S28 respectively. It has been well discussed (Weir, 1992, Devlin and Risch, 1993) that each of these single fragments may belong to one of the following categories: (a) the two fragments are of the same size; (b) the fragments are of different but similar sizes, namely coalesced, and thus appear the same; and (c) the disappearance of a very short fragment due to the limitation of the RFLP technique. Fragments of the first type are called true homozygotes; while those of the other types are apparent homozygotes. It is well known in genetics that substructuring in the population would result in excessive homozygotes even though the sub-groups are in Hardy-Weinberg equilibrium. It is to be noted that the issue of excessive single fragments is a controversial topic in DNA profiling. As we have little idea about the proportion of apparent homozygotes, we try omitting some or all the single fragments for checking the independence hypothesis. Notice that this way of studying independence may not be unrealistic because the single fragments are treated differently in case work. Their random match probabilities are estimated, without assuming independence, using the formula $2p$ which is conservative and gives the benefit of doubt to the accused; while $2pq$ that assumes independence is employed for the double fragments (Weir, 1992), where p and q are the frequencies for the corresponding fragments.

Table 6. *p-values for Geisser-Johnson Z and X tests (within loci)*

Locus	Test	q				
		2	3	4	5	6
D1S7	X	.166	.039	.0012	.078	.103
	Z	.083	.461	.739	.404	.402
D2S44	X	.267	.670	.495	.443	.909
	Z	.134	.246	.375	.674	.268
D4S139	X	.166	.793	.015	.190	.090
	Z	.083	.163	.008	.006	.0002
D10S28	X	.781	.506	.389	.176	.667
	Z	.609	.461	.739	.674	.807

We first have the extreme case of omitting all the single fragments. The p-values using the X statistics are generally greater (not shown), and locus D4S139 is no longer significant. Locus D1S7 still has the dependence structure as it only has five single fragments. The Z statistics do not indicate any dependence structure in the data. However, all but one of the twenty Z values are negative, indicating that the omission of all single fragments is too extreme.

We investigate locus D4S139 further by eliminating the first seven single fragments. The Z statistic for $q = 6$ is significant at the 1% level. It is still significant at the 5% level if 10 single fragments are omitted, but insignificant when 14 fragments are excluded. The associated X values are all insignificant. It is clear that the dependence structure of D4S139 is mainly due to the high proportion of single fragments.

There has been some discussion on the validity of the Geisser-Johnson test for independence in VNTR data. Devlin and Risch (1993) argued that the test is not appropriate for typical VNTR data because it is very sensitive to two physical properties of the data — correlated measurement error and coalescence. They showed using simulations that alleles at a VNTR locus can be independent even when the test gets highly significant deviations from independence. Geisser and Johnson (1993) responded that since the measurement error is approximately only 0.6% of the fragment length and coalescence is not too common in the FBI data, their test is unlikely to be highly affected and would not give a very small p-value if the alleles are independent especially for small q such as $q = 2$ or 3. It seems that their test should be quite robust to the special RFLP electrophoreic properties. However, the results for larger q (such as 5 or 6) may have to be interpreted carefully.

4.2 Between loci

Recently Geisser and Johnson (1995), extending the idea for testing Hardy-Weinberg equilibrium, suggested a test for testing between loci independence. The two fragments at each locus of the random sample are first added as $x_i = x_{i1} + x_{i2}$ and $y_i = y_{i1} + y_{i2}$, $i = 1, \cdots, n$. The q ordered quantiles $Q_1(x), \cdots, Q_q(x)$ can then be formed for the x_i's and similarly $Q_1(y), \cdots, Q_q(y)$ for the y_i's. Let n_{ij} be the number of pairs of (x, y)'s that fall in $(Q_i(x), Q_j(y))$, $i, j = 1, \cdots, q$. It is easy to show that under the linkage equilibrium $E(n_{ij}) = n/q^2$ approximately. Geisser and Johnson (1995) proposed the Pearson χ^2 statistic

$$X^* = \sum_{i=1}^{q} \sum_{j=1}^{q} \frac{(n_{ij} - n/q^2)^2}{n/q^2} \tag{7}$$

that follows a chi-squared distribution with $(q-1)^2$ df asymptotically.

The test is applied to the Hong Kong Chinese data and the results are shown in Table 7. Only $q \leq 4$ is tried since the asymptotic distribution may not be good enough for a larger q with $n = 208$. All the eighteen p-values are greater than 0.05 and thus it indicates that there is no strong departure from linkage equilibrium.

Table 7. *p-values for Geisser-Johnson between loci test* X^*

Loci		q		
		2	3	4
D1,	D2	.166	.205	.581
D1,	D4	.166	.074	.173
D1,	D10	.405	.403	.967
D2,	D4	.782	.621	.740
D2,	D10	.579	.809	.921
D4,	D10	.579	.870	.550

5. Concluding Remarks

A within-group pairwise comparison is carried out for the Hong Kong Chinese DNA profiling database. It is found that the probability of a random match of four loci is extremely small. A battery of tests on independence of fragment lengths has been conducted. The results may indicate some violation of the Hardy-Weinberg law for loci D1S7 and D4S139. However, the cause for the violation of the latter locus is mainly due to the large number of (apparent) homozygotes. The tests also do not indicate any linkage disequilibrium (dependence structure) for pairs of loci. These findings are better than those for other ethnic groups such as in Weir (1992), and Risch and Devlin (1992). The Geisser-Johnson tests indicate strong dependence structure in the FBI's Black, Caucasian, and Hispanic databases, but not in our Chinese database. It seems that the use of the multiplication rule for evaluating the probability of a random match of DNA profiles may not be unjustified in our case. Of course, the actual evaluation of the probability in practice is much more conservative (NRC report, 1992, and Tsui and Wong, 1995).

Further analysis of the independence assumption on the bin levels may have to be done. Although the majority of the Hong Kong population is rather homogeneous (over 95% are Chinese), it may be of interest to study whether 'different' Chinese ethnic groups from Hong Kong, Taiwan and Singapore (or even from Mainland) have different patterns of DNA profiles. The issue of sub-populations could be studied.

Acknowledgements

The author is grateful to the referee's comments and is thankful to Wes Johnson, P. Tsui, D.M. Wong, Seymour Geisser and Jack Lee for assistance. He is indebted to Richard Cowan for introducing the subject. This work was partly supported by CRCG grant A/C 337/017/0003.

References

Berry, D.A. (1990). DNA fingerprinting: What does it prove. *Chance,* **3**, 15–25.

Berry, D.A., Evett, I.W. and Pinchin, R. (1992). Statistical inference in crime investigations using deoxyribonucleic acid profiling (with discussion). *Appl. Statist.*, **41**, 499–531.

Budowle, B., Giusti, A.M., Waye, J.S., Baechtel, F.S., Fourney, R.M., Adams, D.E., Presley, L.A., Deadman, H.A., and Monson, K.L. (1991). Fixed-bin analysis for statistical evaluation of continuous distributions of allelic data from VNTR loci, for use in forensic comparisons. *Am. J. Hum. Genet.*, **48**, 841–855.

Chow, S.T., Tan, W.F., Yap, K.H., and Ng, T.L. (1993). The development of DNA profiling database in an HAE III based RFLP system for Chinese, Malays and Indians in Singapore. *J. Foren. Sci.*, **38**, 874–884.

Chow, S.T., Yap, K.H., and Tan, W.F. (1994). Determination of match criterion for DNA profiling based on casework data. *Unpublished report*. Institute of Science and Forensic Medicine, Singapore.

Devlin, B., and Risch, N. (1993). Physical properties of VNTR data, and their impact on a test of allelic independence. *Am. J. Hum. Genet.*, **53**, 324–329.

Devlin, B., Risch, N., and Roeder, K. (1992). Forensic inference using DNA fingerprints. *J. Am. Statist. Asso.*, **87**, 337-350.

Fisher, R.A. (1958). *Statistical Methods for Research Workers*, Ed. 13. New York: Hafner.

Geisser, S. and Johnson, W. (1992). Testing Hardy-Weinberg equilibrium on allelic data from VNTR loci. *Am. J. Hum. Genet.*, **51**, 1084–1088.

Geisser, S. and Johnson, W. (1993). Testing independence of fragment lengths within VNTR loci. *Am. J. Hum. Genet.*, **53**, 1103–1106.

Geisser, S. and Johnson, W. (1995). Testing independence when the form of the bivariate distribution is unspecified. *Statist. in Med.*, **14**, 1621–1639.

Jeffreys, A.J., Wilson, V., and Thein, S.L. (1985). Individual-specific 'fingerprints' of human DNA. *Nature*, **316**, 76–79.

National Research Council (NRC) (1992). *DNA Technology in Forensic Science*. Washington D.C.: National Academy Press.

Risch, N.J. and Devlin, B. (1992). On the probability of matching DNA fingerprints. *Science*, **255**, 717–720.

Roeder, K. (1994). DNA fingerprinting: A review of the controversy (with discussion). *Statistical Science*, **9**, 222–278.

Tsui, P. and Wong, D.M. (1995). Allele frequencies of four VNTR loci in the Chinese population in Hong Kong. *Paper under revision*.

Waye, J.S. (1989). Forensic analysis of restriction fragment length polymorphism: Theoretical and practical considerations for design and implementation. *Proceedings for the International Symposium on Human Identification*.

Weir, B.S. (1992). Independence of VNTR alleles defined as fixed bins. *Genetics*, **130**, 873–887.

Weir, B.S. (1993). Independence tests for VNTR alleles defined as quantile bins. *Am. J. Hum. Genet.* **53**, 1107–1113.

Bayes Factors for Testing the Equality of Covariance Matrix Eigenvalues

Robert E. McCulloch and Peter. E. Rossi
University of Chicago
Graduate School of Business
Chicago, IL, 60637

Abstract

We use the methods of McCulloch and Rossi (1992) to construct a Bayes Factor for the hypothesis of the equality of eigenvalues of a covariance matrix. If the smallest s eigenvalues of a covariance matrix are equal, then we can think of the p dimensional sample data as arising from a reduced rank model. Tests of this sort are often used to identify the number of components used in a Principal Components analysis. The Bayes Factor approach requires specification of prior distributions for both the restricted and unrestricted covariance matrices. The problem of specifying a prior distribution on the restricted covariance matrix is solved via a projection method. We exploit the duality between sufficient statistics and parameters in exponential families to use the restricted MLE as a projection device. We illustrate this method with simulated and actual data. Our real data example addresses the question of the number of factors underlying stock return data.

Keywords: Monte Carlo, Factor Analysis, Nonlinear.

1 Introduction

A common goal of multivariate analysis is to look for some kind of lower dimensional structure for a random vector X in R^p. Both factor analysis and principal component approaches rely on the existence of a lower dimensional linear subspace such that some component of the distribution of X lies within that space. In factor analysis, we write $X = \mu + Lf + e$ where $f \sim N(0, I)$, L is $p \times (p - s)$, and the covariance matrix of e is diagonal. In this case, the distribution of X is massed closely to the subspace spanned by the columns of the matrix L. If we make the further assumption that the source of the error is iid so that the covariance matrix of e is $\sigma^2 I_p$ and denote the covariance matrix of X as Σ, then $\Sigma = LL' + \sigma^2 I_p$. This structure is equivalent to the hypothesis that the smallest s eigenvalues of Σ are equal (Anderson (1984), p. 485). Thus, the hypothesis of equality of a set of smallest eigenvalues is closely related to factor analysis. In addition, tests of equality of eigenvalues are often used as a pre-test method in principal components analysis.

The standard frequentist approach to testing equality of the smallest eigenvalues is the likelihood ratio test described in Anderson (1984), p. 475. It is well known that the standard chi-squared approximation to the null distribution of the LR statistic is inadequate. Schott (1988) extends the suggestions of Lawley by computing the asymptotic variance of a modified likelihood ratio statistic. The higher-order approximations advanced offer some improvement at the expense of considerable tedious calculations. Even if there were accurate finite sample approximations to the distribution of the LR statistics, power calculations would be impossible without resort to some sort of simulation approach. Furthermore, since no natural pivots are available for this problem, we would have to adopt some sort of parametric bootstrap approach to simulate the null or alternative distribution of the LR statistic.

Given these well-known weaknesses of frequentist approaches to testing hypotheses, it would be desirable to develop a Bayesian Odds Ratio approach for the eigenvalues problem. The odds ratio is the ratio of posterior probabilities of two competing hypotheses and is the product of the prior odds ratio and the Bayes Factor. The odds ratio has the conceptual advantages of providing a natural probability metric for assessing the strength of data and prior beliefs. Because of the difficulties of assessing priors under the nonlinear restrictions and the computation of the required integrals, Bayes factors have not been computed for this problem.

In this paper, we adapt the methods of McCulloch and Rossi (1992) to problem of testing the equality of eigenvalues of a covariance matrix. In section 2, we develop our method and in section 3 we discuss illustrative examples.

2 The Method

2.1 The Projection Approach to Computing Bayes Factors

We briefly review the approach of McCulloch and Rossi (1992) and then specialize the method to the eigenvalue problem. Consider the usual set-up where we have a sampling density, $f(x \mid \omega)$, where x is the observable quantity and ω is the parameter. The general problem consists of testing the hypothesis, $H_o : \omega \in \Omega_o$, versus the hypothesis, $H_A : \omega \in \Omega$, where Ω_o and Ω are two subsets of the parameter space. In this paper, we consider the case where Ω_o is a lower dimensional subspace of Ω defined by some restriction. The likelihood function is $L(\omega) = f(x \mid \omega)$ where x is considered fixed. Given that the prior probability that $\omega \in \Omega_o$ is q, the posterior odds ratio is the product of the priors odds ratio times the Bayes factor (denoted BF in the equation below).

$$\frac{\text{Prob}(\omega \in \Omega_o \mid x)}{\text{Prob}(\omega \in \Omega \mid x)} = \frac{q}{(1-q)}\text{BF} \quad \text{where} \quad \text{BF} = \frac{\int_{\Omega_o} L(\theta) dP_o(\theta)}{\int_{\Omega} L(\omega) dP(\omega)} \tag{1}$$

We use θ to denote an element of Ω_o, and ω to denote a general element of Ω. P_o is a prior on Ω_o and P is a prior on Ω. The Bayes factor compares the average of the likelihood over parameter values satisfying the restriction with the average of the likelihood over unrestricted parameter values. The Bayes factor may also be interpreted as the ratio of the predictive densities. Geisser (1980, 1994) emphasizes the importance of the predictive viewpoint in model selection.

Our goal is to specify the prior P_o so that prior information can be conveniently used and the integral in the numerator can be computed. We assume that we can formulate distributions on the unrestricted parameter space that adequately incorporate available prior information and from which Monte Carlo draws can easily be made. We specify the restricted prior P_o and then draw θ from P_o in two steps:

1. Specify a distribution Q, on the unrestricted parameter space Ω and draw $\omega \sim Q$.
2. Specify a projection function $h : \Omega \to \Omega_o$. Let $\theta = h(\omega)$. The prior P_o is the marginal distribution of $h(\omega)$ for $\omega \sim Q$.

Given draws $\omega_1, \omega_2, \ldots \omega_n \sim Q$, the Bayes factor is then approximated using the sample average of the likelihood evaluated as each draw from the restricted prior.

$$\text{Bayes Factor} \doteq \frac{\frac{1}{n}\sum_{i=1}^{n} L(h(\omega_i))}{\int_{\Omega} L(\omega) dP(\omega)} \tag{2}$$

The integration approach taken for the numerator (simply averaging the likelihood over draws from the restricted prior) can be improved via an importance sampling method while still using the basic projection mechanism (see McCulloch and Rossi (1992) for details).

The projection method used in this paper is based on the interchangeability of parameters and sufficient statistics in the exponential family. Recall that a density function in an exponential family may be written $f(x \mid \omega) = \exp(\omega' x - c(\omega)) f_0(x)$ where c and f_0 are given functions. If $\{x_1, \ldots x_n\}$ is an iid sample from the distribution corrresponding to f, then $f(x_1, x_2, \ldots x_n \mid \omega) = \exp(n(\bar{x} - c(\omega)) \prod f_0(x_i)$, where $\bar{x}$ is the sample mean. Clearly, $\bar{x}$ is a sufficient statistic. Let S be the sample space for the sufficient statistic. The likelihood function is then proportional to, $L(\omega \mid \bar{x}) = \exp(n(\bar{x} - c(\omega))$. There is a one to one correspondence between $\bar{x}$ in S and ω in Ω given by $\bar{x} = \nabla c(\omega)$, where $\nabla c(\omega)$

is the first derivative (the gradient vector) of the function c. The maximum likelihood estimate of ω is given by the solution of the equation $\bar{x} = \nabla c(\omega)$.

We can now define $h(\omega)$ to be the member θ of Ω_o , which maximizes $L(\theta \mid \nabla c(\omega))$:

$$L(h(\omega) \mid \nabla c(\omega)) = \max_{\theta \in \Omega_o} L(\theta \mid \nabla c(\omega)). \tag{3}$$

Thus, $h(\omega)$ is the parameter value in Ω_o which maximizes the likelihood based on a sample with unrestricted maximum likelihood estimate ω .

2.2 The Eigenvalue Problem

Assume we have a random sample of p-variate normal random variables, $x \sim N(0, \Sigma)$. Here we assume the mean $\mu = 0$; it would be straightforward to extend the method to the case of non-zero μ. As discussed in the introduction, we entertain the null hypothesis that the smallest s eigenvalues are equal. This hypothesis is related to the existence of a lower dimensional factor structure for the random vector X. It is convenient to work with Σ^{-1} rather than Σ. In the rest of this paper we let $G = \Sigma^{-1}$. The null hypothesis that the smallest s eigenvalues of Σ are equal then becomes the hypothesis that the largest s eigenvalues of G are equal. If the eigenvalues of G are sorted in descending order and denoted $\lambda_1 \geq \lambda_2 \geq \ldots \geq \lambda_p$, then we are testing the hypothesis:

$$H_o : \lambda_1 = \ldots = \lambda_s \geq \ldots \geq \lambda_p$$

versus the alternative

$$H_A : \lambda_1 \geq \ldots \geq \lambda_p.$$

This hypothesis is nonlinear in the elements of G and linear in the eigenvalues themselves. At first glance, one might be tempted to reparameterize the problem and specify the prior in terms of the eigenvalues and eigenvectors. We have $G = Q'\Lambda Q$ where Q is an orthogonal matrix of eigenvectors and the diagonal of Λ contains the eigenvalues. Specifying a prior distribution over the unique parameters in Q and Λ requires that a prior distribution be assessed over the ordered elements $\lambda_1, \lambda_2, \ldots \lambda_p$ as well as the orthogonal matrix Q. Choosing convenient form for this joint prior and incorporating prior information would be extremely difficult.

In order to employ the method outlined in section 2.1, we need to specify a prior, P, on the unrestricted parameter space, a distribution Q on the unrestricted parameter space, and a projection function, h, mapping unrestricted covariance matrices to restricted matrices. In this paper, we take P and Q to be the same natural conjugate Wishart distribution. It is a simple matter to make Wishart draws and perform the projection which is the key to the success of this method. The projection mapping we use exploits the duality between parameters and sufficient statistics for the exponential family discussed above. For multivariate normal samples from a population with zero mean, the sufficient statistic is $S = \frac{1}{n}\sum x_i x_i'$. The restricted MLE provides a way of mapping unrestricted draws down on H_o. To see this, recall the restricted MLE is defined as the solution to the constrained maximization problem:

$$\begin{aligned} &\max L(G \mid n, S) = (-np/2)log(2\pi) + (n/2)log(|G|) - (n/2)tr(SG) \\ &\text{s.t.}\lambda_1 = \ldots = \lambda_s \geq \lambda_{s+1} \geq \ldots \geq \lambda_p \end{aligned} \tag{4}$$

This nonlinear programming problem maps a positive definite S matrix to the space of positive definite G matrices satisfying H_o. Recognizing the duality between the sufficient statistic S and G^{-1}, we can project any draw G from the unrestricted prior simply by inserting it in the map defined above in place of S^{-1}. In order to impose the eigenvalue constraint, we solve for the restricted MLE with $S = G^{-1}$.

$$\begin{aligned} &\max L(Q'\Lambda Q \mid n, G^{-1}) \\ &\text{s.t. } \lambda_1 = \ldots = \lambda_s \geq \lambda_{s+1} \geq \ldots \geq \lambda_p \\ &q_i'q_j = \delta_{ij} \end{aligned} \tag{5}$$

The restricted draw G^* is constructed as $Q^{*\prime}\Lambda^* Q^*$ where Q^* and Λ^* are solutions to the restricted maximization problem. Even in the $Q'\Lambda Q$ parameterization, it is possible to compute the gradient of the likelihood function w.r.t the unique elements of Q and Λ allowing for the use of analytical gradients and Jacobian of the nonlinear constraints in the maximization routines. To assess the unrestricted Wishart prior, we find it useful to think of the prior as arising from the posterior from a previous sample of data. The Wishart distribution has a scale parameter, ν_0, and a location parameter V. A Wishart prior can arise as the posterior from a previous sample of data with a diffuse prior on G, $p(G) \propto |G|^{-(p+1)/2}$ (as in Geisser(1965)). In this case, G would have a Wishart distribution with parameters, (ν_0, V), with $\nu_0 = n - 1$ and $V = \frac{1}{n}S^{-1}$. n is the size of the data sample on which the prior is based and S is the MLE of Σ for that sample. We pick a location matrix, V, for our prior and simply scale the prior parameters to represent varying degrees of prior sample information.

3 Examples

3.1 Simulated Data

In our examples, we consider samples from a trivariate normal distribution, $x_i \sim N(0, \Sigma)$. The null hypothesis entertained is that the smallest two eigenvalues are equal. This hypothesis is consistent with a one-dimensional factor structure for Σ. As above, the eigenvalues of Σ^{-1} are denoted by λ. We are then testing:

$$\begin{aligned} H_o : \lambda_1 &= \lambda_2 \geq \lambda_3 \\ &\text{vs.} \\ H_A : \lambda_1 &\geq \lambda_2 \geq \lambda_3. \end{aligned} \tag{6}$$

In the first example, a sample of 400 observations from the null distribution, $\Sigma =$ Diag$(5, 5, 10)$, is used. A Wishart prior is used for the unrestricted distribution of G, $G \sim W(0, V)$.

$$p(G) \propto |G|^{\frac{\nu_0-3-1}{2}} \text{etr}(-\frac{1}{2}GV^{-1}) \tag{7}$$

It should be noted that the denominator integral in the Bayes factor can be performed analytically with the Wishart prior. Since many thousands of Wishart draws can be made at extremely low computational cost, we simply used the Monte Carlo estimate of the denominator integral. We shall always choose $q = .5$ so that the Bayes factor and the odds ratio are the same.

If the source of the prior information was a previous sample of data, then the prior scale parameter, ν_0, would be "n"-1 where n is the size of the previous sample. The location parameter of the Wishart distribution, V, would be $\frac{1}{n}S^{-1}$ where S is the MLE of Σ for the previous sample. In this example, a prior which is diffuse relative to the likelihood function is used; we employ priors with the scaling factor $n = 100$ and test data samples of size 400. To assess V in this example, we simulated observations under the null hypothesis and use the restricted MLE of Σ^{-1} scaled by 1/100. The use of the restricted MLE centers the prior over the null hypothesis.

Figure 1 presents the results for our sample of 400 observations from the null in three plots (from left to right): 1. A plot of the smallest two eigenvalues for each draw of Σ from both the restricted and unrestricted priors (the MLE of the lambda is marked with the solid triangle), 2. boxplots of the distributions of the log-likelihood evaluated at the restricted and unrestricted draws, and 3. boxplots of the restricted and unrestricted prior distributions of the normalized likelihood. The high diffusion of the prior is evident from the plots of eigenvalue draws. While there are draws whose first or second eigenvalues are close the MLE, this does not mean that the entire parameter vector is close the MLE. The point here is that it is difficult to explore this likelihood defined over a six dimensional space by looking at any two dimensions. The prior is used to explore the likelihood by focusing attention on a smaller portion of the parameter space. Under this relatively diffuse prior, the restricted draws mass in a region with higher likelihood values than the very diffuse unrestricted draws. Even though no single draw achieved a likelihood value more than 1/2 the maximum, the likelihood distribution for the restricted draws is located around a higher value than the unrestricted distribution. This can be seen most clearly in the center log likelihood distribution plot. The Bayes factor for this prior and sample are 6.3 with a Monte Carlo standard error of .11 reflecting the strong sample evidence in favor of the null. 15,000 draws were used to estimate the Bayes factor.

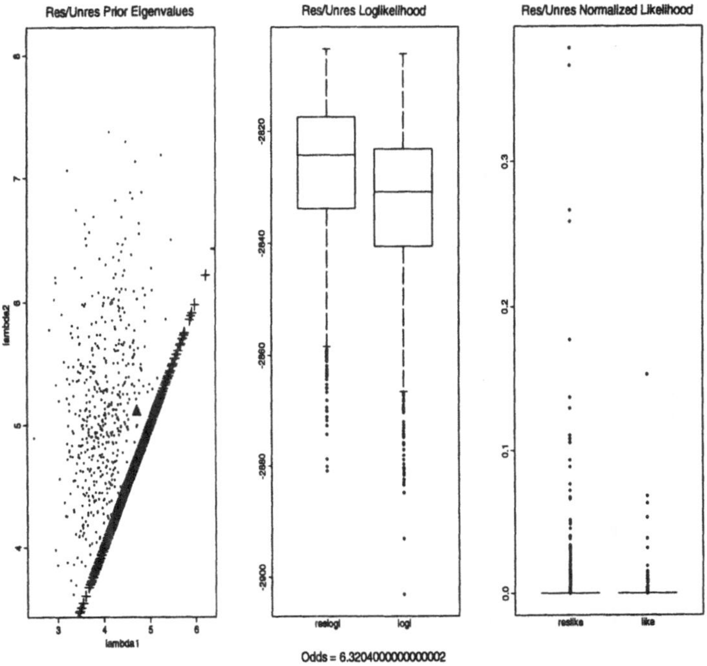

Figure 1: Simulated Eigenvalue Example, Data Generated Under the Null hypothesis.

We consider an alternative to the null hypothesis by generating 400 observations with $\Sigma = \text{diag}(5, 7, 10)$. We assess the prior location matrix of the Wishart by simulating a training sample of 400 observations from the alternative and using the restricted MLE of Σ^{-1} scaled to represent the information in a sample of size 100. Figure 2 displays the results. Draws from both the restricted and unrestricted priors are associated with relatively low values of the likelihood. However, the unrestricted prior put mass on values of the likelihood which are on average higher which is reflected in the small Bayes factor of .0005 with a standard error of .000035.

Notice that the median of the log-likelihood distribution is actually higher under the restricted prior than under the unrestricted prior even though the odds favor the alternative. The restricted prior puts mass on a smaller space and consequently puts less mass on parameter values where the likelihood is small. However only the alternative prior places mass near the maximum of the likelihood. Even though this mass is very small, it dominates the computation of the odds ratio as seen in the likelihood distributions. We believe this scenario is common in odds ratio calculations.

This simulated example shows that the method proposed here is computationally tractible and produces reasonable results. We now consider an application of the method

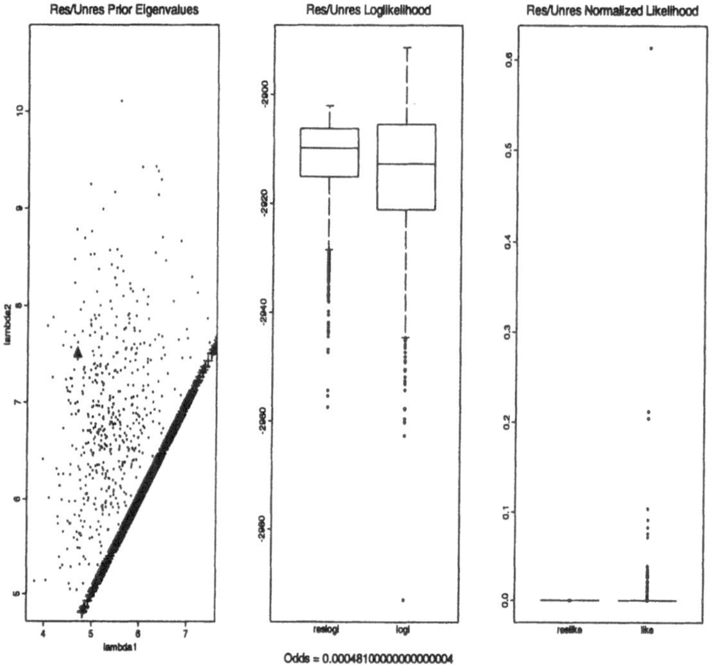

Figure 2: Simulated Eigenvalue Example, Data Generated Under the Alternative Hypothesis.

to the problem of the number of factors present in stock data.

3.2 Stock Return Data

In this section we apply our method to stock market data. A basic goal in financial research is to understand the covariance structure of returns on various assets. The arbitrage pricing theory (and to some extent the Capital Asset Pricing Theory) are both related to the existence of a factor structure for asset returns (Connor and Korajczyk [1986]). In this example, we consider monthly returns on three portfolios and test whether the two smallest eigenvectors are equal. This is equivalent to there being one factor underlying the returns on all three assets (and an iid error). For our three assets, we use returns on the first, fifth, and tenth decile portfolios constructed from the CRSP (center for reaserch in security prices) database at the University of Chicago Graduate School of Business. The decile portfolios are obtained by sorting the firms listed on the New York Stock exchange on the basis of total market capitilization (size) into ten groups. A portfolio is formed from each group by investing an equal amount in each stock within a group. For example, the first decile portfolio is the average return of the 10% smallest stocks. Our choice of three portfolios is designed to capture as much of the overall structure of the market as possible with a small number of portfolios. In addition, much research on the sources of variations in stock returns in the finance literature has used these size-based portfolios to test various multifactor models of returns (c.f. McCulloch and Rossi [1991]).

To compute an odds ratios for the decile portfolio data, we proceed in much the same way as our simulated examples in section 3.1 above. We assess an unrestricted prior in the Wishart form, $G \sim W(0, V)$ were $V = 1/(\nu_0 + 1)S^{-1}$. S is the sample covariance matrix for the testing sample. The testing sample consists of 400 observations (monthly returns from 9/29/61 to 12/30/94). Again, we use $q = .5$. In order to gage the sensitivity of our results to the level of uncertainty in the prior, we tried three different values for ν_0. Figure 3 plots draws from the restricted and unrestricted priors of λ_1 and λ_2 for $\nu_0 = 100$, $\nu_0 = 200$ and $\nu_0 = 300$. Clearly, as ν_0 increases the priors tighten up (all three plots are on the same scale). The corresponding values of the odds ratio are .41, .47, and .70. While .41 favors the alternative slightly and .70 favors the null slightly, basically none of the odds ratios suggest strong evidence for either hypothesis. The conlusion that we are uncertain about the hypotheses given the data, is reasonably insensitive to our level of prior uncertainty.

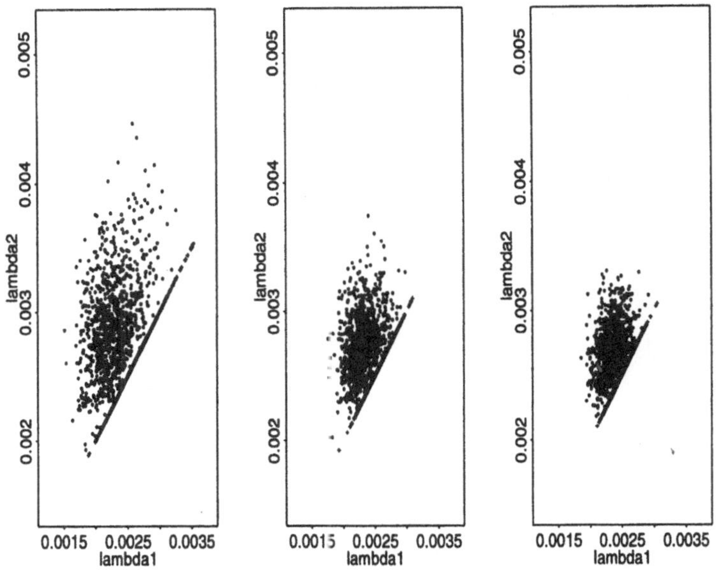

Figure 3: Stock Data Example. Priors with $\nu_0 = 100, 200, 300$ from left to right. Corresponding odds are .41, .47. and .70.

References

Anderson, T. W. (1984), *An Introduction to Multivariate Analysis.* New York: John Wiley and Sons.

Connor, G. and R. Korajczyk (1986), "Performance Measurement with the Arbitrage Pricing Theory: A New Framework for Analysis," *Journal of Financial Economics*, 15, 373-394.

Geisser, S. (1965), "Bayesian Estimation in Multivariate Analysis," *Annals of Mathematical Statistics*, 36, 150-159.

Geisser, S. (1980), "A Predictivistic Primer," in A. Zellner (ed), *Bayesian Analysis in Econometrics and Statistics*, Amsterdam: North-Holland.

Geisser, S. (1994), *Predictive Inference*, New York: Chapman and Hall.

McCulloch, R. and P. Rossi (1991), "A Bayesian Approach to Testing the APT," *Journal of Econometrics*, 49, 141-168.

McCulloch, R. and P. Rossi (1992), "Bayes Factors for Nonlinear Hypotheses and Likelihood Distributions," *Biometrika*, 79, 663-676.

Schott, J. R. (1988), "Testing the Equality of the Smallest Latent Roots of a Correlation Matrix," *Biometrika*, 75, 794-796.

Model Selection in Parapsychology: Psychokinesis, Precognition or Chance?

Jessica Utts
Division of Statistics
University of California, Davis
Davis, CA 95616

Abstract

This paper discusses three models that have been put forward to fit data from a particular type of experiment attempting to test psychic functioning. The experiments were actually designed to test what has been called "micro-psychokinesis," in which participants tried to influence the output of a true random number generator. One of the three models tested is that the participants were capable of such influence, one is that the results were consistent with chance, and one is that the participants were able to anticipate when favorable sequences were about to occur and to start collecting data at that time. Data from over one hundred experiments appear to support this last model.

1 Introduction

For a number of years parapsychologists, who conduct laboratory experiments of psychic functioning, have divided apparent psychic abilities into two basic types. The first type, extrasensory perception, involves abilities in which information is received through unknown means, either from another individual (called telepathy), from the future (called precognition) or from the environment (called clairvoyance). The second type of apparent functioning is called psychokinesis (PK) and is the ability of the mind to actually influence the environment.

In recent years a number of researchers have studied a purported ability called micro-psychokinesis or micro-PK, the "supposed ability to influence small events" (McCrone, 1994, p. 34). In the initial micro-PK studies, conducted by Dr. Helmut Schmidt in the 1970s, subjects attempted to influence the decay of radioactive particles through psychic means. More recent experiments have used random white noise from an electrical diode or other sophisticated microelectronics.

The results of the micro-PK experiments have demonstrated a small non-chance effect, and because the number of trials is huge, the combined results are highly statistically significant. Some researchers accept the database as evidence that psychokinesis is a real phenomenon. In this paper we examine the data and propose another explanation for the results, albeit one that may seem equally unlikely to skeptics of the possibility of psychic functioning.

2 Random Number Generator Experiments

Experiments designed to test micro-PK use various sources of randomness, but they all have the common feature that they produce a supposedly random sequence of Bernoulli random variables, with equal likelihood for zero and one. For instance, devices relying on radioactive decay are designed to measure whether or not a particle is emitted during a fixed, short time period. The random event generator registers a one if a particle is emitted and a zero otherwise, and the time period is chosen so that those two possibilities are equally likely.

In a typical experiment the random number generator (RNG) is producing a continuous string of bits without interference, and a subject is told to push a button to start the collection of data, and then to attempt to influence the device psychically. Sometimes the subject is told to try to "aim high" and produce more ones than zeros, and sometimes to "aim low" and produce more zeros than ones. After the button has been pressed to start the experiment, a predetermined number of bits are collected, constituting the data for that session.

Some experiments also have a "baseline" condition in which the subject is told to push the button to start data collection, but thereafter to try to ignore the device. Those sessions serve as a type of control, to determine whether or not the mere presence of a person pushing the button could somehow be influencing the electronics of the device to change the proportions of zeros and ones.

There are a number of methodological controls to prevent tampering, undetected patterns of nonrandomness in the generators themselves and other potential artifacts. According to Radin and Nelson (1989) who conducted a meta-analysis of micro-PK experiments:

> "Some of the RNGs described in the literature are technically sophisticated, the best devices employing electromagnetic shielding, environmental failsafe mechanisms triggered by deviant voltages, currents, or temperature, automatic computer-based data recording on magnetic media, redundant hard copy output, periodic randomness calibrations, and so on" (p. 1501).

McCrone (1994) in an article in *New Scientist*, gives details about some of the methodological controls used in one type of experiment conducted over many years by Engineering Professor Robert Jahn at Princeton University:

> "To guard against [tampering, patterns, etc.], Jahn has fitted the generator with various warning bells and temperature gauges. But more importantly, the sampling method does not rely on the raw output of the noise diode. Instead, the definition of what counts as a head or tail is alternated with each trial, so a positive signal will be counted as a head on one trial, but a tail the next. This added twist would cancel out any inherent bias that the equipment might develop during the course of an experiment. Switching the polarity criteria a thousand times a second would also seem to rule out any deliberate, or even inadvertent, tampering by subjects" (p. 36).

Depending on the experiment and the random device used, the sequence lengths collected following a single button press have ranged from a low of 16 to a high of many thousands.

Because the strings of bits are assumed to be i.i.d. Bernoulli random variables, the analysis proceeds using standard tests (exact or approximate, depending on the sample size) for binomial random variables. Sometimes those tests are used to determine whether or not the proportion of ones (or zeros) deviated significantly from one-half, whereas sometimes the "high aim" and "low aim" results are compared against each other and/or against the "baseline."

3 The Results of RNG Experiments

Radin and Nelson (1989) conducted a meta-analysis of all of the microelectronic random number generator experiments they could find reported. They located 152 reports, describing 832 studies conducted by 68 different investigators. Of those, 235 were "control" studies, used for comparison, and the remaining 597 were experimental studies.

For each study, the chance hypothesis dictates that $Y \sim Binomial(n, 0.5)$ where n is the length of the entire sequence of zeros and ones and Y is the number of ones. Using the normal approximation to the binomial a $z - score$ is computed as

$$z = \frac{Y - 0.5n}{0.5\sqrt{n}}.$$

Sequence lengths varied widely across studies, so for each study Radin and Nelson computed an estimated effect size measure

$$\text{estimated effect size} = \hat{\epsilon} = \frac{z}{\sqrt{n}}.$$

Notice that two experiments with the same proportion of ones but vastly different sequence lengths would have the same effect size but vastly different z's. The effect size measures the per bit departure from chance (0.5) in terms of the per bit standard deviation of $\sqrt{(.5)(.5)} = .5$. The sign of z and thus of the estimated effect size indicated whether or not the results were in the direction of intent, i.e., a positive score results from more ones in the high aim intent or more zeros in the low aim intent.

For the control studies Radin and Nelson found an average effect size of 0.1×10^{-4} with a standard error of 0.4×10^{-4}, indicating that the results were well within the range of what would be expected by chance (an effect size of zero). On the other hand, for the experimental studies, the average effect size was about 3.2×10^{-4} with a standard error of about 0.15×10^{-4}, far beyond what would be expected by chance.

The analysis by Radin and Nelson included an assessment of quality for each study, using a list of sixteen criteria. Contrary to what might be expected if the results were due to poor quality studies, there was no relationship evident between the quality of a study and its resulting effect size.

They also examined the question of whether or not the results could be due to the "file-drawer problem" in which meta-analysis results are inflated because only successful studies are published. Because parapsychologists recognized this potential problem long ago, their journals have a long-standing practice of not allowing the results to determine whether or not a paper is accepted for publication. Nonetheless, Radin and Nelson conducted numerical analyses showing that a potential "file-drawer" could not explain these results.

By far the largest collection of data from a single laboratory comes from the laboratory run by Professor Jahn at Princeton University. According to McCrone (1994), Jahn has collected over 14 million bits using over 100 subjects. The resulting effect size is about 0.002, meaning that out of 1,000 bits 501 will be the intended outcome and 499 will not, instead of the 500 to 500 split expected by chance. While this is a very small effect, the magnitude of the sample means that the corresponding *p-value* is less than 1/5,000. Also notice that this effect size is almost ten times larger than the average effect size reported in the meta-analysis by Radin and Nelson.

An experiment reported by Hungarian physicist Zoltan Vassy (1990) is worth noting because it resulted in an effect size about ten times larger than that produced by the Princeton Lab, but it did so using a pseudo-random number generator rather than a true random source. Seven subjects participated for a total of 100 runs of 36 bits each, or a total of 3600 bits. This experiment also differed in an aspect that will be seen to be important in considering alternative explanations for all of the RNG data. Rather than collect all 36 bits for each run by pressing the button just once, each bit required a separate press of the button. There were 1,846 bits in the correct direction, 46 more than expected by chance, resulting in a z-score of 1.53 and an effect size of 0.026.

4 Problems with the Psychokinesis Hypothesis

There is little doubt that the results of these experiments are exhibiting something other than simple chance behavior. The researchers who have conducted the majority of these experiments believe that their results demonstrate "micro-psychokinesis," in which the resulting sequence is still Bernoulli, but in which physical interferences increase the probability of producing a one (or a zero for low aim). But there are some inconsistencies with that theory, as evidenced by the data. First, the effect sizes appear to remain relatively constant across different types of random generators using different known forces. To alter the noise diode would require tampering with an electromagnetic force, while to alter radioactive decay would require tampering with the weak nuclear force.

Second, if subjects could actually influence these devices at a constant rate and achieve an additional proportion of ones (or zeros) then the corresponding z-score should grow as a function of the sequence length. Empirically, that did not seem to be the case. Rather, it seemed that it was not possible to obtain a z-score beyond a certain level despite the length of the sequence of bits collected.

Perhaps the most obvious problem is that results similar to those obtained with true random sources were also obtained with pseudo-random number generators. Since they follow a known algorithm, the only way to change their output would be to somehow influence the internal workings of the computer. In the experiment reported above by Vassy and in another reported by Radin and May (1986) the results of the experimental sessions were compared with the known output of the particular pseudo-random number generator and were found to match. In other words, the subjects did not actually physically influence the generator, but perhaps managed to push the button on the computer at a time when an unusual string of bits was coming.

5 An Alternative Explanation: Decision Augmentation Theory

All of these observations led physicist Edwin May and his colleagues to propose a different mechanism to explain the results of these studies. Originally called "Intuitive Data Sorting" (May, Radin, Hubbard, Humphrey and Utts, 1985), and more recently named "Decision Augmentation Theory" (DAT) the idea is that subjects are not able to *influence* the random number generator, but they are able to *predict* when a desirable sequence is about to occur and press the button at just the right time. (See Model 3 in Section 6 for the technical details.)

This mechanism, traditionally called precognition, would explain why results using pseudo-random number generators are similar to those using true random number generators. It would also explain why it would be almost impossible to achieve a very large z-score as the result of a single press of the button. Despite the number of bits being collected, it would rarely be the case that a sequence would appear, even if we could see the future perfectly, that resulted in a z-score beyond about 3. Whether 100 or 100,000 bits were being collected, we should expect to see approximately constant z-scores rather than approximately constant success rates.

6 Three Possible Models

Consider an experiment in which one button press results in n Bernoulli random variables, X_i, $i = 1, ..., n$. The random variable of interest is $Y = \sum X_i$. Assuming n is large, we can construct a test of whether or not the Bernoulli probability is 0.5 using the normal approximation to the binomial distribution and computing the usual z-score.

There is a clear distinction between what is predicted to happen if micro-PK is at work and what is expected to happen if DAT is the explanation. In the first case, we would expect the observed success rate, i.e., the proportion of ones (or zeros, depending on the aim) to remain approximately constant across vastly different sequence lengths, so the z-score should increase as n increases, in fact directly in relation to $\sqrt{n}$. In the second case (DAT) we would expect the final z-score to remain approximately constant across sequence lengths, and thus the success rate would *decline* as n increases.

It is this last observation that allows us to formulate a model to compare possible explanations for the data. We do so by observing the predicted relationship between z^2 and n for the two models. We also include chance, i.e., the absence of any psychic functioning, as a third possible model.

Model 1: Chance Under this model, $X_i \sim \mathit{Bernoulli}(0.5)$. Let μ_0 and σ_0 be the mean and standard deviation of X_i under this model, so $E(X_i) = \mu_0 = 0.5$ and $\sigma_0 = \sqrt{(0.5)(0.5)} = 0.5$. Then $Y \sim \mathit{Binomial}(n, \mu_0)$. Recall that $Z = (Y - n\mu_0)/\sqrt{n}\sigma_0$, and note that $E(Z^2) = 1$ regardless of n. It is approximately true that $Z \sim N(0,1)$.

Model 2: Psychokinesis
Under this model, $X_i \sim \mathit{Bernoulli}(p)$ with $p \neq 0.5$. Using the standard definition of effect size as the per unit effect in number of standard deviations, define an effect size per bit

to be

$$\epsilon = (p - \mu_0)/\sigma_0 = (p - 0.5)/0.5.$$

Then $E(X_i) = \mu_0 + \epsilon\sigma_0$ and $Y \sim Binomial(n, \mu_0 + \epsilon\sigma_0)$. It is approximately true that $Z \sim N(\sqrt{n}\epsilon, (1 - \epsilon^2))$ and $E(Z^2) = 1 - \epsilon^2 + \epsilon^2 n \approx 1 + \epsilon^2 n$, assuming ϵ^2 is small.

Model 3: Precognition or Decision Augmentation Theory (DAT)
Under this model we do not specify the distribution of the individual X_i because the idea is that this ability works to identify an entire sequence of bits that are about to be produced by the generator. It is as if the mind of the subject asks the question "If I push the button *now* will I get a good z-score?" Therefore, we specify the distribution of Z directly and simply to have a constant mean $\mu 0$ and a constant variance $\sigma^2 \geq 1$. In that case, we find $E(Z^2) = \mu^2 + \sigma^2$. Notice that we do not assume anything about the distribution or independence of the individual X_i for this model.

The important feature in comparing these models is the relationship between $E(Z^2)$ and n. Notice that if we were to plot the relationship, we would find a flat line for Models 1 (chance) and 3 (precognition) but a line with a slope of n for Model 2 (psychokinesis). The distinction between models 1 and 3 is in the intercept of the line, which should be 1.0 for Model 1 and greater than 1.0 for Model 3.

To summarize, the chance model predicts an intercept of 1.0 and a slope of 0. An intercept above 1.0 supports Model 3, precognition. A slope greater than 0 supports Model 2, psychokinesis. If we were to find an intercept significantly less than 1.0 but slope close to 0, that would support a version of Model 3 that allowed for smaller variance than 1.0, but we do not consider that to be a likely possibility.

Formulating the 3 models in this way led May, Utts, Spottiswoode and James(1995) to examine the data from all experiments in which enough information was available to determine the values of z and n corresponding to single presses of a button. Recall that Model 3 requires that a single decision be used to collect the data used to compute z. If more than one decision is allowed, then *each* decision could be used to choose an optimal z and the model would not hold.

7 Testing the Three Models

We found 128 studies in which we were able to ascertain the z-score and the sequence length n. In some cases we were able to ascertain only an average z-score for a series of N runs of length n, so that the variance of z^2 was reduced by a factor of $1/N$. Accordingly, we used weighted least squares to account for differing variances, and fit a regression of z^2 versus n.

We found the data to be in support of Model 3, precognition or data augmentation. The slope was not significantly different from zero, with a value of 1.73×10^{-6}. The corresponding t-test for zero slope resulted in $t = 0.543$ with $df = 126$ so $p = 0.295$. In contrast, we found that the intercept was significantly above the value of 1.0 that corresponds to chance. The intercept was 1.036 and the t-test comparing it with the chance value of 1.0 resulted in $t = 9.10, df = 126, p = 4.8 \times 10^{-20}$.

We were somewhat concerned about artifacts, like outliers, influencing the results so we ran a few additional analyses. First, we used only the lower half of the sequence lengths since there were some outlying values of n. (In the original data the median n was 64 but the mean was 566; for the lower half only, the median was 35 and the mean was 34.) Using only the smaller values of n gave basically the same results. The slope was 3.4×10^{-6} resulting in a very small and nonsignificant $t = -0.01$. The intercept was 1.022 with the comparison with the chance value of 1.0 resulting in $t = 3.63, p = 2.9 \times 10^{-4}$.

Finally, there were three outliers in z^2, all above 2.0. When we removed those and reran the analysis we found the intercept to be 1.07, significantly above 1.0 ($t = 4.40$) and the slope to be a nonsignificant 1.071×10^{-5}.

Since the normality assumption required for regression clearly does not hold for the z^2 we were concerned that the skewness in those values might be responsible for the intercept being greater than one. Consequently, we ran a simulation by generating z^2 under the chance hypothesis, using the values of n in the actual data set. We found results consistent with chance; the slope was not significantly different than zero and the intercept was not significantly different from one.

8 Conclusions

We conclude that the data generated by the experiments designed to measure micro-psychokinesis actually support a different hypothesis, namely that the subjects are somehow able to determine an opportune time to start collecting data. This theory is also consistent with other research in parapsychology. Honorton and Ferrari (1989) conducted a meta-analysis of precognition experiments involving card-guessing, in which they reviewed 309 studies involving a total of nearly 50,000 subjects and two million individual guesses. The results showed a small but significant effect size of 0.02, with a standard error of 0.002.

More recently experimenters have conducted "free-response" experiments involving precognition, in which subjects are asked to describe a picture. The correct answer is not selected until *after* the description has been made. Preliminary results show that these experiments are successful as well. Finally, in virtually every experiment conducted in parapsychology the subject is ultimately shown the correct answer. Thus, it could be that precognition is the unifying explanation for all of these experiments, despite the fact that experimenters think they are testing other abilities, such as telepathy (mind-to-mind communication). This theory is consistent with the current state of knowledge in physics, in which time is still a mysterious dimension that is poorly understood.

References

Honorton, C. and D.C. Ferrari (1989). "Future Telling:" A Meta-analysis of forced-choice precognition experiments, 1935-1987, *Journal of Parapsychology*, **53**, 281-308.

May, E.C., D.I. Radin, G.S. Hubbard, B.S. Humphrey and J. Utts (1985). Psi experiments with random number generators: an informational model, *Proceedings of the Parapsychological Association 28th Annual Convention*, 237.266.

May, E.C., J.M. Utts, S.J. Spottiswoode and C.L. James (1995). Applications of Decision Augmentation Theory, *Journal of Parapsychology*, to appear.

McCrone, John (1994). Psychic powers: what are the odds?, *New Scientist*, 26 November 1994, 34-38.

Radin, D.I. and E.C. May (1986). Testing the intuitive data sorting model with pseudo random number generators: A proposed method, *Proceedings of the Parapsychological Association 29th Annual Convention*, 539-554.

Radin, D.I. and R.D. Nelson (1989). Evidence for consciousness-related anomalies in random physical systems, *Foundations of Physics*, **19** 1499-1514.

Vassy, Zoltan (1990). Experimental study of precognitive timing: Indications of a radically noncausal operation, *Journal of Parapsychology*, **54**, 299-320.

Section VI

Modelling and Prediction in Finance

An Empirical Bayesian Approach to Cointegrating Rank Selection and Test of the Present Value Model for Stock Prices

John C. Chao
Department of Economics, University of Maryland
College Park, Maryland 20742

Peter C. B. Phillips
Cowles Foundation for Research in Economics, Yale University
P.O. Box 208281, New Haven, Connecticut 06520–8281

Abstract

This paper provides an empirical Bayesian approach to the problem of jointly estimating the lag order and the cointegrating rank of a partially nonstationary reduced rank regression. The method employed is a variant of the Posterior Information Criterion (PIC) of Phillips and Ploberger (1994, 1995) and is similar to the asymptotic predictive odds version of the PIC criterion given in Phillips (1994). Here, we use a proper (Gaussian) prior whose hyperparameters are estimated from an initial subsample of the data. The form of the prior is suggested by the asymptotic posterior distribution of the parameters of the model, and, hence, the criterion can be interpreted as an approximate predictive odds ratio in the case where the sample size is large. Applying this procedure to the extended Campbell–Shiller data set for stock prices and dividends, we find the present value model for stock prices to be inconsistent with the data.

1 Introduction

Order estimation of the lag dimension of a model is an unavoidable task in applied econometric work involving vector autoregressions (VAR's). In VAR models that are partially nonstationary in the sense of Ahn and Reinsel (1990), it is often desirable to estimate not only the lag order of the model but also the number of linearly independent cointegrating vectors, or the cointegrating rank. In conventional econometric practice, however, the estimation of the lag length of a VAR model is often done with statistical methodologies that are very different from those employed for the determination of the cointegrating rank. While information criteria, such as AIC and BIC, are often favored by researchers for lag estimation (but see Pötscher, 1983, for an alternative method based on LM tests), cointegrating rank determination is most often performed using a sequence of classical tests that come within the Neyman–Pearson tradition (see Johansen, 1992).

Recently, Phillips (1992), Chao and Phillips (1994), and Phillips (1994) have argued that cointegrating rank determination is most naturally a problem of order selection. Applying the Posterior Information Criterion (PIC) of Phillips and Ploberger (1994, 1995), they developed statistical procedures which allow for the joint estimation of the lag order and the cointegration rank in a VAR system. Such procedures are particularly appealing in the light of simulation evidence, presented in Toda and Phillips (1994) and Chao (1995), that shows the performance of classical tests of cointegration, such as those put forth by Johansen (1988, 1992), to be sensitive to autoregressive lag specification. In the present paper, we develop a variant of PIC, which is similar in spirit to the predictive odds ratios of Atkinson (1978), O'Hagan (1991), and Geweke (1994) and to the predictive form of PIC given in Phillips (1994). We derive our criterion using a proper Gaussian prior whose hyperparameters are estimated using an initial subsample of the data. As the form of the prior is suggested by the asymptotic posterior distribution of the parameters of the model, the criterion can be interpreted as an approximate predictive odds ratio in the case where the sample size is large.

The second objective of this paper is to illustrate the use of this model selection criterion through an empirical application that tests the rational expectations present value model for stock prices. An important recent application of the technology of nonstationary time series analysis to empirical economic research has been the work of Campbell and Shiller (1987). In their paper, Campbell and Shiller used classical tests of unit roots and cointegration to address issues, raised by Kleidon (1986) and Marsh and Merton (1986), pertaining to possible nonstationarity in the price and dividend processes. They showed that if both stock prices and dividends are I(1) processes (i.e., stationary in first-order differences), then one implication of the present value model is that these variables are cointegrated. Following the suggestion of Engle and Granger (1987), they use a preliminary estimate of the cointegrating vector to transform their bivariate VAR into a stationary system, where conventional statistical procedures can be applied to test the restrictions implied by the present value model.

An important statistical issue which arises in testing the present value model within the VAR framework of Campbell and Shiller (1987) is the specification of the lag order of the system. Since economic theory offers little guidance in this regard, Campbell and Shiller (1987) used the Akaike Information Criterion (AIC) to pre-select the lag length. However, it is well-known from the results of Shibata (1976) and Sawa (1978) that AIC has a tendency to overestimate the lag order. Moreover, Toda (1991) found classical tests of the present value models to be sensitive to variations in lag selection.

Given the need for pre-selection of the lag order and given the sensitivity of conventional testing procedures to lag specification, the use of our model selection criterion has certain advantages. First, it allows us to jointly select the lag and cointegrating rank order of a VAR and to test the implications of the present value model simultaneously in one coherent framework. Such a joint selection procedure is preferred over a sequential procedure which estimates cointegrating rank conditional on some preliminary lag selection because with sequential procedures, there is always some pre-test bias and associated implications for inference. Moreover, the joint-selection procedure makes comparison across the full array of models and gives consistent order estimates of cointegrating rank and lag order, as shown in Chao and Phillips (1994). A further advantage of our approach is that bar charts and histograms of posterior probabilities for models of different dimensions can be readily constructed so that one may assess the robustness of the inferences with respect to lag length and cointegrating rank.

The paper proceeds as follows. In Section 2, we introduce the time series model to be studied and describe our model selection procedure PIC. Section 3 gives a discussion of the rational expectations present value model for the stock market and the restrictions that it imposes on a VAR in error-correction form. Section 4 presents our empirical results. Finally, we offer some concluding thoughts in the fifth and final section.

2 Order Selection and Hypothesis Testing in VAR Models via Posterior Odds

2.1 The Partially Nonstationary VAR Model

The model framework we consider in this paper is similar to that of Chao and Phillips (1994). The setup is the m-dimensional vector autoregressive model of $(p+1)$-order:

$$y_t = \mu + \sum_{i=1}^{p+1} A_i y_{t-1} + \varepsilon_t \ . \tag{1}$$

It is well-known that equation (1) can be rewritten as an error-correction model (ECM):

$$\Delta y_t = \mu + \sum_{i=1}^{p} A_i^* \Delta y_{t-i} + A_* y_{t-1} + \varepsilon_t \ , \tag{2}$$

where $A_i^* = -\sum_{j=i+1}^{p+1} A_j$ and $A_* = \sum_{i=1}^{p+1} A_i - I_m$. We further assume that the following conditions are applicable to our model:

(i) $\det\left[I_m - \sum_{i=1}^{p+1} A_i L^i\right] = 0$ implies that either $L = 1$ or $|L| > 1$.

(ii) $A_* = \alpha\beta'$, where α and β are $m \times r$ matrices of full column rank r, $0 \leq r \leq m$. (If $r = 0$, we take $\alpha = \beta = 0$, and if $r = m$, we take $\alpha = A_*$ and $\beta = I_m$.)

(iii) $\alpha'_\perp \left(\sum_{i=1}^{p} A_i^* - I_m\right) \beta_\perp$ is nonsingular for $0 \leq r < m$, where $\alpha_\perp$ and $\beta_\perp$ are $m \times (m-r)$ matrices of full column rank $m-r$ such that $\alpha'_\perp \alpha = 0 = \beta'_\perp \beta$. (If $r = 0$, we take $\alpha_\perp = \beta_\perp = I_m$.)

(iv) $\varepsilon \equiv$ i.i.d. $N(0, \Omega)$.

Under these assumptions, $\{y_t\}$ is I(1), but $\beta' y_t$ is I(0) with r linearly independent cointegrating vectors. Thus, in the nomenclature of the literature on cointegration, we say that the multivariate system defined by (2) has a cointegrating rank of r. Note that without further restrictions, α and β are unidentified. To achieve identification, we follow Ahn and Reinsel (1990) in selecting a normalized parameterization in which

$\beta' = [I_r, B]$. We shall refer to equation (2) as a reduced rank regression of order (p, r), or RRR(p, r) for short.

2.2 The Posterior Information Criterion in Asymptotic Predictive Form

Our objective in this paper is to jointly estimate the cointegrating rank r and the lag order p of the model (2). As in Chao and Phillips (1994), the criterion which we use for model selection is a posterior odds ratio. However, unlike the earlier paper where a uniform prior was employed, we use here a Gaussian prior whose hyperparameters are estimated from an initial subsample of the data. To be explicit, let us begin by defining a class of competing models $M_{p,r}(p = 0, 1, ..., \overline{p};\ r = 0, 1, ..., \overline{r})$, given in terms of the parameterized measure $\mathbb{P}_T^{\theta_{p,r}}$ with $\theta_{p,r}$ belonging to some parameter space $\Theta_{p,r}$. Here, $M_{p,r}$ denotes the model with cointegrating rank r and order of lagged differences p. Given the data $\mathbf{y} = \{y_t\}_1^T$, we let $L_T(\theta_{p,r}) = d\mathbb{P}_T^{\theta_{p,r}}/d\nu$ denote the likelihood function of the model $M_{p,r}$ with respect to the Lebesgue measure ν. Suppose we take the prior model probability of $M_{p,r}$ and the prior density of $\theta_{p,r}$ to be $\pi_{p,r}$ and $g(\theta_{p,r}|M_{p,r})$ respectively; then, the posterior odds for the competing models M_{p_0,r_0} and M_{p_1,r_1} is defined as

$$\frac{\Pi_T(M_{p_0,r_0}|\mathbf{y})}{\Pi_T(M_{p_1,r_1}|\mathbf{y})} = \frac{\pi_{p_0,r_0}\overline{L}_T(M_{p_0,r_0})}{\pi_{p_1,r_1}\overline{L}_T(M_{p_1,r_1})}\ , \tag{3}$$

where

$$\overline{L}_T(M_{p,r}) = \int_{\Theta_{p,r}} L_T(\theta_{p,r})g(\theta_{p,r}|M_{p,r})d\theta_{p,r} \tag{4}$$

is the Bayesian data density or the marginalized likelihood. Under the additional assumption of a symmetric loss function, which penalizes Type I and Type II errors equally, we obtain the decision rule:

$$\frac{\Pi_T(M_{p_0,r_0}|\mathbf{y})}{\Pi_T(M_{p_1,r_1}|\mathbf{y})} > 1\ , \quad \text{then decide in favor of } \ M_{p_0,r_0}\ . \tag{5}$$

To implement this posterior odds test for the model in Section 2.1, we must first specify the prior density $g(\theta_{p,r}|M_{p,r})$. Following an approach used in earlier work (e.g., Atkinson (1978), O'Hagan (1991), Geweke (1994), and Phillips (1994)), we divide the sample into two sample periods $[1,\ T_0]$ and $[T_0 + 1,\ T]$, where $T_0 = [\rho T]$ (the integer part of ρT) for some constant $\rho \in (0, 1)$, and use data from the initial sample period $[1,\ T_0]$ to assist in the construction of our prior. Let $\theta_{p,r} = (\phi',\ \omega')'$ $= (\text{vec}(B)',\ \text{vec}(\alpha)',\ \text{vec}(A^*)',\ \omega')'$, where ω is the vector of nonredundant elements of Ω and $A^* = (\mu,\ A_1^*,\ A_2^*,\ ...,\ A_p^*)$, and let the prior densities of the model $M_{p,r}$ be

$$g(\Omega|M_{p,r}) = |\Omega|^{-\frac{1}{2}(m+1)} \tag{6}$$

$$g(\phi|M_{p,r}) = (2\pi)^{-\frac{1}{2}(2mr-r^2+m(mp+1))}|V|^{\frac{1}{2}}\exp\left\{-\tfrac{1}{2}(\phi-\overline{\phi})'V(\phi-\overline{\phi})\right\}\ . \tag{7}$$

The hyperparameters $\overline{\phi}$ and V can be estimated using data from the period $[1,\ T_0]$. More specifically, we estimate ϕ by its maximum likelihood estimate $\widehat{\phi}_{T_0}$

$= (\mathrm{vec}(\widehat{B}_{T_0})', \mathrm{vec}(\widehat{\alpha}_{T_0})', \mathrm{vec}(\widehat{A}^*_{T_0})')'$ and V by the asymptotic formula:

$$\widehat{V}_{T_0} = \begin{pmatrix} (\widehat{\alpha}'_{T_0}\widehat{\Omega}^{-1}_{T_0}\widehat{\alpha}_{T_0} \otimes F'Y'_{-1,T_0}Y_{-1,T_0}F) & 0 & 0 \\ 0 & (\widehat{\Omega}^{-1}_{T_0} \otimes \widehat{\beta}'_{T_0}Y'_{-1,T_0}Y_{-1,T_0}\widehat{\beta}_{T_0}) & (\widehat{\Omega}^{-1}_{T_0} \otimes \widehat{\beta}'_{T_0}Y'_{-1,T_0}W_{T_0}) \\ 0 & (\widehat{\Omega}^{-1}_{T_0} \otimes W'_{T_0}Y_{-1,T_0}\widehat{\beta}_{T_0}) & (\widehat{\Omega}^{-1}_{T_0} \otimes W'_{T_0}W_{T_0}) \end{pmatrix} \tag{8}$$

where $\widehat{\beta}_{T_0} = [I_r, \widehat{B}_{T_0}]'$ and $\widehat{\Omega}_{T_0} = (1/(T_0+m+1))\sum_{t=1}^{T_0}(\Delta y_t - \widehat{A}^*_{T_0}W_t - \widehat{\alpha}_{T_0}\widehat{\beta}'_{T_0}y_{t-1})'$ $(\Delta y_t - \widehat{A}^*_{T_0}W_t - \widehat{\alpha}_{T_0}\widehat{\beta}'_{T_0}y_{t-1})$. Here, we have let $Y_{-1,T_0} = (y_0, \ldots, y_{T_0-1})'$ and $W_{T_0} = (W_1, \ldots, W_{T_0})'$, with $W_t = (1, \Delta y'_{t-1}, \ldots, \Delta y'_{t-p})'$, while $F' = [0, I_{m-r}]$ is an $(m-r) \times m$ matrix. Note that $\widehat{V}_{T_0}$ is an estimator of the precision matrix of the asymptotic posterior distribution of ϕ, and, hence, for large T_0, our prior density,

$$\begin{aligned} g(\phi|M_{p,r}, \widehat{\phi}_{T_0}, \widehat{V}_{T_0}) &= (2\pi)^{-\frac{1}{2}(2mr-r^2+m(mp+1))}|\widehat{V}_{T_0}|^{\frac{1}{2}} \\ &\quad \exp\left\{-\tfrac{1}{2}(\phi-\widehat{\phi}_{T_0})'\widehat{V}_{T_0}(\phi-\widehat{\phi}_{T_0})\right\}, \end{aligned} \tag{9}$$

can be interpreted as the approximate (large sample) posterior distribution for the initial sample period $[1, T_0]$ under a uniform prior.

Given our assumption (iv), the likelihood function for the sample period $[T_0+1, T]$ can be written as

$$\begin{aligned} &L_{T_1}(\theta_{p,r}) \\ &= (2\pi)^{\frac{-mT_1}{2}}|\Omega|^{\frac{-T_1}{2}}\exp\left\{-\tfrac{1}{2}\sum_{t=T_0+1}^{T}(\Delta y_t - A^*W_t - \alpha\beta'y_{t-1})'\Omega^{-1}(\Delta y_t - A^*W_t - \alpha\beta'y_{t-1})\right\}, \end{aligned} \tag{10}$$

where $T_1 = T - T_0 = T - [\rho T]$ and α and β are, as before, $m \times r$ matrices. Combining the likelihood function (10) with the prior densities (6) and (9) and integrating over the parameters $\theta_{p,r} = (\phi, \omega)$ using the Laplace approximation method as in Phillips and Ploberger (1994) and Chao and Phillips (1994), we obtain, up to a multiplicative constant (not involving p and r), the (approximate) Bayesian data density or marginalized likelihood:[1]

$$\begin{aligned} \overline{L}_{T_1}(M_{p,r}) &\sim |\widehat{\Omega}_{T_1}|^{\frac{-T_1}{2}}\left(|\widehat{V}_{T_0} + \widehat{V}_{T_1}|/|\widehat{V}_{T_0}|\right)^{-\frac{1}{2}}\exp\left\{-\tfrac{1}{2}(\widehat{\phi}_{T_1} - \widehat{\phi}_{T_0})'\widehat{V}_{T_0}(\widehat{\phi}_{T_1} - \widehat{\phi}_{T_0})\right\} \\ &= \widehat{\overline{L}}_{T_1}(M_{p,r}) \text{ (say)}, \end{aligned} \tag{11}$$

where

$$\widehat{\Omega}_{T_1} = \frac{1}{T_1+m+1}\sum_{t=T_0+1}^{T}\left(\Delta y_t - \widehat{A}^*_{T_1}W_t - \widehat{\alpha}_{T_1}\widehat{\beta}'_{T_1}y_{t-1}\right)'(\Delta y_t - \widehat{A}^*_{T_1}W_t - \widehat{\alpha}_{T_1}\widehat{\beta}'_{T_1}y_{t-1}), \tag{12}$$

and

$$\widehat{V}_{T_1} = \begin{pmatrix} (\widehat{\alpha}'_{T_1}\widehat{\Omega}^{-1}_{T_1}\widehat{\alpha}_{T_1} \otimes F'Y'_{-1,T_1}Y_{-1,T_1}F) & 0 & 0 \\ 0 & (\widehat{\Omega}^{-1}_{T_1} \otimes \widehat{\beta}'_{T_1}Y'_{-1,T_1}Y_{-1,T_1}\widehat{\beta}_{T_1}) & (\widehat{\Omega}^{-1}_{T_1} \otimes \widehat{\beta}'_{T_1}Y'_{-1,T_1}W_{T_1}) \\ 0 & (\widehat{\Omega}^{-1}_{T_1} \otimes W'_{T_1}Y_{-1,T_1}\widehat{\beta}_{T_1}) & (\widehat{\Omega}^{-1}_{T_1} \otimes W'_{T_1}W_{T_1}) \end{pmatrix}. \tag{13}$$

[1] See the cited references for details of the Laplace derivation.

Here, $Y_{-1,T_1} = [y_{T_0}, ..., y_{T-1}]'$, $W_{T_1} = [W_{T_0+1}, ..., W_T]'$, and the matrices $\widehat{A}^*_{T_1}$, $\widehat{\alpha}_{T_1}$, and $\widehat{\beta}_{T_1}$ are the posterior modes of A^*, α, and β, where the posterior distribution has been updated by the likelihood function (10). A further approximation is possible by taking the transformation

$$-\frac{2}{T_1}\ln\left(\widehat{\widehat{L}}_{T_1}(M_{p,r})\right) \sim \ln|\widehat{\Omega}_{T_1}| + \frac{1}{T_1}\ln\left(|\widehat{V}_{T_0} + \widehat{V}_{T_1}|/|\widehat{V}_{T_0}|\right) , \tag{14}$$

where the term $(1/T_1)(\widehat{\phi}_{T_1} - \widehat{\phi}_{T_0})'\widehat{V}_{T_0}((\widehat{\phi}_{T_1} - \widehat{\phi}_{T_0}) = O_p(T_1^{-1})$ is neglected. If, in addition, we set the prior model probabilities equal across all models so that $\pi_{p,r} = 1/[(\overline{p}+1)(\overline{r}+1)]$, then we can define our information criterion either in terms of expression (11) or in terms of expression (14) as follows:

$$(\widehat{p}, \widehat{r}) = \text{argmax PIC}(p, r) , \tag{15}$$

where

$$\text{PIC}(p, r) = \widehat{\widehat{L}}_{T_1}(M_{p,r})/\widehat{\widehat{L}}_{T_1}(M_{\overline{p},\overline{r}}) \tag{16}$$

with $\widehat{\widehat{L}}_{T_1}(M_{p,r})$ as defined in expression (11). Alternatively, we can write this criterion in terms of (14) as

$$(\widehat{p}, \widehat{r}) = \text{argmin PIC}'(p, r) , \tag{17}$$

where

$$\text{PIC}'(p, r) = \ln|\widehat{\Omega}_{T_1}| + \frac{1}{T_1}\ln(|\widehat{V}_{T_0} + \widehat{V}_{T_1}|/|\widehat{V}_{T_0}|) . \tag{18}$$

Note that for large values of T_0 and T_1, expression (18) and, especially, expression (16) can be interpreted as criteria that are based on the predictive odds ratio. Note also that expression (18) is in a form analogous to other information criteria (for example — AIC, BIC, and the Fisher information criterion (FIC) of Wei (1992)) in that the right-hand side of expression (18) is comprised of two terms, with the first term being a measure of the goodness of fit and the second term being a penalty function reflecting the complexity of the model. This formulation corresponds closely to the asymptotic predictive odds criterion used in Phillips (1994).

2.3 Hypothesis Testing of Linear Restrictions

The procedure we outline in the last subsection can be readily extended to test linear restrictions of the form:

$$H(M^R) \ : \ \text{vec}(A^*) = Sd + s \ \text{ and } \ \text{vec}(\alpha) = Gc + g , \tag{19}$$

where S, G, s, g, d, and c are respectively an $(m + m^2p) \times q_1$ restriction matrix of rank q_1, an $mr \times q_2$ restriction matrix of rank q_2, an $(m + m^2p) \times 1$ vector of known constants, an $mr \times 1$ vector of known constants, a q_1-vector of basic parameters, and a q_2-vector of basic parameters. Under the hypothesis (19), we have the following (restricted) ECM formulation:

$$\Delta y_t = (I_m \otimes W_t')(Sd + s) + \alpha(c)\beta' y_{t-1} + \varepsilon_t \tag{20}$$

or

$$z_t = X_t d + \alpha(c)\beta' y_{t-1} + \varepsilon_t \;, \tag{21}$$

where $X_t = (I_m \otimes W_t')S$ and $z_t = \Delta y_t - (I_m \otimes W_t')s$. Note that equations (20) and (21) describe a cointegrated system with additional restrictions imposed on the coefficients of its stationary components. Analogous to (9), we take the prior density of $\gamma = (\text{vec}(B)', d', c')'$ to be

$$g(\gamma | M_{p,r}^R, \widetilde{\gamma}_{T_0}, \widetilde{P}_{T_0}) = (2\pi)^{-\frac{1}{2}(q_1+q_2)} |\widetilde{P}_{T_0}|^{\frac{1}{2}} \exp\left\{ -\tfrac{1}{2}(\gamma - \widetilde{\gamma}_{T_0})' \widetilde{P}_{T_0} (\gamma - \widetilde{\gamma}_{T_0}) \right\} \;, \tag{22}$$

where $\widetilde{\gamma}_{T_0} = (\text{vec}(\widetilde{B}_{T_0})', \widetilde{d}'_{T_0}, \widetilde{c}'_{T_0})'$ is the posterior mode (or the maximum likelihood estimate) of γ over the sample period $[1, T_0]$ and

$$\widetilde{P}_{T_0} = \begin{pmatrix} (\alpha(\widetilde{c}_{T_0})'\widetilde{\Omega}_{T_0}^{-1}\alpha(\widetilde{c}_{T_0}) \otimes F'Y'_{-1,T_0}Y_{-1,T_0}F) & 0 & 0 \\ 0 & G'(\widetilde{\Omega}_{T_0}^{-1}\otimes\widetilde{\beta}'_{T_0}Y'_{-1,T_0}Y_{-1,T_0}\widetilde{\beta}_{T_0})G & G'(\widetilde{\Omega}_{T_0}^{-1}\otimes\widetilde{\beta}'_{T_0}Y'_{-1,T_0}W_{T_0})S \\ 0 & S'(\widetilde{\Omega}_{T_0}^{-1}\otimes W'_{T_0}Y_{-1,T_0}\widetilde{\beta}_{T_0})G & S'(\widetilde{\Omega}_{T_0}^{-1}\otimes W'_{T_0}W_{T_0})S \end{pmatrix} \tag{23}$$

where $\widetilde{\beta}_{T_0} = [I_r, \widetilde{B}_{T_0}]'$ and $\widetilde{\Omega}_{T_0} = (1/(T_0 + m + 1)) \sum_{t=1}^{T_0} (z_t - X_t \widetilde{d}_t - \alpha(\widetilde{c}_{T_0})\widetilde{\beta}'_{T_0} y_{t-1})'$ $(z_t - X_t \widetilde{d}_{T_0} - \alpha(\widetilde{c}_{T_0})\widetilde{\beta}'_{T_0} y_{t-1})$. As before, we take our prior density for Ω to be (6), but our likelihood function under the restricted model $M_{p,r}^R$ now becomes

$$\begin{aligned} &L_T^R(B, d, c, \omega) \\ &= (2\pi)^{\frac{-mT_1}{2}} |\Omega|^{\frac{-T_1}{2}} \exp\left\{ -\tfrac{1}{2} \sum_{t=T_0+1}^{T} (z_t - X_t d - \alpha(c)\beta' y_{t-1})' \Omega^{-1} (z_t - X_t d - \alpha(c)\beta' y_{t-1}) \right\} . \end{aligned} \tag{24}$$

Combining the likelihood function (24) with the prior densities (6) and (22) and integrating over the parameters (B, d, c, ω) using Laplace's method, we obtain, corresponding to (11), the (approximate) Bayesian data density or marginalized likelihood for the restricted model $M_{p,r}^R$:

$$\begin{aligned} \overline{L}_{T_1}(M_{p,r}^R) &\sim |\widetilde{\Omega}_{T_1}|^{\frac{-T_1}{2}} (|\widetilde{P}_{T_0} + \widetilde{P}_{T_1}| / |\widetilde{P}_{T_0}|)^{-\frac{1}{2}} \exp\left\{ -\tfrac{1}{2} (\widetilde{\gamma}_{T_1} - \widetilde{\gamma}_{T_0})' \widetilde{P}_{T_0} (\widetilde{\gamma}_{T_1} - \widetilde{\gamma}_{T_0}) \right\} \\ &= \widehat{\overline{L}}_{T_1}(M_{p,r}^R) \text{ (say)}, \end{aligned} \tag{25}$$

where

$$\widetilde{\Omega}_{T_1} = \frac{1}{T_1 + m + 1} \sum_{t=T_0+1}^{T} \left(z_t - X_t \widetilde{d}_{T_1} - \alpha(\widetilde{c}_{T_1}) \widetilde{\beta}'_{T_1} y_{t-1} \right)' (z_t - X_t \widetilde{d}_{T_1} - \alpha(\widetilde{c}_{T_1}) \widetilde{\beta}'_{T_1} y_{t-1}) \tag{26}$$

and

$$\widetilde{P}_{T_1} = \begin{pmatrix} (\alpha(\widetilde{c}_{T_1})'\widetilde{\Omega}_{T_1}^{-1}\alpha(\widetilde{c}_{T_1}) \otimes F'Y'_{-1,T_1}Y_{-1,T_1}F) & 0 & 0 \\ 0 & G'(\widetilde{\Omega}_{T_1}^{-1}\otimes\widetilde{\beta}'_{T_1}Y'_{-1,T_1}Y_{-1,T_1}\widetilde{\beta}_{T_1})G & G'(\widetilde{\Omega}_{T_1}^{-1}\otimes\widetilde{\beta}'_{T_1}Y'_{-1,T_1}W_{T_1})S \\ 0 & S'(\widetilde{\Omega}_{T_1}^{-1}\otimes W'_{T_1}Y_{-1,T_1}\widetilde{\beta}_{T_1})G & S'(\widetilde{\Omega}_{T_1}^{-1}\otimes W'_{T_1}W_{T_1})S \end{pmatrix} \tag{27}$$

$\widetilde{\beta}_{T_1}$, $\widetilde{d}_{T_1}$, and $\widetilde{c}_{T_1}$ are the posterior modes of β, d, and c, where the posterior distribution has been updated by the likelihood function (24).

Expression (25) enables us to make decisions on the lag order and the cointegrating rank of the system simultaneously with decisions about the validity of the restrictions represented by (19). Let $H(M^R)$ be the null hypothesis defined by (19) and let $H(M^U)$ be the alternative of an unrestricted ECM as defined in Section 2.2, then the decision rule for choosing M^R over M^U can be stated as:

$$\text{Accept } H(M^R) \text{ in favor of } H(M^U) \text{ if } \widehat{\overline{L}}_{T_1}(M^R_{\widetilde{p},\widetilde{r}})/\widehat{\overline{L}}_{T_1}(M^U_{\widehat{p},\widehat{r}}) > 1 , \tag{28}$$

where

$$\begin{aligned} (\widehat{p},\widehat{r}) &= \text{argmax PIC}U(p,r) , \\ \text{PIC}U(p,r) &= \widehat{\overline{L}}_{T_1}(M^U_{p,r})/\widehat{\overline{L}}_{T_1}(M^U_{\overline{p},\overline{r}}) , \end{aligned} \tag{29}$$

and

$$\begin{aligned} (\widetilde{p},\widetilde{r}) &= \text{argmax PIC}R(p,r) , \\ \text{PIC}R(p,r) &= \widehat{\overline{L}}_{T_1}(M^R_{p,r})/\widehat{\overline{L}}_{T_1}(M^U_{\overline{p},\overline{r}}) . \end{aligned} \tag{30}$$

Note that $\widetilde{p}$ and $\widetilde{r}$ are the estimated lag and cointegrating rank order of a (possibly) cointegrated system having additional restrictions of the form (19) and, thus, may be different from $\widehat{p}$ and $\widehat{r}$, which are the order estimates of a (possibly) cointegrated system having no additional restriction.

3 The Present Value Model and its Testable Implications

In this section, we briefly describe the present value model and discuss its testable implications. Since a detailed discussion of this model is given in Campbell and Shiller (1987), we focus our attention here only on those features of the model which will be relevant for our subsequent empirical analysis. Formally, the present value model can be written as:

$$y_{2t} = \theta(1-\delta)\sum_{i=0}^{\infty} \delta^i E(y_{1t+i}|I_t) + \text{const}, \tag{31}$$

where $E(\cdot|I_t)$ denotes the mathematical expectation conditional on the full public information I_t at time t and where y_{1t} and y_{2t} are, respectively, the dividend and stock price at time t. Here, as elsewhere in this paper, we treat conditional expectations as being equivalent to linear projections on information. Moreover, in the context of the stock market, $\theta = \delta/(1-\delta)$ and const is restricted to be zero so that equation (31) has the simplified form:

$$y_{2t} = \sum_{i=0}^{\infty} \delta^{i+1} E(y_{1t+i}|I_t) . \tag{32}$$

We follow Campbell and Shiller (1987) in defining the random variable $s_t = y_{2t} - \theta y_{1t}$, which they referred to as the "spread." Subtracting θy_{1t} from both sides of equation (32) and rearranging the terms, it is easy to show, as in Campbell and Shiller (1987), that the present value model implies two alternative interpretations of the spread, *viz.*

$$s_t = \theta \sum_{i=1}^{\infty} \delta^i E(\Delta y_{1t+i}|I_t) \ , \quad \text{and} \tag{33}$$

$$s_t = \frac{\delta}{1-\delta} E(\Delta y_{2t+1}|I_t) \ . \tag{34}$$

For our purposes, it is most convenient to work with the relationship (34). To put equation (34) in a more useful form, we multiply both sides by $-(1-\delta)/\delta$ to obtain

$$s_t^* = -E(\Delta y_{2t+1}|I_t) \ , \tag{35}$$

where the left-hand side of equation (35) has the equivalent forms

$$\begin{aligned} s_t^* &= y_{1t} - \frac{1-\delta}{\delta} y_{2t} \\ &= y_{1t} - (1/\theta) y_{2t} \\ &= y_{1t} + b y_{2t} \ \ \text{(say)}. \end{aligned} \tag{36}$$

If Δy_{2t} is stationary, then equation (35) implies that s_t^* is also stationary, from which we deduce (from equation (36)) the cointegration of y_{1t} and y_{2t} with cointegrating vector $(1, b)$.

The statistical model we use to describe the joint dynamics of y_{1t} and y_{2t} is a bivariate version of the general error-correction model (2), which we will rewrite here with the reduced rank restriction imposed:

$$\Delta y_t = \mu + \sum_{i=1}^{p} A_i^* \Delta y_{t-i} + \alpha\beta' y_{t-1} + \varepsilon_t \ . \tag{37}$$

This representation is in line with that of Hansen and Sargent (1981), Campbell and Shiller (1987), Toda (1991), and DeJong and Whiteman (1994), in that it includes the lagged values of not only y_{1t} but also y_{2t} in the information set that is available to the econometrician. Imposing the relationships (35) and (36) on this error-correction model yields the following set of restrictions:

$$a_{i,21}^* = a_{i,22}^* = 0 \ , \quad i = 1, \ldots, p \ , \tag{38}$$

$$\mu_2 = 0 \ , \tag{39}$$

$$\alpha_2 = -1 \ , \tag{40}$$

$$(1, b) = (1, -(1-\delta)/\delta) = (1, -1/\theta) \ , \tag{41}$$

where $(a_{i,21}^*, a_{i,22}^*)$ is the second row of the matrix A_i^* and α_2 is the second element of the vector $\alpha = (\alpha_1, \alpha_2)'$. The restrictions (38)–(40) are of the form (19) and can therefore be tested using the procedure outlined in the last section. Moreover, for a given value of the discount factor δ, equation (41) is also of the form (19). Following Campbell and Shiller (1987), we shall, in the next section of the paper, test the present value model both for the case where the discount factor δ takes on a value implied by the sample mean return on stocks and for the case where the cointegrating vector $(1, b)$ is left unrestricted.

4 Data Description and Empirical Results

In this section, we apply the test procedures discussed in Section 2 to the Campbell–Shiller data set for stock prices and dividends, updated to 1992. A brief discussion of the data is in order. As explained in Campbell and Shiller (1987), the term y_{2t} is the real stock price computed by dividing the Standard and Poor's stock price index for January by the January producer price index normalized so that the 1976 producer price index equals 100. Real dividend y_{1t} on the other hand, is constructed by dividing the nominal dividend series by the annual average producer price index also normalized so that the 1967 producer price index equals 100. It should be noted that the nominal dividend series before and after 1926 were collected from different sources. Since 1926, the nominal dividend series used in the construction is the dividends per share taken from the Standard and Poor's statistical service. Before 1926, the nominal dividend was taken from Cowles (1939).

Note also that a difficulty arises in pairing y_{1t} and y_{2t} since they are not measured contemporaneously. In our data set, y_{2t} is the beginning-of-period stock price while the dividend y_{1t} is paid sometime within period t. Since y_{1t} is not observable at the start of period t, West (1988) and others have argued that treating y_{1t} and y_{2t} as observations from the same period may lead to spurious rejection of the present value model. To circumvent this problem, we follow Campbell and Shiller (1987) and Toda (1991) in writing the VECM (37) in terms of $y_t = (y_{1t-1}, y_{2t})'$, where both variables are now in the information set at the start of time t. Note that cointegration of y_{1t} and y_{2t} implies that y_{1t-1} and y_{2t} are also cointegrated.

The remainder of this section is divided into three subsections, each discussing the results from a different test procedure. The results of unit root tests and tests of cointegration are given in Subsections 4.1 and 4.2, respectively, while Subsection 4.3 presents the results of testing the full set of restrictions implied by the present value model.

4.1 Unit Root Tests

As our setup depends critically on the assumption that both y_{1t} and y_{2t} are I(1) processes, we begin our empirical analysis by testing both the stock price series and the dividend series for unit roots. In their work, Campbell and Shiller (1987) ran Dickey–Fuller regressions on the two series and found that the unit root null hypothesis cannot be rejected at the 10% level for either series. Here, we take a different approach to unit root testing in an effort to bring additional statistical evidence in support of the hypothesis that both real stock prices and real dividends are well-described by I(1) processes. The method we use is closely related to the model selection criterion PICF detailed in Phillips (1992, 1995) and is, in fact, the univariate version of the multivariate test procedure we outlined in Section 2. To test unit root models against alternatives which may be trend stationary, we compare a general autoregressive model with trend (written in difference form), *viz.*,

$$H(M_{p,\ell}^{REF}) \ : \ \Delta y_t = a_0 y_{t-1} + \sum_{i=1}^{p-1} a_i \Delta y_{t-i} + \sum_{j=0}^{\ell} b_j t^j + \varepsilon_t \tag{42}$$

with one which explicitly incorporates a unit root

$$H(M_{p,\ell}^{UR}) \ : \ \Delta y_t = \sum_{i=1}^{p-1} a_i \Delta y_{t-i} + \sum_{j=0}^{\ell} b_j t^j + \varepsilon_t \ . \tag{43}$$

Decisions about unit roots can then be made on the basis of the criterion:

$$\text{Decide in favor of unit root if } \widehat{\overline{L}}_{T_1}(M_{\widetilde{p},\widetilde{\ell}}^{UR})/\widehat{\overline{L}}_{T_1}(M_{\widehat{p},\widehat{r}}^{REF}) > 1 \ , \tag{44}$$

where

$$(\widehat{p},\widehat{\ell}) = \text{argmax PIC}^{REF}(p,\ell) \ , \tag{45}$$
$$\text{PIC}^{REF}(p,\ell) = \widehat{\overline{L}}_{T_1}(M_{p,\ell}^{REF})/\widehat{\overline{L}}_{T_1}(M_{\overline{p},\overline{\ell}}^{REF}) \ ,$$

and

$$(\widetilde{p},\widetilde{\ell}) = \text{argmax PIC}^{UR}(p,\ell) \ , \tag{46}$$
$$\text{PIC}^{UR}(p,\ell) = \widehat{\overline{L}}_{T_1}(M_{p,\ell}^{UR})/\widehat{\overline{L}}_{T_1}(M_{\overline{p},\overline{\ell}}^{REF}) \ ,$$

The formulae for the (approximate) marginalized likelihoods, $\widehat{\overline{L}}_{T_1}(M_{p,\ell}^{UR})$ and $\widehat{\overline{L}}_{T_1}(M_{p,\ell}^{REF})$, are analogous to their multivariate counterparts presented in Section 2 (see equations (11) and (25)). Hence for brevity, we will not state them here.

Table 1: Unit Root Tests for the Sample Period 1871–1992

Variable	Initialization T_0	Lag selected[a] under $H(M^{UR})$	Trend selected[b] under $H(M^{UR})$	Posterior odds in favor of a unit root
	0 (uniform prior)	1	0[c]	42.662
	22	1	0	121.750
	26	1	0	778.628
y_{1t}	30	1	0	170.532
	34	1	0	8.880
	38	3	0	5.179
	42	1	0	1.540
	0 (uniform prior)	1	0	33.357
	22	1	0	357.095
	26	4	0	7691.930
y_{2t}	30	4	0	476.178
	34	1	0	23.647
	38	1	0	46.517
	42	1	0	27.456

[a] The maximum lag length $\overline{p}$ is set equal to 9.
[b] The maximum trend degree $\overline{\ell}$ is set equal to 1.
[c] 0 denotes the inclusion of a constant term but not a linear trend.

Table 1 documents the results of unit root tests using the test criterion (44). As our empirical Bayesian approach inevitably involves a subjective choice of the sample split point T_0, the results in Table 1 and beyond will always be reported for several different values of T_0 so as to give an indication of the sensitivity of our results to the choice of the split point. Note also that there is a tradeoff in the choice of T_0: as T_0 increases, the hyperparameters of the prior distribution are estimated more precisely, but T_1 decreases so a smaller portion of the sample is being used for model comparison.

The results presented in Table 1 corroborate those obtained by Campbell and Shiller (1987) in that both dividends and stock prices are found to have a unit root specification, although the strength of the evidence in favor of a unit root (as measured by the posterior odds) varies with different values of T_0. Our criterion, however, does not favor a linear trend specification for either series.

4.2 Estimation of the Lag Order and the Cointegrating Rank

Sometimes, the question of whether dividends and stock prices are cointegrated is of independent interest. In particular, it can be quite independent of any interest in the validity of the present value model. For instance, we may only wish to obtain an appropriate time series representation for the variables y_{1t} and y_{2t} for forecasting purposes. Hence, in this section, we set out to estimate the lag order and the cointegrating rank of the model (37) using the test criterion (15) given in Section 2. Note that the criterion (15) selects the mode amongst possible PIC values. Alternatively, one could also construct point estimates of p and r by taking a weighted average using the PIC values as weights:

$$(p^m, r^m) = \text{round}\left\{\left(\sum_{p,r} \text{PIC}(p,r)\right)^{-1} \sum_{p,r} [\text{PIC}(p,r) \times (p,r)]\right\} . \tag{47}$$

An advantage of using a mean criterion like (47) in addition to the modal criterion (15) is that it is affected by and therefore alerts the investigator to cases (i.e., order combinations) where an appreciable mass of PIC values may occur in regions away from the mode.

Table 2 reports order estimates (p,r) from both the modal criterion (15) and the mean criterion (46) for different values of T_0, the last observation used to construct the prior. From Table 2, we see that a cointegrating rank of zero was selected by both criteria regardless of initialization. These findings are roughly in accord with previous results obtained by Campbell and Shiller (1987), Phillips and Ouliaris (1988), and Toda (1991) for the shorter version of the same data set covering the period 1871–1986. Phillips and Ouliaris (1988) and Toda (1991) found no evidence of cointegration. Campbell and Shiller (1987), on the other hand, rejected the null hypothesis of no cointegration using a Dickey–Fuller regression but failed to reject the same null hypothesis when an Augmented Dickey–Fuller regression was used.

The sharpness of our inference on the lag order and the cointegrating rank is portrayed in Figures 1(a)–(g), where we depict bar charts of PIC values in (p,r) space for different choices of T_0. Note that in each of these figures, our selection of the RRR(1,0) specification is well-determined in the sense that it has far and away the

Table 2: Estimation of Cointegrating Rank and Order of Lagged Differences for the Sample, Period 1871–1992

Initialization T_0	$\widehat{p}$	$\widehat{r}$	p^m	r^m
0 (uniform prior)	1	0	1	0
22	1	0	1	0
26	1	0	2	0
30	1	0	1	0
34	1	0	1	0
38	1	0	1	0
42	1	0	1	0

Notes: $\widehat{p}$, $\widehat{r}$, p^m, and r^m are as defined in expressions (15) and (47).

highest PIC value amongst competing models. That our data strongly favors the RRR(1,0) specification is also reflected in the close agreement between the order estimates from the modal criterion and those from the mean criterion in Table 2, with the only exception being the selection of 2 lags by the mean criterion in the case $T_0 = 26$.

4.3 Tests of the Present Value Model

We proceed now to test the restrictions (38)–(41) of the present value model using the test criterion (28). To help summarize our results, we define the following statistics:

$$\tau_1 = \widehat{L}_{T_1}(M^{PV}_{\widetilde{p},1})/\widehat{L}_{T_1}(M^{U}_{\widetilde{p},\widetilde{r}})$$

and

$$\tau_2 = \widehat{L}_{T_1}(M^{PV}_{\widetilde{p},1})/\widehat{L}_{T_1}(M^{U}_{\widetilde{p},1}) \; .$$

Here, M^{PV} and M^U denote, respectively, the null model which satisfies the present value restrictions and the unrestricted VECM given by equation (2). The statistic τ_1 compares the restricted model of the chosen lag order $\widetilde{p}$ with the model having the highest density amongst those in the class of unrestricted VECM's while τ_2 compares the same null model with an unrestricted model of the same order $(\widetilde{p}, 1)$. We test the present value relations both for the case where $(1 - \delta)/\delta = R = .085$ and the case where R is left unrestricted. The number .085 is the sample mean return on stocks for the period 1871–1992 and is used here as a possible discount rate.

Tables 3 and 4 summarize our results which do not seem to be sensitive in any substantive way to whether R is taken to be .085 or left unrestricted. Focusing on the τ_1 statistic, we see that our criterion favors a RRR(1,0) specification over the present value model uniformly over the different choices of T_0. Rejection of the present value model by our procedure is on the whole consistent with the results of Campbell and Shiller (1987) and Toda (1991), who also rejected the full set of present value

Table 3: Model Selection Test of the Present Value Restrictions ($R = .085$)

Initialization T_0	$\widehat{p}$ under $H(M^{PV})$	p^m under $H(M^{PV})$	τ_1	τ_2
0 (uniform prior)	1	1	0.198	1.194
22	1	2	0.018	0.113
26	1	1	0.017	0.115
30	1	1	0.008	4.552
34	1	1	0.035	3.131
38	1	1	0.057	1.562
42	1	1	0.030	7.066

Table 4: Model Selection Test of the Present Value Restriction (R unrestricted)

Initialization T_0	$\widehat{p}$ under $H(M^{PV})$	p^m under $H(M^{PV})$	τ_1	τ_2
0 (uniform prior)	1	1	0.362	1.823
22	1	1	0.307	1.942
26	1	1	0.085	0.569
30	1	1	0.090	51.434
34	1	1	0.049	4.353
38	1	1	0.055	1.511
42	1	2	0.051	12.070

restrictions using the classical Wald test. Only with the exclusion of the restriction corresponding to our equation (39) did Campbell and Shiller (1987) find favorable evidence for the present value model. In addition, our results are in agreement with those of DeJong and Whiteman (1994) in the case where a relatively tight Minnesota prior was used. We note, however, that those authors found more favorable evidence for the present value model when they allowed their priors to be more diffuse.

Looking at the τ_2 statistics, we see that for a majority of the cases under consideration, the present value model of the chosen lag order $\widetilde{p}$ compares favorably with a reduced rank regression of the same lag order and one cointegrating vector. This suggests that in most cases the rejection of the present value model by our criterion is primarily a rejection of the hypothesis that dividends and stock prices are cointegrated, which we showed in Section 3 to be an implication of the present value model when the data are well-described as integrated processes.

To assess the sensitivity of our results to lag specification, we turn our attention to Figures 2(a)–(g) which, for different choices of the initialization T_0, plot PIC values for four models (VECM with $r = 0$, VECM with $r = 1$, present value model with

$R = .085$, and present value model with R unrestricted) against different lag specifications. Note that for the cases where $T_0 \leq 34$, our results are only mildly sensitive to variations in the lag length. In these cases, our choice of an unrestricted model with no cointegration over the present value model can be overturned only if one decides to condition upon lag orders that are extremely small ($p = 0$) or moderately large ($p \geq 5$). On the other hand, in the two cases where the initial sample size is taken to be 38 and 42, the choice between the same two models depends more critically on the lag order selected. Interesting enough, these cases where our results are most sensitive to lag specification are also cases where our inference on the lag order is very sharp, as is evident from Figures 2(f) and 2(g). Hence, even in these cases, there is a clear choice in favor of an unrestricted model with no cointegration.

5 Conclusion

This paper argues for and illustrates a Bayesian approach to the joint estimation of the order of lagged differences and the cointegrating rank in a vector error-correction model (VECM). Our method is a variant of the Posterior Information Criterion (PIC), developed and analyzed in Phillips and Ploberger (1994, 1995), and is very similar to the asymptotic predictive odds version of the PIC criterion given in Phillips (1994). In the formulation of the PIC criterion here, we use a proper (Gaussian) prior whose hyperparameters are estimated from an initial subsample (of length T_0) of the data. As the form of the prior is suggested by the asymptotic posterior distribution of the parameters of our model, our criterion can be interpreted, in large samples, as an approximate predictive odds ratio. As in our earlier work (see Chao and Phillips, 1994), this procedure delivers consistent estimates of the lag order and cointegrating rank of a VECM.

Our procedure also has the advantage that it enables us to select the lag and cointegrating rank order of a VECM at the same time as it tests restrictions implied by economic theory, and it does so in the same coherent framework. Hence, the difficulties of accounting for the uncertainty associated with preliminary lag selection that arise with other methods of inference are avoided here. In addition, bar charts of PIC values for models of different dimensions can be readily constructed so that one may assess the sharpness of the inferences with respect to lag length and cointegrating rank.

We applied this method in an empirical analysis of the Campbell–Shiller stock market data, extended through to 1992. Using our procedure to compare models of different lag lengths and cointegrating rank, we consistently select a model with no cointegration for different choices of the initial value T_0. Further examination of the distribution of PIC values finds the lag and rank estimates to be sharply determined in every case.

Finally, we test the full set of present value restrictions using our criterion. We find models which satisfy these restrictions to be less plausible than time series models with no cointegration. These results are by and large consistent with the results of Campbell and Shiller (1987) and Toda (1991) using classical methodologies and with the Bayesian results of DeJong and Whiteman (1994) in the case where a relatively tight Minnesota prior was used.

6 Acknowledgements

We thank Don Andrews, Chris Sims, participants of the Conference on Forecasting, Prediction and Modeling in Statistics and Econometrics, particularly John Geweke and Arnold Zellner, and two anonymous referees for their many helpful comments and suggestions during the evolution of this paper to its present form. An expression of gratitude is also due to Robert Shiller for supplying the data used here. In addition, PCBP thanks the NSF for research support under Grant No. SES 9122142. Last but not least, we are grateful to Glena Ames for expertly keyboarding the manuscript. All the usual disclaimers apply.

References

Ahn, S. K. and G. C. Reinsel (1990). "Estimation for Partially Nonstationary Multivariate Autoregressive Models," *Journal of the American Statistical Association*, 85, 813–823.

Atkinson, A. C. (1978). "Posterior Probabilities for Choosing a Regression Model," *Biometrika*, 65, 39–48.

Berger, J. O. (1985). *Statistical Decision Theory and Bayesian Analysis*. New York: Springer–Verlag.

Campbell, J. Y. and R. J. Shiller (1987). "Cointegration and Tests of Present Value Models," *Journal of Political Economy*, 95, 1062–1088.

Chao, J. (1995). "Some Simulation Results of Likelihood Ratio Tests for Cointegrating Ranks in Misspecified Vector Autoregressions," in preparation.

Chao, J. C. and P. C. B. Phillips (1994). "Bayesian Model Selection in Partially Nonstationary Vector Autoregressive Processes with Reduced Rank Structure," mimeographed, Yale University.

Cowles, A. (1939). *Common Stock Indexes*. Bloomington, IN: Principia.

DeJong, D. N. and C. H. Whiteman (1992). "More Unsettling Evidence on the Perfect Markets Hypothesis: Trend-Stationarity Revisited," *Federal Reserve Bank of Atlanta Economic Review*, 77, 1–13.

DeJong, D. N. and C. H. Whiteman (1994). "Modeling Stock Prices without Knowing How to Induce Stationarity," *Econometric Theory*, 10, 701–719.

Engle, R. F. and C. W. J. Granger (1987). "Cointegration and Error-correction: Representation, Estimation, and Testing," *Econometrica*, 55, 251–276.

Geweke, J. (1994). "Bayesian Comparison of Econometric Models," Federal Reserve Bank of Minneapolis Research Department working paper 532, July.

Hansen, L. P. and T. Sargent (1981). "Exact Linear Rational Expectations Models: Specification and Estimation," Staff Report No. 71, Minneapolis: Federal Reserve Bank.

Johansen, S. (1988). "Statistical Analysis of Cointegrating Vectors," *Journal of Economic Dynamic and Control, 12*, 231–254.

Johansen, S. (1991). "Estimation and Hypothesis Testing of Cointegrating Vectors in Gaussian Vector Autoregressive Models," *Econometrica*, 59, 1551–1580.

Johansen, S. (1992). "Determination of Cointegrating Rank in the Presence of a Linear Trend," *Oxford Bulletin of Economics and Statistics*, 54, 383–397.

Kleidon, A. W. (1986). "Variance Bounds Test and Stock Price Valuation Models," *Journal of Political Economy*, 94, 953–1001.

Marsh, T. A. and R. C. Merton (1986). "Dividend Variability and Variance Bounds Tests for the Rationality of Stock Prices," *American Economic Review*, 46, 483–498.

O'Hagan, A. (1991). "Discussion on Posterior Bayes Factor (by M. Aitkin)," *Journal of the Royal Statistical Society Series B*, 53, 136.

Phillips, P. C. B. (1992). "Bayes Methods for Trending Multiple Time Series with an Empirical Application to the U.S. Economy," Cowles Foundation Discussion Paper No. 1025, Yale University.

Phillips, P. C. B. (1994). "Model Determination and Macroeconomic Activity," mimeographed, Yale University.

Phillips, P. C. B. (1995). "Bayesian Model Selection and Prediction with Empirical Applications," *Journal of Econometrics* (in press).

Phillips, P. C. B. and W. Ploberger (1994). "Posterior Odds Testing for a Unit Root with Data-based Model Selection," *Econometric Theory*, 10, 774–808.

Phillips, P. C. B. and W. Ploberger (1995). "An Asymptotic Theory of Bayesian Inference for Time Series," *Econometrica* (forthcoming).

Phillips, P. C. B. and S. Ouliaris (1988). "Testing for Cointegration Using Principal Components Methods," *Journal of Economic Dynamics and Control*, 12, 205–230.

Pötscher, B. M. (1983), "Order Estimation in ARMA Models by Lagrange Multiplier Tests," *Annals of Statistics*, 11, 872–885.

Sawa, T. (1978). "Information Criteria for Discriminating among Alternative Regression Models," *Econometrica*, 46, 1273–1291.

Shibata, R. (1976). "Selection of the Order of an Autoregressive Model by Akaike's Information Criterion," *Biometrika*, 63, 117–126.

Toda, H. Y. (1991). "An ECM Approach to Tests of Present Value Models," mimeographed, Yale University.

Toda, H. Y. and P. C. B. Phillips (1994). "Vector Autoregression and Causality: A Theoretical Overview and Simulation Study," *Econometric Reviews*, 13, 259–285.

Tsay, R. S. (1984). "Order Selection in Nonstationary Autoregressive Models," *Annals of Statistics*, 12, 1425–1433.

Wei, C. Z. (1992). "On Predictive Least Squares Principles," *Annals of Statistics*, 20, 1–42.

West, K. D. (1988). "Dividend Innovations and Stock Market Volatility," *Econometrica*, 56, 37–62.

Zellner, A. (1971). *An Introduction to Bayesian Inference to Econometrics*. New York: John Wiley and Sons.

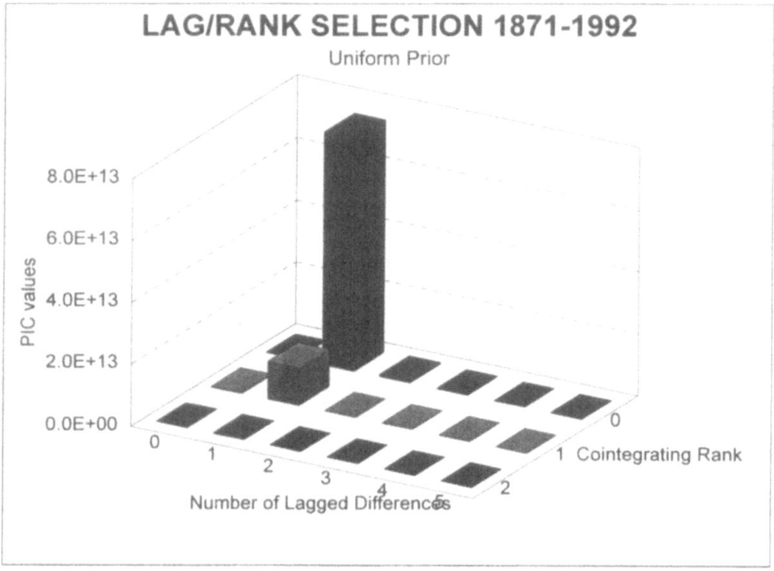

Figure 1(a)

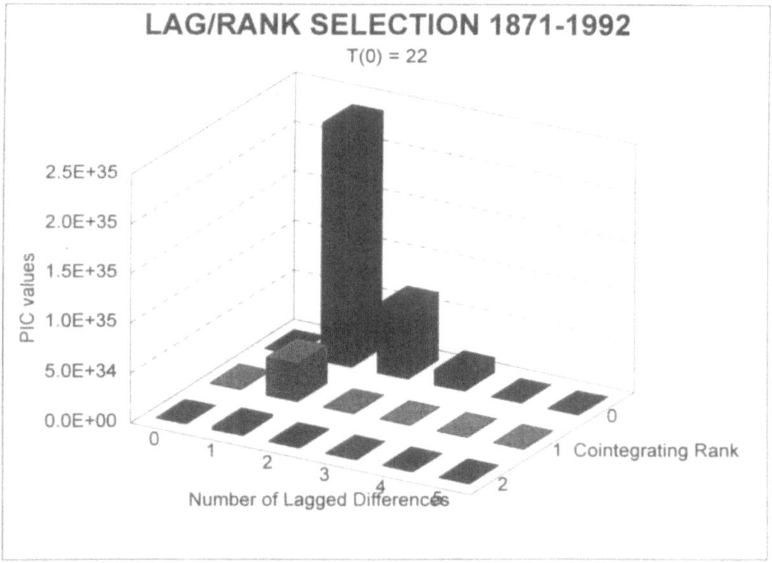

Figure 1(b)

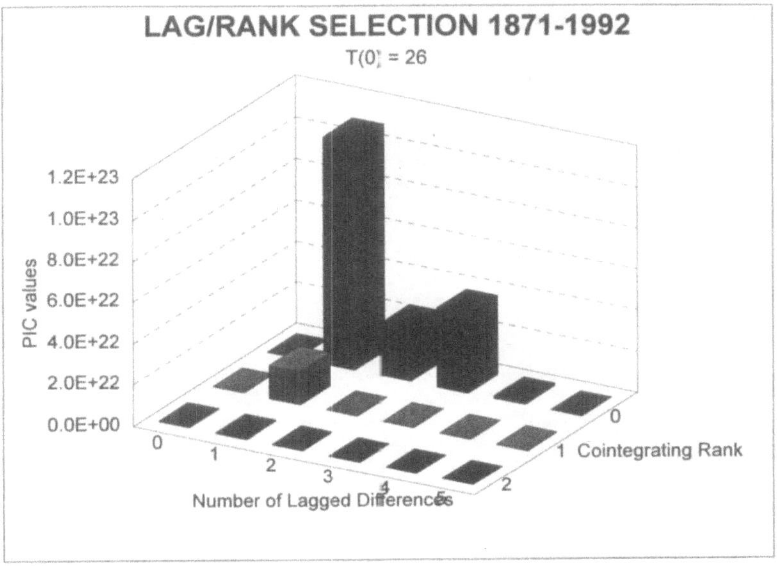

Figure 1(c)

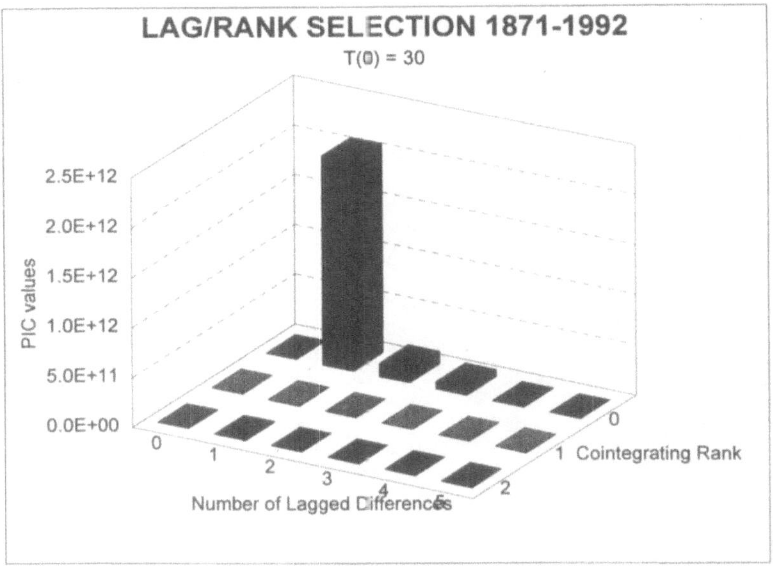

Figure 1(d)

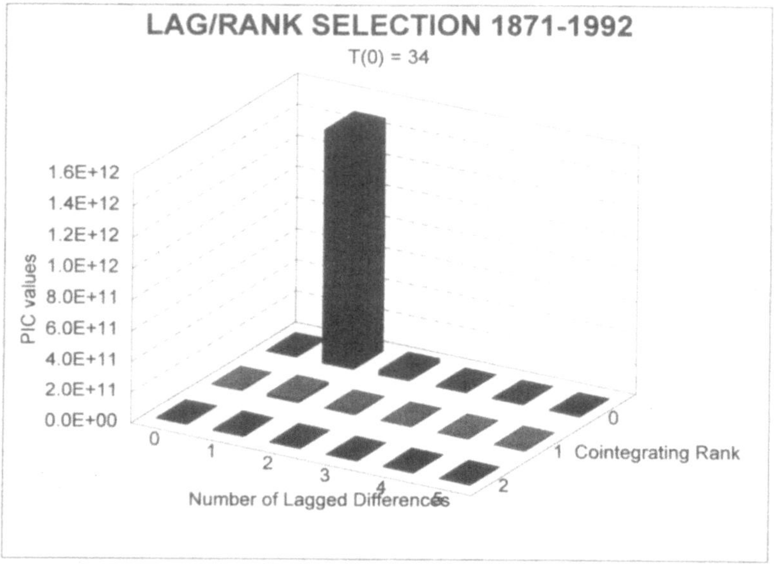

Figure 1(e)

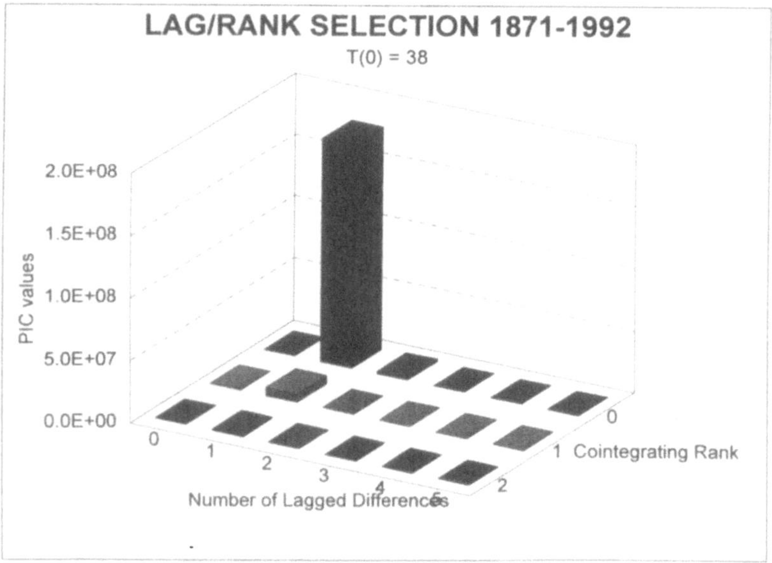

Figure 1(f)

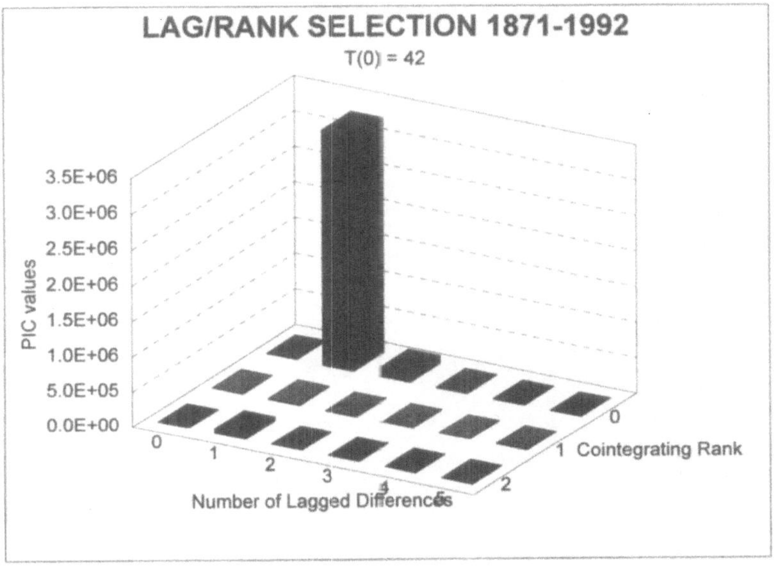
LAG/RANK SELECTION 1871-1992
T(0) = 42
PIC values
3.5E+06
3.0E+06
2.5E+06
2.0E+06
1.5E+06
1.0E+06
5.0E+05
0.0E+00
0
1
2
3
4
5
Number of Lagged Differences
0
1
2
Cointegrating Rank

Figure 1(g)

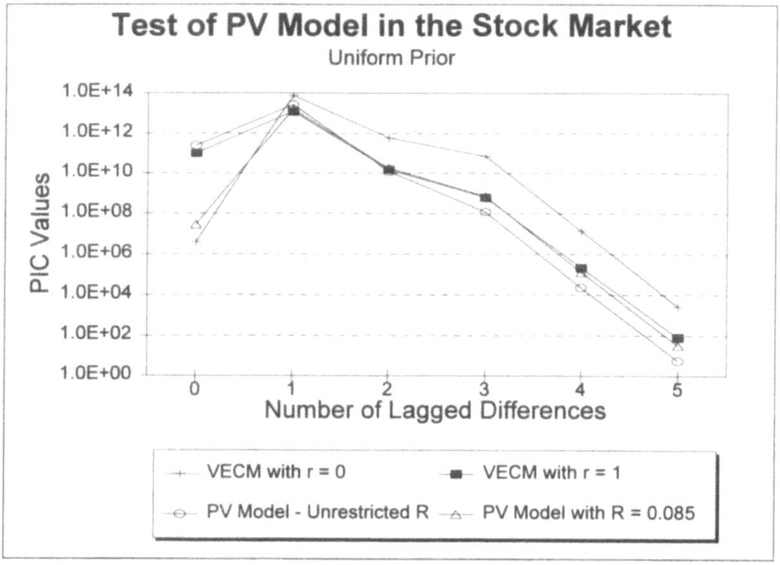

Figure 2(a)

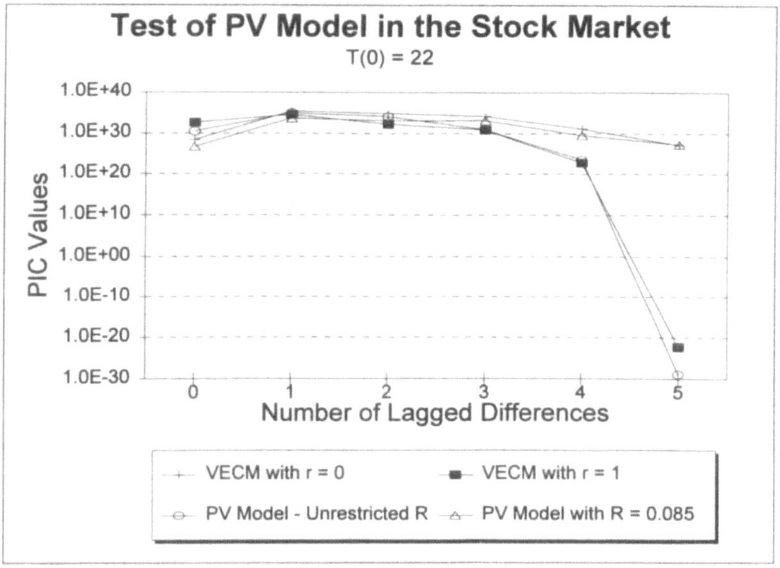

Figure 2(b)

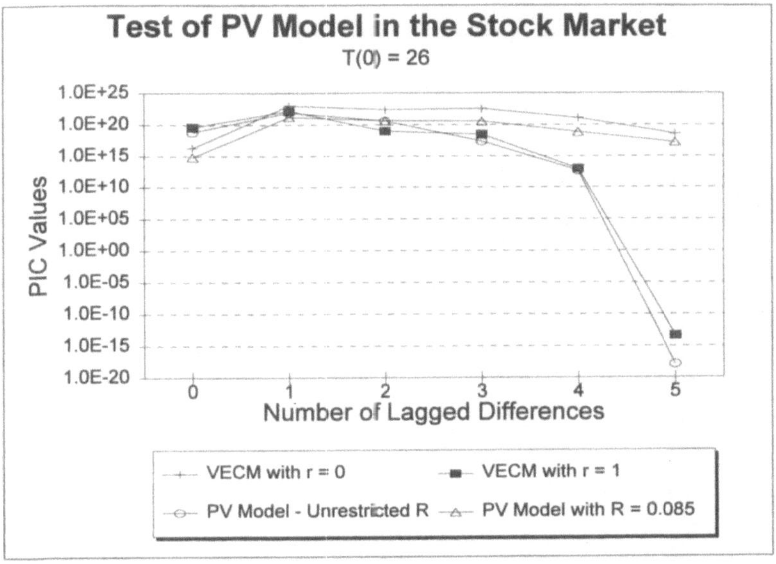

Figure 2(c)

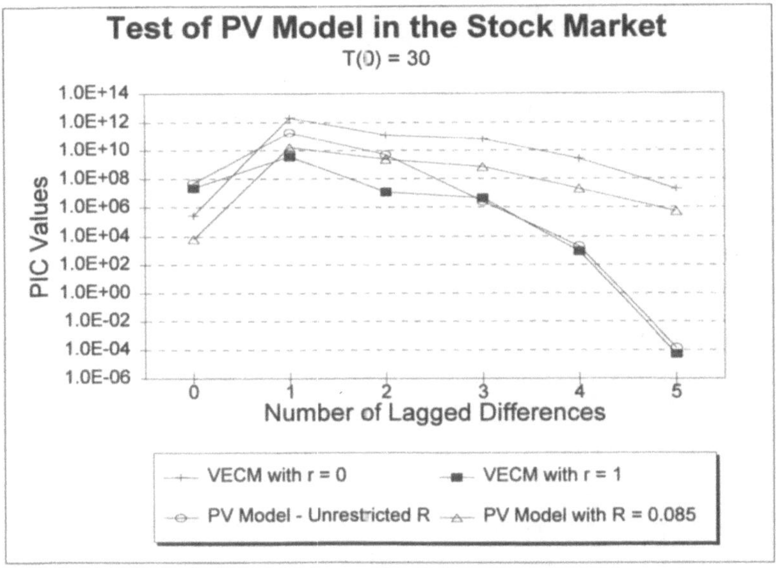

Figure 2(d)

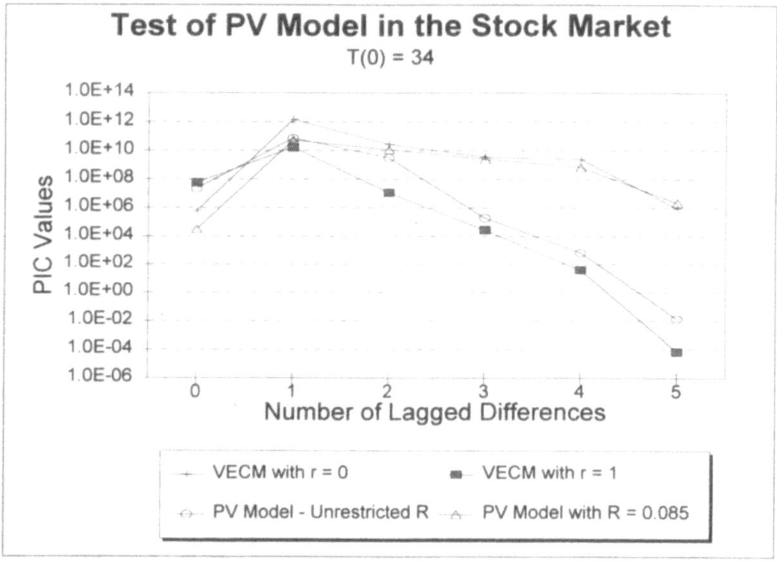

Figure 2(e)

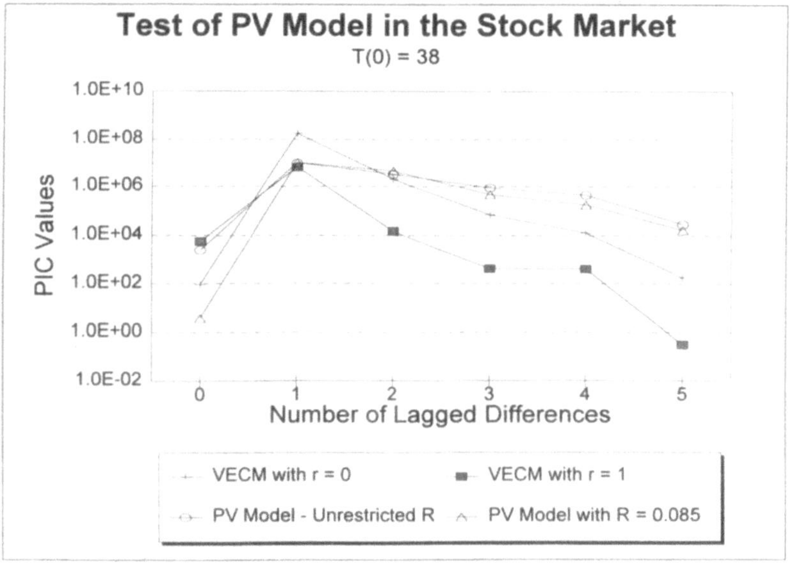

Figure 2(f)

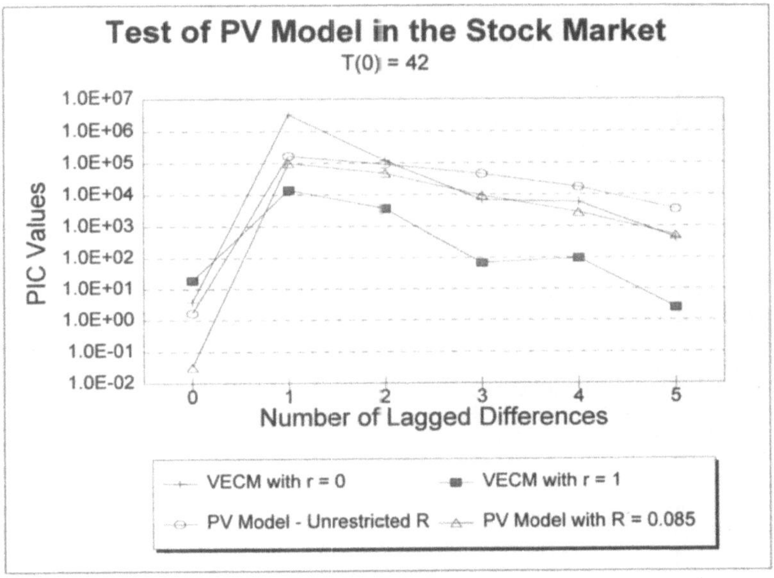
Test of PV Model in the Stock Market
T(0) = 42
1.0E+07
1.0E+06
1.0E+05
1.0E+04
1.0E+03
1.0E+02
1.0E+01
1.0E+00
1.0E-01
1.0E-02
PIC Values
0
1
2
3
4
5
Number of Lagged Differences
VECM with r = 0
VECM with r = 1
PV Model - Unrestricted R
PV Model with R = 0.085

Figure 2(g)

Prediction of Individual Bond Prices via the TDM Model

Takeaki KARIYA
Institute of Economic Research
Hitotsubashi University
Kunitachi, Tokyo 186, JAPAN

Hiroshi TSUDA
The NLI Research Institute
1-1-1 Yuraku-cho
Chiyoda-ku, Tokyo 100, JAPAN

Abstract

In Kariya and Tsuda (1994), the TDM (Time Dependent Markov) bond pricing model is shown to be of great in-sample performance. In fact, the standard errors of the model are almost all less than 0.5 yen among 120 models where the face value of a bond is 100 yen. In this paper, the TDM model is applied to the prediction of monthly individual bond prices and it is shown that the predictive power of the model is rather good for two recent years.

1 Introduction

This paper will test the predictive performance of the TDM (Time Dependent Markov) bond pricing model Kariya and Tsuda (1994) (abbreviated as TK (1994) below) proposed as an extension of the CSM (Cross-Sectional Market) model that Kariya (1993) formulated. The in-sample performance of the TDM model, when applied to the monthly prices of individual JG (Japanese Government) bonds with initial maturities of 10 years, is amazingly good even though the model has only 5 or 6 parameters. In fact, among 120 monthly models tested from 1983.1 through 1992. 2, there are only 7 models with standard errors that are greater than 0.5 yen, where the face value of a JG bond is 100 yen and there are 45 to 90 JG bonds for each month which were issued at different times. Hence, the next step is to test the predictive power of the model. In this paper, we show that the TDM model is even effective in predicting individual JG bond prices at least for recent years.

In prediction, we need to predict the 4 time dependent parameters involved in the discount functions which depend on the coupon rate and maturity as bond attributes. The VAR (vector autoregressive) model is used for the parameter prediction with the AIC (Akaike Information Criterion). Unfortunately the VAR models thus selected do not fit the parameter movements up to 1990.12. After 1991.1, the VAR models fit well. This fact is reflected in the predictive performance of the TDM models for prices. In fact, the predictive standard error for the period 1988.1 through 1990.12 is 2 yen on the average, while it is 0.8 yen for the period 1991.1 through 1992.12. Therefore, we focus on the latter period and investigate the predictive performance of individual prices, i.e., term structure of prices. There remain some patterns in the term structure of the differences between realized prices and their predicted prices. But the term structures of realized and predicted returns will show that the predictive TDM model is effective in forming portfolios. A comparison of the predictive performance is also made for the CSM model versus the TDM model. In Section 2, the models are described briefly.

2 CSM and TDM

In this section, the CSM and TDM models that Kariya (1993) and KT (1994) formulated are described. Let $P_{it}(0)$ denote the market price of the i-th bond at time t and let

$$s(i)_j \qquad (j=1,\cdots,M(i): i=1,\cdots,N) \tag{2.1}$$

denote the terms (periods) to the j-th cash flow of the i-th bond from time t. All these terms in N bonds are enumerated in an increasing order as

$$s_{*1} < s_{*2} < \cdots < s_{*M}, \quad M=\max M(i). \tag{2.2}$$

Hence s_{*k} is one of the terms in which at least one of the N bonds yields a cash flow. Let $C_{it}(s)$ be the cash flow function of the i-th bond defined on $0 \leqq s \leqq s_M$. Of course, $C_{it}(s)=0$ unless s is one of $s(i)_j$'s. In the CSM model, a realization of price $P_{it}(0)$ which is random is viewed as equivalent to a whole realization of the stochastic process

$$\{D_{it}(s): 0 \leqq s \leqq s_M\} \tag{2.3}$$

of the discount function of the i-th bond in the model

$$P_{it}(0) = \sum_{j=1}^{M} C_{it}(s_{*j}) D(s_{*j}) = C_{it}' D_{it}, \tag{2.4}$$

where

$$C_{it} = (C_{it}(s_{*1}),\cdots, C_{it}(s_{*M}))': M\times 1 \quad \text{and} \tag{2.5}$$

$$D_{it} = (D_{it}(s_{*1}),\cdots, D_{it}(s_{*M}))': M\times 1.$$

Then letting

$$\overline{D}_{it}(s) = E[D_{it}(s)] \tag{2.6}$$

be the mean discount function of the i-th bond, the model (2.4) becomes

(2.7) $$P_{it}(0) = C_{it}'\bar{D}_{it} + \eta_{it},$$

where

(2.8) $$\bar{D}_{it} = E(D_{it}), \quad \eta_{it} = C_{it}'\nu_{it} \quad \text{and} \quad \nu_{it} = D_{it} - \bar{D}_{it}.$$

It is noted that $C_{it}'\bar{D}_{it}$ is the sum of the cash flows discounted by the mean discount function, η_{it} is the sum of the cash flows randomly discounted by the random discount factor ν_{it}. The mean discount function in (2.6) is specified as

(2.9) $$\bar{D}_{it}(s) = 1 + \sum_{k=1}^{p} \delta_k(z_{it})\, s^k \quad \text{with}$$

(2.10) $$\delta_k(z_{it}) = \sum_{j=0}^{q} \delta_{kjt}\, z_{ijt},$$

where $z_{it} = (z_{i1t}, \cdots, z_{iqt})'$ is a vector of the attributed variables of the i-th bond. On the other hand, the covariance structure of η_{it}'s is specified as

(2.11) $$\mathrm{Cov}(\eta_{it}, \eta_{kt}) = \lambda_{ikt} C_{it}' \Phi_{ikt} C_{kt},$$

where

(2.12) $$\begin{aligned} \lambda_{ikt} &= \sigma^2 a_{iit} \\ &= \sigma^2 \rho\, a_{ikt} \qquad (i \neq k), \\ a_{ikt} &= \min(s_{M(i)}^{\theta}, \; s_{M(k)}^{\theta}) \exp(-\mid s_{M(i)} - s_{M(k)} \mid) \end{aligned}$$

and

(2.13) $$\Phi_{ikt} = (\phi_{ikt\cdot jr}) = (\exp(-\mid s_{\cdot j} - s_{\cdot r} \mid)).$$

In the specification in (2.11), the following variational features of bond prices are taken into account;

(a) prices between a short term bond and a long term bond are less correlated,
(b) random discount factors for cash flows occurring close by are more correlated, and
(c) bond prices with shorter maturities fluctuate less.

market) model. It is expressed as a regression model and estimated by the GLS (generalized least squares) method (see KT (1994) for details).

TDM model

The stochastic process of the discount function in (2.3) evolves with t. In other words, the term structure of the stochastic discount function changes over time. The time series evolution is specified in KT (1994) as

$$(2.14)\qquad D_{it}(s) = \overline{D}_{it}(s)+\xi_t\nu_{it-h}(s+h)+\tau_{it}(s) \quad \text{or equivalently}$$
$$\nu_{it}(s) = \xi_t\nu_{it-1}(s+h)+\tau_{it}(s), \quad \nu_{it-h}$$

where

$$\nu_{it}(s) = D_{it}(s)-\overline{D}_{it}(s)$$

and for each i

$$\{\tau_{it-kh}(s+kh):\quad k=1,2,\cdots,K\} \quad \text{with} \quad Kh\leqq t$$

is a white noise for given t, h and s. Here the relation of $t-h$ and $s+h$ in $\nu_{it-h}(s+h)$ describes the fact that s years from t is equal to $s+h$ years from $t-h$. In the sequal, we set $h=1$ by measuring t and s in a specific unit of time. Let

$$(2.15)\qquad D_{it}(s_a^t) = (D_{it}(s_{a1}^t),\cdots,D_{it}(s_{aM}^t))',$$
$$\nu_{it}(s_a^t) = (\nu_{it}(s_{a1}^t),\cdots,\nu_{it}(s_{aM}^t))', \quad \text{and}$$
$$\tau_{it}(s_a^t) = (\tau_{it}(s_{a1}^t),\cdots,\tau_{it}(s_{aM}^t))',$$

where

$$(2.16)\qquad s_a^t=(s_{a1}^t,\cdots,s_{aM}^t)'.$$

Then (2.14) is expressed as

$$(2.17)\qquad D_{it}(s_a^t) = \overline{D}_{it}(s_a^t)+\xi_t\nu_{it-1}(s_a^{t-1})+\tau_{it}(s_a^t),$$

where $s_{aj}^t+1=s_{aj}^{t-1}$ and

$$(2.18)\qquad \overline{D}_{it}(s_a^t) = (\overline{D}_{it}(s_{a1}^t),\cdots,\overline{D}_{it}(s_{aM}^t))'.$$

Also noting $C_{it-h}(s+h)=C_{it}(s)$, let

$$(2.19) \qquad C_{it}(s_{\cdot}^{t}) = (C_{it}(s_{\cdot 1}^{t}), \cdots, C_{it}(s_{\cdot M}^{t}))'$$

Then the model is rewritten as

$$(2.20a) \qquad P_{it}(0) = C_{it}(s_{\cdot}^{t})' \bar{D}_{it}(s_{\cdot}^{t}) + \eta_{it}(s_{\cdot}^{t}) \qquad \text{with}$$

$$(2.20b) \qquad \eta_{it}(s_{\cdot}^{t}) = \xi_{t} \eta_{it-1}(s_{\cdot}^{t-1}) + \varepsilon_{it}(s_{\cdot}^{t}),$$

where

$$(2.21) \qquad \eta_{it-1}(s_{\cdot}^{t-1}) = C_{it-1}(s_{\cdot}^{t-1})' \nu_{it-1}(s_{\cdot}^{t-1}) \qquad \text{and}$$

$$\varepsilon_{it}(s_{\cdot}^{t}) = C_{it}(s_{\cdot}^{t})' \tau_{it}(s_{\cdot}^{t}) .$$

The model (2.20) with (2.21) is called the TDM model, which is equivalent to the CSM model except for the fact that the random part $\eta_{it}(s_{\cdot}^{t})$ of price $P_{it}(0)$ follows a time dependent Markov model as in (2.20b). The same cross-sectional covariance structure as (2.11) is assumed for $\tau_{it}(s_{\cdot j})$'s for given t, and then the TDM model is estimated by the GLS method (see KT (1994)).

3 Prediction of bond prices

In KT (1944), the CSM and TDM models are applied to monthly data of JG bond prices for the period 1991.1～1992.12, where the initial maturities of these bonds are 10 years, there are 44 to 95 bonds for each month, which were issued at different times and the face value of a bond here is 100 yen. The in-sample performance of the TDM model is very good and the standard errors of the 120 models estimated are almost all less than 0.5 yen. There the mean discount function is specified as a polynomial of order 2;

$$(3.1) \qquad \bar{D}_{it}(s_{\cdot j}) = 1 + (\delta_{11t} z_{i1t} + \delta_{12t} z_{i2t}) s_{\cdot j} + (\delta_{21t} z_{i1t} + \delta_{22t} z_{i2t}) s_{\cdot j}^{2},$$

where z_{i1t} and z_{i2t} are respectively the coupon rate and term to maturity of the i-th bond at t.

In this section, one month ahead ($t+1$) prices of individual bonds are predicted based on

$$(3.2) \qquad \hat{P}_{it+1}(0) = C_{it+1}(s_{\cdot}^{t+1})' \hat{\bar{D}}_{it+1}(s_{\cdot}^{t+1}) + \xi_{t+1} \hat{\eta}_{it}(s_{\cdot}^{t})$$

where we assume $\xi_{t+1} = 1$ based on the result of TK (1994) for prediction and $\hat{\eta}_{it}(s_i^t)$ is the GLS residual of the CSM model. We also assume $\theta = 0$ in (2.12). When t is replaced by $t+1$ in (3.1), the unknown parameters δ_{kjt+1}'s need to be predicted so that $\hat{\bar{D}}_{it+1}$'s are predicted. For this purpose, we apply a VAR model to estimate parameters $\hat{\delta}_r$'s $(r=1,\cdots,t)$, where

$$\hat{\delta}_r = (\hat{\delta}_{11r}, \hat{\delta}_{21r}, \hat{\delta}_{21r}, \hat{\delta}_{22r})'.$$

The movement of $\hat{\delta}_t$'s, which are estimated by the TDM model with $\xi_t = 1$ is given in Fig. 3-1.

Fig. 3-1, Fig. 3-2

The AIC selects VAR(1) models for $\hat{\delta}_{kj}$'s till December of 1989 and VAR(2) models afterwords. In Fig. 3-2, we also plot one month ahead predicted values of δ_{kjt+1}'s based on the selected VAR models, where $t = 1988.1, \cdots, 1992.12$.

Although the δ_{11t+1}'s are well predicted over all the period, the other δ_{kjt+1}'s are not well predicted before 1991. In particular, the predictive performance of the model for δ_{22t+1}'s is not good till the middle of 1991. This is reflected in the prediction of prices. The estimated prediction standard deviations in terms of prices defined by

$$v_{t+1} = \left[\frac{1}{N}\sum_{i=1}^{N}(P_{it+1}(0) - \hat{P}_{it+1}(0))^2\right]^{1/2}$$

are plotted for $t+1 = 1988.1, \cdots, 1992.12$ in Fig. 3-3 and tabulated in Table 3-1.

Fig. 3-3, Table 3-1

From the graph and table it is observed that the prices of the months of 1991 and 1992 are better predicted than the prices of the earlier months. In fact, for the period from 1991.1 through 1992.12, the estimated standard deviations of prediction are less than 1.2 yen except for $v_{1991.3} = 1.83$, $v_{1991.6} = 1.59$ and $v_{1992.4} = 2.06$. The mean and standard deviation of the estimated prediction standard deviations for the period of 1988.1~1990.12 are respectively 1.691 and 0.690 while for the period of 1991.1~1992.12 they are respectively 0.811 and 0.454. Hence in the recent two years the predictive performance is rather good on the average. In particular, it is even better for the months of 1992. The stimated prediction standard deviations of prices via the CSM models are also plotted in Fig. 3-3. The prediction performance of the TDM model is not far diferent from that of the CSM model for the period up to 1990.12, but for the period of 1991.1~1992.12, the TDM model performs better. In fact, the mean and standard deviation of the prediction standard deviations of the CSM model are respectively 1.149 and 0.441 for the period of 1991.1~1992.12. In Fig.3-4, the predicted and realized prices of individual bonds are plotted for the months of 1992.1, 1992.6 and 1992.12, their residuals $P_{it+1}(0)-\hat{P}_{it+1}(0)$ are plotted in Fig.3-5 and their predicted returns $[\hat{P}_{it+1}(0)-P_{it}(0)]/P_{it}(0)\times 100$ and realized returns are plotted in Fig.3-6. From Fig.3-5, it is observed that the residuals follow some patterns in the term structure. In particular, they exhibit bigger deviations for longer term bonds, which may indicate that the covariance structure in (2.12) for $\tau_{it}(s_{\cdot j})$'s with $\theta=0$ is not good enough to delete the feature that prices of longer term bonds fluctuate more than those of shorter term bond. However, in terms of returns, the patterns of the term structure of the realized returns are well captured by those of the predicted returns, and hence using this information, we will be able to make investment decisions that longer term bonds should be included in a portfolio for 1992.12.

References

Heath,D., Jarrow,R. and Morton (1990). Bond pricing and the term structure of interest rates: a discrete time approach, *Journal of Financial and Quantitative Analysis* 25, 419-440.

Heath,D., Jarrow,R. and Morton (1990). Bond pricing and the term structure of interest rates: a new methodology for contigent claims valuation, *Econometrica* 60, 77-105.

Ho,T.S. and S.Lee (1986). Term structure movements and pricing interest rate contigent claims, *Journal of Finance* 41, 1011-1028.

Kariya,T. (1993). *Quantitative Methods for Portfolio Analysis*, Kluwer Academic Publisher.

Kariya,T. and Tsuda,H. (1994). New bond peicing models with applications to Japanese data, *Financial Engineering and the Japanese Markets* 1, 1-20, Kluwer Academic Publishers.

McCulloch,J.H. (1971). Measuring the term structures of interest rates, *Journal of Business*, 19-31.

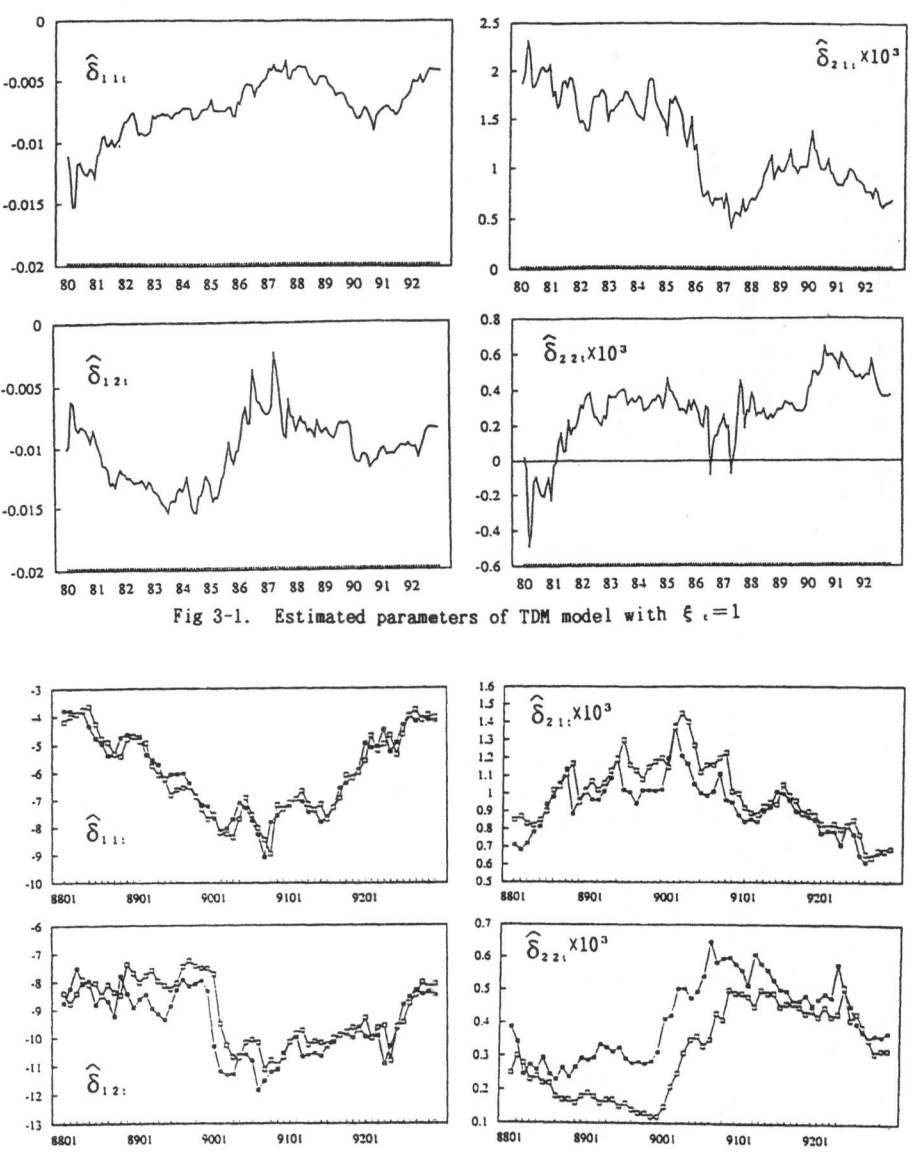

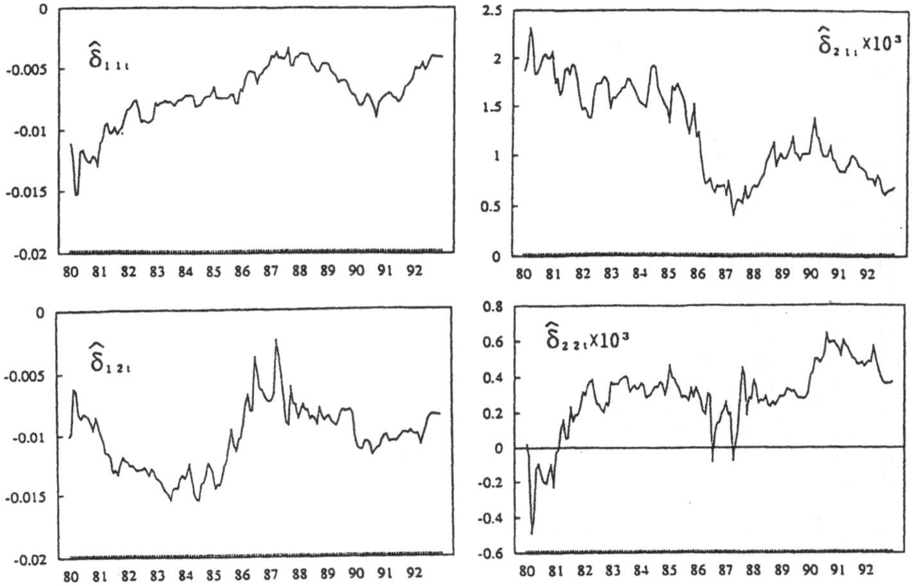

Fig 3-1. Estimated parameters of TDM model with $\xi_t = 1$

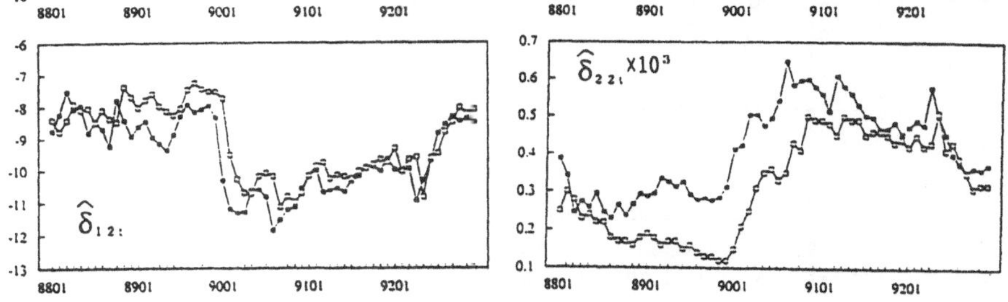

Fig 3-2. Prediction of estimated parameters via VAR model
Here ■ and □ respectively cenote realized values and one month ahead predicted values of estimated parameters.

Table 3-1 Estimated standard deviations of prediction of the TDM model *N* and *v* denote respectively the number of bonds and standard deviation (Yen).

	88		89		90		91		92	
	N	*v*	*N*	*v*	*N*	*v*	*N*	*v*	*N*	*v*
1	83	2.181	83	1.607	84	2.812	85	0.698	86	0.842
2	81	1.113	80	2.458	82	1.628	83	0.446	84	1.127
3	82	0.763	82	1.244	83	2.182	83	1.833	84	0.273
4	83	0.656	82	0.963	84	1.742	83	0.901	83	2.059
5	81	2.821	82	1.240	82	2.523	83	0.503	84	0.642
6	84	2.233	82	1.289	84	1.941	84	1.585	83	0.764
7	84	0.848	82	2.751	85	1.968	84	0.771	80	0.674
8	82	2.696	82	1.068	82	2.786	83	0.241	79	0.809
9	83	0.630	83	1.242	83	1.857	85	0.506	80	0.529
10	83	1.615	83	1.465	84	2.746	84	1.118	79	0.381
11	80	0.450	82	1.807	83	1.165	85	0.618	79	0.449
12	83	1.404	82	1.710	84	1.296	86	1.061	76	0.646
$\bar{v}$		1.451		1.570		2.054		0.857		0.766

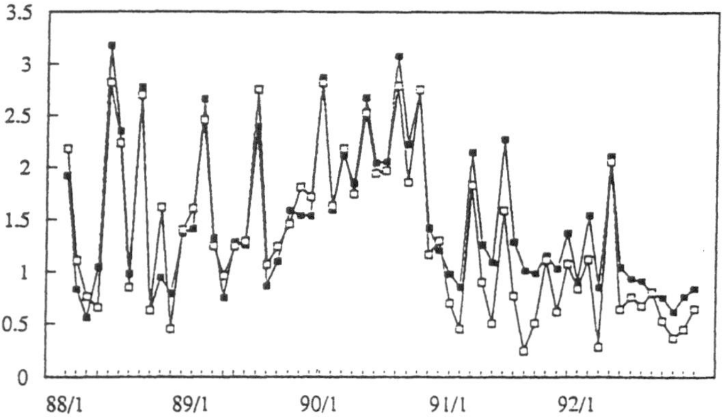

Fig 3-3. Prediction standard deviations (v_{t+1})
Here ■ and □ respectively denote the prediction standard deviations of CSM and TDM model.

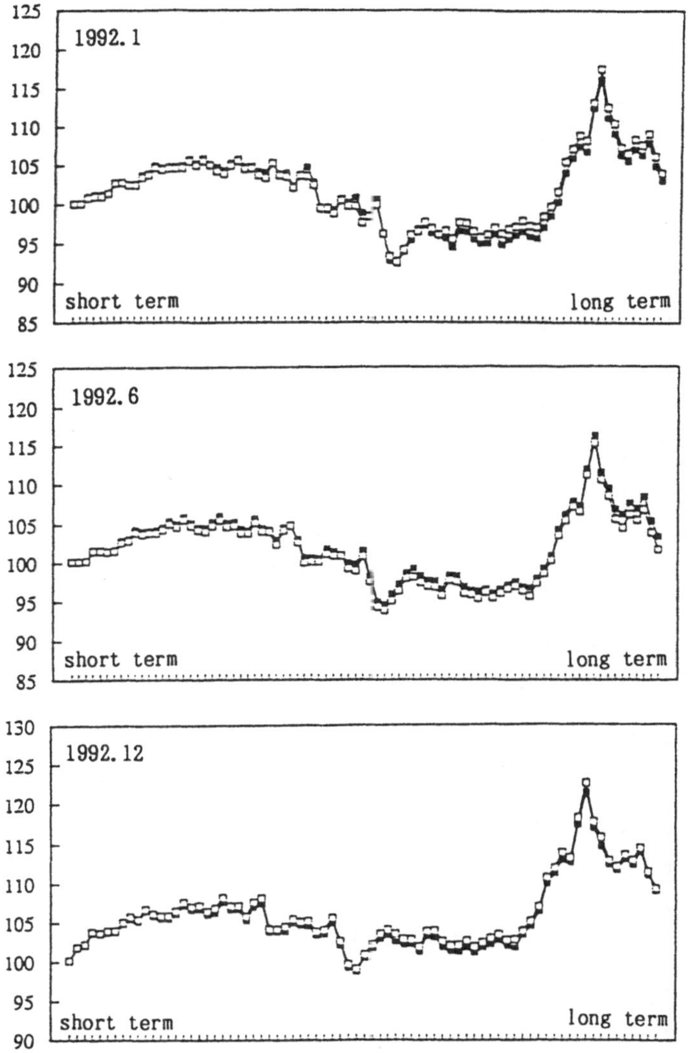

Fig 3-4. Realized values (■) and predicted values (□) of individual prices (yen)

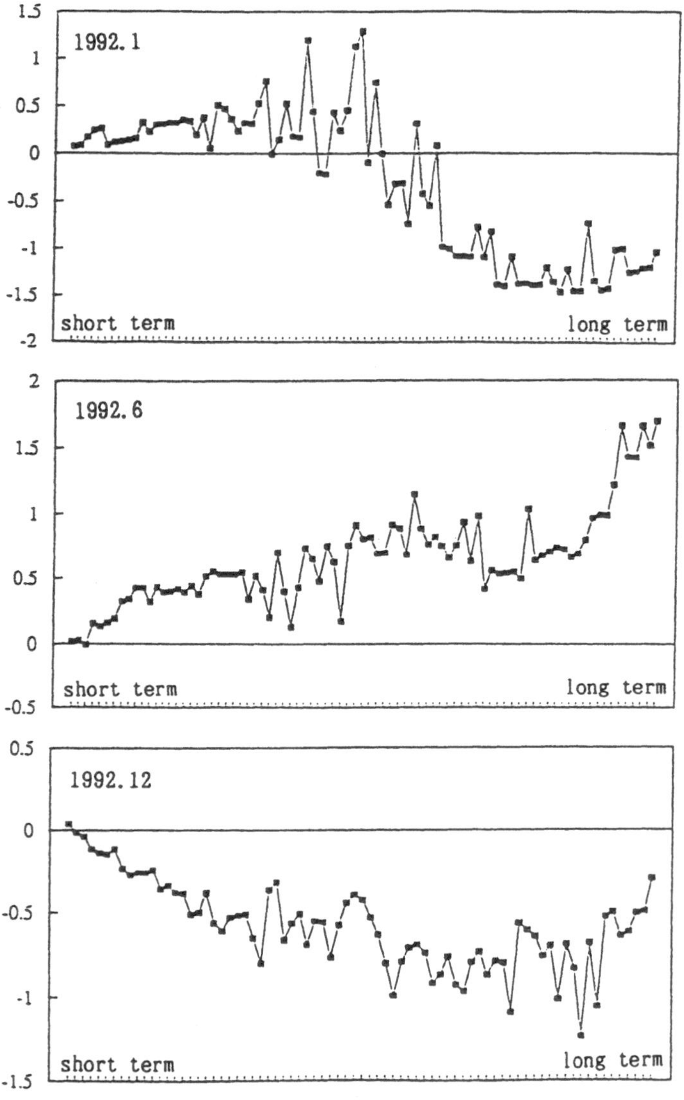

Fig 3-5. Term structure of predicted errors (yen)

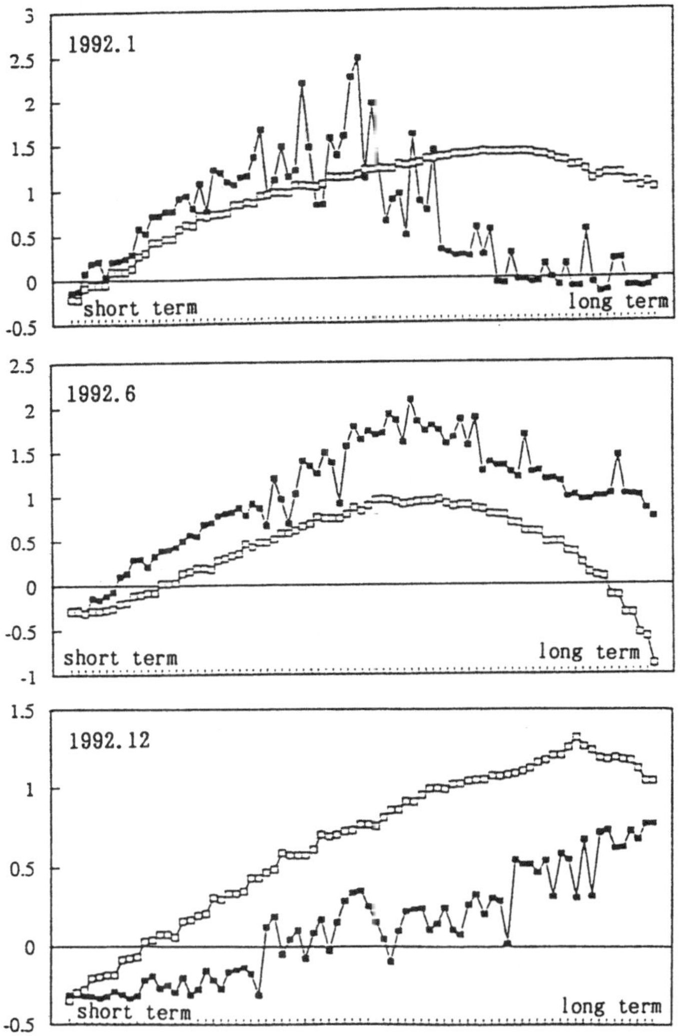

Fig 3-6. The term structure of realized returns (■) and predicted returns (□) (%)

Section VII

Time Series Modelling and Applications

On the Use of Canonical Correlation Analysis in Testing Common Trends

N.H. Chan
Department of Statistics
Carnegie Mellon University
Pittsburgh, PA 15213

Ruey S. Tsay
Graduate School of Business
University of Chicago
Chicago, IL 60637

Abstract

Motivated by the asymptotic uncorrelatedness between the stationary and nonstationary components of a vector time series, a statistic is constructed from the canonical correlations of these components to test for the number of common trends and, hence, the presence of co-integration. For univariate series, such a test statistic possesses direct relationships with the classical Dickey-Fuller test. An iterative testing procedure is then proposed which can handle unit roots of higher multiplicities as well as seasonal co-integrations. In applications, both bootstrap and simulation are used to obtain the empirical critical values of the test statistic. The proposed procedure is illustrated by two real examples.

1 Introduction

Recently, considerable attention has been given in the econometric literature to the issue of testing for common trends in a co-integrated vector time series, e.g. Engle and Granger (1987). To introduce the concept of co-integration, it is best to consider the simple case of two related time series X_{1t} and X_{2t} of interest. Individually, both time series are non-stationary of order d, that is, their d-th differences, $(1-B)^d X_{it}$, are stationary, where $d > 0$ and B is the usual backshift operator. The two series are said to be *co-integrated* (of order $b > 0$) if there exists a non-null linear combination $Z_t = \alpha_1 X_{1t} + \alpha_2 X_{2t}$ such that $(1-B)^{d-b} Z_t$ is a stationary process, where $d - b \geq 0$. The vector $(\alpha_1, \alpha_2)'$ is called a *co-integrating vector* and the vector process $X_t = (X_{1t}, X_{2t})'$ a co-integrated series. From a practical point of view, the interesting feature is that for $d = b$, co-integration says that some linear combination of individually non-stationary time series is stationary. This phenomenon was discussed by Box and Tiao (1977) via the well-known U.S. Hog Data, where $d = b = 1$. But only recently, primarily due to the development in the study of unit-root time series, such possibilities of introducing stationarity are rigorously investigated.

In the above two series example, there must exist another linear combination $Y_t = \beta_1 X_{1t} + \beta_2 X_{2t}$, which jointly with Z_t, contains the full information of X_{1t} and X_{2t} and is an integrated process of order d. This process represents the common non-stationary component of X_t. Consequently, that X_t is a co-integrated process means that the non-stationarity of X_{it}'s is attributed to a single source (or in general to a source of dimension lower than that of the observed vector process X_t). In the literature, such a Y_t process is referred to as a "common trend" of the original series X_{it}, because jointly there is only a single unit root in the system. Stock and Watson (1988) gave some accounts on why studying common trends is important both from a theoretical and a practical point of view. In terms of methodologies, several different procedures have been proposed to test for the presence of co-integrations, see for example, Engle and Granger (1987), Stock and Watson (1988), Johansen (1988), and Reinsel and Ahn (1992) among others. Gonzalo (1990) gave a review and comparison on some of these methods. Some reservations, however, remain as to the practical usefulness of co-integration tests (see Tiao, Tsay and Wang (1993)).

A main feature in handling co-integrated systems is the persistency of the unit-root phenomenon. This persistent phenomenon could result in nonstandard limiting distributions for some commonly used statistics, see Chan and Wei (1988) and Tsay and Tiao (1990) among others, for the asymptotic properties of the least squares estimators for non-stationary time series.

A second feature of a co-integrated system is the presence of common trends. Mathematically, the number of common trends in a co-integrated series is the number of independent non-stationary components in the joint series. Knowing this number is equivalent to knowing the number of (independent) co-integrating vectors in the system. Such information could have interesting implications in studying the joint behavior of a vector series. For instance, it provides the necessary information for understanding the eventual forecasting function of a vector time series. In econometrics, the common trends also provide information on the number of persistent innovations (or shocks) to the system under study.

Naturally, to investigate co-integration, one needs to make use of the unit-root characteristics of the system. This is also the approach taken in this paper. Motivated by the asymptotic uncorrelatedness between stationary and nonstationary components, cf. Theorem 3.4.1 of Chan and Wei (1988), one might expect that statistics constructed from canonical correlations between the differenced series and the lagged original process could reveal the number of common trends and the associated co-integration factors, if any. Canonical correlation analysis has been used extensively in multivariate time series analysis, especially in specifying parsimonious models for stationary as well as unit-root nonstationary vector time series, see Tiao and Tsay (1989) and the references therein. Box and Tiao (1977) use canonical correlation analysis to study the dynamical structure of a vector time series.

The main purpose of this paper is to provide a statistical procedure that makes use of the canonical correlations between the differenced series and the lagged original process to detect the number of common trends. It turns out that the proposed procedure can be easily extended to deal with unit roots of higher multiplicities as well as complex roots on the unit circle. Similar ideas of using canonical correlations to test for common trends are also discussed in Johansen (1988) and Reinsel and Ahn (1992) among others. Most of these results, however, consider mainly for the simple unit root cases.

Another goal of this paper is to relax the dependence of co-integration tests on the normality assumption and asymptotics. Most of the co-integration tests available in the literature use percentiles obtained by large scale simulations that rely on using Gaussian errors in the generating process. In practice, there seems to be a need to address the issue of non-normality and finite sample properties of a test statistic. In this paper, we use bootstrap and simulation to obtain finite sample percentiles of the test statistic.

Simulation provides finite sample results whereas bootstrap further relaxes the normality assumption.

This paper is organized as follows. Section 2 describes the proposed test procedure, the asymptotic properties of the test statistic, and two methods for obtaining finite sample percentiles of the test statistic. Sections 3 applies the procedure to two interest-rate data sets; one from the U.S. and the other from Taiwan. Finally, some discussions and extensions of the proposed procedure are given in Section 4.

2 A Test Prodecure

In this section we consider a test procedure for determining the number of unit-root common trends in a k-dimensional time series X_t based on the observations $\{X_t\}_{t=1}^n$, where n is the sample size. The null hypothesis under study is that there is no co-integration, i.e. X_t has k unit roots. The section is divided into subsections.

2.1 A test statistic

As mentioned earlier, the main idea of the proposed method lies in exploiting the asymptotic uncorrelatedness of the stationary component $\{\Delta X_t\}$, where $\Delta X_t = X_t - X_{t-1}$, and the nonstationary part of $\{X_{t-1}\}$. To this end, consider the canonical correlations between X_{t-1} and ΔX_t. Define the following moment matrices:

$$\begin{aligned} S_{00} &= \sum_{t=1}^{n} X_{t-1}X_{t-1}', \quad S_{01} = \sum_{t=1}^{n} X_{t-1}\Delta X_t', \\ S_{10} &= S_{01}' = \sum_{t=1}^{n} \Delta X_t X_{t-1}', \quad \text{and} \quad S_{11} = \sum_{t=1}^{n} \Delta X_t \Delta X_t'. \end{aligned}$$

Let $A = S_{00}^{-1}S_{01}S_{11}^{-1}S_{10}$, where the inverses S_{00}^{-1} and S_{11}^{-1} exist almost surely under the conditions given in the next subsection. Further, let $\lambda_1 < \ldots < \lambda_k$ be the eigenvalues of A and $v_1, \ldots, v_k$ be the corresponding (right) eigenvectors. Observe that the λ_i's are the squares of the canonical correlations of X_{t-1} and ΔX_t and the v_i's provide the canonical variables, upon a scale factor. For further information on canonical correlation analysis, see Chapter 12 of Anderson (1984). The test statistic employed in this paper is

$$\Lambda = n \sum_{i=1}^{k} \lambda_i. \tag{1}$$

2.2 Some asymptotic results

As will be seen later, to detect the number of common trends in X_t via the proposed procedure, it suffices to consider the properties of the test statistic Λ in (1) for the case $d = 1$. The case of unit roots with higher multiplicities can be handled by iterations. Following Engle and Granger (1987), a useful way to express a co-integrated series X_t is to use the moving average representation of the differenced series

$$\Delta X_t = C(B)\epsilon_t, \quad \sum_{j=1}^{\infty} j|C_j| < \infty, \tag{2}$$

where $C(z) = \sum_{i=0}^{\infty} C_i z^i$ with $C(0) = I_k$, $\Delta = (I_k - I_k B)$, and $\{\epsilon_t\}$ is a sequence of independent and identically distributed random variables with mean zero and positive-definite covariance matrix Σ. The assumption of $\Sigma > 0$ ensures the existence of the matrix A defined in Subsection 2.1. Engle and Granger (1987) show that if X_t is co-integrated, then $C(1)$ is singular. Further discussions on this property can be found in Stock(1987). Let $e_t = C(B)\epsilon_t$ and $X_n(v) = n^{-\frac{1}{2}} \sum_{i=1}^{[nv]} e_i$, where $[nv]$ is the integer part of nv with $v \in [0,1]$. Suppose that the sequence $\{e_t\}$ has a continuous spectral density function $f_{ee}(\omega)$ and satisfies the following conditions as $n \to \infty$:

$$E(X_n(1)X_n'(1)) \to \Omega_k, \tag{3}$$

$$n^{-1}(\sum_{t=1}^{n} e_t e_t') \to \Omega_e \quad \text{in probability,} \tag{4}$$

for some positive definite matrices Ω_k and Ω_e. Then, we have the following result due to Phillips (1988).

Proposition 1. Under (2) to (4), as $n \to \infty$,

$$X_n(v) \Rightarrow B(v) \text{ for } v \in [0,1],$$

where $B(v) = \Omega_k^{\frac{1}{2}} W(v)$, $W(v)$ is a k-dimensional Brownian motion, $\Omega_k = 2\pi f_{ee}(0) = \Omega_e + \Gamma + \Gamma'$, $\Omega_e = E(e_1 e_1')$, $\Gamma = \sum_{k=1}^{\infty} E(\epsilon_0 e_k')$, and

$$n^{-1} \sum_{t=1}^{n} X_{t-1} e_t' \Rightarrow \int_0^1 B(v) dB'(v) + \Gamma,$$

where "$\Rightarrow$" designates convergence in distribution.

Conditions (2) to (4) are standard assumptions for multivariate functional central limit theorem (Herrndorf, 1984) which are used extensively in econometrics (Phillips, 1988). In particular, under these conditions, the series $\{e_t\}$ encompasses the commonly used stationary and invertible vector ARMA processes. Using Proposition 1 and Lemma 3.1 of Phillips and Durlauf (1986), we have the following asymptotic result for Λ in (1).

Theorem 1. Let $\{X_t\}$ satisfy (2) with $X_0 = 0$ and $\{e_t\}$ satisfy the assumptions (3) and (4). Under the hypothesis that the rank of $C(1)$ is k, then as $n \to \infty$,

$$\Lambda \Rightarrow \text{trace } (\xi),$$

where

$$\xi = \Omega_e^{-1/2}(\int_0^1 B(v)dB'(v) + \Gamma)'(\int_0^1 B(v)B'(v)dv)^{-1}(\int_0^1 B(v)dB'(v) + \Gamma)\Omega_e^{-1/2}.$$

Note that when $C(B) = I_k$, $\Gamma = 0$ and the above result reduces to that of Johansen (1988). In general, the limiting distribution of Theorem 1 depends on the nuisance parameters Ω_e and Γ. There are many ways to handle these nuisance parameters. One possible approach is to adjust the effect of these nuisance parameters on test statistics, e.g. Stock and Watson (1988). Another approach is to assume a vector AR(p) model for X_t and use regression techniques to remove the nuisance parameters, e.g. Johansen (1988) and Reinsel and Ahn (1992). One advantage of these two adjusting methods is that asymptotic critical values of the test statistics involved can be tabulated. A third approach is to use the idea of canonical correlation analysis given in Park (1992) to transform the data.

A different approach is adopted in this paper. Instead of adjusting or removing the effects of nuisance parameters from the test statistic, we estimate the nuisance parameters to obtain the finite sample critical values of the test statistic. Specifically, the empirical percentiles of the test statistic Λ are obtained by using either a bootstrap method or simulation. An advantage of this approach is that finite sample percentiles are used instead of those of the limiting distribution. This advantage is gained by imposing an additional assumption that $\{e_t\}$ follows a parametric model with a finite number of parameters, e.g. a vector ARMA model. Details of the bootstrap method and simulation used are given in the next subsection.

2.3 Finite sample percentiles

Suppose that under the null hypothesis of no co-integration, the differenced series $e_t = \Delta X_t$ follows adequately the vector ARMA model

$$e_t = \sum_{i=1}^{p} \hat{\Phi}_i e_{t-i} + a_t - \sum_{j=1}^{q} \hat{\Theta}_j a_{t-j}, \tag{5}$$

where p and q are non-negative integers, $\hat{\Phi}_i$ and $\hat{\Theta}_j$ are $k \times k$ matrices, and $\{a_t\}_{t=2}^{n}$ is the residual series with covariance matrix $\hat{\Sigma}$. Adequacy here means that the residual series $\{a_t\}_{t=2}^{n}$ is close to white noise, i.e. its sample autocorrelation matrices are all small. The ARMA model in (5) is used for convenience. Other parametric models can be used as long as e_t can be generated from a_t and the lagged values e_{t-i} and a_{t-j}.

A bootstrap method: A bootstrap sample of the series X_t can be constructed as follows:

- Generate a bootstrap residual series $\{a_t^*\}_{t=2}^{n}$ from the fitted residuals $\{a_t\}_{t=2}^{n}$ by using simple random sampling with replacement. That is, each a_t^* is an independent random draw with equal probability from the set $\{a_t\}_{t=2}^{n}$.
- Generate a bootstrap sample $\{e_t^*\}_{t=2}^{n}$ of the differenced series e_t by

  $$e_t^* = \sum_{i=1}^{p} \hat{\Phi}_i e_{t-i}^* + a_t^* - \sum_{j=1}^{q} \hat{\Theta}_j a_{t-j}^*$$

  where $e_t^* = e_t$ for $t = 2, \cdots, 2 + p - 1$ if $p > 0$, $\hat{\Phi}_i$ and $\hat{\Theta}_j$ are given in (5).
- Construct a bootstrap sample $\{X_t^*\}_{t=1}^{n}$ of X_t by $X_t^* = X_{t-1}^* + e_t^*$ with $X_1^* = X_1$.

The bootstrap percentiles of the test statistic Λ are then obtained by (a) generating M independent bootstrap samples of X_t, (b) computing the test statistic Λ for each bootstrap sample, and (c) obtaining the percentiles of these Λ values. In this paper, we use $M = 3{,}000$. The percentiles appear to be stable even when M is increased to 10,000; see Table 3 of Example 4 below. In the proposed bootstrap method, we use the estimated coefficients $\hat{\Phi}_i$ and $\hat{\Theta}_j$ because the true coefficients are unknown. To investigate the effect of such substitutions, we compare the critical values of the test statistic via simulation. Consider a 3-dimensional time series X_t such that the first difference, $e_t = X_t - X_{t-1}$, follows the ARMA(1,1) model $e_t = \Phi e_{t-1} + a_t - \Theta a_{t-1}$ with $\text{cov}(a_t) = \Sigma$, where

$$\Phi = \begin{bmatrix} .2 & .3 & 0. \\ -.6 & 1.1 & 0. \\ .4 & .5 & .7 \end{bmatrix}, \quad \Theta = \begin{bmatrix} -.5 & 0. & .4 \\ 0. & -.5 & .5 \\ 0. & 0. & -.5 \end{bmatrix}, \quad \Sigma = \begin{bmatrix} 1. & .1 & .1 \\ .1 & 1. & .1 \\ .1 & .1 & 1. \end{bmatrix}.$$

This vector ARMA(1,1) model has strong serial correlations and represents the situation in which the convergence of the test statistic to its limiting distribution of Theorem 1 might be slow. Two hundred observations of X_t were generated from which the follwoing estimates were obtained:

$$\hat{\Phi} = \begin{bmatrix} .0 & .35 .03 & \\ -.62 & 1.12 & .01 \\ .46 & .50 & .68 \end{bmatrix}, \quad \hat{\Theta} = \begin{bmatrix} -.63 & .07 & .48 \\ -.06 & -.45 & .55 \\ -.05 & .03 & -.41 \end{bmatrix}, \quad \hat{\Sigma} = \begin{bmatrix} .89 & -.04 & .11 \\ -.04 & .91 & .11 \\ .11 & .11 & 1.15 \end{bmatrix}.$$

We then applied the proposed bootstrap method using both the "true" and "estimated" coefficients with 3000 iterations. The resulting percentiles of the test statistic are close to each other. For example, the 10%, 5% and 2.5% critical values of the test statistic Λ are 210.61, 223.67, 232.85 and 213.23, 224.98, 234.33, respectively, for the "true" and "estimated" coefficients. Thus, using the estimated coefficients in the proposed bootstrap method does not substantially alter the finite-sample critical values of the co-integration test.

A simulation method: If one further assumes that the residual series $\{a_t\}$ of model (5) is a Gaussian white noise sequence, then one can use simulation to obtain the finite sample percentiles of Λ. The procedure is the same as that of the bootstrap method except that the residual sample $\{a_t^*\}_{t=2}^n$ is generated independently from a multivariate Gaussian distribution with mean zero and covariance-matrix $\hat{\Sigma}$ in (5).

In practice, if the normality assumption is reasonable, one would expect the simulation method to work better. On the other hand, the bootstrap method is preferred when the Gaussian assumption is skeptical.

2.4 A test procedure

By repeatedly using the test statistic Λ and the preceding methods for obtaining its finite sample critical values, the following two-step procedure to detect the number of common trends in X_t is proposed.

- **Step 1.** Perform a canonical correlation analysis between X_{t-1} and ΔX_t to obtain the sample test statistic $\hat{\Lambda}$. Build an adequate vector ARMA model for ΔX_t and obtain the finite sample critical values of Λ by using the bootstrap method or simulation. Test the null hypothesis that rank $(C(1)) = k$, the dimension of X_t. If the null hypothesis is rejected, go to the next step. Otherwise, stop and no co-integration is detected, because the hypothesis that all of the components of X_t are non-stationary cannot be rejected.

- **Step 2.** Step 1 indicates that there exists at least one linear combination of X_{t-1} being co-integrated, so there exist vectors α and β such that $\alpha' X_{t-1}$, being stationary, is related to $\beta' \Delta X_t$. This is revealed by λ_k, the largest squared canonical correlations between X_{t-1} and ΔX_t. Delete this component and construct $Y_t = V' X_t$, where $V = (v_1, \ldots, v_{k-1})$, which is orthogonal to v_k if the largest eigenvalue λ_k is different from the other eigenvalues. This transformation leaves out the largest canonical correlation component. Go to Step 1 with X_t replaced by the transformed series Y_t.

The essential feature of the preceding procedure is to test for the full-rank model of the $C(1)$ matrix sequentially, i.e. testing rank $(C(1)) = \ell$, for $\ell = k, k-1, \cdots, 1$ repeatedly until the null hypothesis cannot be rejected or $\ell = 1$. The canonical transformation V deletes the most correlated component from the data each time and reduces its dimension by 1. If $k = 1, \Lambda$ has a direct meaning as can be seen from the following example.

Example 1. Let $\{X_t\}$ be the 1-dimensional random walk model $\Delta X_t = \epsilon_t$. Let ρ be the sample correlation between X_{t-1} and ΔX_t and $\hat{\beta}$ be the least squares estimate of the autoregressive coefficient in $X_t = \beta X_{t-1} + \epsilon_t$. Then

$$\begin{aligned} \rho^2 &= (\textstyle\sum_{t=1}^n X_{t-1} \Delta X_t)^2 / (\sum_{t=1}^n X_{t-1}^2)(\sum_{t=1}^n \Delta X_t^2) \\ &= (\textstyle\sum_{t=1}^n X_{t-1} \epsilon_t)^2 / \sum_{t=1}^n X_{t-1}^2 \sum_{t=1}^n \epsilon_t^2, \end{aligned}$$

which implies

$$(\sum_{t=1}^n \epsilon_t^2) \rho^2 = (\sum_{t=1}^n X_{t-1}^2)(\hat{\beta} - 1)^2.$$

Thus,

$$\Lambda = n\rho^2 \Rightarrow \tau^2 = \{\int_0^1 W(v) dW(v)\}^2 / \int_0^1 W^2(v) dv.$$

In this special case, Λ is simply the square of the correlation between X_{t-1} and ϵ_t which in turn is the square of the well-known Dickey-Fuller statistic. Hence, the limiting distribution of Λ is the same as the square of the standard unit root test, namely, τ^2. □

The test procedure also allows one to deal with unit roots of higher multiplicities. For example, if instead of (2), $\{X_t\}$ has a component such that $\Delta^h X_t^{(i)} = e_t^{(i)}$ for some i in $\{1, \ldots, k\}$, then one can start the proposed test procedure with $Y_t = \Delta^{h-1} X_t$. In this case, the canonical correlations between the nonstationary component $Y_{t-1} = \Delta^{h-1} X_{t-1}$ and the stationary part $\Delta Y_t = \Delta^h X_t$ is extracted. Once the highest multiplicity is completed, repeat the test procedure on $(\Delta^{h-2} X_{t-1}, \Delta^{h-1} X_t)$ and keep doing the analysis down to $(X_{t-1}, \Delta X_t)$. To summarize, the idea is to repeat the canonical correlation analysis in steps 1-2 on the pairs $(\Delta^{\ell-1} X_{t-1}, \Delta^{\ell} X_t)$ for $\ell = h, h-1, h-2, \ldots, 1$. To understand this idea more concretely, consider the following simple but illustrative example.

Example 2. Let a 3–dimensional time series $X_t = (X_t^{(1)}, X_t^{(2)}, X_t^{(3)})'$ be such that $(1-B)^2 X_t^{(1)} = e_t^{(1)}, (1-B) X_t^{(2)} = e_t^{(2)}$, and $X_t^{(3)} = e_t^{(3)}$ where $\{e_t^{(1)}\}$, $\{e_t^{(2)}\}$, and $\{e_t^{(3)}\}$ are sequences of independent and identically distributed random variables with mean zeros and unit variances. Since canonical correlation analysis is invariant under linear transformation of X_t, the observed process, in effect, could be $Z_t = HX_t$, where H is a non-singular 3×3 matrix. Since $h = 2$ in this case, we first analyze the canonical correlations on the pair $(Y_{t-1}, \Delta Y_t)$ where $Y_t = \Delta X_t$. In practice, $h = 2$ can be inferred from the data by testing the number of unit roots in each component. Since Y_t contains a single unit root, the proposed test procedure would require three iterations to identify a single common trend. Here, of course, a unit root in Y_t means a unit root with multiplicity 2 in X_t. Consider next the possibility of the existence of common trends of multiplicity less than two by analyzing the pair $(X_{t-1}, \Delta X_t)$. Observe that $X_{t-1} = (X_{t-1}^{(1)}, X_{t-1}^{(2)}, e_{t-1}^{(3)})'$, and $\Delta X_t = (Y_t^{(1)}, e_t^{(2)}, e_t^{(3)} - e_{t-1}^{(3)})'$. So, the first components of both X_{t-1} and ΔX_t are nonstationary whereas the last components of them are stationary. Since these two components violate the conditions of the null hypothesis, the test statistic Λ does not follow the null distribution when they are in the system. Consequently, they are deleted sequentially in Step 2 until we are left with $X_{t-1}^{(2)}$ and $e_t^{(2)}$. Thus, the presence of a unit root in the second component is recovered.

On the other hand, if instead of having a unit root in the second component of X_t, $X_t^{(2)} = e_t^{(2)}$, then the canonical correlations between X_{t-1} and ΔX_t are large and the hypothesis that there are common trends of multiplicity less than two is rejected. The procedure then detects only a double unit root in the first component. □

It is possible to construct test statistics based on the ordered eigenvalues of the matrix A so that the two steps of the proposed procedure can be combined when all the unit roots are of multiplicity 1. However, the two-step procedure can handle unit roots with higher and different multiplicities.

It should be pointed out that the same methodology can be applied to processes with complex unit roots. In that case, the limiting result of Theorem 1 will be replaced by more complicated forms of Brownian functionals such as those in Theorem 3.5.1 of Chan and Wei (1988). But the proposed method is sufficient to handle this more general form of nonstationarity.

3 Examples

The test procedure of Section 2 is applied to two real examples in this section. The first example consists of three U.S. monthly interest-rate series over the period 1960.1 through 1979.12. The second example considers three monthly time-deposit interest rates of Taiwan from 1961.3 to 1989.7. Because the interest rates were controlled by the government in Taiwan during the sample span, differences between the two examples might shed some light on the differences between a free and a controlled interest-rate market.

Example 3. The three US monthly interest-rate series considered are the Federal Fund Rate, the 90-day Treasury Bill Rate, and the one-year Treasury Bill Rate. These data have been analyzed by Stock and Watson (1988), who used the original series, and by Reinsel and Ahn (1992), who employed the logged series. Both analyses used asymptotic critical values and suggested a common trend, i.e. two co-integrated series in the data. The evidence for the second co-integrated series is not strong, however.

Following Reinsel and Ahn, we also used the logged series in our analysis. Table 1 gives the test results of the proposed procedure of Section 2 in which Part (a) shows that each individual component is unit-root nonstationary and Part (b) gives the results of joint tests. The finite sample percentiles in Table 1 were obtained by the two methods of Subsection 2.3. Specifically, they are percentiles of the test statistic Λ based on 3,000 replications of the bootstrap method or simulation. For the first iteration of the proposed

Table 1: Summary of Unit-Root Tests via Canonical Correlation Analysis for the U.S. Interest-Rate Series in Example 3 where $Q(w)$ Denotes the finite sample w-th Percentile of the Test Statistic Λ.

(a) Individual Component		
Series	Eigenvalue	Λ
Federal	.00064	.15
90-day	.00003	.01
one-year	.00063	.15

(b) Joint test				Bootstrap			Simulation		
Iter	k	Eigenvalues $\times 10^2$	Λ	$Q(90)$	$Q(95)$	$Q(97.5)$	$Q(90)$	$Q(95)$	$Q(97.5)$
1	3	.02 8.16 13.32	51.4	35.7	40.0	44.0	34.5	38.9	43.4
2	2	.02 8.16	19.6	18.3	22.0	25.5	20.0	24.4	28.5
3	1	.01	.02	4.09	5.34	6.61	4.63	6.20	7.86

procedure, the vector ARMA model used for the differenced series e_t is

$$e_t = \begin{bmatrix} .076 & -.136 & .435 \\ .212 & -.152 & .271 \\ .132 & -.068 & .235 \end{bmatrix} e_{t-1} + a_t - \begin{bmatrix} -.192 & -.071 & -.062 \\ -.035 & .091 & -.113 \\ -.044 & .245 & -.184 \end{bmatrix} a_{t-3},$$

with residual covariance matrix

$$\hat{\Sigma} \times 10^2 = \begin{bmatrix} .465 & .186 & .173 \\ .186 & .321 & .269 \\ .173 & .269 & .298 \end{bmatrix}.$$

This model appears to be adequate, because its residual autocorrelation matrices are all small compared with their sample standard errors. See Table 2 for some residual autocorrelation matrices. Because the null hypothesis of three unit roots was rejected, we deleted the canonical variate of the largest eigenvalue and iterated the proposed test procedure with a 2-dimensional series given by the two remaining canonical variates.

For the second iteration of the proposed procedure, the differenced series is again identified as a vector ARMA(1,3) model with parameter estimates

$$e_t = \begin{bmatrix} .316 & -.491 \\ .003 & -.015 \end{bmatrix} e_{t-1} + a_t - \begin{bmatrix} -.288 & .251 \\ -.098 & .001 \end{bmatrix} a_{t-3}, \quad \hat{\Sigma} \times 10^2 = \begin{bmatrix} .11 & .03 \\ .03 & .04 \end{bmatrix}.$$

Residual analysis shows that this model is also adequate. For example, consider the first 12 lags of the residual autocorrelation matrices. Only three correlations out of 48 are beyond their asymptotic 2 standard errors, and the maximum correlation is only 0.18 at the (2,2)-position of lag 4.

For the last iteration, the differenced series is identified as a univariate ARMA(1,3) model given by

$$e_t = .202 e_{t-1} + a_t + .258 a_{t-3}, \quad \hat{\sigma}^2 = .00118.$$

Table 2: Residual autocorrelation Matrices of a Vector ARMA(1,3) Model for Example 3.

Lag											
1			2			3			4		
-.02	.03	.04	-.04	-.01	.04	.04	.03	.03	.02	.09	.08
.02	.02	.04	.03	-.09	-.06	.06	.05	.04	.00	.03	.06
.05	.04	.05	.01	-.10	-.09	.04	.04	.03	-.04	-.05	-.06
Lag											
5			6			7			8		
-.01	.04	.01	.08	.06	.04	-.11	-.12	-.11	-.03	.04	-.06
-.02	-.01	.00	-.02	-.11	-.08	-.08	-.09	-.11	.08	.05	-.02
-.05	.03	.00	-.08	-.13	-.11	-.15	-.11	-.16	.07	.11	.02

The residual autocorrelations of this model are all small with the Box-Ljung Q-statiatic for the first 12 lags at 8.8, which, compared with a chi-square distribution with 10 degrees of freedom, is significant at any reasonable significance level, indicating that there are essentially no serial correlations in the rsiduals. In Reinsel and Ahn (1992) a vector AR(2) model was used for the logged interest-rate series. A lag-3 term in the moving-average part seems to improve the fit.

In computing the test statistic Λ for the data and for simulation, the sample mean is removed before obtaining the matrix A. From Table 1, it is seen clearly that (a) each individual series contains a unit root, (b) there is a co-integration in the data, and (c) there is some evidence of a second common trend in the data. However, this second common trend is rejected for a one-sided test at the 10% significance level based on the bootstrap percentiles. In summary, results here are in general agreement with those of the previous analyses. □

In this particular example, the differenced series of all three iterations follow an ARMA(1,3) model. This is just a coincidence. In general, dropping a component in a vector time series may change the vector ARMA specification. As indicated in Step-1 of the proposed procedure, there is a need to re-identify the model for the differenced series in each iteration of the proposed procedure.

Example 4. In this example, the monthly time-deposit interest rates with 1-, 3-, and 6-month maturities of Taiwan from March 1961 to July 1989 are considered. There are 340 observations. Time-plots of the data show that the three interest rates are closely related. In particular, they moved together over time with strong co-movement. This data set was used in Tiao, Tsay, and Wang (1993) to illustrate the usefulness of linear transformations in multivariate time series analysis. Using the co-integration tests of Stock and Watson (1988), Engle and Granger (1987) and Johansen (1988), these authors found that there is no co-integration in the data.

Because interest rates were controlled by the government in Taiwan during the data span, the normality assumption appears to be questionable. In fact, the interest rates only assumed about 25 different values over the data span, and the time plots of the data show patterns of step-functions. Therefore, we re-analyze the series by using critical values of

Table 3: Finite Sample Percentiles of The Test Statistic Λ for Example 4. The Null Hypothesis is That the Series Has Three Unit Roots, Where $Q(w)$ Denotes the w-th Percentile.

Method	$Q(2.5)$	$Q(5)$	$Q(10)$	$Q(90)$	$Q(95)$	$Q(97.5)$
	(a) 3,000 replications					
Bootstrap	9.03	10.4	12.4	42.0	48.3	55.6
Simulation	9.77	11.2	13.0	39.8	45.7	51.7
	(b) 10,000 replications					
Bootstrap	8.98	10.3	12.2	40.9	47.4	54.1
Simulation	9.64	11.2	13.1	40.6	46.9	53.4

Λ obtained by the proposed bootstrap method. Table 3 gives the finite sample percentiles of Λ based on 3,000 and 10,000 replications of the two methods proposed in Section 2.3. These percentiles are stable. For the data, the test statistic under the null hypothesis of three unit roots is $\hat{\Lambda} = 15.92$ which, compared with the critical values in Table 3, cannot reject the null hypothesis. The testing procedure is terminated in one iteration. Consequently, the proposed test procedure further confirms the lack of co-integration in Taiwan's monthly time-deposit interest rates.

The finite sample percentiles of Table 3 were based on the vector AR(3) model of the differenced series

$$e_t = \begin{bmatrix} .261 & .016 & .113 \\ .095 & .162 & .126 \\ .062 & .017 & .295 \end{bmatrix} e_{t-1} + \begin{bmatrix} .093 & .049 & .073 \\ .102 & .044 & .080 \\ .113 & .032 & .087 \end{bmatrix} e_{t-3} + a_t$$

with residual covariance matrix

$$\hat{\Sigma} \times 10 = \begin{bmatrix} .820 & .793 & .738 \\ .793 & .798 & .765 \\ .738 & .765 & .798 \end{bmatrix}.$$

The residual autocorrelation matrices of this model are all small. In fact, all elements of the autocorrelation matrices from lag 1 to lag 12 are less than their asymptotic two standard errors in modulus under the white noise assumption for the residuals. □

It is interesting to compare the finite sample percentiles of the bootstrap method and simulation in Table 3. The results of simulation seem to be more concentrated, suggesting that imposing the normality assumption in this particular instance underestimates the tail probability of the test statistic. Consequently, overlooking the non-normality of interest rates might result in underestimating the number of unit roots in the system.

4 Extension and Discussion

An extension: The test procedure of Section 2 can be extended to processes with complex roots on the unit circle such as seasonal co-integrations in a seasonal vector time series. The problem of testing for seasonal co-integration was investigated by Engle, et al. (1990). Here we only consider the special case in which co-integrations exist throughout the seasonal frequencies. Let s be the period of a seasonal time series X_t and $\Delta_s X_t = X_t - X_{t-s}$ denote the seasonally differenced series. Define the moment matrices:

$$\begin{aligned} S_{00}^s &= \sum_{t=1}^n X_{t-s} X_{t-s}', \quad S_{01}^s = \sum_{t=1}^n X_{t-s} \Delta_s X_t', \\ S_{10}^s &= (S_{01}^s)', \quad S_{11}^s = \sum_{t=1}^n \Delta_s X_t \Delta_s X_t'. \end{aligned}$$

Let $A^s = (S_{00}^s)^{-1} S_{01}^s (S_{11}^s)^{-1} S_{10}^s$ and $\lambda_1^s < \lambda_2^s < \ldots < \lambda_k^s$ be the eigenvalues of A^s. Again, the λ_i^s's are the squared canonical correlations of X_{t-s} and $\Delta_s X_t$. The test statistic then becomes $\Lambda^s = n \sum_{i=1}^k \lambda_i^s$. It can be shown that under conditions similar to Theorem 1, the following result holds.

Theorem 2.

$$\Lambda^s \Rightarrow \text{trace}\,(\xi^s) \quad \text{as} \quad n \to \infty, \quad \text{where}$$

$$\begin{aligned} \xi^s = {} & (\textstyle\sum_{i=1}^s \Omega_{ei})^{-1/2} (\sum_{i=1}^s \int_0^1 B_i(v) dB_i'(v) + \Gamma_i)' (\sum_{i=1}^s \int_0^1 B_i(v) B_i'(v) dv)^{-1} \\ & (\textstyle\sum_{i=1}^s \int_0^1 B_i(v) dB_i'(v) + \Gamma_i)(\sum_{i=1}^s \Omega_{ei})^{-1/2}, \end{aligned}$$

$B_1(v), \ldots, B_s(v)$ are s independent k-dimensional Brownian motions each with covariance matrix Ω_{ki}, with $\Omega_{ki} = \Omega_{ei} + \Gamma_i + \Gamma_i'$, Ω_{ei} and Γ_i are defined as in Proposition 1 for the *ith* block of the e_t's. Both the bootstrap method and simulation of Subsection 2.3 can also be extended to obtain the finite sample percentiles of the test statistic Λ^s.

Discussion: A procedure for testing co-integrations in a vector time series is considered in this paper. The test statistic employed is an unadjusted version of Reinsel and Ahn (1992). It is simple and easily computable. Its limiting distribution, however, depends on the nuisance parameters that are often unknown in practice. To circumvent this difficulty, a bootstrap method and a simulation procedure for calculating finite sample critical values of the test statistic are proposed. This is achieved by assuming that the differenced series follows a parametric model under the null hypothesis, yielding a new approach to co-integration testing. Contrary to most of the commonly used co-integration tests that require adjustments for nuisance parameters and use asymptotic results, the finite sample critical values of the proposed test can be obtained by taking advantage of modern computing power. For the examples considered in Section 3, the computing time for 3,000 replications of bootstrap samples or simulation require only 2 to 4 minutes on a SUN SPARCstation 10.

Acknowledgement: This research is supported in part by the National Science Foundation Grants DMS-8902177, DMS-9003324, by the Office of Naval Research Grant N00014-89-J-1851, and by the Institute of Mathematics and Applications with funds provided by the National Science Foundation while the authors were in residence at the Institute during the Summer of 1990. We thank a referee and Dr. S.K. Ahn for helpful comments.

REFERENCES

Anderson, T.W. (1984). *An Introduction to Multivariate Statistical Analysis, 2nd Ed.*, New York: Wiley.

Box, G. E. P. and Tiao, G. C. (1977). A Canonical Analysis of Multiple Time Series. *Biometrika*, **64**, 355-365.

Chan, N.H. and Wei, C.Z. (1988). Limiting Distributions of Least Squares Estimates of Unstable Autoregressive Processes. *The Annals of Statistics*, **16**, 367-401.

Engle, R.F. and Granger, C.W.J. (1987). Co-integration and Error Correction: Representation, Estimation, and Testing. *Econometrica*, **55**, 251-276.

Engle, R.F., Granger, C.W.J., Hylleberg, S. and Lee, H.S. (1990). Seasonal Co-integration: The Japanese Consumption Function 1961.1 - 1987.4. UCSD Discussion Paper.

Gonzalo, J. (1990). Comparison of Five Alternative Methods of Estimating Long Run Equilibrium Relationships. UCSD Discussion Paper.

Herrndorf, N. (1984). A Functional Central Limit Theorem for Weakly Dependent Sequences of Random Variables. *The Annals of Probability*, **12**, 141-153.

Johansen, S. (1988). Statistical Analysis of Co-integration Vectors. *Journal of Economic Dynamics and Control*, **12**, 231-254.

Park, J.Y. (1992). Canonical Cointegrating Regressions. *Econometrica*, **60**, 119-143.

Phillips, P.C.B. (1988). Weak Convergence of Sample Covariance Matrices to Stochastic Integrals Via Martingale Approximations. *Econometric Theory*, **4**, 528-533.

Phillips, P.C.B. and Durlauf, S.N. (1986). Multiple time series regression with integrated processes. *Review of Economic Studies*, **53**, 473-495.

Reinsel, G.C. and Ahn, S.K. (1992). Vector AR Models With Unit Roots and Reduced Rank Structure: Estimation, Likelihood Ratio Test, and Forecasting. *Journal of Time Series Analysis*, **13**, 353-375.

Stock, J.H. (1987). Asymptotic Properties of Least Squares Estimators of Cointegrating Vectors, *Econometrica*, **55**, 1035-1056.

Stock, J.H. and Watson, M.W. (1988). Testing for Common Trends. *Journal of the American Statistical Association*, **83**, 1097-1107.

Tiao, G.C. and Tsay, R.S. (1989). Model Specifications in Multivariate Time Series (with discussion). *Journal of the Royal Statistical Association, Series B*, **51**, 157-213.

Tiao, G.C., Tsay, R.S., and Wang, T. (1993). Usefulness of linear transformation in multivariate time series analysis. *Empirical Economics*, **18**, 567-593.

Tsay, R.S. and Tiao, G.C. (1990). Asymptotic Properties of Multivariate Nonstationary Processes with Applications to Autoregressions. *The Annals of Statistics*, **18**, 220-250.

Conjugate Processes*

Clive W. J. Granger
Department of Economics
University of California, San Diego
La Jolla, CA 92093-0508
and
Jin-Lung Lin
Institute of Economics, Academia Sinica
Nankang, Taipei, Taiwan, 11529

Abstract

Two series are said to be conjugate if each of them has dynamic structure, yet they add up to a white noise. It imposes constraints on the parameters of the data generation processes and has implications for estimation and forecasting. Simulations are carried out to study the size of efficiency gain in estimation from this additional piece of information and an empirical application is considered. In addition, the relation between conjugate processes and cointegration is investigated.

1 Introduction

It is generally the case that if two stochastic processes each have fairly simple dynamic structure, then their sum will have a more complicated structure. For example, if X_t and Y_t are independent and are each generated by different AR(p) models then generally the sum will be ARMA(2p,p). However, this simple to complex effect of adding processes is not inevitable and it is possible to go from complex to simple. This paper explores a particular case of this second possibility, where X_t and Y_t have dynamic structure, yet they add to a white noise. When this occurs, X_t and Y_t will be said to be conjugate processes. It might be noted that when it is said that X_t and Y_t are "added", this can usually be equally well interpreted as meaning that a "linear combination of X_t and Y_t is formed."

Depending if the component series are independent, there are two types of conjugate processes: common factor conjugate process and independent conjugate process. For the former

*The authors would like to thank an anonymous referee for helpful comments

case, each component series has a common factor which can have any form of dynamic structure including deterministic trend and random walk but is cancelled out in the process of adding. For the latter case, component series are mutually independent but their own dynamics offset each other when combined linearly. These two types of conjugate processes impose different types of constraints on the parameters of the data generating processes and have different implications for forecasting. Independent conjugate process is discussed in next section and then the following two sections considered a pair of application. The first asks about if a pair of random walks can be cointegrated and the second looks into linear models possibly to be used to forecast a white noise.

2 Independent Conjugate Processes

If X_t and Y_t are independent, having power spectra $f_x(\omega), f_y(\omega)$ respectively, then to be conjugate these spectra must obey the constraint

$$f_x(\omega) + f_y(\omega) = \text{constant} = \frac{\sigma_z^2}{2\pi} \tag{1}$$

where σ_z^2 is the variance of the white noise Z_t such that $X_t + Y_t = Z_t$. Assuming σ_z^2 is finite, it follows that neither X_t nor Y_t can have a spectrum which is infinite at some frequency. Thus, X_t or Y_t cannot be a random walk or any form of integrated process.

For a given process X_t, the conjugate condition determines the properties of the conjugate process Y_t but not the process itself. If X_t, Y_t are independent and conjugate, so will X_t and Y_{t-k} for any k, as Y_t and Y_{t-k} have the same spectrum. Further, X_t and $Y_{t-k} + a_t$ will be conjugate, where a_t is a white noise independent of Y_t and of X_t. Now the sum will be a white noise but with a larger variance. In what follows, it is assumed that X_t and Y_t are conjugate when they add to a white noise having the smallest possible variance, except when stated otherwise. If Y_{1t} and Y_{2t} are both conjugate processes with a given X_t it now follows that they will have the same spectral shape and consequently Y_{1t}, Y_{2t} cannot themselves be conjugate with each other, except for the trivial case when X_t is itself white noise.

The following examples suffice to illustrate some extreme forms of independent conjugate processes. Let

$$\begin{aligned} X_t &= \epsilon_t + b\epsilon_{t-1} \\ Y_t &= e_t + \beta e_{t-1} \end{aligned}$$

where ϵ_t, e_t are a pair of zero mean, independent white noises.

Let $Z_t = X_t + Y_t$, then note that

$$\text{covariance}(Z_t, Z_{t-1}) = b\sigma_\epsilon^2 + \beta\sigma_e^2$$

and this is zero if $\beta = \frac{-b\sigma_\epsilon^2}{\sigma_e^2}$

As $\text{cov}(Z_t, Z_{t-k}) = 0$, all $k > 1$, because the components of Z_t are MA(1), it follows that Z_t has all non-zero lag autocorrelations equal to zero and is therefore a white noise. Thus, independent series can be conjugate processes. It may be noted that any pair of independent MA(1)s will be conjugate if weighted sums are used, provided that β and b have opposite signs.

It is easy to show that if X_t is MA(q) then processes conjugate with it will always exist. The proof starts by noting that the MA(r) process $W_{rt} = Q_t - Q_{t-r}$ where Q_t is a white noise has spectrum

$$f_r^w(\omega) = c(1 - cos(r\omega))$$

where $c = \frac{1}{2\pi}$, if var(Q) = 1. It follows that the spectrum of the weighted sum of independent MA(r) components, $r = 0, 1, 2, 3, \ldots,$ q can give a series with spectrum of form

$$f(\omega) = \sum_{j=0}^{q} a_j cos(j\omega)$$

As any MA(q) will have a spectrum of this form, the spectrum of the conjugate process will also be of this form, because of (1) and so there will always be a conjugate to an MA(q) process. As stated before, this process is not unique, but will always be MA(q). An example is if

$$X_t = \epsilon_t + \epsilon_{t-1} + \epsilon_{t-2}, \quad \text{var}(\epsilon) = 1 \tag{2}$$

then a conjugate is

$$Y_t = (1 - B)(1 + \lambda B)e_t \tag{3}$$

where $\lambda = 2 - \sqrt{3}$, $\text{var}(e) = \frac{1}{\lambda}$, indicating that it is not always easy to find an explicit representation for the conjugate process.

If X_t is an autoregressive process of order p, ie. $X_t \sim AR(p)$, with spectrum bounded above at all frequencies, it will always have a conjugate but the conjugate is not in general also an AR(p) process. A method for constructing a conjugate is as follows: Suppose that X_t is generated by

$$a(B)X_t = \epsilon_t \tag{4}$$

where ϵ_t is white noise and all the root of a(z)=0 lie outside the unit circle. Suppose that Y_t is generated by

$$a(B)Y_t = b(B)e_t \tag{5}$$

where e_t is also a white noise, independent of the series, ϵ_t. Then, Z_t given by $X_t + Y_t$ has representation

$$a(B)Z_t = \epsilon_t + b(B)e_t \tag{6}$$

If the right hand side can be represented by $a(B)\eta_t$ - that is it has the same temporal properties, with η_t white noise, then $Z_t = \eta_t$ and so will be white noise, making X_t, Y_t conjugate. The question then becomes - can $b(B)$ and var(e_t) be chosen so that the spectrum of $b(B)e_t$ equals the spectrum of $a(B)\eta_t$ minus $\frac{\sigma_\epsilon^2}{2\pi}$? This can always be done, using the same argument as for the MA(q) case and by picking var(η_t) big enough so that spectrum $(a(B)\eta_t - \frac{\sigma_\epsilon^2}{2\pi})$ is positive for all frequencies. This can be done provided $a(z) > 0, z = e^{i\omega}$, for all frequencies ω, which is always true from the condition that the spectrum of X_t is bounded. Thus, if x_t is AR(p) a conjugate process Y_t can be formed that is ARMA(p,p).

It is clear that if X_t is ARMA(p,q) with a bounded spectrum, then an independent conjugate process will exist, by a simple generalization of these arguments.

3 Common Factor Conjugate Processes

A simple case suffices to illustrate the basic concept of common factor conjugate processes. Suppose that

$$
\begin{aligned}
X_t &= W_t + \epsilon_{xt} \\
Y_t &= -W_t + \epsilon_{yt}
\end{aligned}
$$

where $\epsilon_{xt}, \epsilon_{yt}$ are mutually independent white noises. Then

$$
\begin{aligned}
Z_t &= X_t + Y_t \\
&= \epsilon_{xt} + \epsilon_{yt} \\
&= \text{white noise}
\end{aligned}
$$

As W_t can have any dynamic structure, it follows that separately X_t and Y_t can have any dynamic structure. In this case X_t, Y_t are dependent via the common factor W_t. It should be noted that the literature is currently unclear if W_t should be called a common factor or a common feature. Here no distinction is made between these two descriptions.

X_t may be called integrated of order one, denoted $X_t \sim I(1)$, if its first difference is an ARMA process with bounded spectrum $f(\omega)$, so that $0 < f(\omega) < \infty$, all ω. A pair of I(1) series X_t, Y_t are said to be cointegrated if there is a linear combination $Z_t = X_t - AY_t$ which is stationary, here denoted I(0). The consequences and uses of this concept are discussed in Granger(1986) and in the collection of readings by Engle and Granger(1991). A known consequence of cointegration is that the series must have a common factor representation

$$
\begin{aligned}
X_t &= AW_t + \epsilon_{xt} \\
Y_t &= W_t + \epsilon_{yt}
\end{aligned}
$$

where W_t is I(1) and $\epsilon_{xt}, \epsilon_{yt}$ are each I(0).

Note that $Z_t = \epsilon_{xt} - A\epsilon_{yt}$ and that generally, a linear combination of I(0) processes is I(0). In what follows, it will be assumed that $A = 1$, which corresponds to using a scaled X_t. In economics, and particular in finance, many series are thought to be random walks, according to some theory, examples being stock and commodity prices, exchange rates in a floating rate period, interest rates when not directly controlled and real aggregate nondurable consumption. The question often asked is if a pair of these random walk can be cointegrated. A problem arises from the common factor representation and the fact that the sum of a random walk and an independent white noise gives an IMA(1,1) series, so that it is integrated and its change is MA(1). thus, in the common factor representation, X_t cannot be a random walk, and W_t also be a random walk with ϵ_{xt} independent of W_t. To show that a common factor representation is possible, suppose that W_t is IMA(1,1) given by

$$(1 - B)W_t = e_t + be_{t-1}$$

$\epsilon_{xt} = \epsilon_t$ is white noise, where ϵ_t, e_t are independent processes and let $A = 1$. Then

$$X_t = W_t + \epsilon_t$$

and

$$(1 - B)X_t = e_t + be_{t-1} + (1 - B)\epsilon_t$$

If and only if $e_t + be_{t-1}$ and $(1-B)\epsilon_t$ are conjugate, so that their sum is a white noise, a_t, will X_t be a random walk given by $(1-B)X_t = a_t$. A required condition is that $\sigma_\epsilon^2 = b\sigma_e^2$ using the theory of conjugate MA(1) processes, As Y_t is also assumed to be a random walk, it follows (with cointegration parameter A=1) that if $Y_t = W_t + a_t$ with a_t white noise, then also $\sigma_\epsilon^2 = b\sigma_e^2$ and so one gets the rather strong result that the two I(0) components in the common factor representation will have identical variance, in the case where W_t is MA(1,1) and $A = 1$.

More generally, if W_t is IMA(1,q), where $X_t = W_t + \epsilon_{xt}$, with ϵ_{xt}, W_t independent, for X_t to be a random walk, ΔW_t and $\Delta\epsilon_t$ will be conjugate, and ϵ_t will be MA(q-1). The spectrum of ϵ_{xt} and ϵ_{yt} will be identical. Further, as the spectrum of $\Delta\epsilon_{xt}$ is zero at zero frequency, it follows directly that var(ΔX_t) $= 2\pi$ spectrum(ΔW) (at $\omega = 0$) and so $\Delta X_t, \Delta Y_t$ have the same variance. The same results hold if $W_t \sim$ ARMA(p,1,q) except that ϵ_{xt} and ϵ_{yt} will now be ARMA. If p=0, q=1, so that W_t is IMA(1,1) then Z_t will be white noise but this is not generally true.

If $Z_t = X_t - Y_t$ and with the additional assumption that $\epsilon_{xt}, \epsilon_{yt}$ are independent, it follows that spectrum(Z_t)= 2 spectrum (ϵ_{xt}). Thus, in this case the spectrum of ΔW_t can be determined and, at least in principle the values of p,q identified. If ϵ_{xt} and ϵ_{yt} are not independent, p and q cannot usually be determined, even in theory.

It is easy to consider a more general framework, such as having

$$Y_t = a(B)W_t + \epsilon_{yt} \tag{7}$$

where $a(B)$ is a lag polynomial with $a(1) = A$. Now the cointegrated process $\epsilon_{xt}, \epsilon_{yt}$ will be different and generally Z_t will not be white noise.

There is an interesting difference between these two types of conjugate processes. In the common factor case, Z_t cannot be forecast either using the information set $I_t : Z_{t-j}, j \geq 0$ or using the wider information set $J_t : X_{t-j}, Y_{t-j}, j \geq 0$. However, in the independent case, Z_t can be forecast using J_t, as X_t, Y_t can individually, and independently, be forecast.

There is currently interest in pairs of series that individually have a temporal structure, so that they have a non-flat spectrum, but there is a linear combination that is a white noise. That such a property occurs with important economic variables is shown in Engle and Kozicki (1993). They explain the result by the existence of common features (ie. factors), so that they use the common factor conjugate processes model. However, an implication of this model is that the spectra of X_t and Y_t differ only by a constant which may sometimes be considered to be too stringent for reality. If, in the model above, ϵ_{xt} and ϵ_{yt} are replaced by a pair of independent conjugate processes, then a wider class of actual series may have the property that a linear combination is white noise. Thus, actual data could combine the common factor conjugate process model and the independent conjugate process model.

4 Estimation Properties

Consider the simple regression

$$Y_t = \beta_1 X_t + \beta_2 X_{t-1} + e_t \tag{8}$$

where the coefficients can be estimated by various techniques. If one is given the extra information that the dependent variable is white noise, say, and possibly additionally that the explanatory variable is also white noise, then most estimation techniques are not affected, so that this additional information is not utilized. There are occasions when this extra information is available, or that a researcher would like to conduct the estimation under the assumption that this assumption is correct. Examples are price changes, or rates of return for assets in speculative markets.

Suppose that Y_t is actually generated by

$$Y_t = \beta_1 X_t + \beta_2 X_{t-1} + \epsilon_t - \alpha\epsilon_{t-1} \tag{9}$$

where X_t, ϵ_t are independent iid processes, with zero mean and unit variance and with

$$\alpha = \beta_1\beta_2 \tag{10}$$

so that the conjugate process theory holds and Y_t is white noise. Three methods of estimation have been considered in a small simulation study:

Method 1: Moving average residuals, unconstrained

The method is proposed by Balestra(1980) in which (9) is transformed into a relationship between new series

$$\begin{aligned} Y^* &= (1-\alpha B)^{-1}Y_t \\ X^* &= (1-\alpha B)^{-1}X_t \end{aligned}$$

using an approximated truncated transformation

$$(1-\alpha B)^{-1} \cong \sum_{j=0}^{T} \alpha^j B^j \tag{11}$$

Maximum likelihood estimates of β_1, β_2 and α are obtained in the usual fashion.

Method 2: Moving average residual, constrained

As method 1 but imposing the constraint that $\alpha = \beta_1\beta_2$

Method 3: Ordinary least squares, not using the information that the residual is MA(1)

Table 1 concentrates on the estimates of β_2, the crucial dynamic parameter. The simulation used time series of length 100, a true β_1 value of 1, true β_2 values of 0.1, 0.2, 0.3, etc up to 0.9 and 100 simulations. The table shows the mean $\hat{\beta}_2$, the associated bias, the root mean square error of $\hat{\beta}_2$, and a theoretical standard deviation taken from chapter 6 of Judge et.al.(1985) and the ratio of root mean squared error of method 2 to that of method 1. Note that X_t and ϵ_t are normal iid generated separately using the multiplicative congruential schemes and Box-Muller transformation.

Table 1: Simulation Results For $\hat{\beta}_2$

True β_2	0.1	0.2	0.3	0.4	0.5	0.6	0.7	0.8	0.9
Estimation Method 1									
Mean $\hat{\beta}_2$	.1048	.2079	.3060	.4155	.5126	.6087	.7041	.8095	.9179
bias $\hat{\beta}_2$	.0048	.0079	.0060	.0155	.0126	.0087	.0041	.0095	.0179
Root mean square error	.0993	.0985	.1027	.1009	.1022	.1031	.0923	.0914	.0983
Theoretical s.d.	.1022	.1022	.1022	.1022	.1022	.1022	.1021	.0968	.0709
Estimation Method 2									
mean $\hat{\beta}_2$	.0904	.1900	.2724	.3767	.4716	.5668	.6803	.7982	.8983
Bias $\hat{\beta}_2$	-.0096	-.0100	-.0276	-.0233	-.0284	-.0232	-.0197	-.0018	-.0017
Root mean square error	.0638	.0705	.0767	.0797	.0855	.0970	.0869	.0854	.1054
Theoretical s.d.	.0713	.0735	.0739	.0762	.0790	.0820	.0859	.0857	.0639
Ratio mean square error	.41	.56	.56	.58	.69	.88	.88	.86	1.14
Estimation Method 3 (OLS)									
mean $\hat{\beta}_2$	.0903	.1913	.2924	.3934	.4945	.5954	.6964	.7974	.8985
Bias $\hat{\beta}_2$	-.0097	-.0087	-.0076	-.0066	-.0055	-.0046	-.0036	-.0026	-.0015
Root mean square error	.0965	.0970	.0985	.1011	.1047	.1091	.1143	.1201	.1265

A few features of table 1 are (i) method 1 always produces a small positive bias but the other two estimation procedures produce negative biases. (ii) for true β_2 value up to 0.8 the constrained nonlinear estimation method 2 produces more efficient estimates of β_2 than the other two methods, with gain of around 50% in terms of reduction in the variation of $\hat{\beta}_2$ for true β_2 value range 0.1 and 0.5. The gain were substantially lower for the true β_2 values between 0.6 and 0.8 and there is no gain at $\beta_2 = 0.9$. (iii) there is little difference, in term of root mean square error of $\hat{\beta}_2$, between OLS and the unconstrained moving average estimation. (iv) the empirical results agree well with the theoretical variance of the estimates.

Overall it is seen that using the conjugate process constraints is generally helpful in the estimation of β_2. This is not true for the estimation of β_1, as shown in table 4 which just shows biases and root mean squared errors for $\hat{\beta}_1$ for various true β_2 values. Overall, method 1 provides the better estimates, although these are usually similar to those from method 3 (OLS), with the constrained moving average method being somewhat worse.

Equation (8) is one-way causal specification, with X_t causing Y_t both lagged and instantaneously. An equivalent specification would include just X_{t-1} and X_{t-2}, say, but now Y_t will be more forecastable from the X_t series, in general. It is possible to have a feedback system, with (8) being associated with a second equation such as

$$X_t = \gamma_1 Y_{t-1} + \gamma_2 Y_{t-1} + e_{xt} \tag{12}$$

Table 2: Simulation Results For $\hat{\beta}_1$

True $\hat{\beta}_1$	1.0	,1.0	1.0	1.0	1.0	1.0	1.0	1.0	1.0
Estimation Method 1									
Mean $\hat{\beta}_1$	1.0114	1.0042	1.0118	1.0191	1.0190	1.0190	1.0158	.9904	.9802
Bias $\hat{\beta}_1$	.0114	.0042	.0118	.0191	.0190	.0190	.0158	-.0096	-.0198
Root mean square error	.1094	.1067	.1047	.1119	.1119	.1121	.0989	.1076	.1179
Theoretical s.d.	.1010	.1010	.1010	.1010	1010	.1011	.1012	.0964	.0705
Estimation Method 2									
Mean $\hat{\beta}_1$	1.0125	1.0039	1.0187	1.0191	1.0230	1.0282	1.0183	.9877	.9891
Bias $\hat{\beta}_1$	.0125	.0039	.0187	.0191	.0230	.0282	.0183	-.0123	-.0109
Root mean square error	.1170	.1183	.1308	.1263	.1258	.1362	.1088	.1078	.1175
Theoretical s.d.	.1005	.1005	.1004	.1003	.1001	.0998	.0996	.0930	.0651
Estimation Method 3 (OLS)									
mean $\hat{\beta}_1$	.9915	.9918	.9921	.9925	.9928	.9931	.9935	.9938	.9941
Bias $\hat{\beta}_1$	-.0085	-.0082	-.0079	-.0075	-.0072	-.0069	-.0065	-.0062	-.0059
Root mean square error	.1058	.1071	.1092	.1121	.1156	.1198	.1245	.1297	.1353

Table 3: Estimated ACF and PACF for 2-month and 8-month Interest Rates

	Δ log (8-month interest rate)		Δ log (2-month interest rate)	
	Estimated ACF	Estimated PACF	Estimated ACF	Estimated PACF
Lag 1	.15	.15	.10	.10
Lag 2	.01	-.01	-.11	-.12
Lag 3	-.06	-.06	.01	.03
Lag 4	-.06	-.05	-.00	-.02
Lag 5	-.04	-.03	-.06	-.06
Lag 6	-.12	-.11	-.16	-.15
Lag 7	-.17	-.15	-.03	.00
Lag 8	.04	.09	.09	.06
Lag 9	.09	.06	.01	-.01
Lag 10	.07	.02	.06	.08
Lag 11	.15	.14	.08	.05
Lag 12	.03	-.02	.03	.01
Lag 13	.02	.00	-.03	-.02
Lag 14	.03	.04	-.01	.03
Lag 15	-.06	-.01	-.02	-.02

e_{xt} has to be a conjugate process to the MA(1) in Y_t, lagged, to ensure that X_t, is white noise. The properties of this system has not been considered at this time.

The results in this section can easily extended to multivariate models. Suppose, for example, that Y_t is generated by

$$Y_t = \beta_1 X_t + \beta_2 X_{t-1} + \gamma_1 \phi_t + \gamma_2 \phi_{t-1} + e_t \tag{13}$$

where Y, X, ϕ and e are all white noise series and with X, e and ϕ being independent, then the conjugate conditions is that

$$\beta_1 \beta_2 \text{var}(X) + \gamma_1 \gamma_2 \text{var}(\phi) = 0$$

5 Empirical Examples: A Pair of Interest Rates

To produce an example of conjugate processes, two series, yields on two-month and eight-month treasury bills, are taken from the CRSP bond file. Both series are monthly data from July 1964 to December 1988 with sample size being 293 observations. The original series are transformed by taking logarithms, then taking difference and finally subtracting their respective means. The transformed series look like white noise as shown by their autocorrelations and partial autocorrelations as reported in table 3.

Before fitting (9) and applying the estimation methods mentioned in session 4, it is worth noting that in the simulation study, the variances of X_t and ϵ_t are set to be one in advance, which is clearly not the case when using real data. To get the appropriate constraint implied by the conjugate process, square both sides of (9), then take expectations and make use of the fact that

Table 4: Estimation Results for Interest Rates
$Y = \Delta$ log (2-month interest rate); $X = \Delta$ log (8-month interest rate)

	Estimation method 1	Estimation Method 2
$\hat{\beta}_1$	.932	.937
Theoretical s.d.	.0405	.0382
$\hat{\beta}_2$	.039	.114
Theoretical s.d.	.0406	.0196
$\hat{\alpha}$	.257	.218
Theoretical s.d.	.0564	.0574

all three series, X_t, Y_t and ϵ_t are all white noises to get

$$\alpha V_y = \alpha(\beta_1^2 + \beta_2^2)V_x + (1 + \alpha^2)\beta_1\beta_2 V_x \tag{14}$$

where V_x and V_y are the variance of Y, and X respectively, which can be directly estimated.

The estimation results are summarized in table 4. Two methods of estimator were used, method 1 as discussed in section 4 and method 2 uses the constraint (14). The parameter estimates, $\hat{\beta}_1, \hat{\beta}_2$ and $\hat{\alpha}$ do not vary much across methods 1 and method 2. However, the estimated theoretical standard deviation of $\hat{\beta}_2$ using method 2 is much smaller than that using method 1, which is consistent with the simulation results. Note that -2 log likelihood ratio is 0.8408 which is far below the critical value $\chi_{0.1}(1) = 2.70554$, indicating that the data fail to reject the null hypothesis of the constraint (14) being correct. It follows that for this pair of series model (9) with constraint (14) seems to be an appropriate one.

6 Conclusion

This paper has introduced the concept of conjugate processes which might be expected to occur in economics. It adds an additional piece of information, which can be used to sharpen estimates. It is planned to investigate how to extend this idea to the multivariate case and how to make use of these properties to improve forecasts. If a series is found to be white noise, a consequence of this theory is that non-white series can possibly be used to forecast it, so that the search for forecastability should not stop.

The case considered is just when the summation of correlated variables is white noise. The theory can be easily extended to cases where this variable has any given generating mechanism, such as AR(1) or random walk.

References

[1] Balestra, P (1980), "A note on the exact transformation associated with the first-order moving process," **Journal of Econometrics**, 14, 381-394.

[2] Engle, R. & C.W.J. Granger (1991) **Long-run economic relationships: Readings in Cointegration** Oxford University Press

[3] Engle, R. & S. Kozicki (1991), "Testing for common features," **Journal of Business and Economic Statistics,** 11, 369-80.

[4] Granger, C.W.J. (1986) "Developments in the Study of Cointegrated Economic Variables,"**Oxford Bulletin of Economics and Statistics** 48, 213-218.

[5] Judge, G. G. W.E. Griffihs, R.C. Hill, H. Lutkepohl & T.C Lee (1985) **The theory and Practice of Econometrics,** 2nd. New York: Wiley

Nonlinear Time Series - Bifurcation, Chaos, and Stationarity

Wolfgang Kliemann
Department of Mathematics
Iowa State University
Ames, Iowa 50011
USA

Abstract

The qualitative behavior of time series (like stationarity and ergodicity) is often assumed to be known for time series analysis (like estimation and prediction). For nonlinear time series, the qualitative behavior is often difficult to study. Using associated control systems, we describe a method to determine the system behavior that is successful for series with bounded random term. In the process, we also analyze the relation between deterministic dynamical systems and their random counterparts. This includes bifurcation behavior depending on the parameters of the system, and random perturbations of chaotic attractors.

1 Introduction

Time series as models of 'real world systems' usually contain a variety of parameters that can be structural, fixed (estimated), time-varying (e.g. stochastic processes), or controlled. Forecasting, prediction, and the assessment of policy influences in a time series then depend often crucially on the nature of these parameters and on the assumptions made about them. While the qualitative theory of linear time series with Gaussian randomness is quite well understood, nonlinear models still pose various challenges for estimation, prediction, qualitative behavior, and control. Recent advances in the theory of perturbed and controlled dynamical systems allow us today to analyze and understand some of the basic phenomena that can occur in nonlinear time series models. The purpose of this paper is to describe a set-up in which some of these phenomena can be formulated and analyzed rigorously, together with implications for the validity of conclusions drawn from such a model.

Our starting point are nonlinear time series of the form

$$x_{n+1} = f(x_n, \alpha, \xi_n, u_n) \text{ in } \mathbb{R}^d, d \geq 1 \tag{1}$$

where x is the d-dimensional state vector. $\alpha \in \mathbb{R}^p$ is a vector of parameters whose values are assumed to be constant, i.e. either structure parameters, or parameters that have been estimated and that are set at the estimated value. $\xi_n \in \mathbb{R}^q$ is a (time-varying) random process describing the uncertainties in the model, e.g. measurement errors, parameter or dynamic uncertainties. $u_n \in \mathbb{R}^r$ is a vector of controls, i.e. of possible external intervention, where u_n can be either constant (usually called policy tuning), open loop (i.e. dependent on time and on the initial state), or feedback (i.e. a function on the entire state space). By starting from the time series (1) we assume that modeling (i.e. choosing a functional form for f), estimation of the (time-invariant) parameters in α, assumptions about the stochastic process ξ_n, and choice of a policy range for the control u_n have been performed. Note that this step may include various crucial tasks, like model reconstruction techniques from observations, comparison of different estimation procedures, decisions about stochastic vs. deterministic perturbation models, and analysis of policy tools and ranges. These problems are addressed in the statistical literature, and we point to the monograph [T] as a reference for the statistics of nonlinear time series, which also contains some results concerning the qualitative behavior from a distribution point of view.

The focus of this paper is the analysis of the possible system behavior starting from some initial value $x_0 \in \mathbb{R}^d$. This includes concepts like stationarity, multistability, transient behavior, stability, influence of the parameter α on the possible system behavior, influence of the random process ξ_n when compared to the unperturbed system with $\xi_n \equiv 0$. The precise set-up for the random process ξ_n is as follows:

Let $\{\eta_n, n \in \mathbb{N}\}$ be an ergodic Markov chain on a finite dimensional manifold N, whose one-step transition probability has a continuous density. The function $h : N \to U \subset \mathbb{R}^q$, $h(\eta) = \xi$ maps the 'background noise' to the system uncertainty ξ_n, where U is the uncertainty range (compact, convex, with $0 \in U$). In this set-up, the pair process $(\eta_n,\ x_n)$ is a Markov chain, and we assume an absolute continuity condition for the distribution of the x-component (see e.g. [Ho], Condition (E) on p. 536).

This set-up includes all models whose uncertainties are bounded functions of some ergodic Markov chain, which is realistic in applications as well as for computations.

Standard unbounded noise models, like ARMA processes, should be treated in our context by confining them to some large set $V \subset \mathbb{R}^q$, outside of which their distribution is smaller than some preset value $\gamma << 1$.

2 Some Modeling Issues

When modeling the influence of system parameters that are not structural, i.e. those parameters that need to be estimated from data, one may choose to model them as fixed (constants) or as time varying (random processes). Two possible effects have to be taken into account:

If the parameter is constant, does a small change in the parameter value change the system behavior drastically? And if an (in reality) time varying parameter is modeled as a constant, can the system behavior be substantially different for small time varying influences?

Let us look at two simple examples:

1. Example

Consider the system

$$x_{n+1} = \alpha(x_n + c)^2 \quad \text{in} \quad \mathbb{R}, \tag{2}$$

where c is a structural parameter, which we set to $c = 1$ for the purpose of this example. Assume that we have estimated $\hat{\alpha} = 0.2$ with standard deviation 0.05 from given data. The system $x_{n+1} = \hat{\alpha}(x + c)^2$ has, for x-values in the range $[0, 1]$ one equilibrium $x^* = \frac{3}{2} - \frac{1}{2}\sqrt{5} \sim 0.382$, and the system will converge for all initial values $x_0 \in [0, 1]$ towards the equilibrium x^*. However, for $\alpha > 0.25$ the system does not possess any equilibrium, and all solutions will blow up to $+\infty$. The bifurcation diagram in α captures the situation.

In Figure 1., the arrows indicate the (increasing or decreasing) direction of the solution of (2) for the corresponding (α, x)-combination.

2. Example

Here we consider a system $x_{n+1} = f(x_n, \xi_n)$, with $\xi_n \in [\alpha - \rho, \alpha + \rho]$, where the corresponding bifurcation diagram in (α, x) is given in Figure 2.

The lower line of equilibria will be approached by the system for initial values x_0 below the upper line. In particular, for $x_0 < 1$, the system will approach the lower equibrium for <u>each constant</u> α in the α-range shown in Figure 2. If, however, the parameter α is time-varying as described in Section I, with range $[\alpha_0 - \rho, \alpha_0 + \rho]$, $\alpha_0 = 1$, $\rho > 0$, then the system will blow up to $+\infty$ for all initial values $x_0 \geq 0$. The same remains true in this case for other combinations of α and ρ, depending on the explicit form of the lines of equilibria in Figure 2.

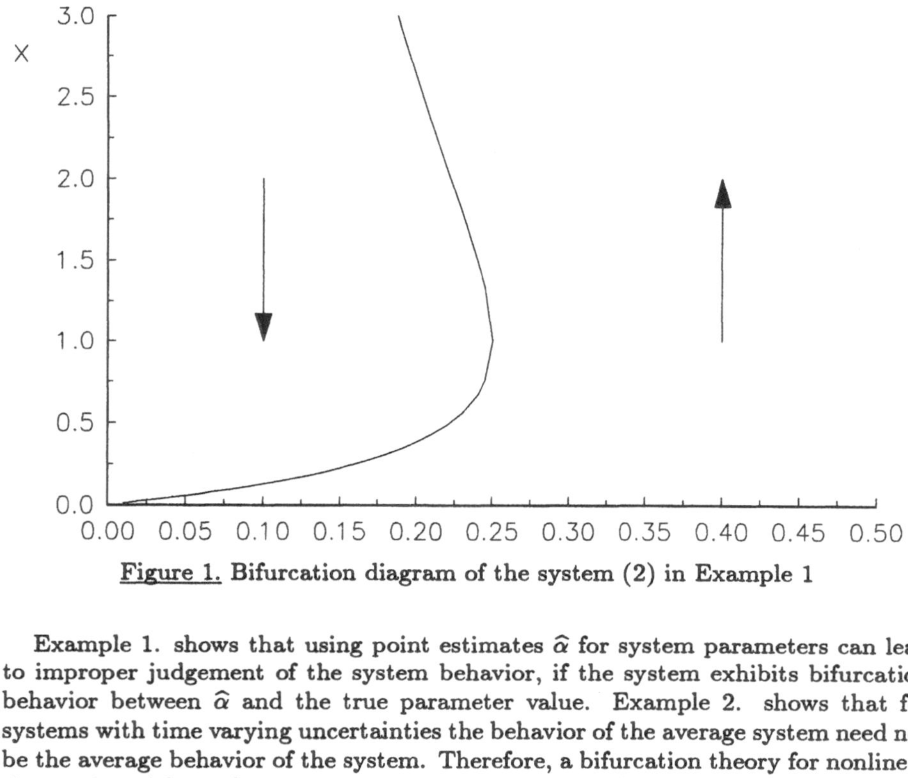

Figure 1. Bifurcation diagram of the system (2) in Example 1

Example 1. shows that using point estimates $\hat{\alpha}$ for system parameters can lead to improper judgement of the system behavior, if the system exhibits bifurcation behavior between $\hat{\alpha}$ and the true parameter value. Example 2. shows that for systems with time varying uncertainties the behavior of the average system need not be the average behavior of the system. Therefore, a bifurcation theory for nonlinear time series with random process uncertainties is needed. The next sections outline the beginnings of such a theory, drawing heavily on existing concepts and results for time continuous systems.

3 Qualitative Behavior of Nonlinear Time Series

Qualitative theory describes the behavior of (random) dynamical systems when no explicit solutions to the system equations are available. It is based on the idea to first find the limit sets of the system, i.e. those sets of points to which the system converges a time $n \to \infty$. These sets can be equilibria, periodic solutions, or more complex sets, such as chaotic attractors, and their stochastic counter parts. For Markov processes, these limit sets are the supports of invariant distributions, i.e. they correspond to stationary solutions. Next the domains of attraction of the limit sets are analyzed, i.e. those points from which the system converges to the different limit sets. In the Markovian context, we distinguish between points,

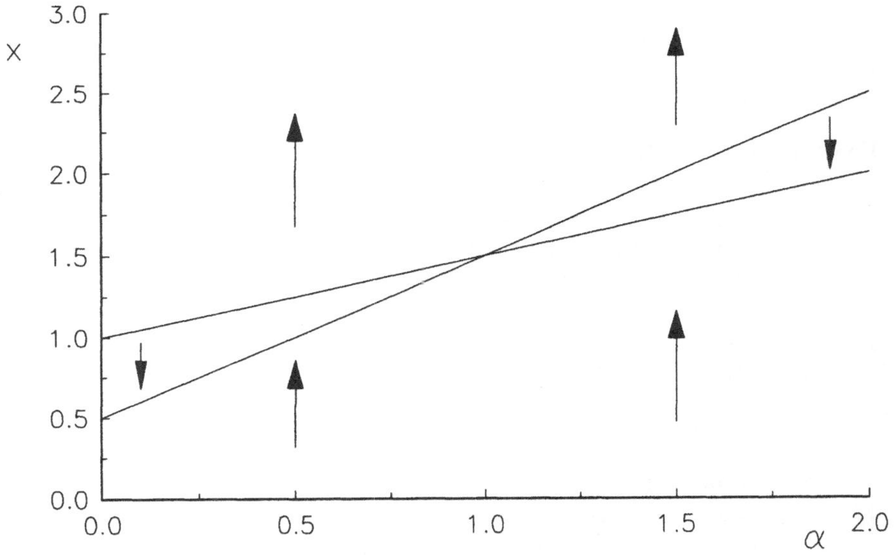

Figure 2. Bifurcation diagram of the system in Example 2.

for which a law of large numbers holds (i.e. initial values from which the system converges towards exactly one stationary solution), and multistable points, from which different limit sets are approached with positive probability. This gives an overview over the system behavior for large times. Concepts related to transient behavior (like exit times from sets, sojurn times etc.), and precise local behavior around limit sets (linearization, Lyapunov exponents, stable and unstable manifolds etc.) are beyond the scope of this short paper.

One standard way to obtain results on qualitative theory of time series of the form

$$x_{n+1} = f(x_n, \xi_n) \text{ in } \mathbb{R}^d \tag{3}$$

under the assumptions described in Section I. is to associate to (3) a control system of the form

$$x_{n+1} = f(x_n, u_n) \tag{4}$$

where $u_n \in \mathcal{U} = \{u : \mathbb{N} \to U\}$ is the set of admissible controls. We define the regions of controllability of (4), which will give information about stationarity and multistability of (3).

We denote the solutions of the control system (4) for a specified initial value $x \in \mathbb{R}^d$ and a specified control sequence $\{u_n, u \in \mathbb{N}\}$ at time $n \in \mathbb{N}$ by $\varphi(n, x, u)$. For a point $x \in \mathbb{R}^d$, the totality of all these solutions determine the reachable set

$\mathcal{O}^+(x)$, i.e. all points through which there exists a solution, starting from x, for some admissible control $u \in \mathcal{U}$: $\mathcal{O}^+(x) = \{z \in \mathbb{R}^d$; there exist $u \in \mathcal{U}$ and $n \in \mathbb{N}$ with $\varphi(n, x, u) = z\}$. Moreover, we denote the (topological) closure of a set $A \subset \mathbb{R}^d$ by $c\ell A$. With these preparations we are ready to define 'control sets'.

3. Definition A set $D \subset \mathbb{R}^d$ is a control set of (4), if

(i) for all $x, y \in D$ it holds that $y \in c\ell\mathcal{O}^+(x)$,
(ii) for all $x \in D$ there exists $u \in \mathcal{U}$ with $\varphi(n, x, u) \in D$ for all $n \in \mathbb{N}$,
(iii) D is maximal with respect to the properties (i) and (ii). A control set C is said to be invariant, if $c\ell\mathcal{O}^+(x) \subset c\ell C$ for all $x \in C$.

We refer the reader to [Ho] for a detailed discussion of control sets for nonlinear time series. Under a weak regularity condition on (4) (see e.g. [MC], [Ho], [AS], [B]) we obtain the following result (compare [Ho]).

4. Theorem *(i) All invariant distributions of the time series (3) are attained on invariant*
control sets of (4).
(ii) *Each compact invariant control set C of (4) carries exactly one stationary solution of (3), the corresponding invariant distribution μ has support* $\text{supp}\mu = C$.
(iii) *In each compact invariant set of (4) there exists at least one, and at most finitely many invariant control sets.*

Theorem 4. characterizes, in compact invariant sets, exactly the possible stationary solutions of (3): There exists one stationary solution on each invariant control set of (4). This result explains e.g. the behavior in Example 2.

5. Example 2. revisited
In one-dimensional systems the control sets are determined by the fixed points of the system, i.e. by points $x \in \mathbb{R}$ with $f(x, u) = x$ for $x \in U$, compare [CK1]. Figure 3. shows, for each α and ρ as indicated, the corresponding control sets. The sets in area A are invariant, hence carry stationary solutions, those in area B are variant. Around the point α_0 we have an area of variant control sets, hence no stationary solution can exist according to Theorem 4(i), and the system blows up to ∞ for all initial values $x_0 \geq 0$.

To describe convergence towards stationary solutions, we need two new concepts:

6. Definition For a control set D of (4), its domain of attraction is given by $\mathcal{A}(D) = \{y \in \mathbb{R}^d; \mathcal{O}^+(y) \cap D \neq \phi\}$.

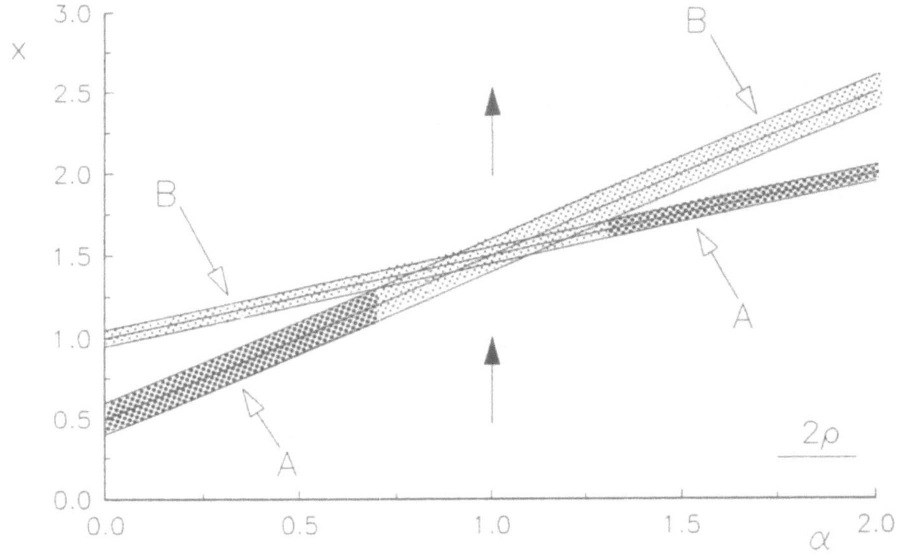

Figure 3. Control sets of the system in Example 2.

7. Definition On the control sets of (4) we define an order, $<$, via $D_1 < D_2$ if there exists $x \in D_1$ with $\mathcal{O}^+(x) \cap D_2 \neq \phi$.

We note that invariant control sets are maximal w.r.t. the order defined in Definition 7. The next important class of control sets are the relatively invariant ones:

8. Definition A control set $E \subset \mathbb{R}^d$ is called relatively invariant, if there exist (at least two) invariant control sets C_1 and C_2 such that $E < C_1$, $E < C_2$, and E is maximal among the control sets with this property.

These concepts allow us to describe convergence towards stationary solutions, again for simplicity in compact invariant sets of (4):

9. Theorem *(i) Let C be a compact control set carrying a stationary solution of the*

time series (3). Then $P\{\varphi(n,x,\xi) \to C\} > 0$ iff $x \in \mathcal{A}(C)$.

(ii) *$P\{\varphi(n,x,\xi) \to C\} < 1$ iff $x \in \mathcal{A}(E)$ for some relatively invariant control set E, otherwise this probability is equal to 1.*

(iii) *There exist at most finitely many relatively invariant control sets.*

Together with Theorem 4., this result characterizes the ergodic behavior of nonlinear time series in compact invariant sets: If there exists exactly one invariant control set, then Equation (3) has a unique ergodic solution and all other solutions converge in distribution to the ergodic one. If there exist more than one invariant control set, then the limit distribution from an initial value x is a convex combination of the stationary distributions in those invariant control sets C that satisfy $x \in \mathcal{A}(C)$, compare the next example.

10. Example

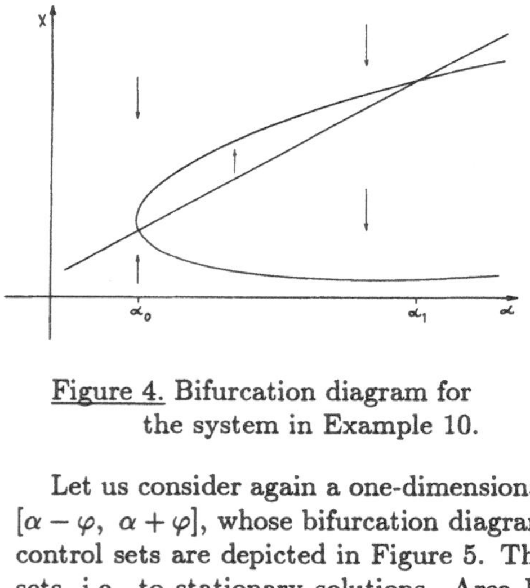

Figure 4. Bifurcation diagram for the system in Example 10.

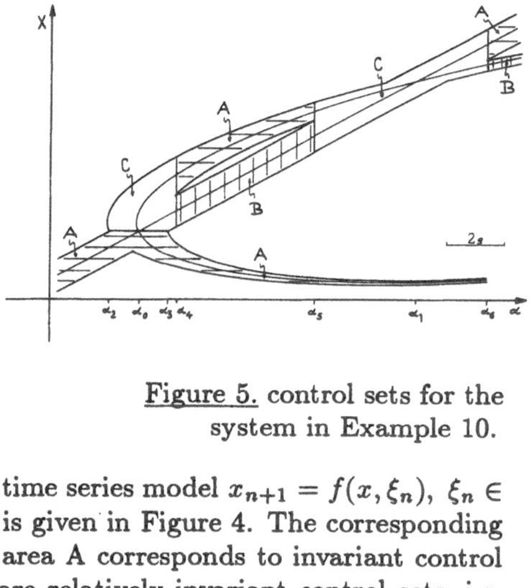

Figure 5. control sets for the system in Example 10.

Let us consider again a one-dimensional time series model $x_{n+1} = f(x, \xi_n)$, $\xi_n \in [\alpha - \varphi,\ \alpha + \varphi]$, whose bifurcation diagram is given in Figure 4. The corresponding control sets are depicted in Figure 5. The area A corresponds to invariant control sets, i.e. to stationary solutions. Area B are relatively invariant control sets, i.e. starting in these sets the system will, with positive probability, more towards each of the invariant sets in area A above and below area B, and will approach there the invariant distribution. The control sets in area C are sojourn sets, into which the time series enters at one boundary and exists at the other boundary according to the direction indicated by the arrows.

It is worthwhile to note that the control sets of (4) correspond exactly to the chaotic sets of the flow induced by the time series (3), compare [CK2] and [AS].

4 Nonlinear Time Series with Varying Uncertainty Range

In this section we discuss the behavior of nonlinear time series

$$x_{n+1} = f(x_n, \xi_n, \rho) \text{ in } \mathbb{R}^d \tag{5}$$

where ρ is a parameter that corresponds to the range of the process ξ_n. More precisely, we use the set-up in Section I. and consider a family of onto functions

$$h^\rho : N \to U^\rho,\ U^\rho = \rho \cdot U,\ \rho \geq 0$$

that map the 'background noise' η_n into the increasing family of noise ranges U^ρ. This allows us to study time series with increasing uncertainty, and gives additional information on stationarity and multistability. The precise theory is based on concepts and results from topological dynamics (see e.g. [CK3] or [B]) which are beyond the scope of this paper. We will describe intuitively the main ideas.

We need a regularity condition on the functions h^ρ, which says that a solution $\varphi(n, x, \xi, \rho)$ is contained at a later time $m > n$ in the interior of the orbit $\mathcal{O}^+_{\leq m}(x, \rho')$ for $\rho < \rho'$. This means that the increased uncertainty range does have an additional effect on the system. Furthermore, we denote by $M_1, \ldots, M_k$ the components of the limit sets of the deterministic, noise free system $x_{n+1} = f(x_n, \xi_n, 0)$, which is just a deterministic difference equation. The sets $M_1, \ldots, M_k$ can be equilibria, periodic trajectories, chaotic attractors, etc. The following result describes the behavior of the time series under increasing uncertainty.

11. Theorem *(i) Each limit set M_i, $i = 1 \ldots k$ of the deterministic system for $\rho = 0$ is contained in a control set of (5) for $\rho > 0$.*

(ii) *Attractive limit sets (for $\rho = 0$) are contained in invariant control sets for $\rho > 0$ small enough.*

(iii) *The control sets of (5) depend continuously on ρ, except for at most finitely many ρ-points, at which control sets can merge together.*

This theorem, the proof of which can be found in [B], has a variety of consequences. First of all, stable limit sets of the deterministic system turn into stationary solutions of the time series (for small uncertainty range). In particular, chaotic attractors are enlarged to areas with invariant distribution. The following example of the Hénon system explains this fact (compare [B]).

12. Example

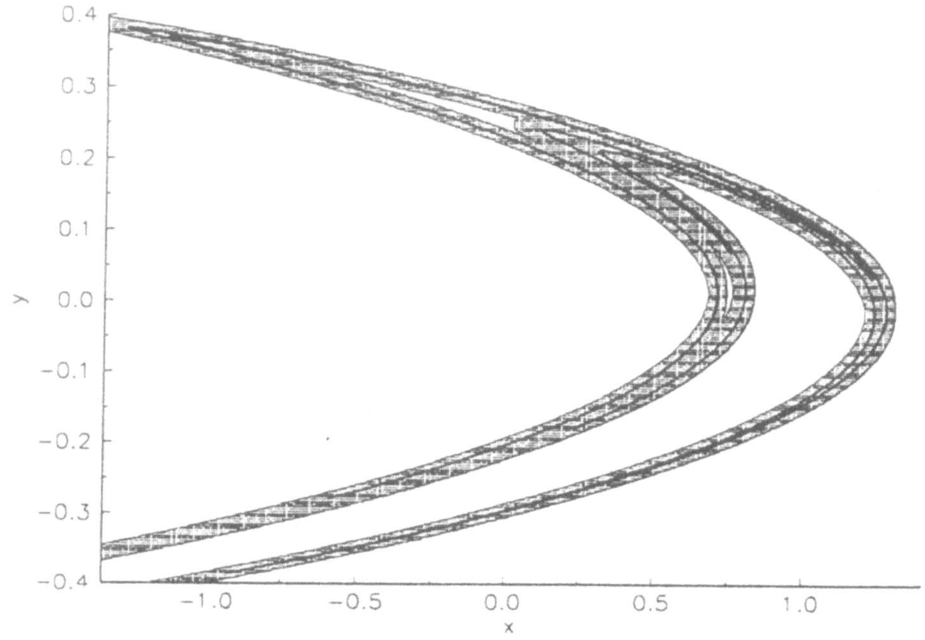

Figure 6. Support of the stationary distribution in Example 12.

Consider the two dimensional system

$$\begin{pmatrix} x_{n+1} \\ y_{n+1} \end{pmatrix} = \begin{pmatrix} 1 + y_n - \alpha x_n^2 \\ \beta x_n \end{pmatrix} + \begin{pmatrix} \xi_n^1 \\ \xi_n^2 \end{pmatrix} \tag{6}$$

with structure parameters $\alpha = 1.4$ and $\beta = 0.3$. For $\xi^1 \equiv \xi^2 \equiv 0$ this system exhibits a chaotic attractor (see e.g. [GH] and [BC]).

With random uncertainty $(\xi_n^1, \xi_n^2) \in [-3\rho, 3\rho] \times [-\rho, \rho], \rho > 0$ this chaotic attractor turns into a stationary solution, which is attractive for almost all initial values (x_0, y_0). For ρ small, the support of the stationary solution traces the chaotic attractor, as shown in Figure 6. for $\rho = 0.006$.

Secondly, there may exist (finitely many) ρ-values, for which control sets merge, and hence they may change their character between invariant, relatively invariant, or variant. This can lead to a drastic change in the behavior of the system, as explained in Theorems 4. and 9. Therefore the ρ-discontinuity points of the control sets may constitute bifurcation points for the nonlinear time series (5), describing the bifurcation behavior depending on the uncertainty range, as the next example illustrates.

13. Example 10. revisited
Consider again Example 10, but now with enlarged uncertainty range. The corresponding control sets, and hence the qualitative behavior, is illustrated in Figure 7. Note that in this case only one region A with invariant distributions 'survives', and the ergodic theorem (Theorem 9) now holds for all initial values $x_0 \in \mathbb{R}$ over the entire α-range.

Finally, Theorem 11. can serve as the basis for reliable numerics of stationarity solutions, multistability etc. for nonlinear time series. Since control sets can only be computed explicitly for one-dimensional systems (see e.g. [CK1]), numerical algorithms are required for higher dimensional systems. If the limit sets of the deterministic (undisturbed) equation $x_{n+1} = f(x_n, \xi_n, 0)$ are known, they can be used according to this theorem as a starting point for the computation of the orbits of the associated control systems for $\rho > 0$. The software packages CS (compare [Ha]) and DISORBIT (compare [B]) are based on this idea. These packages allow for a very precise computation of supports, bistability sets, and domains of attraction. It is worthwhile to point out, that for $\rho \downarrow 0$ the control sets of (4) converge toward the sets $M_i, i = 1 \ldots k$ of the deterministic system. Hence, for decreasing $\rho \downarrow 0$, these algorithms allow for a numerical outer approximation of e.g. chaotic attractors, while customary dynamical systems software is based on single trajectories, i.e. some inner approximation.

5 Bifurcations in Nonlinear Times Series

Bifurcation theory for nonlinear time series can be based on different concepts. In the Physics literature, a concept based on the changing shape of invariant densities (e.g. from unimodal to bimodal) is quite common (compare e.g. [HL]). To obtain strong correspondences with deterministic concepts, one can also consider bifurcations based on the Lyapunov exponents of stochastic flows associated to time series. This concept requires the analysis of objects that do not exist within the Markovian context. A third range of ideas is based on the Markov theory of nonlinear time series and considers stationary solutions and bistability regions as described above. While these results can be formulated strictly for Markov processes, some of the basic concepts and proofs of this approach also make use of more general dynamical systems (stochastic and control flows).

In the latter approach, the idea is roughly to consider time series

$$x_{n+1} = f(x_n, \xi_n, \rho, \alpha), \ x \in \mathbb{R}^d \tag{7}$$

where α is a bifurcation parameter of the underlying deterministic system (for $\rho = 0$), and $\rho \geq 0$ is the range of the stochastic term as in Equation (5). The initial

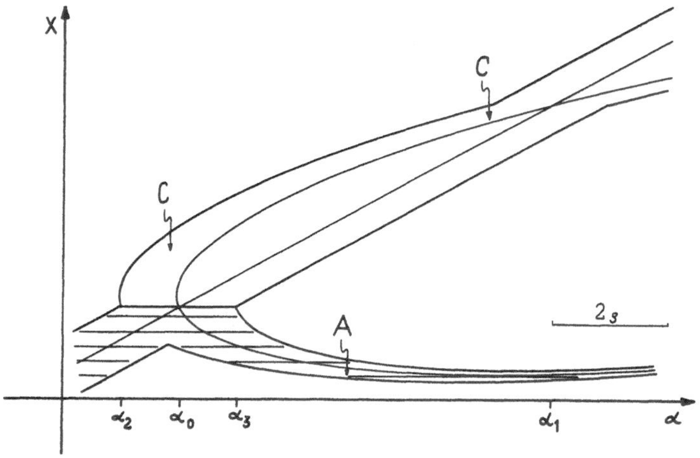

Figure 7. Control sets of the system in Example 13.

result here, based on Theorem 11, is that if α° is not a bifurcation value of the deterministic system, then there exists a value $\rho^\circ > 0$ such that for all uncertainty ranges $\rho < \rho^\circ$ the time series (7) 'reflects' the behavior of the deterministic system in the sense that the stochastic behavior expressed by the control sets corresponds to the deterministic behavior. If α° is a bifurcation point for $\rho = 0$, then the stochastic time series (for $\rho > 0$) may or may not exhibit a similar behavior. Precise results depend on the nature of the deterministic bifurcation scenario and on the way the uncertainty ξ_n enters into the system (7). Examples 10 and 13 above can be interpreted in this way as random bifurcation effects. For more details along these lines, and also for bifurcations from a pair $(\alpha^\circ, \rho^\circ)$ with $\rho^\circ > 0$, we refer the reader to e.g. [K], [CK4], [HS].

Acknowledgement Figure 6. was produced by Claudia Bauer, Universität Augsburg.

References

[AS] Albertini, F., Sontag, E.D., *Some connections between chaotic dynamical systems and control systems*, Proceedings European Control Conference, Grenoble 1, (1991), 158–163.

[B] Bauer, C., *Limit behavior and control sets of discrete-time systems* (1994), submitted.

[BC] Benedicks, M., Carleson, L., *The dynamics of the Hénon map*, Annals of Mathematics, Second Series **133** (1991), 73–169.

[CK1] Colonius, F., Kliemann, W., *Remarks on ergodic theory of stochastic flows and control flows*, Diffusion Processes and Related Problems in Analysis (1992), Vol II, Pinsky, M., Wihstutz, V (eds.), Birkhäuser, 203–240.

[CK2] Colonius, F., Kliemann, W., *Some aspects of control systems as dynamical systems*, J. Dynamics Diff. Equa. **5** (1993), 469–494.

[CK3] Colonius, F., Kliemann, W., *Limit behavior and genericity for nonlinear control systems*, J. Differential Equations **109** (1994), 8–41.

[GH] Guckenheimer, J., Holmes, P., *Nonlinear oscillations, dynamical systems and bifurcation of vector fields*, Springer–Verlag, Berlin, 1986, (2nd printing).

[Ha] Häckl, G., *Numerical approximation of reachable sets and control sets*, Random and Computational Dynamics **1** (1992–93), 371–394.

[HS] Häckl, G., Schneider, K.R., *Controllability near a Takens-Bogdanov-bifurcation*, Systems and Networks: Mathematical Theory and Applications, Vol II (Helmke, U., Mennicken, R., Saurer, J., eds), Akademie Verlag (1994), 193–196.

[Ho] Homble, P., *Ergodicity conditions for nonlinear discrete time stochastic dynamical systems with Markovian noise*, Stochastic Analysis and Applications **11** (1993), 513–568.

[HL] Horsthemke, W., Lefever, R., *Noise-induced transitions*, Springer– Verlag, 1984.

[K] Kliemann, W., *Analysis of nonlinear stochastic systems* in: Analysis and Estimation of Stochastic Mechanical Systems, W. Schiehlen and W. Wedig, Eds., Springer–Verlag, (1988), 43–102.

[MC] Meyn, S.P., Caines, P.E., *Asymptotic behavior of stochastic systems possessing Markovian realizations*, SIAM J. Control Optim. **29** (1991), 535– 561.

[T] Tong, H., *Non-linear time series: a dynamical systems approach*, Oxford University Press, 1990.

The VAR–VARCH model: A Bayesian approach

Wolfgang Polasek*
and
Hideo Kozumi†

Abstract

In this paper, we develop a combined Bayesian vector autoregressive and conditional heteroskedasticity (VAR–VARCH) models. A Gibbs sampling approach is suggested for the univariate and multivariate VAR–VARCH model. Using a random coefficient formulation it is shown that full conditional distributions are derived in closed analytical forms. The method is applied to monthly exchange rate series, the Swiss Franc, and the Deutsch Mark to the U.S. Dollar.

Keywords: exchange rates, full conditional distributions, Gibbs sampling, random coefficients, VAR–VARCH models.

1 Introduction

ARCH (autoregressive conditional heteroskedasticity) models have obtained considerably attention in the analysis of financial time series since their introduction by Engle (1982). They are used to capture the phenomenon of volatility clustering, i.e., the tendency of large (small) price changes to be followed by other large (small) price changes. There are now more than several hundred papers discussing theoretical properties of ARCH models as well as empirical applications (see Bollerslev *et al.*, 1992 for a good review).

From a Bayesian point of view, Geweke (1988, 1989a, 1989b) analyzed univariate ARCH models via Monte Carlo integration. Instead of employing the Monte Carlo

*Institute for Statistics and Econometrics, University of Basel, Holbeinstrasse 12, 4051 Basel, Switzerland

†Institute for Statistics and Econometrics, University of Basel, Holbeinstrasse 12, 4051 Basel, Switzerland and Faculty of Economics & Business Administration, Hokkaido University, Kita-ku, Kita 9 Nishi 7, Sapporo 060, Japan

integration method, Korn (1993), and Polasek (1993) applied the Gibbs sampler to univariate ARCH models using the random coefficient (RC) formulation of Tsay (1988).

As an extension of univariate results, it is sometimes necessary to analyze volatility clustering within a multivariate framework. Along this line, several authors have proposed vector ARCH (VARCH) models (see, for example, Diebold and Nerlove, 1989 and Bollerslev, 1990). However, there are only few articles using Bayesian inference in this field.

In this paper, we develop a vector autoregressive-VARCH (VAR-VARCH) model which is suitable for a Markov chain Monte Carlo simulation. Using the random coefficient model as a parameterization for the ARCH models, we can derive all full conditional distributions in closed analytical forms, and thus it is easy to apply the Gibbs sampler. The plan of this paper is as follows. In section 2, we describe a model using a RC formulation. In section 3, full conditional distributions are derived, which are necessary to apply the Gibbs samplers. Section 4 illustrates our model with foreign exchange rate data. Conclusions are given in section 5.

2 The models

2.1 The AR-ARCH (p, q) model

Univariate ARCH models with AR structure in the mean are written as

$$y_t = \beta_0 + \beta_1 y_{t-1} + \cdots + \beta_p y_{t-p} + e_t, \quad e_t | I_t \sim N(0, h_t), \tag{1}$$
$$h_t = \alpha_0 + \alpha_1 e_{t-1}^2 + \cdots + \alpha_q e_{t-q}^2, \tag{2}$$

where y_t is the observation at time t ($t = 1, \ldots, T$), and I_t is the information set available at time t. The AR-ARCH model in (1) and (2) has an observationally equivalent RC representation of the form

$$y_t = x_t'\beta + z_t'\gamma_t + u_t, \quad u_t \sim N(0, \omega), \tag{3}$$

where $x_t = (1, y_{t-1}, \ldots, y_{t-p})'$, $\beta = (\beta_0, \ldots, \beta_p)'$, $z_t = (e_{t-1}, \ldots, e_{t-q})'$, and $\gamma_t \sim N(0, \Sigma)$. Also, the error terms u_t and the random coefficients γ_t are assumed to be mutually independent. It is easily verified that the model (3) has the following moments

$$\mathrm{E}(e_t | z_t) = 0, \tag{4}$$
$$\mathrm{var}(e_t | z_t) = \omega + z_t' \Sigma z_t. \tag{5}$$

If Σ is a full symmetric matrix, the model is called an augmented ARCH model (see Bera *et al.*, 1992 and Polasek, 1993). When Σ is a diagonal matrix, the model (3) reduces to the usual ARCH model in (1) and (2), and in such a case $\omega = \alpha_0$ and $\sigma_{ii} = \alpha_i$ ($i = 1, \ldots, p$) where σ_{ii} is the i-th diagonal element of Σ. The model (3) was analyzed by Korn (1993) and Polasek (1993).

2.2 The VAR–VARCH (p, q) model

This section extends the univariate AR–ARCH model given by (3) to a multivariate VAR–VARCH model. Let us consider a VAR(p) model of dimension of M (see, e.g., Lütkepohl, 1993) with time dependent (heteroskedastic) covariance matrices:

$$y_t = a_0 + A_1 y_{t-1} + \cdots + A_p y_{t-p} + e_t, \quad e_t | I_t \sim N(0, H_t), \tag{6}$$

where I_t is the information set, $y_t = (y_{1t}, \ldots, y_{Mt})'$ is an $M \times 1$ vector of observed time series at time t, A_i $(i = 1, \ldots, p)$ are fixed $M \times M$ coefficient matrices, $a_0 = (a_{01}, \ldots, a_{0M})'$ is a fixed $M \times 1$ vector of intercept terms, and $e_t = (e_{1t}, \ldots, e_{Mt})'$ is an $M \times 1$ vector of error terms.

The above model is rewritten compactly as

$$y_t = B'x_t + e_t, \tag{7}$$

where $x_t = (1, y'_{t-1}, \ldots, y'_{t-p})'$ is a $(Mp + 1) \times 1$ vector of regressor variables,

$$B = \begin{pmatrix} a_{10} & \cdots & a_{M0} \\ a_{11} & \cdots & a_{M1} \\ \cdots & \cdots & \cdots \\ a_{1p} & \cdots & a_{Mp} \end{pmatrix},$$

and a_{mi} $(i = 1, \ldots, p;\ m = 1, \ldots, M)$ is the m-th $M \times 1$ vector of $A_i = (a_{1i}, \ldots, a_{Mi})'$. In order to extend univariate ARCH models to a multivariate setting, various parameterizations in H_t are proposed (see, for example, Bollerslev *et al.*, 1988 and Bollerslev, 1990). However, a general definition does not exist. Therefore, for simplicity, we assume a diagonal structure for H_t:

$$H_t = \text{diag}(h_{1t}, \ldots, h_{Mt}),$$

and the variance elements depend on the past q residuals of all M time series. This simplification will be modified in the following subsection.

2.3 The VAR–VARCH(p, q) model in RC form

We write the above formulated VAR–VARCH model in a random coefficient form:

$$\begin{aligned} y_t &= B'x_t + e_t, \\ &= B'x_t + \Gamma'_t z_t + u_t, \end{aligned} \tag{8}$$

where $B'x_t$ stands for the VAR component of the VAR–VARCH model and $\Gamma'_t z_t$ for the RC component defined by

$$\begin{aligned} z_t &= (e'_{t-1}, \ldots, e'_{t-q})', \\ \Gamma_t &= (\gamma_{1t}, \ldots, \gamma_{Mt}), \end{aligned}$$

and q is the order of the VARCH model The error term is normally distributed with seemingly unrelated regression (SUR) type covariance matrix:

$$u_t \sim N(0, \Omega), \tag{9}$$

where Ω is an $M \times M$ positive definite matrix. For the random coefficient matrix Γ_t we assume the stochastic structure

$$\text{vec}\Gamma_t \sim N(0, \Sigma),$$

where Σ is an $M^2q \times M^2q$ block diagonal matrices:

$$\Sigma = \begin{pmatrix} \Sigma_1 & & 0 \\ & \ddots & \\ 0 & & \Sigma_M \end{pmatrix}.$$

For a parsimonious model, we further assume that each Σ_m is a block diagonal matrix of dimension $Mq \times Mq$, i.e.,

$$\Sigma_m = \begin{pmatrix} \Sigma_{m1} & & 0 \\ & \ddots & \\ 0 & & \Sigma_{mq} \end{pmatrix},$$

where the sub-blocks Σ_{mi} are $M \times M$ positive definite matrices.

Thus, the RC-component of model (8) has the moments

$$\begin{aligned} \text{E}(\Gamma_t' z_t | z_t) &= 0, \\ \text{var}(\Gamma_t' z_t | z_t) &= \text{diag}\Big(\sum_{i=1}^{q} e_{t-i}' \Sigma_{1i} e_{t-i}, \ldots, \sum_{i=1}^{q} e_{t-i}' \Sigma_{Mi} e_{t-i}\Big). \end{aligned}$$

It should be noted that the distribution of the error vector u_t in (9) captures the SUR type correlation in the VAR model, while each RC-component models the volatility movement in each time series. The conditional variance matrix is always guaranteed to be positive definite. Moreover, when $M = 1$ this model reduces to the univariate ARCH model given by (3).

3 Full conditional distributions

We now show the proposed RC model has a convenient conditional structure needed to apply the Gibbs sampler. The Bayesian VAR-VARCH(p, q) model in RC form is given as

$$y_t = B' z_t + \Gamma_t' z_t + u_t, \tag{10}$$

and prior distributions are chosen from the families of normal and Wishart distributions, hence

$$\begin{aligned} \text{vec}B &\sim N(\beta_*, H_*), \\ \Omega^{-1} &\sim W_M(\Omega_*, n_*), \\ \text{vec}\Gamma_t &\sim N(0, \Sigma), \\ \Sigma_{mi}^{-1} &\sim W_M(\Sigma_{mi*}, \nu_{mi*}), \end{aligned} \tag{11}$$

where all of the hyper-parameters (which are denoted with a star) are known a priori. Thus, we can derive the following full conditional distributions (f.c.d.'s) for the Gibbs simulation process.

a) The f.c.d. for the VAR coefficients vec$B$:
The model (10) can be rewritten as

$$
\begin{aligned}
u_t &= y_t - (\mathrm{I}_M \otimes x_t')\mathrm{vec}B - \begin{pmatrix} \gamma_{1t}' \\ \vdots \\ \gamma_{Mt}' \end{pmatrix}\begin{pmatrix} y_{t-1} \\ \vdots \\ y_{t-q} \end{pmatrix} + \begin{pmatrix} \gamma_{1t}' \\ \vdots \\ \gamma_{Mt}' \end{pmatrix}\begin{pmatrix} \mathrm{I}_M \otimes x_{t-1}' \\ \vdots \\ \mathrm{I}_M \otimes x_{t-q}' \end{pmatrix}\mathrm{vec}B, \\
&= y_t - \begin{pmatrix} \gamma_{1t}' \\ \vdots \\ \gamma_{Mt}' \end{pmatrix}\begin{pmatrix} y_{t-1} \\ \vdots \\ y_{t-q} \end{pmatrix} - \left[(\mathrm{I}_M \otimes x_t') - \begin{pmatrix} \gamma_{1t}' \\ \vdots \\ \gamma_{Mt}' \end{pmatrix}\begin{pmatrix} \mathrm{I}_M \otimes x_{t-1}' \\ \vdots \\ \mathrm{I}_M \otimes x_{t-q}' \end{pmatrix}\right]\mathrm{vec}B, \\
&= \tilde{y}_t - \tilde{x}_t'\mathrm{vec}B,
\end{aligned}
$$

where $\otimes$ denotes the Kronecker product. Using the notations $\tilde{Y} = (\tilde{y}_1', \ldots, \tilde{y}_T')'$ and $\tilde{X} = (\tilde{x}_1, \ldots, \tilde{x}_T)'$, we can derive the f.c.d. as

$$
p(\mathrm{vec}B|\theta^c, data) = N(\beta_{**}, H_{**}), \tag{12}
$$

where

$$
\begin{aligned}
H_{**}^{-1} &= \tilde{X}'(\mathrm{I}_T \otimes \Omega)^{-1}\tilde{X} + H_*^{-1}, \\
\beta_{**} &= H_{**}\{\tilde{X}'(\mathrm{I}_T \otimes \Omega)^{-1}\tilde{Y} + H_*^{-1}\beta_*\},
\end{aligned}
$$

and θ^c denotes all parameters with the argument of the f.c.d. removed.

b) The f.c.d. for the precision matrix Ω^{-1}:
By standard results for the Bayesian normal-Wishart regression model we find for the f.c.d.

$$
\begin{aligned}
p(\Omega^{-1}|\theta^c, data) &\propto |\mathrm{I}_T \otimes \Omega|^{-\frac{1}{2}} \exp\Big\{-\frac{1}{2}\sum_{t=1}^{T} u_t'\Omega^{-1}u_t\Big\} \\
&\quad \times |\Omega^{-1}|^{\frac{1}{2}(n_*-M-1)}\mathrm{etr}\Big\{-\frac{1}{2}\Omega_*\Omega^{-1}\Big\}, \\
&= W_M(\Omega_{**}, n_{**}),
\end{aligned} \tag{13}
$$

where the parameters of the M-dimensional Wishart distribution are

$$
\begin{aligned}
n_{**} &= n_* + T, \\
\Omega_{**} &= \Omega_* + \sum_{t=1}^{T} u_t u_t',
\end{aligned}
$$

and etr$(\cdot)$ stands for exp(tr$(\cdot)$).

c) The f.c.d. for the random coefficients vecΓ_t:
From the independency of u_t and Γ_t, we have for $t = 1, \ldots, T$:

$$\begin{aligned} p(\text{vec}\Gamma_t|\theta^c, data) &= \exp\Big\{-\frac{1}{2}(y_t - B'x_t - \Gamma_t'z_t)'\Omega^{-1}(y_t - B'x_t - \Gamma_t'z_t)\Big\} \\ &\quad \times \exp\Big\{-\frac{1}{2}(\text{vec}\Gamma_t)'\Sigma^{-1}(\text{vec}\Gamma_t)\Big\}, \\ &= N(\Gamma_{t**}, G_{t**}), \end{aligned} \tag{14}$$

where

$$\begin{aligned} G_{t**}^{-1} &= (\mathrm{I}_M \otimes z_t')'\Omega^{-1}(\mathrm{I}_M \otimes z_t') + \Sigma^{-1}, \\ \Gamma_{t**} &= G_{t**}(\mathrm{I}_M \otimes z_t')'\Omega^{-1}(y_t - B'x_t). \end{aligned}$$

d) The f.c.d. for the ARCH coefficient matrix Σ_{mi}^{-1}:
Exploiting the block diagonal structure we find for $m = 1, \ldots, M$ and $i = 1, \ldots, q$

$$\begin{aligned} p(\Sigma_{mi}^{-1}|\theta^c, data) &\propto |\mathrm{I}_T \otimes \Sigma_{mi}|^{-\frac{1}{2}} \exp\Big\{-\frac{1}{2}\sum_{t=1}^{T} \gamma_{mt,i}'\Sigma_{mi}^{-1}\gamma_{mt,i}\Big\} \\ &\quad \times |\Sigma_{mi}^{-1}|^{\frac{1}{2}(\nu_{mi*}-M-1)} \text{etr}\Big\{-\frac{1}{2}\Sigma_{mi*}\Sigma_{mi}^{-1}\Big\}, \\ &= W_M(\Sigma_{mi**}, \nu_{mi**}), \end{aligned} \tag{15}$$

where

$$\begin{aligned} \nu_{mi**} &= \nu_{mi*} + T, \\ \Sigma_{mi**} &= \Sigma_{mi*} + \sum_{t=1}^{T} \gamma_{mt,i}\gamma_{mt,i}', \end{aligned}$$

and $\gamma_{mt} = (\gamma_{mt,1}', \ldots, \gamma_{mt,q}')'$.

The Gibbs sampling technique is explained in Gelfand and Smith (1990). The iteration process consists of drawing random numbers from the distributions in (14), (13), (15) and (12) in that order. The choice of the starting values is explained in the examples.

4 Illustrative examples

We present two examples to illustrate the univariate and the multivariate ARCH models. In the first example, we analyze a data set of monthly exchange rates for the Swiss Franc/Dollar (SFr/USD) and the Deutsch Mark/Dollar (DM/USD) for the period October 1973 to March 1988 using univariate ARCH models (i.e., $M = 1$). (The SFr/USD and DM/USD are shown in Figure 1.) We estimate VAR-VARCH models for the same two exchange rates in the second example. All time series are in log difference form, giving 186 observations of monthly growth rates.

The results given below are based on the following Gibbs sampling scheme. We used multiple parallel chains with an overdispersed initial sample to be able to apply the convergence diagnostic proposed by Gelman and Rubin (1992). This diagnostic $\sqrt{\hat{R}}$ estimates the potential scale reduction, and $\hat{R}$ is the ratio of the current variance estimate to the within-sequence variance with a factor to account for the extra variance of the Student's t distribution. To initialize the simulation, we first ran a single chain with 1,000 iterations, computed for each parameters (θ_i) approximate posterior means $\bar{\theta}_i$ and standard deviations σ_i based on the last 500 iterations, and then generated initial values from a $t(\bar{\theta}_i, \sigma_i^2)$ distribution with 4 degrees of freedom. After obtaining initial values, we ran 10 parallel chains with 6,000 iterations, and the last 1,000 samples for each chain are used for the parameter estimation.

The specification of the hyper-parameters in (11) is rather noninformative. The prior parameters of the VAR component are

$$\beta_* = 0, \qquad H_*^{-1} = \mathrm{diag}(0.01, \dots, 0.01),$$

which can be viewed as a tightness prior around zero. The prior for the SUR–type covariance matrix in the VAR component is

$$n_* = M + 2, \qquad \Omega_* = \mathrm{diag}(0.01, \dots, 0.01).$$

The prior parameters for ARCH coefficients matrix Σ_{mi} are

$$\nu_{mi*} = M + 2, \qquad \Sigma_{mi*} = \mathrm{diag}\left(\frac{0.8(q+1-i)}{0.5q(q+1)}, \dots, \frac{0.8(q+1-i)}{0.5q(q+1)}\right).$$

This specification follows a linear decay pattern with the properties $\mathrm{E}(\Sigma_{mi}) = \Sigma_{mi*}$ and $\sum_{i=1}^{q} \frac{0.8(q+1-i)}{0.5q(q+1)} = 0.8$ for each of the $m = 1, \dots, M$ components.

4.1 Example 1: Univariate AR–ARCH model

In the first example, we estimated the model (3) without the intercept for both exchange rates, SFr/USD and DM/USD, and for various combination of orders, p and q. For model selection, we have calculated the conditional predictive ordinate (CPO) based on the forecasting predictive density for all time points $t = 1, \dots, T$. The CPO is explained in the appendix. Since the CPO's for SFr/USD are very similar to those for DM/USD, the CPO–plots only for SFr/USD up to order (4,4) are shown in Figure 2. From Figure 2, it is seen that the medians of the CPO–plots are very similar, and that the distribution of the CPO's is skewed to the left. Also, we see that the ARCH models with high order tend to produce large values of CPO's. These large values occur for the period 1980–1982 and 1985. As seen from Figure 1, exchange rates are rather unstable and volatile in these periods. It may be considered that ARCH models with high order are suitable for these periods. However, since the ARCH parameters for high orders are not "significant", we chose an AR–ARCH(1,1) model as the sufficient model. ("Significant" in our interpretation means that the posterior mean is more than two posterior standard deviations away from zero.)

The estimated potential scale reductions of the Gelman–Rubin diagnostics are reported in Table 1. Since all values are close to 1.0 we conclude that the simulated values are close to the target distribution.

Posterior means and standard deviations are shown in Table 2. It can be seen that ARCH parameters σ_{11} are significant both in SFr/USD and DM/USD. Using monthly exchange rate data from March 1980 to January 1985, Baillie and Bollerslev (1989) found that there are no significant ARCH effects. Compared with our results, we conclude that their results are possibly due to the small sample size. As noted by Domowitz and Hakkio (1985) and Baillie and Bollerslev (1989), ARCH effects tend to weaken with less frequently sampled data. Since the posterior means of ARCH parameters are 0.1365 in SFr/USD and 0.1223 in DM/USD respectively, their findings are confirmed from our example. Moreover, the estimates for SFr/USD and DM/USD are very similar. This result is consistent with a visual inspection of Figure 1. Figure 5 shows that the conditional variances of the DM/USD are smaller than those of the SFr/USD.

The kernel density estimates for the marginal posterior distributions are shown in Figure 3. The distributions for β_1 and ω appear to be fairly symmetric. Not surprisingly, the distributions for σ_{11} are skewed to the right (reflecting the contribution to the conditional variance of the time series).

4.2 Example 2: A bivariate VAR–VARCH model

After the previous univariate results, we estimated the bivariate VAR–VARCH(1,1) model using the SFr/USD and DM/USD exchange rates. The CPO-plots of the VAR–VARCH model up to order $(p,q) = (2,2)$ are shown in Figure 4. As in the previous example, it can be seen that the predictive performance of the VAR–VARCH(1,1) model is about the same as that for the VAR–VARCH(2,2) model. There is no unique answer for the model choice from the CPO-plots.

The estimated potential scale reductions for the Gelman–Rubin diagnostics are given in Table 3, indicating that all values are close to one. The parameter estimates for the VAR–VARCH(1,1) model are for the VAR component:

$$\begin{aligned}
\text{SFr/USD}_t &= \underset{(0.1532)}{0.4338}\,\text{SFr/USD}_{t-1} - \underset{(0.1723)}{0.1595}\,\text{DM/USD}_{t-1}, \\
\text{DM/USD}_t &= \underset{(0.1506)}{0.1025}\,\text{SFr/USD}_{t-1} + \underset{(0.1641)}{0.2082}\,\text{DM/USD}_{t-1},
\end{aligned}$$

with the SUR covariance matrix

$$\Omega \times 10^3 = \begin{pmatrix} \underset{(0.0987)}{0.8757} & \\ \underset{(0.0816)}{0.6798} & \underset{(0.0819)}{0.7431} \end{pmatrix},$$

and for the VARCH component (the covariance matrices of the RC component):

$$\Sigma_1 = \begin{pmatrix} 0.1345 & \\ (0.0663) & \\ -0.0857 & 0.1373 \\ (0.0637) & (0.0704) \end{pmatrix}, \qquad \Sigma_2 = \begin{pmatrix} 0.1052 & \\ (0.0462) & \\ -0.0645 & 0.1102 \\ (0.0451) & (0.0517) \end{pmatrix}.$$

See (numerical) posterior standard deviations are given in parentheses below the posterior means of the coefficients.

The estimates of the VAR component of the model show that only the AR coefficient for SFr/USD$_{t-1}$ is "significant". Again "significant" means that the posterior t-value (i.e., the ratio of posterior mean and standard deviation) for this coefficient is larger than 2. Moreover, all elements of Ω are highly significant, and a high correlation is found. This finding agrees with a visual inspection from Figure 1.

The estimated coefficients of the VARCH component show that both diagonal elements of Σ_2 are significant and their magnitudes are almost the same. This finding implies that the SFr/USD helps to explain the volatiliy movement of the DM/USD. However, the second diagonal element of Σ_1 is not significant, and this means that the DM/USD does not explain the volatility movement of the SFr/USD exchange rate on a monthly basis. Off-diagonal elements of Σ_1 and Σ_2 are negative, but not significant.

In Figure 5, the conditional variances both for univariate and multivariate models are plotted. It can be seen that the conditional variance for univariate model of the SFr/USD rate is slightly larger than that for multivariate model. In contrast to the SFr/USD exchange rate, the result for the DM/USD shows that the conditional variance for the univariate model is smaller than that for the multivariate model over the whole time period.

Figure 6 displays the kernel density estimates for the marginal posterior distributions. Note that ARCH coefficients are skewed to the right for diagonal elements, and to the left for off-diagonal elements with all modal values distinctly different from zero.

Since it is interesting to see how the VAR–VARCH model performs in prediction, we carried out a 4 step prediction exercise. To evaluate the predictive performance, we computed the mean square error (MSE) and the mean absolute error (MAE) which are defined as, respectively,

$$\begin{aligned} \text{MSE} &= \frac{1}{4}\sum_{i=1}^{4}(z_i - \hat{z}_i)^2, \\ \text{MAE} &= \frac{1}{4}\sum_{i=1}^{4}|z_i - \hat{z}_i|, \end{aligned}$$

where z_i denotes the i-th observation while $\hat{z}_i$ denotes the posterior mean from the i-th marginal predictive distribution. The results are shown in Table 4. For comparison, the results for the VAR(1) model with the same prior specification as the VAR–VARCH(1,1) model are also reported. From Table 4, it can be seen that the VAR–VARCH model outperforms the VAR model in terms of prediction. In case of the MSE, the reduction is about 30% and in case of the MAE, the reduction is about 20%.

5 Conclusions

In this paper, we have analyzed Bayesian VAR–VARCH models for exchange rates using a random coefficient formulation. We have shown that all full conditional distributions can be derived in closed form. This allows an easy implementation of the Gibbs sampling routine since random variables can be easily drawn from the full conditional distributions.

Using monthly exchange rate data, we have compared models with different orders both for univariate and multivariate models. The univariate analysis shows that the ARCH effects can be clearly seen for the SFr/USD and the DM/USD series, but that the effects are rather small. Also, the posterior distributions for the ARCH parameters are skewed to the right. The multivariate analysis shows that the SFr/USD helps to explain the volatility movement of the DM/USD, but the reverse is not true. We conclude that a VAR–VARCH model reduces and re-allocates the effects of conditional variances. Furthermore, we could show that the VAR–VARCH model outperforms the VAR model in terms of short term predictions.

Appendix

The conditional predictive ordinate (CPO) based on the forecasting predictive distribution of y_t given $(y_1, \ldots, y_{t-1})$ is defined as

$$\begin{aligned} f(y_t|y_1, \ldots, y_{t-1}) &= \frac{f(y_1, \ldots, y_t)}{f(y_1, \ldots, y_{t-1})}, \\ &= \frac{\int \frac{f(y_1,\ldots,y_t|\theta)}{f(y_1,\ldots,y_T|\theta)} \pi(\theta|y_1,\ldots,y_T) d\theta}{\int \frac{f(y_1,\ldots,y_{t-1}|\theta)}{f(y_1,\ldots,y_T|\theta)} \pi(\theta|y_1,\ldots,y_T) d\theta}, \end{aligned} \tag{16}$$

where θ is a parameter vector, $f(\cdot|\theta)$ is the likelihood function, and $\pi(\theta|\cdot)$ is the posterior density function. The CPO given by (16) shows what values of y_t are likely, given that the model was fitted to the observations $y_1, \ldots, y_{t-1}$, and it is possible to see whether the observation supports the model. Using the output from the Gibbs sampler, $\theta^{(i)}$, $i = 1, \ldots, N$, an approximation to (16) can be obtained by Monte Carlo integration

$$\hat{f}(y_t|y_1, \ldots, y_{t-1}) = \frac{\frac{1}{N}\sum_{i=1}^N \frac{f(y_1,\ldots,y_t|\theta^{(i)})}{f(y_1,\ldots,y_T|\theta^{(i)})}}{\frac{1}{N}\sum_{i=1}^N \frac{f(y_1,\ldots,y_{t-1}|\theta^{(i)})}{f(y_1,\ldots,y_T|\theta^{(i)})}}.$$

References

[1] Baillie, R.T. and Bollerslev, T. (1989). The message in daily exchange rates: A conditional-variance tale, *Journal of Business and Economic Statistics 7*, 297–305.

[2] Bera, A.K., Higgins, M.L., and Lee, S. (1992). Interaction between autocorrelation and conditional heteroscedasticity: a random-coefficient approach, *Journal of Business and Economic Statistics 10*, 133–142.

[3] Bollerslev, T. (1990). Modelling the coherence in short-run nominal exchange rates: A multivariate generalized ARCH approach, *Review of Economics and Statistics 72*, 498–505.

[4] Bollerslev, T., Engle, R.F., and Wooldridge, J.M. (1988). A capital asset pricing model with time varying covariances, *Journal of Political Economy 96*, 116–131.

[5] Bollerslev, T., Chou, R.Y., and Kroner, K.F. (1992). ARCH modeling in finance, *Journal of Econometrics 52*, 5–59.

[6] Diebold, F.X. and Nerlove, M. (1989). The dynamics of exchange rate volatility: A multivariate latent factor ARCH model, *Journal of Applied Econometrics 4*, 1–21.

[7] Domowitz, I. and Hakkio, C.S. (1985). Conditional variance and the risk premium in the foreign exchange market, *Journal of International Economics 19*, 47–66.

[8] Engle, R.F. (1982). Autoregressive conditional heteroskedasticity with estimates of the variance of U.K. inflation, *Econometrica 50*, 987–1008.

[9] Gelfand, A.E. and Smith, A.F.M. (1990). Sampling based approaches to calculating marginal densities, *Journal of the American Statistical Association 85*, 398–409.

[10] Gelman, A. and Rubin, D.B. (1992). Inference from iterative simulation using multiple sequences, *Statistical Science 7*, 63–86.

[11] Geweke, J. (1988). Exact inference in models with autoregressive conditional heteroskedasticity, in :E. Berndt, H. White, and W. Barnett, eds., *Dynamic econometric modeling*, (Cambridge University Press, Cambridge) 73–104.

[12] Geweke, J. (1989a). Exact predictive densities in linear models with ARCH disturbances, *Journal of Econometrics 44*, 307–325.

[13] Geweke, J. (1989b). Bayesian inference in econometric models using Monte Carlo integration, *Econometrica 57*, 1317–1339.

[14] Korn, O. (1993). A Gibbs sampler for Bayesian ARCH models, WWZ discussion paper, University of Basel.

[15] Lütkepohl, H. (1993). Introduction to multiple time series analysis 2nd ed., Springer–Verlag, Berlin.

[16] Polasek, W. (1993). Bayesian augmented ARCH models, WWZ discussion paper, University of Basel.

[17] Tsay, R.S. (1987). Conditional heteroscedastic time series models, *Journal of the American Statistical Association 82*, 590–604.

Table 1: Gelman–Rubin diagnostics $\sqrt{\hat{R}}$ for example 1.

	β_1	ω	σ_{11}
SFr/USD	1.0006	1.0003	1.0169
DM/USD	1.0004	1.0002	1.0145

Table 2: Parameter estimates for example 1.

	SFr/USD		DM/USD	
	mean	sd.v	mean	sd.v
β_1	0.3522	0.0792	0.3291	0.0796
$\omega \times 10^3$	0.8833	0.1014	0.7206	0.0809
σ_{11}	0.1365	0.0599	0.1223	0.0512

Table 3: Gelman–Rubin diagnostics $\sqrt{\hat{R}}$ for example 2.

$B(1,1)$	$B(1,2)$	$B(2,1)$	$B(2,2)$	$\Omega(1,1)$	$\Omega(1,2)$	$\Omega(2,2)$
1.0006	1.0005	1.0007	1.0005	1.0003	1.0001	1.0001

$\Sigma_1(1,1)$	$\Sigma_1(1,2)$	$\Sigma_1(2,2)$	$\Sigma_2(1,1)$	$\Sigma_2(1,2)$	$\Sigma_2(2,2)$
1.0276	1.0339	1.0293	1.0195	1.0219	1.0153

Table 4: MSE and MAE for example 2.

	VAR-VARCH(1,1)		VAR(1)	
	SFr/USD	DM/USD	SFr/USD	DM/USD
MSE $\times 10^2$	0.114	0.071	0.169	0.110
MAE $\times 10^2$	2.889	2.150	3.695	2.924

Figure 1: SFr/USD and DM/USD rates.

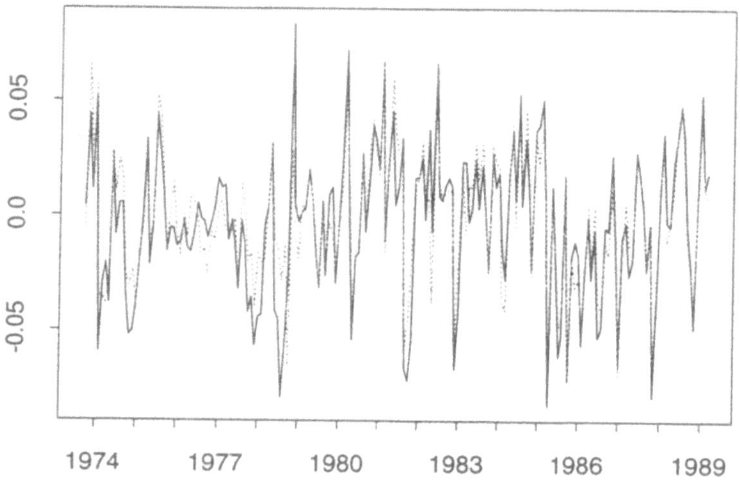

SFr/USD (solid line) and DM/USD (dotted line).

Figure 2: CPO box-plots for AR–ARCH models.

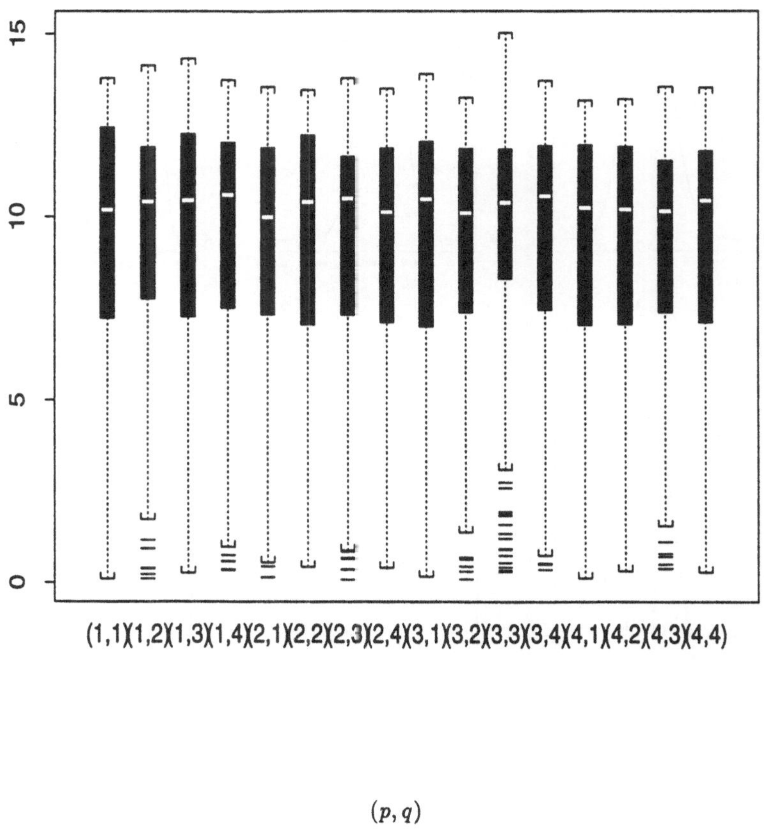

Figure 3: Marginal posterior densities for the AR–ARCH(1,1) model.

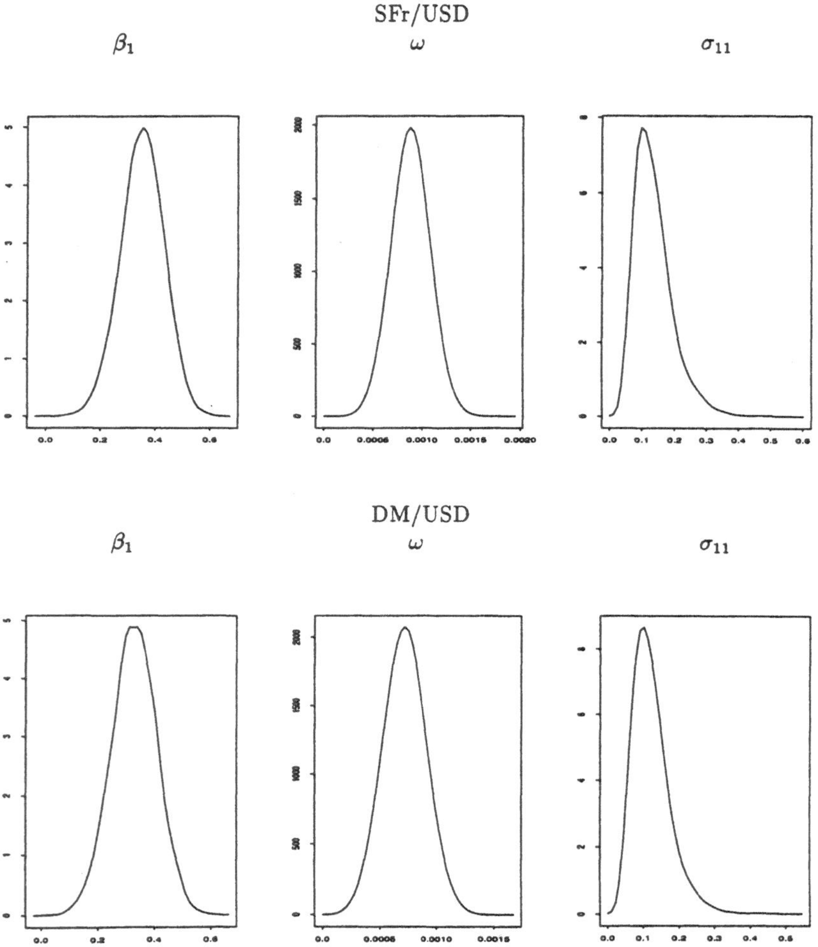

Figure 4: CPO box–plots for VAR–VARCH models.

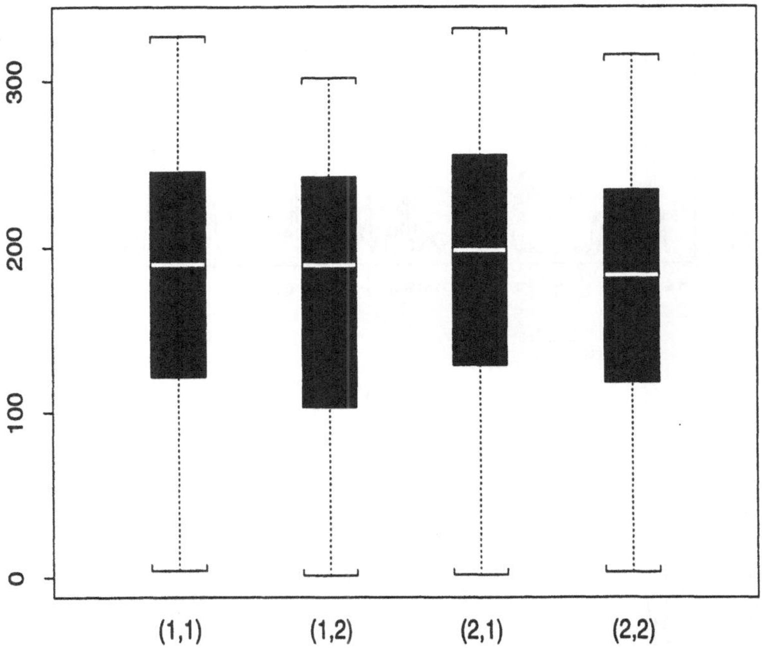

(p, q)

Figure 5: Conditional variances for univariate and multivariate models.

SFr/USD

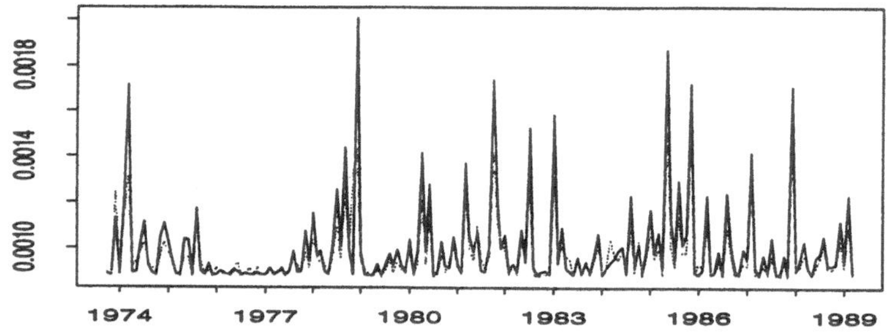

DM/USD

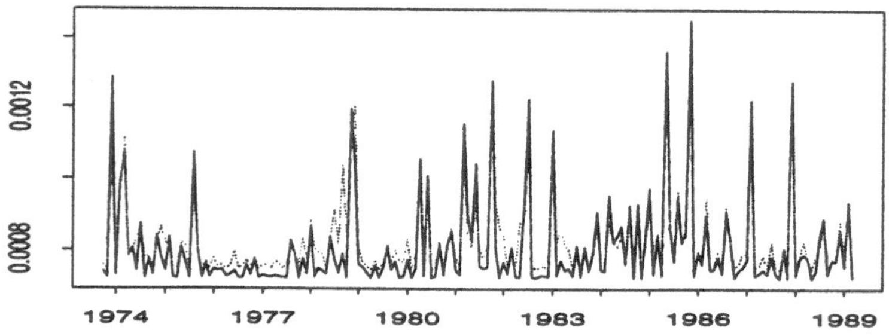

Univariate model (solid line) and multivariate model (dotted line).

Figure 6: Marginal posterior densities for the VAR–VARCH(1,1) model.

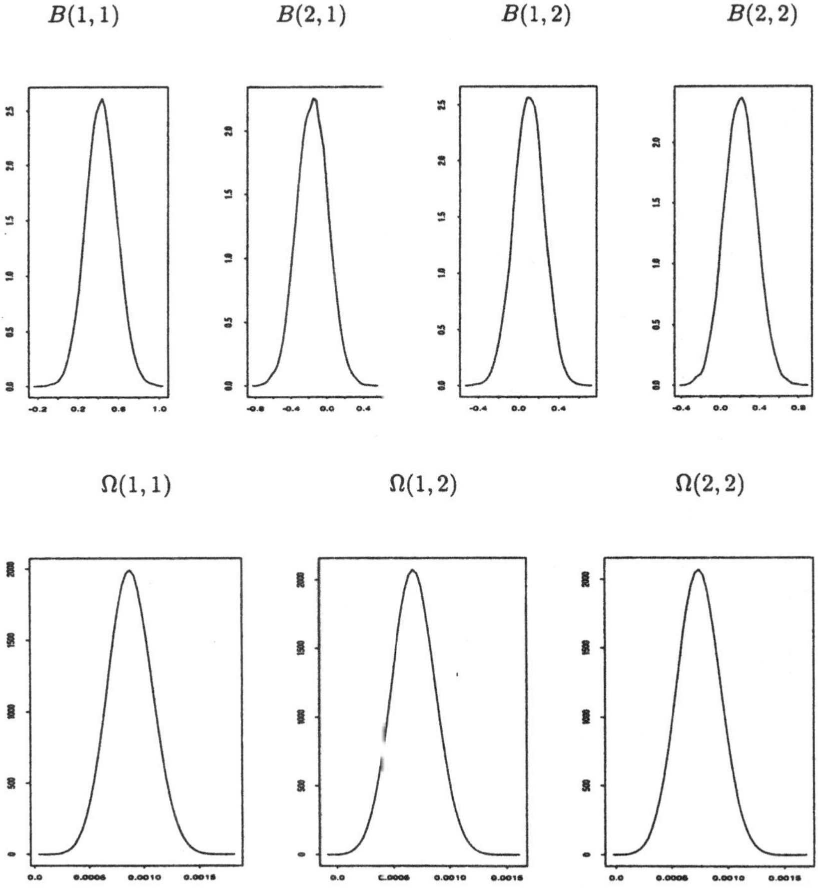

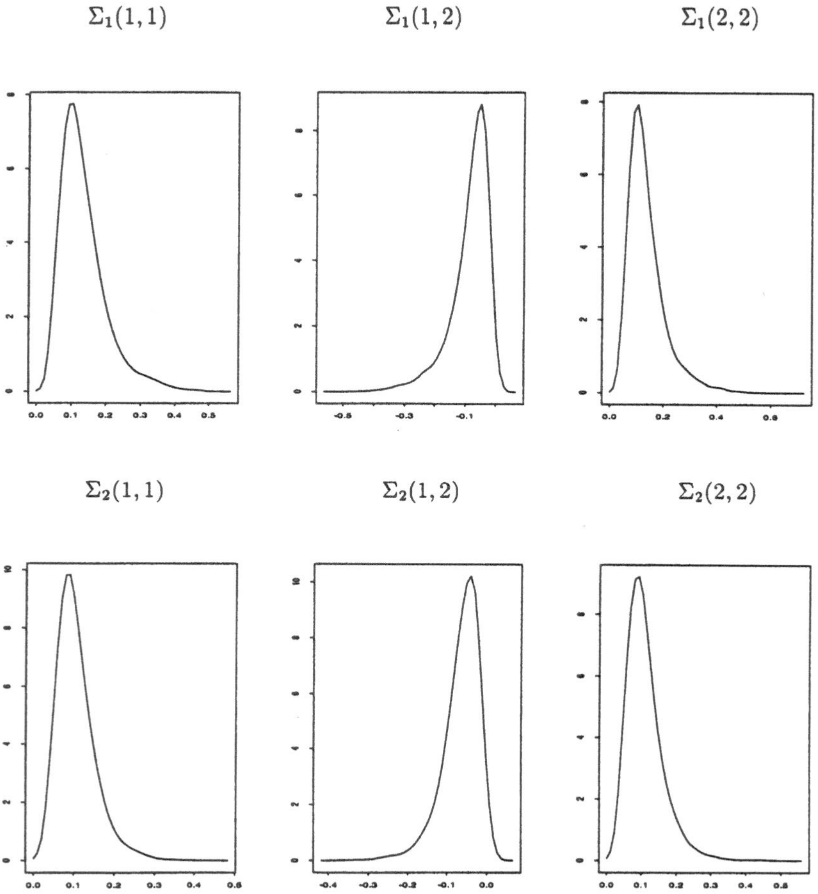
$\Sigma_1(1,1)$
$\Sigma_1(1,2)$
$\Sigma_1(2,2)$
$\Sigma_2(1,1)$
$\Sigma_2(1,2)$
$\Sigma_2(2,2)$

Statistical Assessment of Atmospheric Ozone Data for Depletion

Gregory C. Reinsel and George C. Tiao
University of Wisconsin-Madison and University of Chicago

Abstract

A seasonal trend analysis of Dobson station total ozone data from a network of 56 stations has been performed using data from 1964 through 1991. Random effects models for the individual station seasonal trend estimates, to allow for individual station and regional trend variations, are used to combine individual station trend estimates to obtain overall trend estimates in ozone for different seasons of the year as a function of latitude. The trend results indicate significant negative trends in ozone since 1970, of the order of −2.5% per decade, during the winter and spring seasons in the higher northern latitudes (40°N–65°N). A similar seasonal trend analysis over the more recent shorter time period November 1978 through 1991 is also performed using total ozone data from the TOMS satellite experiment, and results are compared with trend analysis of the Dobson ground station data for the comparable time period.

1 Introduction

The issue of depletion of the earth's atmospheric ozone layer due to the release of chlorofluorocarbons (CFCs) was first addressed from a photochemical viewpoint by Cicerone, Stolarski, and Walters (1974), Molina and Rowland (1974), and Stolarski and Cicerone (1974). Subsequent calculations of the effects of CFCs on ozone were made by many atmospheric scientists (e.g., Wuebbles, Luther, and Penner, 1983) using theoretical models of the photochemistry of the atmosphere. A useful review article that summarizes the basic ozone depletion theory and corresponding earlier chemical model predictions is by Stolarski (1982). Over the last 10 years, since the occurrence of the substantial ozone hole over Antarctica during the austral spring months was first noticed in 1985, there has been increased concern about the possible depletion, on a global scale, of the ozone layer due to the release of CFCs. Consequently, analyses of trends in atmospheric total ozone have received substantial attention in recent years, using both ground-based and satellite measurements. Recent statistical trend analyses have been reported by Bojkov et al. (1990), Stolarski et al. (1991, 1992), and Reinsel et al. (1994). We will review some of the statistical methods and results from this last reference.

Monthly average time series data on atmospheric total column ozone from a collection of 56 Dobson ground-based recording stations over the globe and from the total ozone mapping spectrometer (TOMS) satellite experiment are analyzed for the detection of trend in total ozone that may be associated with the effects of the release of CFCs. The ground station data used in the analysis have recently been critically re-evaluated and revised for some stations. For the ground station data, regression time series models that include components for seasonal variations, linear trend over different months of the year (seasonal trend), solar flux variations, and quasi-biennial oscillations (QBO), are estimated from the data at each individual station for the period starting from 1964 through 1991. Random effects models for the individual station trend estimates, to allow for individual station and regional trend variations, are used to combine individual trend estimates to obtain overall trend estimates in ozone for different seasons of the year and year-round as a function of latitude. A related seasonal trend analysis over the more recent shorter time period November 1978 through December 1991 is also performed using zonal average time series data from the TOMS satellite experiment, and results are compared with trend analysis of the Dobson ground station data for the comparable time period.

2 Trend Analysis of Ground Station Total Ozone Data

2.1 Statistical Trend Model for Individual Station Total Ozone Data

Total ozone data are available for trend analysis from 56 Dobson ground stations located around the globe. However, the large majority of stations (43) are located in the Northern Hemisphere, and these stations tend to be concentrated mainly in the regions of North America (10 stations), Europe (14 stations), Japan (4 stations), and India (6 stations). As an illustration, Figure 1 displays time series plots of monthly averages of the total ozone data for four selected Dobson stations. The data are measured as a thickness of the total amount of ozone in a vertical column of the atmosphere in units of milli-atmosphere-centimeters (also known as Dobson units). Prominent characteristics of the total ozone data that are exhibited in Figure 1 are the strong annual seasonal pattern and the increase in both mean amount and amplitude of the seasonal variation with an increase in latitude.

To investigate the seasonal nature of ozone trends in the ground station total ozone data, a statistical trend model that allows for a different trend for each month of the year was considered, and the model was estimated separately for each of the 56 stations. (The allowance for the seasonal trend structure differs from earlier trend analysis studies of Reinsel and Tiao, 1987, and Reinsel et al., 1987, which assumed a constant or uniform trend over different months.) The seasonal trend model considered for an individual station monthly total ozone series Y_t is given by

$$Y_t = \sum_{m=1}^{12} \mu_m I_{mt} + \sum_{m=1}^{12} \omega_m I_{mt} X_t + \gamma_1 Z_{1,t} + \gamma_2 Z_{2,t-k} + N_t , \qquad (1)$$

where I_{mt}, $m = 1, \ldots, 12$, denotes an indicator series for the mth month of the year which equals 1 if t corresponds to month m of the year and 0 otherwise; μ_m denotes the mean ozone amount in month m; X_t denotes a linear trend function starting in January 1970, with $X_t = (t - T_0)/12$ for $t > T_0 =$ Dec. 1969 and $X_t = 0$ for $t \leq T_0$; ω_m denotes the trend or change (in Dobson units per year) starting in January 1970 for month m; $Z_{1,t}$ denotes the 10.7 cm solar radio flux series of measurements from Ottawa; $Z_{2,t-k}$ denotes a quasi-biennial oscillation (QBO) equatorial 50 mb zonal wind series of measurements at Singapore lagged k months with the lag depending on the latitude of the individual Dobson station; and N_t is a noise series modeled as a first-order autoregressive (AR(1)) model, $N_t = \phi N_{t-1} + e_t$.

The linear trend starting in 1970 corresponds approximately to the form of the overall effects on ozone globally due to CFCs as calculated from photochemical model theory. The inclusion of the solar flux and QBO terms in (1) is to account for certain natural variations in atmospheric ozone. Specifically, through the photochemistry of ozone formation and destruction, ozone amounts vary to some extent with variations in the output of the sun, which has prominent longer-term changes over an approximate 11-year cycle. Although the dependence of ozone is on the solar flux in the UV region of the spectrum, measurements of UV solar flux over these important wavelengths are not available over the full ozone data time period, and the closely related 10.7 cm radio flux is used as a proxy solar variable. In addition, there is much evidence that ozone is dependent on the stratospheric circulational changes related to QBO in equatorial stratospheric winds, and the 50 mb equatorial wind measurements are used as an indicator to represent this behavior. To account for differences in the variability of ozone over different months of the year, in the AR(1) model it is assumed that $\mathrm{var}(e_t) = \sigma_t^2 \equiv \sigma_m^2$ depends on the month m of the time t and so is allowed to differ over the months of the year.

In the trend analysis, intervention mean level shift adjustments to account for instrument calibrations and other instrument effects on the ozone measurements were also used in the statistical model (1) for some stations. For these stations an additional term $\gamma_3 Z_{3,t}$ is included in the regression model (equation (1)), where $Z_{3,t}$ is an indicator variable which is equal to 0 before the beginning date of the intervention and is equal to 1 after the date of the intervention.

A slight complication to the analysis of model (1) is that for most stations there are some months for which the total ozone data are not observed. Employing a maximum likelihood procedure, time series models of the preceding form were estimated from the available total ozone data at each of the 56 stations. The parameter estimates of (1) and the AR(1) noise model were obtained by an iteratively reweighted least squares procedure to minimize $\sum_{t=1}^{T} e_t^2/\sigma_m^2$, with adjustment in the weights for the first time point $t = 1$ and any time point t immediately preceded by a missing observation. The variance terms σ_m^2, $m = 1, 2, \ldots, 12$, were estimated in the usual manner as the sample sum of squares of those $\hat{e}_t$ corresponding to the appropriate month m divided by the number of squared terms in the sum, again with adjustment for time points preceded by a missing value. More specifically, suppose monthly observations are available only for the (integer) times $t_1 < t_2 < \cdots < t_n$, and express model (1) as $Y_t = \boldsymbol{X}_t' \boldsymbol{\gamma} + N_t$. Then for given value of ϕ and the σ_m^2, let $Y_{t_i}^* = Y_{t_i} - \phi^{\Delta_i} Y_{t_{i-1}}$ and $\boldsymbol{X}_{t_i}^* = \boldsymbol{X}_{t_i} - \phi^{\Delta_i} \boldsymbol{X}_{t_{i-1}}$ with $\Delta_i = t_i - t_{i-1}$, and let $V_{t_i} = \sum_{j=0}^{\Delta_i - 1} \phi^{2j} \sigma_{t_i - j}^2$. Then the weighted least squares estimator of $\boldsymbol{\gamma}$ at that stage is

$$\hat{\boldsymbol{\gamma}} = \left(\sum_{i=1}^{n} \boldsymbol{X}_{t_i}^* \boldsymbol{X}_{t_i}^{*\prime} / V_{t_i} \right)^{-1} \sum_{i=1}^{n} \boldsymbol{X}_{t_i}^* Y_{t_i}^* / V_{t_i} . \tag{2}$$

The estimator of ϕ is obtained at a given stage similar to the method described in Reinsel and Wincek (1987), with slight modification to account for unequal variances of the e_t.

2.2 Random Effects Model for Individual Station Seasonal Trend Estimates

The individual station trend estimates $\hat{\omega}_m$ of the monthly trend parameters ω_m in (1) are of main interest, and are retained from the estimation results for further analysis. An investigation of variations in these estimated trends over seasons and as a function of latitude is performed for the ground station total ozone data. The monthly trend estimates of an individual station are grouped into four seasons, which are defined as December, January, and February (DJF), March, April, and May (MAM), June, July, and August (JJA), and September, October, and November (SON). Based on estimation results of the model (1) fit to the data at each station, the four seasonal trend values are obtained, expressed in percent per decade, by averaging the three monthly trend estimates $\hat{\omega}_m$ (in Dobson units per year) within each season, dividing by the average of the monthly mean ozone amounts $\hat{\mu}_m$ within the season, and then scaling to percentage per decade. These percentage trends will be referred to as the seasonal trend estimates. Similarly, an annual trend estimate is formed for each individual station from the average of the 12 monthly trend estimates, also scaled to percent per decade. These individual station trend estimates for each season are plotted against the latitude of the station in Figure 2.

The dependence of these seasonal trends on latitude is then considered, and trend summaries for each of 7 latitude zones are obtained. The latitude zones are indicated in Table 1. To obtain overall trend estimates for each of these latitude zones, a random effects model is considered for the collection of individual station trend estimates for each season. The stations are grouped into $A = 9$ distinct geographic regions (e.g., North America, Europe, Japan, and so on), and we let $\hat{\omega}^*_{ij}$ denote the trend estimate of the jth station within the ith region. Then the model considered for the $\hat{\omega}^*_{ij}$ is given by

$$\hat{\omega}^*_{ij} = \delta(L_{ij}) + \alpha_i + \beta_{ij} + \varepsilon_{ij}, \quad j = 1, \ldots, n_i, \quad i = 1, \ldots, A, \tag{3}$$

with $A = 9$, $n_1 = 10$, $n_2 = 14$, $n_3 = 4$, and so on, where L_{ij} denotes the latitude of the individual station, and the mean function $\delta(L_{ij}) = \delta_k$ is assumed to be a constant for all stations within the kth latitude zone but possibly different across latitude zones, for $k = 1, \ldots, 7$. In (3), α_i represents a random effect term corresponding to a regional effect and is common to all stations in a given region, distributed with mean 0 and variance σ^2_α, β_{ij} denotes an individual station random effect corresponding to station to station variations, distributed with mean 0 and variance σ^2_β, and ε_{ij} represents the estimation error (within station) in obtaining the estimates $\hat{\omega}^*_{ij}$, distributed with mean 0 and variance σ^2_{ij}. All the random terms in (3) are also assumed to be mutually independently and normally distributed. The particular model (3) assumes the trend estimates $\hat{\omega}^*_{ij}$ have the simple mean structure $\delta(L_{ij})$ as a function of latitude, but no special form of structure (such as a linear function of latitude) is assumed for the mean of trends between different latitude zones. Note that the presence of the regional random effect term α_i in (3) allows for correlation between trends at different stations within the same region, of a rather special structure.

Given the individual trend estimates $\hat{\omega}^*_{ij}$ and the estimate $\hat{\sigma}^2_{ij}$ of $\text{var}(\varepsilon_{ij})$, which is directly available from the fitting of the individual station trend model (1), the unknown variance components σ^2_α and σ^2_β and the parameters of the mean function in (3) are estimated by maximum likelihood. The variance components are estimated using the EM algorithm approach, treating the random effects α_i and β_{ij} as "missing data". In general, assume the mean function in (3) is represented by a linear regression model as $\delta(L_{ij}) = \boldsymbol{u}'_{ij}\boldsymbol{\delta}$, where $\boldsymbol{\delta} = (\delta_1, \ldots, \delta_r)'$, and set $U_i = (\boldsymbol{u}_{i1}, \ldots, \boldsymbol{u}_{in_i})'$ and $\hat{\boldsymbol{\omega}}^*_i = (\hat{\omega}^*_{i1}, \ldots, \hat{\omega}^*_{in_i})'$. Then it follows that $\text{Cov}(\hat{\boldsymbol{\omega}}^*_i) = V_i = \sigma^2_\alpha \mathbf{1}\mathbf{1}' + \sigma^2_\beta I + D_i = \sigma^2_\alpha \mathbf{1}\mathbf{1}' + A_i$, where $\mathbf{1}$ denotes a vector of ones, $A_i = \sigma^2_\beta I + D_i$ and $D_i = \text{Diag}(\sigma^2_{i1}, \ldots, \sigma^2_{in_i})$, and $V_i^{-1} = A_i^{-1} - (\sigma^2_\alpha d_i) A_i^{-1}\mathbf{1}\mathbf{1}'A_i^{-1}$, where $d_i = [1 + \sigma^2_\alpha \sum_{j=1}^{n_i} (1/(\sigma^2_\beta + \sigma^2_{ij}))]^{-1}$. The EM algorithm approach (e.g., Laird and Ware, 1982) then gives estimates of the variance components at each iteration as

$$\hat{\sigma}_\alpha^2 = (1/A)\sum_{i=1}^A (\tilde{\alpha}_i^2 + R_i^2), \qquad \hat{\sigma}_\beta^2 = (1/n)\sum_{i=1}^A \sum_{j=1}^{n_i} (\tilde{\beta}_{ij}^2 + P_{ij}^2), \tag{4}$$

where $\tilde{\alpha}_i = E(\alpha_i \mid \hat{\boldsymbol{\omega}}_i^*) = \sigma_\alpha^2 \mathbf{1}' V_i^{-1}(\hat{\boldsymbol{\omega}}_i^* - U_i \boldsymbol{\delta})$ and $R_i^2 = \text{var}(\alpha_i \mid \hat{\boldsymbol{\omega}}_i^*) = \sigma_\alpha^2 d_i$, and $\tilde{\boldsymbol{\beta}}_i = (\tilde{\beta}_{i1}, \ldots, \tilde{\beta}_{in_i})' = \sigma_\beta^2 V_i^{-1}(\hat{\boldsymbol{\omega}}_i^* - U_i \boldsymbol{\delta})$ and $\sum_{j=1}^{n_i} P_{ij}^2 = \sum_{j=1}^{n_i} \text{var}(\beta_{ij} \mid \hat{\boldsymbol{\omega}}_i^*) = \sigma_\beta^2 [n_i - \sigma_\beta^2 \text{tr}(V_i^{-1})]$, which are all evaluated at parameter estimates from the previous iteration. The maximum likelihood or generalized least squares estimator of $\boldsymbol{\delta}$ corresponding to the maximum likelihood estimates of the variance components is given by

$$\begin{aligned}\hat{\boldsymbol{\delta}} &= \{ \textstyle\sum_{i=1}^A U_i' \hat{V}_i^{-1} U_i \}^{-1} \sum_{i=1}^A U_i' \hat{V}_i^{-1} \hat{\boldsymbol{\omega}}_i^* \\ &= \{ \textstyle\sum_{i=1}^A [U_i' \hat{A}_i^{-1} U_i - \hat{\sigma}_\alpha^2 \hat{d}_i (U_i' \hat{A}_i^{-1} \mathbf{1})(\mathbf{1}' \hat{A}_i^{-1} U_i)] \}^{-1} \\ &\qquad \textstyle\sum_{i=1}^A [U_i' \hat{A}_i^{-1} \hat{\boldsymbol{\omega}}_i^* - \hat{\sigma}_\alpha^2 \hat{d}_i (U_i' \hat{A}_i^{-1} \mathbf{1})(\mathbf{1}' \hat{A}_i^{-1} \hat{\boldsymbol{\omega}}_i^*)],\end{aligned} \tag{5}$$

with the covariance matrix of $\hat{\boldsymbol{\delta}}$ estimated by

$$\text{Cov}(\hat{\boldsymbol{\delta}}) = \{ \textstyle\sum_{i=1}^A U_i' \hat{V}_i^{-1} U_i \}^{-1} = \{ \sum_{i=1}^A [U_i' \hat{A}_i^{-1} U_i - \hat{\sigma}_\alpha^2 \hat{d}_i (U_i' \hat{A}_i^{-1} \mathbf{1})(\mathbf{1}' \hat{A}_i^{-1} U_i)] \}^{-1}. \tag{6}$$

Using the above model (3), trend estimates were obtained by the preceding estimation procedure for each of the 7 latitude zones and each season. The results are summarized in Table 1(a), with trend estimates expressed as percent changes per decade during the period 1970–1991. One of the main features seen from Table 1(a) is the substantial negative trend in the winter season at higher northern latitudes, particularly in the 40°N–50°N zone. It is also noted that for the 0°N–20°N and 0°S–30°S zones the year-round trends are close to zero and the differences in trends over seasons are small. The trends in the 20°N–30°N latitude zone are only slightly negative (around –0.4% per decade), and the seasonal trends for this zone also do not exhibit much difference across the seasons. For the higher southern latitude zone (30°S–55°S), trends are more negative over all seasons than in the tropical latitude zones, but there is not too much evidence of variation in trends over the seasons in this zone. Analysis of the residuals $\hat{\varepsilon}_{ij} = \hat{\omega}_{ij}^* - \hat{\delta}(L_{ij}) - \tilde{\alpha}_i - \tilde{\beta}_{ij}$ from the fitting of model (3) is also considered. Taking into account that differences in variances of the ε_{ij} are allowed, examination of various residual plots, such as plots of standardized forms of the $\hat{\varepsilon}_{ij}$ versus station latitude, does not indicate any outliers, unusual features, or obvious violations of the assumptions of model (3).

Overall, the general trend findings for the analysis over this period 1964–1991 have similarity to the trends obtained in previous studies of trend for published total ozone data through 1986, e.g., see Bojkov et al. (1990). In particular, substantial negative trends are reconfirmed for the winter season at higher northern latitudes, with average wintertime trend in the 30°N–65°N latitude range of about −2.5% per decade. However, we also now find some substantial negative trends in the spring, summer and fall seasons, in the 40°N–50°N latitude zone, with average trend value of about −1.5% per decade in that latitude zone, but considerably smaller and nonsignificant trend (with average trend value around –0.2% per decade) in spring, summer, and fall seasons for the northern latitude zones below 40°N. Consequently, the year-round or annual trend is also considerably more negative in the 40°N–50°N latitude zone than in northern latitudes below 40°N, with an average annual trend of about −2% per decade in the 40°N–50°N latitude zone.

2.3 Sensitivity of Trend Results by Comparison of Trends Over Two Different Data Periods

A similar seasonal trend analysis of Dobson ground station data was performed for the shorter, more recent, data period of November 1978 through 1991, for comparison with trend results available from the TOMS satellite total ozone data for this same period (see Section 3). A summary of the seasonal and annual trend-fitting results, in terms of latitude zonal trend estimates based on the model of equation (3), is presented in Table 1(b) for the data period November 1978–1991. Based on the trend results summarized in Table 1(a-b), we make the following comparisons of trends over the shorter data period November 1978–1991 relative to trends over the longer data period 1964–1991. First, in the latitude range of 30°S–20°N there is very little difference in trend results between the data periods, and if anything, trends for the period November 1978–1991 are slightly more positive than those for the period 1964–1991 for this latitude range. However, in the 20°N–30°N, 30°N–40°N, and 40°N–50°N latitude zones there are some moderate trend differences. Specifically, the trends in these latitudes are more negative over the shorter data period November 1978–1991 than over the longer period during the winter season, by about −1% per decade, and the spring season, by about −2% per decade; they are less negative over the shorter data period in the fall season (only in the 30°N–40°N and 40°N–50°N latitude zones) by about 1% per decade, and the trends are fairly similar for the summer season between the shorter time period November 1978–1991 and the longer data period. On the other hand, for the latitude range of 50°N–65°N the trends are much more negative over the shorter data period November 1978–1991 than over the longer period during the summer season, as well as in the spring season, by about −2% per decade, and there is less trend difference between the data periods for the fall and winter seasons in this latitude zone. For the southern midlatitude zone of 30°S–55°S the trends are more negative over the shorter data period November 1978–1991 than the longer period for all seasons, with the most substantial trend differences of about −1.5% per decade occurring during the MAM and JJA seasons (fall and wintertime for SH) between the shorter data period and the longer period.

Comparison of these overall seasonal trend results for the latitude zones over the two different data periods are summarized graphically in Figure 3. Figure 3 does provide the overall impression that trends for the data period November 1978–1991 have become generally more negative for the spring and summer seasons in the 50°N–65°N latitude zone, more negative for the winter and spring seasons in the 20°N–30°N, 30°N–40°N, and 40°N–50°N latitude zones, less negative for the fall season in the 30°N–50°N latitude range, and more negative for all seasons (but especially the MAM and JJA seasons) in the 30°S–55°S latitude zone. The most prominent differences in trends between the two data periods occur in the spring season for latitude zones of 20°N and greater. The statistical significance of the differences can be investigated by pairwise comparison of the individual station trend estimates $\hat{\omega}_{ij}^{*}$ over the two data periods. To illustrate, in Figure 4 the individual station differences in trend estimates between the shorter data period November 1978–1991 and the longer period are plotted against latitude for the spring (MAM) season trends. From this figure we see rather strong graphical evidence for more negative trends over the shorter data period for latitudes of 20°N and greater. Under the assumptions of model (1), it can be established that the (within station) variance of the difference in individual station trend estimates between the shorter and longer data periods can be approximated by the difference in the corresponding variances $\hat{\sigma}_{ij}^{2}$ of the trend estimates between the shorter and longer periods. Using this approximation, the individual station trend differences can be analyzed on the basis of model (3) in a similar way as the analysis which led to the results in Table 1. From this method of analysis, the trend differences for the spring (MAM) season in each of the latitude zones 20°N–30°N, 30°N–40°N, 40°N–50°N, and 50°N–65°N are found to have estimated mean values of the order of –2% per decade with standard errors of about 0.5% per decade. Hence these overall mean trend differences between the two data periods are statistically significant. From this same type of analysis applied to each of the four seasons, the only other statistically significant differences between trends over the two data periods are found to occur for the summer season in the 50°N–65°N latitude zone and for the winter (DJF) and spring (MAM) seasons in the 30°S–55°S latitude zone.

2.4 Modification of the Random Effects Model for Correlation Across Stations

We note that the above model (3) has assumed that the ε_{ij} from individual stations are independent across stations, which is closely related to the assumption that the noise series e_t associated with model (1) are independent across stations. In fact, examination of the residuals $\hat{e}_t$ shows that the residuals are generally positively cross-correlated across stations within a given region, but have negligible cross-correlations between stations in different regions. Thus we have also considered a modification of model (3) to allow the ε_{ij} to be correlated across stations within a given region, with the correlations being estimated (approximately) from the cross-correlations of the (standardized) residuals $\hat{e}_t/\hat{\sigma}_t$ from the fitted time series models between different stations.

Examination of the cross-correlation structure of the (standardized) residuals $\hat{e}_t/\hat{\sigma}_t$ is of interest for other purposes as well, for example, in the selection of station locations for effective monitoring purposes. Hence, for illustration, in Figure 5 we display the cross-correlations of standardized residuals among all pairs of stations within each of three main regions, North America, Europe and Japan, as a function of the distance between stations. These cross-correlations display a well-developed pattern with respect to distance, as expected, with correlations becoming negligible for distances greater than about 1800 km.

The maximum likelihood estimation procedure for this modification of model (3), in which the ε_{ij} are assumed to be correlated across stations within each region with the cross-correlations estimated from the cross-correlations of the (standardized) residuals $\hat{e}_t/\hat{\sigma}_t$ from the time series models, is similar to the estimation procedure described previously. In particular, the maximum likelihood estimator of $\boldsymbol{\delta}$ and its estimated covariance matrix are again given by expressions in (5) and (6), except that the matrices A_i are no longer diagonal because the D_i are not diagonal matrices in the modified model. Applying the estimation procedures for this model yielded results similar to those obtained in Table 1, however, and hence they are not reported here. Qualitatively, when the (positive) correlation among the ε_{ij} across stations within each region is allowed for, the latitudinal trend estimates remain similar to those in Table 1 but their standard errors are increased slightly. There is also a tendency for the between station within region variance component estimate $\hat{\sigma}_\beta^2$ to become larger whereas the between region variance component estimate $\hat{\sigma}_\alpha^2$ is smaller. More details of the comparative results from a similar study covering an earlier data period are presented in Bojkov et al. (1990).

3 Comparison With Trend Analysis of TOMS Ozone Satellite Data

For comparison with the above trends obtained from the Dobson total ozone data over the period November 1978–1991, we also consider trend analysis of Nimbus 7 TOMS satellite total ozone data (version 6, see Herman et al. (1991)) available for the same data period. For the TOMS data trend analysis we use monthly averages of latitudinal zonal average data obtained for each 10° latitude zone in the range 60°S–60°N. In our analysis, the globe was partitioned into 10° latitude zones, and the TOMS total ozone data were aggregated to form monthly averages for each of the 12 zones between 60°N and 60°S. This procedure is motivated by the fact that variations in ozone tend to be much more dependent on latitude than on longitude. The advantage of the satellite data is that they provide nearly global data coverage over the latitude range 60°S–60°N, although they cover a shorter time period than the ground station data.

For each 10° latitudinal monthly average time series the seasonal trend model (equation (1)) was estimated for the TOMS data period November 1978–1991. The resulting seasonal trend estimates for each of the four seasons as well as the annual or year-round trend and the solar effect estimates are presented in Table 2 for each of the 12 latitude zones between 60°S and 60°N. These trends seem to have generally similar features as obtained previously in similar trend analyses of the TOMS data by Stolarski et al. (1991, 1992). Specifically, these features include the large negative trends in the northern midlatitudes (30°N–60°N) during the winter and spring seasons, the generally much smaller trends in the tropical latitudes (30°S–30°N) during all seasons, and the large negative trends in the higher southern latitudes (40°S–60°S) during all seasons but especially the JJA season (wintertime for SH).

For comparison of these TOMS zonal average data trends with trends from the Dobson data for the same data period, the overall seasonal trend estimation results for the latitude zones are presented graphically in Figure 6 for both the Dobson data and the TOMS data for the data period November 1978–1991. Generally, we see similar trend results obtained from the analyses of the two data sets. The two sets of trends are most similar in the MAM and JJA seasons, while the TOMS trends are more negative in the DJF season for all latitude zones, by about −1% per decade for zones below 30°N and about −2% per decade for the higher northern latitudes 30°N–65°N, and they are also more negative in the fall (SON) season for the higher northern latitudes by about −2% per decade.

4 Summary and Concluding Remarks

A seasonal trend analysis has been performed for Dobson total ozone data from 56 Dobson stations in the northern and southern hemispheres, using two different data periods, 1964–1991 and November 1978–1991. Over the longer data period, significant negative trends have been found in the higher northern latitudes (above 30°N) during the winter (DJF) and spring (MAM) seasons. There is also some evidence of negative trend in the summer (JJA) and fall (SON) seasons, especially in the 40°N–50°N latitude zone. Moderate negative trends have also been estimated in the higher southern latitudes (30°S–55°S) during all seasons, based on a much smaller number of Dobson stations for the southern hemisphere region. The trend results for the higher northern latitudes are reasonably similar to recent results of Bojkov et al. (1990) and Stolarski et al. (1992) that were based on Dobson data provisionally revised by R. Bojkov, although the trend results from this study generally seem to be somewhat more negative than those of Bojkov et al. (1990) and slightly less negative than the corresponding trend results of Stolarski et al. (1992).

For the shorter time period November 1978–1991 there is an indication that trends have become more negative in the higher northern latitudes, especially during the winter and spring seasons, and also in the higher southern latitudes in all seasons. An implication of these more negative trend findings for the shorter time period is that over the longer time period, some mild nonlinearity in the trend with time may be present, and this should be given consideration in future analyses of the long-term trends in total ozone. For the present analysis, the linear trend for the longer time period is a reasonable approximation which is fairly accurate and useful for descriptive summary purposes. The negative trend results from Dobson data for the shorter time period are qualitatively in agreement with results based on recent trend analyses of satellite-based total ozone data, such as TOMS and SAGE, over comparable data periods. However, trend analysis of the SAGE column ozone data (McCormick et al., 1992) has shown a somewhat more substantial negative trend in the tropical latitudes than the TOMS total ozone trend results or our Dobson data trend results. Our own trend analysis of the TOMS data for the same data period indicates that the satellite data trends seem to be more negative than the Dobson data trends for the higher northern latitude range of 30°N–65°N, with the differences of the order of –2% per decade mainly in the winter and fall seasons. A potential source of some difference between the trends from Dobson and TOMS data is the well-recognized lack of sensitivity of the satellite observations to ozone changes in the lower troposphere (Mateer et al., 1971), coupled with the expected positive trend behavior in tropospheric ozone. The extent to which this effect can account for the seasonal and latitudinal pattern of differences noted in Figure 6 is not certain.

Overall, the Dobson station total ozone data are found to be extremely valuable for trend analysis because of their long-term nature, affording the opportunity for comparisons of trends over different data periods. Another valuable attribute of the ground-based Dobson network is that since the stations use different instruments, the maintenance of instrument calibration over time, while important, is less critical than for a single instrument satellite system. It will be important that the efforts of individual operating stations and agencies to perform detailed reevaluations of their total ozone data using the calibration history of the instrument be continued and maintained so that the highest quality Dobson data be available for long-term trend analyses and for comparisons with current and future satellite data.

The above analyses reveal many statistical and other issues that require further investigation. These include the development and proper evaluation of methods to combine individual station information, with proper accounting for spatial correlation, for the ground station data, and for proper comparison of data and trends from different data sources (e.g., ground-based versus satellite). Other issues involve the accuracy and reliability of the seasonal trend model results, especially their dependence on the length and particular time period analyzed, and their dependence on the influence of rather unusual behavior in the total ozone over many parts of the globe during 1983 and 1985 related to strong natural circulation patterns associated with the El Chichon volcanic eruption in 1982 and the strong El Nino of 1983 combined with QBO variations during this period (Bojkov, 1987).

References

Bojkov, R. D., The 1983 and 1985 anomalies in ozone distribution in perspective, *Mon. Weath. Rev., 115*, 2187–2201, 1987.

Bojkov, R., L. Bishop, W. J. Hill, G. C. Reinsel, and G. C. Tiao, A statistical trend analysis of revised Dobson total ozone data over the northern hemisphere, *J. Geophys. Res., 95*, 9785–9807, 1990.

Cicerone, R. J., R. S. Stolarski, and S. Walters, Stratospheric ozone destruction by man-made chlorofluoromethanes, *Science, 185*, 1165–1167, 1974.

Herman, J. R., R. Hudson, R. McPeters, R. Stolarski, Z. Ahmad, X.-Y. Gu, S. Taylor, and C. Wellemeyer, A new self-calibration method applied to TOMS and SBUV backscattered ultraviolet data to determine long-term global ozone change, *J. Geophys. Res., 96*, 7531–7545, 1991.

Laird, N. M., and J. H. Ware, Random-effects models for longitudinal data, *Biometrics, 38*, 963–974, 1982.

Mateer, C. L., D. F. Heath, and A. J. Krueger, Estimation of total ozone from satellite measurements of backscatter ultraviolet earth radiance, *J. Atmos. Sci., 28*, 1307–1311, 1971.

McCormick, M. P., R. E. Veiga, and W. P. Chu, Stratospheric ozone profile and total ozone trends derived from the SAGE I and SAGE II data, *Geophys. Res. Lett., 19*, 269–272, 1992.

Molina, M. J., and F. S. Rowland, Stratospheric sink for chlorofluoromethanes: Chlorine atom catalyzed destruction of ozone, *Nature, 249*, 810–812, 1974.

Reinsel, G. C., G. C. Tiao, A. J. Miller, D. J. Wuebbles, P. S. Connell, C. L. Mateer, and J. J. DeLuisi, Statistical analysis of total ozone and stratospheric Umkehr data for trends and solar cycle relationship, *J. Geophys. Res., 92*, 2201–2209, 1987.

Reinsel, G. C., and G. C. Tiao, Impact of chlorofluoromethanes on stratospheric ozone: A statistical analysis of ozone data for trends, *J. Amer. Statist. Assoc., 82*, 20–30, 1987.

Reinsel, G. C., G. C. Tiao, D. J. Wuebbles, J. B. Kerr, A. J. Miller, R. M. Nagatani, L. Bishop, and L. H. Ying, Seasonal trend analysis of published ground-based and TOMS total ozone data through 1991, *J. Geophys. Res., 99*, 5449–5464, 1994.

Reinsel, G. C., and M. A. Wincek, Asymptotic distribution of parameter estimators for nonconsecutively observed time series, *Biometrika, 74*, 115–124, 1987.

Stolarski, R. S., Fluorocarbons and stratospheric ozone: a review of current knowledge, *Amer. Statist., 36*, 303–311, 1982.

Stolarski, R. S., P. Bloomfield, and R. D. McPeters, Total ozone trends deduced from Nimbus 7 TOMS data, *Geophys. Res. Lett., 18*, 1015–1018, 1991.

Stolarski, R., R. Bojkov, L. Bishop, C. Zerefos, J. Staehelin, and J. Zawodny, Measured trends in stratospheric ozone, *Science, 256*, 342–349, 1992.

Stolarski, R. S., and R. J. Cicerone, Stratospheric chlorine: A possible sink for ozone, *Can. J. Chem., 52*, 1610–1615, 1974.

World Meteorological Organization, Atmospheric ozone 1985: Assessment of our understanding of the processes controlling its present distribution and change, Global Ozone Res. and Monit. Proj., *Rep. 16*, Geneva, 1985.

Wuebbles, D. J., F. M. Luther, and J. E. Penner, Effect of coupled anthropogenic perturbations on stratospheric ozone, *J. Geophys. Res., 88*, 1444–1456, 1983.

Table 1. Latitudinal Zonal Year-Round (Annual) and Seasonal Trend Summaries (in % per Decade) and Solar Cycle Effect Summary (in % per 100 Units of 10.7 Flux) Based on Dobson Total Ozone Data for Two Different Data Periods. The Zonal Trend Estimation Results are Based on Model (3). (Standard errors of trend estimates are given in parentheses.)

(a) Trend results using data over the data period 1964–1991.

Zone	Year-round	DJF	MAM	JJA	SON	Solar
50°–65°N	−1.37(0.34)	−2.45(0.47)	−1.73(0.40)	−0.46(0.36)	−0.70(0.34)	1.02(0.21)
40°–50°N	−2.02(0.37)	−3.34(0.47)	−2.47(0.47)	−1.19(0.37)	−0.98(0.38)	0.78(0.23)
30°–40°N	−0.86(0.38)	−1.85(0.47)	−0.85(0.49)	−0.50(0.39)	−0.27(0.38)	1.10(0.23)
20°–30°N	−0.37(0.40)	−0.83(0.49)	−0.50(0.52)	0.01(0.42)	−0.14(0.37)	0.66(0.23)
0°–20°N	0.01(0.46)	−0.51(0.56)	0.09(0.59)	−0.01(0.49)	0.26(0.44)	0.34(0.24)
0°–30°S	−0.19(0.44)	−0.46(0.49)	−0.13(0.52)	−0.31(0.48)	−0.09(0.45)	0.51(0.23)
30°–55°S	−1.31(0.47)	−1.41(0.52)	−1.19(0.54)	−1.66(0.54)	−1.03(0.48)	0.73(0.27)

(b) Trend results using data over the data period November 1978–1991.

Zone	Year-round	DJF	MAM	JJA	SON	Solar
50°–65°N	−2.71(0.43)	−2.83(0.86)	−3.99(0.62)	−2.85(0.77)	−1.54(0.60)	0.62(0.24)
40°–50°N	−2.67(0.45)	−4.30(0.86)	−4.20(0.68)	−1.44(0.76)	0.09(0.61)	0.83(0.26)
30°–40°N	−1.43(0.47)	−2.94(0.75)	−2.52(0.76)	0.01(0.78)	0.56(0.61)	1.13(0.27)
20°–30°N	−1.76(0.47)	−2.04(0.75)	−2.82(0.68)	−0.77(0.76)	−0.79(0.50)	0.97(0.25)
0°–20°N	0.16(0.48)	−0.40(0.75)	0.70(0.65)	0.26(0.77)	−0.07(0.55)	0.69(0.27)
0°–30°S	0.20(0.47)	0.05(0.59)	0.36(0.53)	−0.27(0.85)	0.27(0.57)	0.63(0.27)
30°–55°S	−2.51(0.48)	−2.36(0.62)	−2.85(0.58)	−3.28(1.06)	−1.85(0.72)	0.97(0.29)

Table 2. Latitudinal Zonal Year-Round (Annual) and Seasonal Trend Estimates (in % per Decade) and Solar Cycle Effect Estimates (in % per 100 Units of 10.7 Flux) Based on Nimbus 7 Satellite TOMS Total Ozone Data for the Period November 1978–1991. (Standard errors of trend estimates are given in parentheses.)

Zone	Year-round	DJF	MAM	JJA	SON	Solar
50°–60°N	−3.96(1.15)	−5.03(1.46)	−4.99(1.64)	−1.96(1.19)	−3.44(1.00)	−0.42(0.39)
40°–50°N	−3.74(0.94)	−5.91(1.29)	−4.72(1.46)	−1.72(1.03)	−1.95(0.74)	0.24(0.29)
30°–40°N	−2.93(0.90)	−4.92(1.15)	−3.63(1.36)	−1.49(1.01)	−1.39(0.68)	0.75(0.28)
20°–30°N	−1.61(0.98)	−2.96(1.21)	−1.45(1.33)	−0.78(1.03)	−1.33(0.82)	1.14(0.30)
10°–20°N	−0.91(0.95)	−2.09(1.18)	−0.07(1.21)	−0.44(1.02)	−1.11(0.87)	0.87(0.32)
0°–10°S	−0.57(1.01)	−1.13(1.16)	0.60(1.27)	−0.48(1.09)	−1.24(1.02)	0.49(0.33)
0°–10°S	−0.45(0.86)	−0.58(1.00)	0.82(1.19)	−0.63(0.88)	−1.34(0.92)	0.70(0.30)
10°–20°S	−0.44(0.48)	−0.75(0.67)	−0.28(0.77)	−0.71(0.65)	−0.03(0.53)	0.99(0.24)
20°–30°S	−1.00(1.06)	−1.57(1.12)	−1.42(1.20)	−0.87(1.29)	−0.19(1.10)	0.54(0.37)
30°–40°S	−2.04(1.70)	−2.58(1.99)	−2.41(1.75)	−2.19(1.77)	−1.09(1.69)	0.07(0.38)
40°–50°S	−3.86(1.03)	−4.24(1.24)	−3.67(1.03)	−4.74(1.26)	−2.85(1.29)	1.74(0.38)
50°–60°S	−5.94(1.03)	−5.31(1.14)	−4.83(0.96)	−7.04(1.39)	−6.42(1.56)	1.38(0.39)

Fig. 1. Monthly Averages of Total Atmospheric Ozone from 4 Selected Dobson Stations

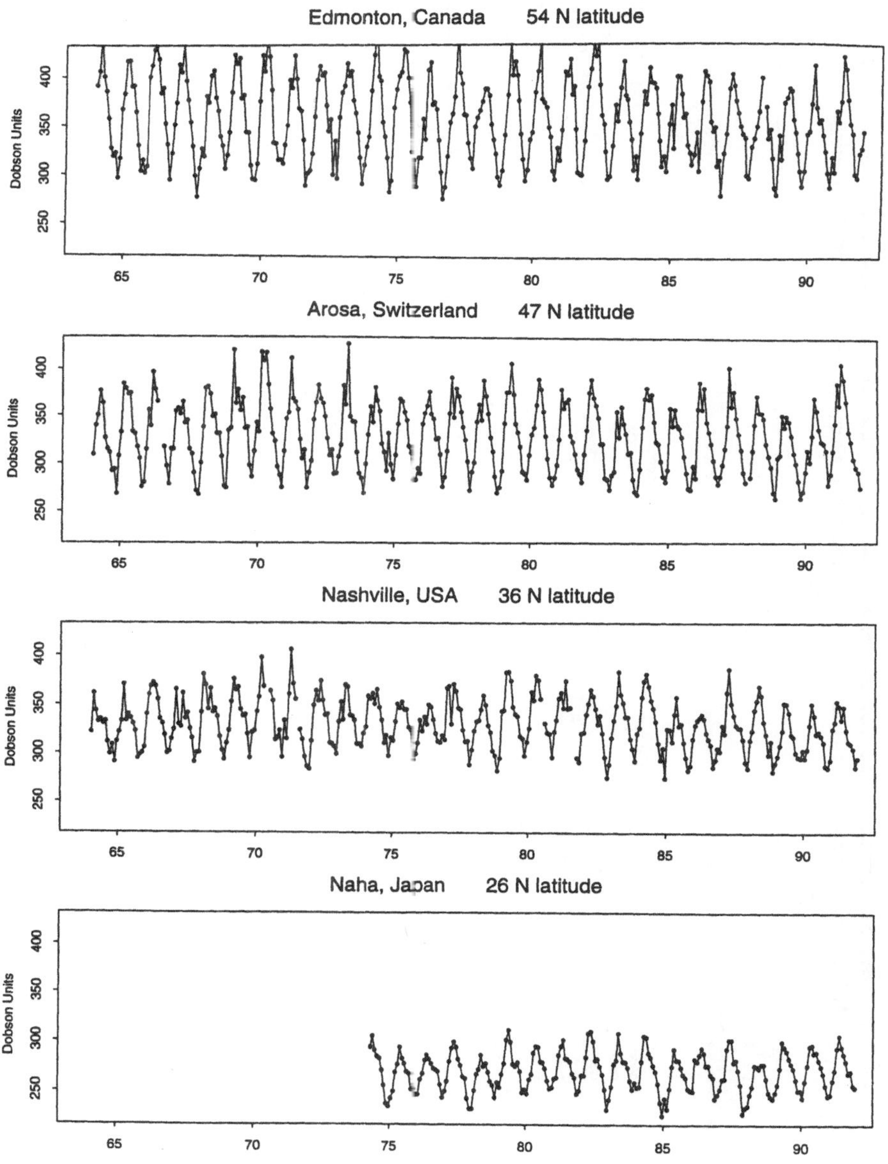

Fig. 2. Individual Dobson station estimates of seasonal trend (% per decade) from total ozone data for 56 stations versus latitude, data period 1964-1991. N, North America; E, Europe; J, Japan; I, India; F, Far East; A, Australia; S, South America; C, Cairo; m, Mahe; P, Samoa; L, Mauna Loa; M, Mexico City; H, St. Helena

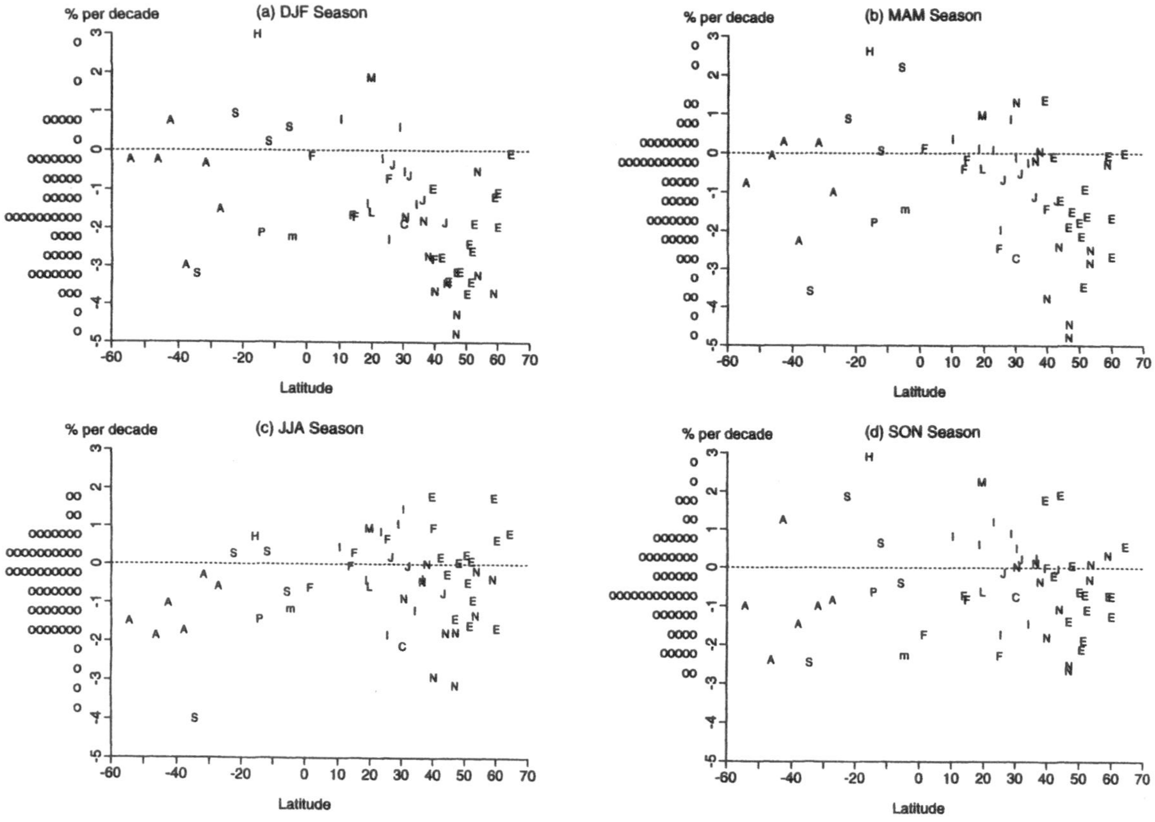

Fig. 3. Seasonal total ozone trends (in % per decade) for 7 different latitude zones obtained from individual station trends using model (3), for 2 data periods: L = 1964-1991, S = 11/1978-1991

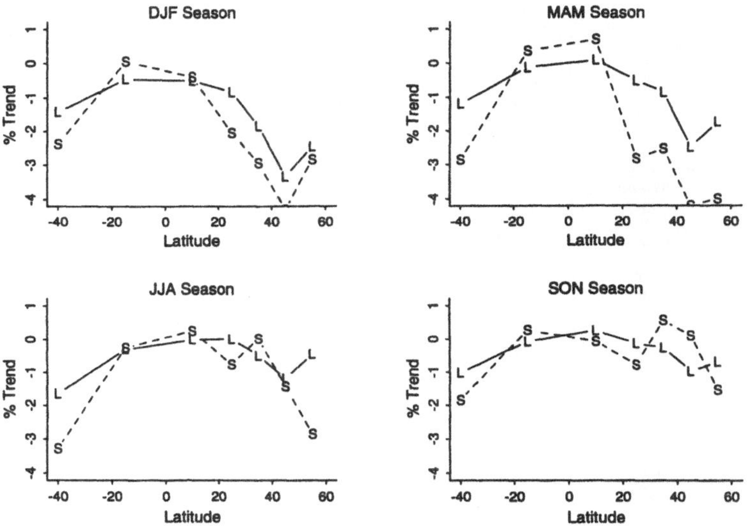

Fig. 4. Individual station differences in spring (MAM) season trend estimates (% per decade) between data periods 11/1978-1991 and 1964-1991. N, North America; E, Europe; J, Japan; I, India; F, Far East; A, Australia; S, South America; C, Cairo; m, Mahe; P, Samoa; L, Mauna Loa; M, Mexico City; H, St. Helena

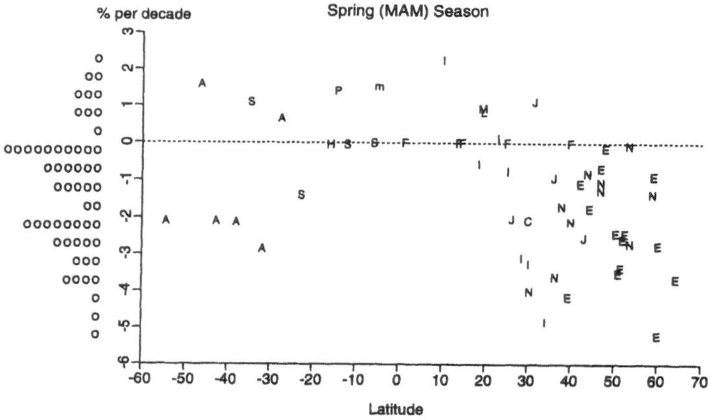

Fig. 5. Plot of cross-correlations of standardized residuals from fitted time series models versus distance

Cross-correlations are between different stations within each region

Region: 1 = North America; 2 = Europe; 3 = Japan

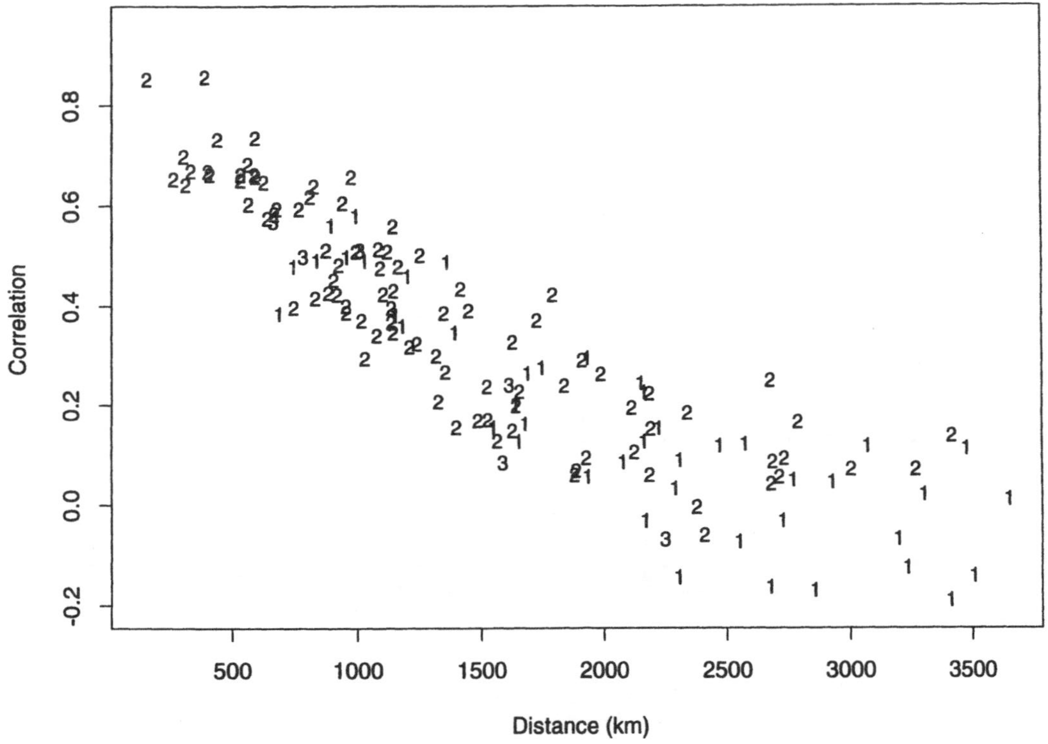

Fig. 6. Seasonal total ozone trends (in % per decade) of Dobson and TOMS total ozone data versus latitude zone for the period November 1978-1991: D = Dobson data, T = TOMS data

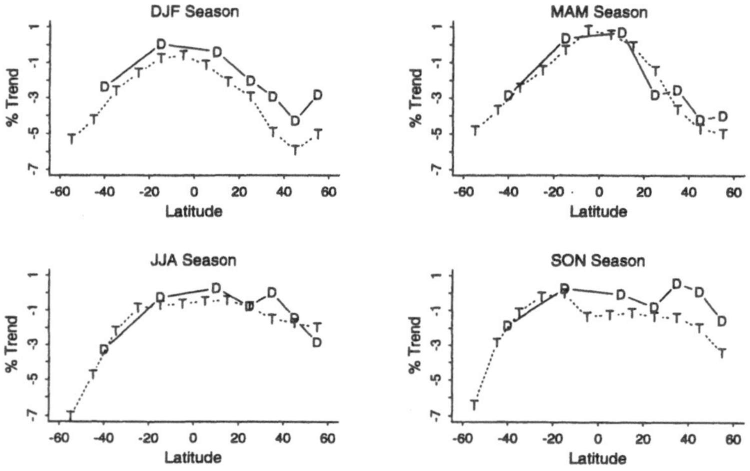

Author Index

Subject Index